Published by
SCIENTIFIC PUBLISHERS (INDIA)

Jodhpur –
5 A, New Pali Road
P.O. Box 91
Jodhpur - 342 001 INDIA
Phone : 0291-2433323
Fax : 0291-2624154
E-mail : info@scientificpub.com

Delhi –
4806/24, Ansari Road
Daryaganj
New Delhi - 110 002 INDIA

ISBN: 978-93-88043-08-3

Visit the Scientific Publishers (India) website at
http://www.scientificpub.com
Printed in India

About the Author

Professor (Dr.) T.S. Chouhan, former Senior Guest Faculty Member in the Department of Geography, University of Rajasthan, Jaipur (India). He is engaged in teaching and research in the fields of environmental geography, remote sensing and its applications, geography of arid lands, agricultural geography, cartography, GIS, GPS Applications, Geospatial Technology, Geoinformatics and Integrated Area Development etc. Dr. Chouhan has over 35 years of experience in teaching and research in interdisciplinary subject.

Dr. Chouhan has participated in number of International Conferences and Presented Research Paper and Chaired the Technical Sessions. He has visited Brazil, Morocco, Saudi Arabia, Tunisia, Argentina, Germany and France etc.

Dr. Chouhan has worked intensively on the following project:

(i) Ecology of Tropical Deserts in Special Reference to Arid Land Physiology (UNESCO Project Encyclopaedia of Life Support System).

(ii) Desert Eco System in India (UNESCO Project Encyclopaedia of Life Support System).

(iii) Ecology of Tropical Deserts in Special Reference to Biogeography and Evolution of Desert Animals (UNESCO Project Encyclopaedia of Life Support System).

He has also completed one UGC major Research Project on "Geo-informatics for Combating Land Degradation and Desertification in fragile Aravalli foothills eco system of Rajasthan".

He has to his credit more than 70 research papers published in reputed journals in India and abroad and is the author of more than two dozen well acclaimed books. He has bean guiding research work for the last more than two decades and supervised more than fifty doctoral theses.

Dr. Chouhan is a recognized consultant/expert to the Government of Rajasthan, Planning Department (Gr. IV), Jaipur in various fields like Wasteland Development, Forestry, Land Management, Water Conservation, Remote Sensing and Cartography. Presently he is working on the following research projects : Application of Satellite Remote Sensing Technology in Planning Natural Resources of Rajasthan; Environmental Protection and Management of Biodiversity in Rajasthan; Development of Wastelands on Watershed basis in Rajasthan; District Planning Atlas of Rajasthan (Based on GIS Applications) and Natural Disaster Management in Rajasthan using Remote Sensing and GIS.

Dr. Chouhan holds the membership of various International, National and Regional Academic Societies contributing in most valuable manner.

Foreword

Dr. J. K. JAIN

Ex. Head, Department of Geography
Jai Narayan Vyas University, Jodhpur, India

It is generally believed that desertification and land degradation are primarily due to misuse and degradation of natural resources by human-induced efforts and their domesticated animals in arid, semi-arid and dry sub-humid regions of the world. There are. however, two conflicting views on the subject of trends towards increasing aridity. According to the first view, the era of long period droughts as a consequential impact is likely to further worsen the situation. This hypothesis is based on the expansion of the cold air from the polar zones, which has caused equatorword shifting of the major high-pressure belts that limits the advances of the Inter-tropical front from the equatorial regions. The second aspect, to which most biologists and agriculturalists now subscribe, that desert expansion in recent years has been entirely due to human interference with the ecosystem.

The world is presently facing two severe problems, viz., decrease in the productivity as the resultant impact of desertification and land degradation and its feared impact on reduction of food production. The other problem is the unchecked growth of population. The production of food/grains increases slower than the increase in the population as the present trend reveals and its further acceleration due to desertification and land degradation could result in devastation. Therefore, the problem needs to be addressed in the earnest manner to avoid hue and cry for food and survival of the humanity. The social and human dimensions of desertification are equally relevant with stress on physical degradation and resultant social changes as impact.

Efforts made by the international agencies and country governments to arrest the problem could not yield requisite success due to various reasons explained in the book implicitly and explicitly. The situation highlighted through the publication is evident that about 40% of the global land is arid, semi-arid and dry sub-humid, which is severely prone to desertification and land degradation and the population inhabited in such regions is more than 47% of the total world population. Although the arid, semi-arid and dry sub-humid regions respectively cover 12%, 18% and 10% of the global area but the inter-continental and inter countries variations significantly very. Further the specific situations need specific treatment of the problem keeping in view the geo-physical and climatic conditions. The citations given in the book reveal that there cannot be a single universal system of combating desertification and land degradation problems. The efforts made by international and country governments for replication of success stories of one region to other parts could not yield similar results as no efforts could be made to assess the local acceptability of area and the people. The lessons need to be learnt, as the process of combating desertification is cost-effective and time-consuming.

Many International agencies and developed countries have come forward for helping the developing and underdeveloped countries for treatment of the problems. Technical and financial assistance has been liberally provided but the results could not be visible in view of various factors. The major factor for unsuccessful or insignificant success was lack of awareness generation among the people and their active association in formulation and implementation of programmes. There is necessity to learn lessons from the past and more in right perspective as the people in dryland areas are facing severe problems and have no courage to wait for further duration in view of faulty implementation process.

Desertification and land degradation are inversely related to poverty and recurrent droughts are the resultant impacts responsible to worsen the situation. The assistance for facing the drought-hit population is human cause for survival of the people but the affected countries need to develop conditions to face their problems as assistance is not the ultimate remedy of the problem This adversely affects the developmental efforts of the country governments and the primary need becomes to save the affected population. The examples quoted are lessons to initiate remedial measures to save the fertility of the land and survival of vegetative cover. The climatic surfaces of the dryland areas need to be focused on aridity determination with potential impact of climatic variations on desertification. The mitigation measures of droughts need to be incorporated in the context as these have significant impact on the nature and dynamics of vegetation growth and resilience in the drylands.

The population growth in dryland regions is alarming which hampers the efforts of developmental process as the growth of net domestic product is mitigated by the population growth. The measures suggested for combating desertification and land degradation need to be applied keeping in view of the areas-specific conditions. There is an urgent need for developing political will to implement the suggested measures in a well defined timeframe.

Water management is the prime necessity in dryland regions as water is very scarce. Study of water resources of dryland areas is very relevant and this needs to be taken as an indicator of social and economic situation of the people of these area; Positive and negative impacts of desertification need to be assessed with stocktaking the impact of combating measures for changes in the dryland areas. The national governments need to derive strategy for efficient water management available from rainfall and underground sources. There is necessity to enforce the measures for water saving devices as well as adoption of suitable cropping management for specific regions. Adoption of biotechnology is the ultimate answer to solve various problems of land degradation.

The cultivators of the dryland areas are very knowledgeable of the problems, but they possess traditional practices. Their knowledge need to be equipped with technological support as this is the best mode of application of requisite measures. Every national and local government need to develop its own research system and feasibility studies need to incorporate the cultivators, as their experiences are quite relevant. This can help in developing extension efforts for encouraging sustainable development of dryland areas.

The book is useful for the country governments, planners, geographers and agronomists, and ultimately to the ordinary people who are the victim of the problems. The publication is the signpost pointing towards the exit from the trap. The future of the human species will depend upon whether or not the advice is taken.

J. K. JAIN

Date: June 5, 2018
Place: Jodhpur

Preface

The term "desertification" was employed in 1949 by the French forester Aubreville, who used the term to refer to the displacement of tropical rain-forest by secondary savannah and scrub in those parts of Africa, where forest was being cleared and burned to provide land for cultivation (WMO-UNEP, 1996). Aubreville concluded that the process was especially active in the sub-humid tropics of Africa and was akin to the creation of deserts in the formerly forested areas. The extent of accelerated soil erosion induced by indiscriminate felling of forests and woodlands in Africa and changes in the soil-water budget and hydrological cycle were understood as some of the factors leading to land degradation. There was also a growing recognition of the part played by human activities and climate changes such as prolonged and frequent droughts aggravating land degradation This led to formally defining desertification as "land degradation in arid, semi-arid and dry sub-humid areas resulting from various factors, including climatic variations and human activities" (UNCED. 1992), which is used as the basis of the UNCCD.

Desertification and land degradation are admittedly impressive terms and have unusual relationship. This is a worldwide problem but in view of its very slow effect, very insignificant attention is paid especially in developing countries. The people living in the affected areas are facing various problems and recurrent droughts are the resultant impact, causing misery to humans, livestock and environmental deterioration. Declining productivity, combined with world population expansion, fluctuations in climate and increasing encroachment on productive lands by roads and cities during drought period constitute an alarming threat to the global life support system. In addition, there are social and human dimensions of desertification, the physical degradation and social change.

Desertification directly affects and puts at risk the livelihoods of more than one billion people, who are directly dependent on land for their survival. Desertification is the degradation of productive land with variable impacts over drylands in the form of droughts, and in irrigated areas as water-logging, salinity and alkalinity. The intensity of the problem lies in the fact that on one side the productivity of land is decreasing constantly and on the other side the population growth is alarmingly high. Main problematic areas are the world's drylands including the Savannas of Africa, the Great Plains and the Pampas of the Americas, the steppes of Asia, the outback of Australia and the margins of the Mediterranean. Desertification is occurring to such a degree that some lands can no longer sustain life.

Drylands of today, have been central in the evolution of humankind, These are the lands that sustained our transition from hunting to pastoralism and agriculture. They still provide much of our grain and livestock and provide the habitat that supports most of the remaining big game animals. They also support a burgeoning human population, but with increasing insecurity as the available land per capita diminish. In this sense, the world is proceeding on the path of destruction and alarms for immediate corrective measures to rectify the problem until it is too late.

Desertification and land degradation affect mainly the arid and semi-arid areas of the earth, but though it may be accelerated by droughts and famines, it is rarely caused by it. On the contrary, their causes are man-made. Desertification is a symptom of the disease of under-development. It results from the combination of social and economic factors, including overgrazing, deforestation, expansion of intensive cash cropping on the marginal land more suitable for pastoralism, poor management of boreholes to water stock and the settling of previously nomadic people, nature and dynamics of vegetation growth and resilience in the drylands.

Desertification also tends to occur more commonly in semi-arid than in arid areas. When rainfall is at least moderately reliable, the desperation to grow more food is higher; and when the ground cover begins to disappear, rains may be heavy and frequent enough to do real damage to the exposed topsoil. The natural causes of desertification and land degradation are eminent and no efforts can succeed on natural disasters. However, the intensity of severity can be minimized with sincere efforts. Human induced effects are very severe, even more than natural hazards; the stoppage of further deterioration is the immediate action to be taken by the human beings, societies, national and local governments. The measures to combat desertification are cost effective and require planned efforts. The on-site and off-site economics need to be ascertained and economic impact and relationship with desertification are the relevant factors of application mode.

The causes of desertification and land degradation cannot be generalized as the reasons for each area are certainly different, but the general issues, which are common in all the regions, need to be addressed with international, national and local efforts. Desertified land can recover, provided enough good soil remains in place and provided local climates have not changed too radically, the land recovers slowly, Once the process to recover is initiated, the impact starts regaining the fertility and ecology.

Therefore, desertification and land degradation need to be understood through various facets, like the causes and severity of the problem particularly in different areas of the world, the natural and manmade reasons for the problem. Unless the problem is not understood fully, further course of actions is completely futile. The second part of the problem is stoppage of the problems identified for each area and making all out efforts at all the levels stalling from the people at gross root upto the international levels. The third and ultimate issue is rectification of the problems to make the world prosperous

Desertification is a complex phenomena resulting from the factors of physical, biological, socio-economic, cultural and political nature. Desertification is not confined to the desert areas of the arid region, but relates to the land degradation in significant area falling within the arid, semi-arid and dry sub-humid regions. Land degradation has a direct impact on land and other natural resources, which results in reduced agricultural productivity, loss of bio-diversity and vegetation cover, decline in groundwater and availability of water in the affected regions. All these lead to a decline in the quality of life, eventually affecting the socio-economic status of the region. About 40% of the total global land is degraded in some form or the other and the problems of land degradation have reached serious proportions threatening the very existence of the people.

United Nations Convention to Combat Desertification (UNCCD) was adopted on 17th June 1994, which stressed the need for integrated efforts and long term strategies on cross-sectoral issues such as environmental conservation, agricultural productivity, sustainable energy and fodder production and use, efficient management of land, water and other natural resources, developmental activities for the local communities to improve their living standards. The UNCCD has provided a platform for addressing these issues in the global context. The study of water in dryland areas is the need of time. For assessment of the impact of combating measures of desertification need to incorporate social and economic indicators to assess and monitor changes in the dryland areas.

The International Organizations have yielded in alarming the world community to take suitable measures including technological and financial assistance for combating the desertification, but the ultimate approach is to be formed by the national governments keeping in view the local conditions as neither common yardstick can be framed and nor it is applicable in every part of the world. The climatic conditions and area specific situations need to be kept in view while formulating any strategy by national or local governments for combating the desertification and land degradation.

Efforts have been made to site various examples of applications and it is an accepted fact that the success stories of one region cannot be applied in similar manner in other parts of the world, without suitable changes in view of area specific conditions and without taking whole hearted support of the local community in view of their socio-economic situations. Therefore, country governments are required to formulate suitable action plans by associating their own experts of all related disciplines and the people who are facing the problems as well as well aware with the implementation strategy for their own benefits.

Combating desertification and land degradation is a costly process and requires reasonable time to rectify the situations. The people of the area especially the cultivators are well conversed with the situations their traditional knowledge need to be upgraded with technological developments relevant to area and such practices for combating desertification are the best way to initiate the process. There is also necessity of concurrent evaluation with the flexible approach to rectify the practices for further development.

There is need for intensive and extensive research to evolve strategy for reversal of the course of desertification through feasibility studies. The technical know-how relevant to local area can encourage people for sustainable management of drylands. It is therefore, necessary to take the earnest efforts for combating desertification for the well-being of the mankind keeping in view the present and future, where the world is likely to face serious challenges of population increase.

T. S. Chouhan

Acknowledgements

The author is grateful to the International Organizations and National Governments for then earnest efforts made for combating desertification and land degradation. In addition, the contributions made by various learned personnels in this field also deserve due recognition. I express my heartiest gratitudes for their sincere efforts for awakening the world community with their views and warning about the problems. The efforts made by all the concerns are not to depict the worst situation ahead but to alarm them with the problems and convince for the future consequences to take adequate measures.

I am thankful to our colleagues for inspiring us for initiating the tedious task and express our gratitude for encouraging us in one way or the other to publish this book.

I shall be failing in my duty, if I do not express my sincere thanks to my wife Vidhya Chouhan, who strived hard during preparation of this publication.

T. S. Chouhan

Contents

Foreword iii

Preface vii

Acknowledgements x

1. Introduction 1

1. Desertification and Land Degradation 1
   - 1.1 Definition of Desertification and Land Degradation 4
   - 1.2 Desertification and land Degradation: The Background: 5
   - 1.3 Sustainable Land Use 9
   - 1.4 Variability in Drylands 9
   - 1.5 Implications for the Study of Desertification 10

2. Status of Drylands, Desertification and Land Degradation 12

2.1 Desertification, Land Degradation and Poverty in Drylands 12
   1. Estimated Dryland and Population Affected 14
   2. Estimates of Corrective Measures 15
   3. Poverty in Drylands 16

2.2 Extent and Distribution of Drylands 20
   1. Criteria for Assessment of Desertification 24
   2. Estimates of Extent of Desertification 25
   3. Social Impact of Desertification 28
   4. Estimates of Rates of Desertification 29

2.3 Land-use in Different Dryland Regions of the World 31
   - (i) Africa 31
   - (ii) Asia 32
   - (iii) Mediterranean Europe 33
   - (iv) Australia 33
   - (v) North America and Mexico 33
   - (vi) South America 34

2.4 Extent of Desertification and Land Degradation in Asia Continent 34
   1. Afghanistan 35
   2. Bangladesh 37
   3. China 38
   4. Indonesia 39
   5. Iran 41

6. Mongolia 43
7. Nepal 44
8. Thailand 44
9. Vietnam 45
2.5 Status of Drylands and Desertification and Land Degradation in India 46
1. General Topography 46
2. Bio-Climatic Regions 51
3. Type and Extent of Land Degradation 51
4. Climate and Desertification 52
5. Major Factors Causing Desertification 53
6. Impacts of Desertification 54

3. Integrated Indicators of Desertification and Land Degradation **56**
3.1 Hydrological Indicators 61
3.2 Physical Indicators 65
1. Soils 68
2. Water Indicators 72
3. Relative Reflectance Indicator 74
3.3 Ecological and Environmental Indicators 74
1. Vegetation 80
2. Animal Indicators 82
3.4 Social and Economic Indicators 89
1. Types of and Changes in Land and Water use 89
2. Settlement Patterns 93
3. Human / Biological Indicators 95
4. Social Process Parameters 97

4. Processes and Causes of Desertification and Land Degradation **99**
4.1 Climatic Surface and Designation of Aridity Zones 99
4.2 Climatic Variability and Change in Drylands 107
4.3 Soil Degradation/ Land Degradation 119
4.4 Soil Degradation/ Land Degradation in Drylands 127
4.5 Soil Erosion 135
A. Water Erosion 138
B. Wind Erosion 143
4.6 Soil Deterioration 148
(A) Processes of Chemical Deterioration 148
(B) Processes of Physical Deterioration 154
4.7 Causes of Soil Degradation /Land Degradation 158

4.8 Soil Degradation and Vegetation 168
(A) Characterizing Vegetation 171
(B) A Global Vegetation Assessment: NDVI and GVI 172
(C) Interpreting the Soil Degradation and Vegetation Map 172
(D) Spatial Patterns Within Drylands 173

5. Consequences of Desertification and Land Degradation 175
5.1 Geographical Regions 178
1. Dryland types 179
2. Problems of Delineating Dryland Boundaries 180
5.2 Soil Degradation / Land Degradation 186
1. Severity and Distribution of Desertification and Land Degradation 186
2. Soil Degradation in Susceptible Drylands 191
5.3 Water Erosion 194
5.4 Wind Erosion 197
5.5 Chemical Deterioration 200
5.6 Physical Deterioration 203
5.7 Causes of Soil Degradation and Land Degradation 206
5.8 Soil Degradation Severity and Vegetation 212
1. Characterizing Vegetation 212
2. Global Vegetation Assessment 213
3. Interpreting Soil Degradation and Vegetation Map 214
4. Spatial Patterns of Drylands 215

6. Combating Processes of Desertification and Land Degradation 216
6.1 Improving Rainfed Cropping 218
6.2 Improving Irrigated Cropping 227
6.3 Controlling Desertification by Improved Livestock Raising 235
1. Improving Animal Quality 236
2. Reducing Stock Levels 238
3. Improving Rangelands 239
4. Digging of More Wells 242
5. Regulating Nomads 243
6. Prospects for Livestock Development 246
7. Livestock Dilemma 247
6.4 Controlling Desertification by Planting Trees 247
1. Large-Scale Fuel-wood Plantations 247
2. Social Forestry 248
3. Community Forestry 251

7. Desertification and Land Degration in India 256

PART–A

7.1 Geographical Background 256
1. Geology 257
2. Physiography 258
7.2 Problems of Desertification and Land Degradation 260
1. Regions of absolute water deficit 266
2. Regions of relative water deficit 267
3. Regions provided with water resources but use less optimally 267
4. Population demography 268
5. Livestock 269
7.3 Processes and Causes of Desertification in India 274
1. Unsustainable agricultural practices 275
2. Unsustainable water management practices 275
3. Land use change 275
4. Deforestation 275
5. Industrial and mining activities 275
6. Demographic pressures 276
7.4 Processes of Desertification 277
1. Wind Erosion 277
2. Water Erosion 277
3. Soil Salinity and Alkalinity 278
4. Water Logging 279
7.5 Desertification Problems in thar Desert 279
7.6 Studies and efforts made by Government of India 281
7.7 Institutional Framework 285
7.8 Assessment of Desertification and Land Degradation 289

PART–B

Degree, Extent and Treatment of Desertification Hazards in India 294

7.9 Methodology 295
7.10 Degree and Extent of Desertification Hazard 297
(A) Degree of Desertification Hazard 301
(B) Vulnerability of Land to Desertification Processes 302
(C) Bio-climatic Zones of the Problematic States 304
(D) Magnitude of the Problem and its Treatment 306
(E) Conclusion 307
1. Approach for Arid Regions 307
2. Semi- Arid Regions 308
3. Sub-Humid Regions 309

PART–C

Desertification and Human Response - Indian Experience **333**

7.11 Extent and Causes of Desertification 335
7.12 Extent of Desertification 335
7.13 Causes of Desertification 338
7.14 Physical Factors 338
1. Variability in Rainfall Pattern 338
2. Climatic Shifts 339
3. Moisture Factor 339
4. Albedo Factor 339
5. Carbon-di-oxide Factor 340
6. Dust Particles Factor 340
7. Accumulation and Accretion of Sand and Sand Dunes 340
8. Geological and Geo-morphological Factors 340
9. Soil Characteristics 343
10. Recurrence of Droughts 349
11. Wind Erosion 349

7.15 Anthropogenic Factors 351
1. Pressure of Human Population 351
2. Changes in Land use Pattern 354
3. Deforestation 360
4. Land Degradation and Desertification 361
5. Water Management Problems 361
6. Use of Chemical Fertilizers and Chemical Pesticides 369
7. Deterioration in Ethnic Values 369
8. Flora and Fauna 369

7.16 Results and Distribution of Desertification in India 369
7.17 Combating Measures 372
7.18 Conclusions and Suggestions 372
1. Management of Land Resources 373
2. Management of Water Resources 374
3. Human Resource Management 375
4. Controlling Human Interface 375

8. Bio-Diversity in Susceptible Drylands **412**

8.1 Human Impact on Biodeversity 414
1. Ecological Consequences of Forest Destruction 415
2. The Problem of Environmental Degradation 415
3. Desertification and Land Degradation by Human Impact 416
4. Desertification and Land Degradation Through Livestock 417
5. Drought Proofing Measures 417

8.2 Valuing Biodiversity 418
8.3 Policy Responses 420
1. Policy Measures to Mitigate Land Degradation and Desertification 421
2. Livestock Breed Improvement Programme 422
3. Bio- Diversity Imbalance Through Chemical Fertilizers and Pesticides 422
8.4 Conclusions 424

9. Dryland Plants and their uses 423
9.1 Plant Communities and Coping Strategies 426
1. Plant Communities in Various Drylands 426
2. Coping Strategies 427
9.2 Centres of Dryland Plant Diversity 429
9.3 Use of Dryland Plants 433
9.4 Conservation Analysis 441
9.5 Conclusions 441

10. Carbon Sequestration in Drylands 442
10.1 Rationale for Sequestering Carbon in Drylands 443
10.2 The Carbon Story 445
10.3 Evidence of Carbon Storage 446
10.4 Possible effects of Climate change on Carbon Storage in Dryland Soils 448
10.5 Feasibility of Carbon Storage in Drylands 450
1. Treating the Problem 451
2. Interpreting the Information for Forecast 452
3. Importance of Weather Forecasts in Dryland Regions 452
4. Strategy for Future 452
10.6 Conclusions 454

11. Saline Soils in the Drylands: Extent of the Problem and Prospects of Utilization 456
11.1 Global Extent of Salt-Affected Soils 456
11.2 Land-use Practices that lead to the Salinization of Cropland 463
11.3 A Crisis for Dryland Irrigated Agriculture 464
1. Areas of Concerns 466
2. Soil Resilience 467
3. Rate of New Soil Formation 468
4. Soil Degradation Rate 468
11.4 Methods of Preventing and Repairing Salt Damaged Soils 469
11.5 New Solutions to Salinity Problems 470

1. Inputs and Improved Technology 472
2. Time Scale 473
3. Preventive Measures 473
4. Curative Measures 474
11.6 Conclusions 475

12. Population and Desertification **476**
12.1 Expressing Population Characteristics 477
1. Economic Development 479
2. Desertification and underdevelopment 480
3. Role of Government 482
12.2 Population in Global Drylands: Data Soucres and Methods 486
12.3 Population in Global Susceptible Drylands 486

13. Poverty and Environmental Degradation **490**
13.1 Poverty, Natural Capital and Investment 492
1. Poverty and Population Growth 494
2. Poverty and Desertification 495
3. Poverty and Famine 495
4. Linkage Between Desertification and Poverty in Drylands of the World 496
13.2 Mining the Future 502
1. Efforts by Government to treat degraded lands 502
2. Micro Level Efforts 504
13.3 Poverty Mitigation and Degradation 505
1. Poverty Eradication 506
2. Introduction of Self-employment Activities for Poverty Mitigation 508
13.4 Poverty as a Disabling Factor 508
1. Accelerating the Economy 509
2. Idle Working Force 509
13.5 Conclusion 510

14. Programmes for Combating Desertification and Land Degradation **511**
14.1 Efforts of UNCCD so far for Combating Deserti-fication and Land Degradation 512
14.2 Activities of UNCCD at Global and Inter-regional Level 513
14.3 New Initiatives for Combating Desertification 527
14.4 Information Dissemination and Awareness Raising 535
14.5 Social sector and Community development programmes 536
14.6 Programmes specifically for Poverty eradication 539

14.7 Technologies for Combating Desertification 545
14.8 Role of Non-Governmental Organization in Combating desertification and land degradation 554

Bibliography **559**

Chapter 1

Introduction

1. DESERTIFICATION AND LAND DEGRADATION

Desertification is a concurrent phenomenon involving climate, soils, flora, fauna and human beings. It may be regarded as a form of deterioration of the eco-system due to the change in the pressure of human use system. Desertification occurs from the fragility of the dryland eco-systems. which under excessive pressure of human use lose productivity and ability to recover. Stability and resilience are major factors of the viability of all the systems- social and cultural as well as physical and biological.

When these systems break down due to effects of desertification; it may be difficult to constitute another viable human use system, so as to restore the primary productivity. As a result, while the productivity of the natural resources is partially lost through desertification, this productivity may be totally lost to the human community. For these reasons, desertification need to be diagnosed and monitored through the whole rang of physical, biological and social processes relevant to eco-system and human use system of dry lands. Therefore, on the one side, people at large responsible for desertification for irrational use of land and on the other side social and economic conditions are equally responsible for indifferent attitude for combating by diagnosing and arresting it.

Desertification directly affects, or puts at risk, the livelihoods of more than one billion people, who are directly dependent on the land for their survival. Desertification is the degradation of productive dry-lands, including the Savannas of Africa, the Great Plains and the Pampas of the Americas, the Steppes of Asia, the "outback" of Australia and the margins of the Mediterranean. Desertification is occurring to such a degree that some land no longer sustain life.

Drylands have been central in the evolution of humankind. These are the lands that sustained our transition from hunting to pastoralism and agriculture. They still provide much of our grains and livestock and provide habitat that supports most of the remaining big game animals. They also support a burgeoning human population, but with increasing insecurity as the available land per-capital diminishes. Desertification is land degradation occurring in the arid, semi-arid and dry sub-humid areas of the globe. These are also referred to as 'the susceptible drylands' and they cover 40 percent of the world's land surface.

One fifth of the world's population is threatened by the impact of global desertification. Its effects are visible all over the world, be in Asia, the Sahel, Latin America, throughout North America or along the Mediterranean. Today, a third of the earth's surface is threatened by desertification, which adds up to an area of over four billion hectares of the planet.

Desertification is a land degradation process that happens in all regions of the world. Its causes are brought about by human-induced factors and climate change. Desertification occurs slowly, as different areas of degraded land spread and merge together, rather than through advancing desert. It is comparable to a slowly but clearly progressing "skin disease" over the planet. Fertile topsoil takes centuries to form, but it can be washed or blown away in a few seasons. Human activities such as over-cultivation, deforestation and poor irrigation practices combined with climate change are turning once fertile soils into barren patches of land.

Arable land per person is shrinking throughout the world, threatening food security, particularly in poor rural areas and triggering humanitarian and economic crises. Since 1990. it is estimated that some six million hectares of productive land has been lost every year due to land degradation. This in turn has caused income losses worldwide of US$ 42 billion per year. Yet the costs associated with inaction in regards to desertification are estimated at one to three percent of developing countries' GDP. In most cases, investment in combating desertification is one order of magnitude below the amount.

Desertification and recurrent drought force people to leave their home in search of a better life. It is estimated that 135 million people, the combined population of France and Germany, are at risk of being displaced by desertification. The problem appears to be most severe in sub-Saharan Africa, the Sahel and the Horn of Africa. Some 60 million are estimated to eventually move from the desertified areas of sub-Saharan Africa towards Northern Africa and Europe by the year 2020. By contrast, in the past 20 years, nearly half of the total population in Mali has migrated at least once to neighboring African countries (96 percent) or to Europe (2.7 percent). In Burkina Faso, desertification can be identified as the cause for 60 percent of swelling of main urban centers. In Peru, farmers on the southern coastal areas face a similar situation. Recurrent droughts and the effects of EI Nino has forced one out of every five to move from rural areas into the main towns. In Mexico, 47 percent of its land is affected by desertification causing vast population movements. Out of the total population of 91 millions, an estimated 8 to 10 million Mexicans migrated to United States every year in search of more prosperous life.

Poverty and desertification are closely linked. Two-thirds of the world's hungry people live in rural areas of developing countries. As highlighted in the 2003 Human Development Report, about half live in farm households on migrated and where environmental degradation threatens agricultural production. Forced to take as much as they can from the land for food, energy housing and income, the poor are both the causes and the victims of desertification. In reverse, desertification is both a cause and a consequence of poverty. With no arable land to grow crops or raise livestock, many have no choice but to leave and resettle in a more fertile area.

Without access to sustainable land use practices, institutional services, credit and technology, most poor farmers are obliged to cultivate degraded land that is unable to meet their needs This constant pressure on the land causes a decline in food production that further aggravates poverty. Eradicating rural poverty is one of the first steps in fighting desertification. No long-term strategy of poverty eradication can succeed in the face of the environmental forces that promote persistent erosion of the soil poor people depend on. No programme for protecting the environment can succeed without alleviating the day-to-day reality of poverty.

India is endowed with a wide variety of climate, ecological regions and land and water resources. However, with barely 2.4 percent of the total land area of the world, our country has to support 16.7 percent of total human population and about 18 percent of the total livestock population of the world. This had put enormous pressure on our natural resources. Ecosystems are highly complex systems relating to a number of factors both– biotic and abiotic, governing them. Natural ecosystems by and large have a high resilience for stability and regeneration. However, continued interference and relentless pressures on utilization of resources lead to an upset of this balance. If these issues are not effectively and adequately addressed in a holistic manner, they can lead to major environmental problems such as depletion of vegetation cover, increase in soil erosion, decline of water table and loss of biodiversity all of which have direct impact on our survival. Thus, measures for conservation of soil and other natural resources, watershed development and efficient water management are the key to sustainable development of the country. The socio-economic aspects of human activities form an important dimension to the issue of conservation and protection of natural resources. The measures should not only conclude rehabilitation of degraded lands but also to ensure that the living conditions of the local communities are improved.

Desertification is a complex phenomena resulting from factors of physical, biological, socio-economic, cultural and political nature. Desertification is not confined to the desert areas or to the arid regions, but relates to land degradation in about two-thirds of our country's geographical area falling within arid, semi-arid and dry sub-humid regions. Land degradation has a direct impact on land and other natural resources, which results in reduced agricultural productivity, loss of bio-diversity and vegetation cover, decline of ground water and availability of water in the affected regions. All these lead to a decline in the quality of life, eventually affecting the socio-economic status of a region. A major part of our country is degraded in some form or the other. In some regions, problems of land degradation have reached serious proportions threatening the very existence of the people.

Desertification is a term that refers to the process of land degradation that ultimately leads to the transformation of productive land into an ecological desert. The causative agent over the last one to two hundred years or more has been man. Over the much larger period of time, climatic changes have produced deserts, where they once were not present. The Sahara is an example of latter situation.

Land degradation by man is initiated when the vegetation cover is reduced or destroyed by grazing, cultivation, mining, road construction, irrigation, urbanization, recreation and other activities that disturb the natural conditions. Subsequently water and wind erosion accelerate the degradation process. Finally, conditions deteriorate to the point where land is abandoned. Recovery of the land, whether natural or under man's management, after deterioration may take a few years, decades or centuries, depending upon the severity of degradation and the climate of the region.

1.1 Definition of Desertification and Land Degradation

The word "desertification" is derived from various similar words like desert, desertic. deserted and desertify which possesses all the situations inhibited in each of the word. Desertification is the situation, which initiates from the desert. The formation of desert is caused due to continuous water deficiency, which converted the cultivated and irrigated land into desert. The second term in consequence is desertic and interprets for the situation that the desert was preceding to desertic situation in view of continued absence of positive ingredients. The term deserted is a situation where the degradation process had reached to the proven level and further efforts to treat the land were treated as impossible. The concluding word 'desertify' is the process of expansion of land degradation process in the surrounding cultivated land and its spread was treated as a concern for area and all the ingredients were proved in vain. Thus, the process of desertification from the cultivated potential land to fully degraded wasteland indicates the efforts for its treatment proving no result and the conclusion has been drawn to regularly detorioting the upfront and the speed of increase of desertification is the degree where the alarming situation of warning the people, planners, geographers, decision makers and similar experts with failure of all possible alternatives.

Desertification arises from the fragility of the dry-land eco-system, which under excessive pressure of human use lose productivity and ability to recover. Although desertification develops from natural cause alone in nay climatic zone, the UN Conference at Nairobi in 1977 mainly focused attention on desertification that derives from interaction of human use system with natural eco-system in arid, semi-arid, sub-humid and humid zones. Therefore, desertification can be defined as a package of processes which brings about certain basic changes in particular eco-system and converts it from a relatively non-desertic to a desertic terrain. This involves an interplay of climatic, edaphic and biotic factors, which may (i) work through time starting with a climatic accident and ultimately resulting in the formation of a desert, or (ii) accentuate arid conditions within an already desertic terrain, or (iii) cause or creep desertic conditions within a desertic tract towards its adjoining areas (Mann, 1977).

The basic cause of desertification is the inter-relationships between physio-environmental system and human adoption. This is because the environment, when defined as the sum total of the conditions that control human activities, are under stress of human needs of adoption, when adoption is defined as the various means of the human being adapt to survive themselves. Thus, these are the human activities that primarily initiate desertification process. Some of the main human and biotic interactions causing desertification are hereunder:

1. Uneconomic land use cultivation.
2. Intensive and uneconomic cultivation or over-exploitation of water resources leading to depletion of water table, seepage problems and increased salinity.
3. Uncontrolled grazing and indiscriminate cutting of trees or over-exploitation of grazing lands and forests resources leading to drought, soil erosion, deterioration of soil fertility and stunted growth of trees.

1.2 Desertification and land Degradation: The Background

Desertification has long been recognized as a major environmental problem affecting the living conditions of the people in the affected regions in many countries of the world. The term "desertification" was employed in 1949 for the first time by the French botanist and ecologist, Aubreville, who used the term to refer to the displacement of the tropical rainforest by secondary savannah and scrub in those parts of Africa, where forests was being cleared and burned to provide land for cultivation (World Meteorological Organization - United Nations Environment Programme, 1996). Aubreville concluded that the process was especially active in the sub-humid tropics of Africa and was akin to the creation of deserts in the formerly forested areas. The extent of accelerated soil erosion induced by in discriminate felling and burning of forest and woodland in Africa and changes in the soil-water budget and hydrological cycle were understood as some of the factors leading to land degradation. There was also a growing recognition of the part played by human activities and climate changes such as prolonged and frequent droughts aggravating land degradation. Desertification reached a wider audience in 1970s, when international attention was focused on the plight of the drought-stricken Sahel of Africa, south of Sahara.

This term was raised as a major environmental issue at the United Nations Conference on Human Environment, convened in Stockholm in 1972. at which the United Nations Environment Programme (UNEP) was formally established and formally defining desertification as "land degradation in arid, semi-arid and dry sub-humid areas resulting from various factors, including climatic variations and human activities," which became the basis for further interactions.

In 1972, the initiative for the "Cambridge Meeting on Desertification" came from the International Geographical Working Group set up in Montreal and it planned a conference on problems in the development and conservation of desert and semi-desert lands, which took place in Ashkabad from July 20 to 26. 1976. But the phenomenon popularly known as desertification received widespread attention only after the United Nations Conference on Desertification (UNCOD) organized at Nairobi in 1977 (UN 1977 (a) and (b). A world map of land degradation by desertification and a number of studies from all the continents were prepared for the Conference to illustrate the phenomenon (UNESCO, 1977. 1980). These studies indicated that desertification was not just happening in Africa, but in the dry-lands the world over. Following the conference. UNEP was charged with coordinating the implementation of the United Nations Plan of Action to Combat Desertification, which was adopted by the conference and endorsed by the United Nations General Assembly.

Desertification involves degradation of vegetative cover, water and wind erosion, soil crusting, soil compaction, loss of soil fertility, salinization and water logging. A definition of desertification should encompass the concept that desertification is a land degradation process involving a continuum of change, from no degradation to very severe degradation and that it is the result of man's activities (Glantz and Orlovsky, 1983) the definition used here is the following:

"Desertification is the impoverishment of terrestrial ecosystems under the impact of man. It is a process of deterioration in these ecosystems that can be measured by reduced productivity of desirable plants, undesirable alterations in the biomass and the diversity of the micro and macro fauna and flora, accelerated soil deterioration and increased hazards for human occupancy" (Dregne, 1978).

In 1984, a Desertification Hazard Map was produced in a co-operative venture between UNEP, the Food and Agriculture Organization (FAO) of the United Nations, the United Nations Educational Scientific and Cultural Organization (UNESCO) and the World Meteorological Organization (WMO). This was accompanied by a Provisional Methodology for the Assessment and Mapping of Desertification (UNEP, 1984). At the same lime, the revised definition of the problem was produced, one of more than a hundred such definitions that have been used both in scientific and political circles (Glantz and Orlovasky, 1983). The lack of an agreed definition of desertification, an inevitable consequence of the issue's complexity, has been characterized by controversy, conflicting opinions and differing interpretations (Hellden 1991. Thomas and Middleton 1994. Stiles 1995).

Discussions in academic circles apart, the reason for interest in desertification is clear: degradation in drylands adversely affects the people, who live there. To better address the issue, the 1992 United nations Conference on Environment and Development (UNCED) in Rio de Janeiro requested the United Nations General Assembly to establish an inter-governmental committee to negotiate the Convention to Combat Desertification, which entered into force on 26th December. 1996. Land degradation implies reduction of resource potential by one or a combination of processes acting on the land. Arid, semi-arid and dry sub-humid climatic zones are collectively referred to as the 'susceptible dry-lands'. such area have been depicted on the world. (Figure 1)

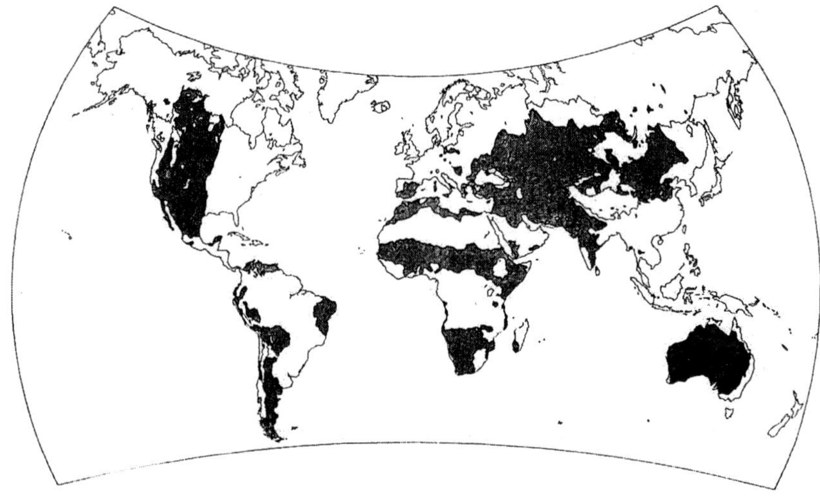

***Figure 1** : Susceptible drylands*

The United Nations Convention to Combat Desertification (UNCCD) was adopted on 17th June 1994. The convention stressed the need for integrating efforts and long term strategies on cross sectoral issues such as environmental conservation, agricultural productivity, sustainable energy and fodder production and use, efficient management of land, water and other natural resources, developmental activities for the local communities to improve their living standards. Thus the UNCCD provided a platform for addressing the issues not only in the national but also in the global context.

In 1992, UNEP produced a World Atlas of Desertification (UNEP 1992-b). The studies indicated that over the preceding 20 years, the problem of land degradation had continued to worsen. In the past, dry-lands recovered easily following long droughts and dry periods. Under modern conditions, however, they tend to lose their biological and economic productivity quickly unless they are managed in a sustainable manner.

For stocktaking the area infested by desertification, the aridity was defined and accordingly the aridity of a region is categorized by the ratio of P= Mean Annual Precipitation to PE = Mean Annual Potential Evapotranspiration, using modified Thornthwaite formula. As per this, the aridity zones are classified as given Table 1 to understand worldwide situation:

Table 1 : Situation of Desertification on Aridity Scale

Climate Zone	R/PE Ratio	Percent of world covered
1. Hyper -arid	< 0.05	7.5
2. Arid	0.05 -0.20	12.5
3. Semi arid	0.21 -0.50	17.5
4. Dry sub-humid	0.51-0.65	9.9
5. Humid	> 0.65	39.0
6. Cold	> 0.65	13.6

Source: WMO-UNEP Report (1996): Interactions of Desertification and Climate.

Land degradation is a process of deterioration of fertility. This contains all the lands lying unutilized or which are not being used to their optimum potential due to different constraints. The degraded lands are those lands which can be brought under vegetative cover, with reasonable efforts and which are currently lying unutilized, which are deteriorated for lack of appropriate water and soil management or on account of natural causes. These can result from inherent and imposed disabilities, such as location, environment, chemical and physical properties of the soil or financial or management constraints. Land degradation is basically of two broad categories: culturable and non-culturable lands. The culturable degraded lands are capable of or have the potential for development for agricultural or pasture purposes, or even can be afforestated. They are not being used, at present, due to such constraints of water for irrigation, salinity or alkanity in the soil, soil erosion, water-logging, an unfavourable physiographic position or human neglect. All these lands can be put to some productive use after proper treatment and reclamation measures.

Non-culturable degraded lands, on the other hand, are barren lands, which cannot be put to any productive use, either for agriculture or to develop forest cover. Barren rocky outcrops are one of the examples of such lands. Degraded lands have been classified into following categories:

1. Gullied and / or ravinous lands,
2. Undulating uplands with or without scrub,
3. Surface water-logged lands and marshes,
4. Shilling cultivated areas,
5. Salt affected lands,
6. Degraded forestlands,
7. Degraded pasture / grazing lands,
8. Demaded non-forest plantation lands,
9. Strip lands,
10. Sands,
11. Mining/ industrial degraded lands,
12. Barren rocky / stony/ sheet rock areas,
13. Steep slopping areas.

Out of the above, first nine categories of degraded lands are treatable as culturable and the remaining four categories are unfit to be treated. Controlling desertification and land degradation involves all aspects of environmental management. The role of water management in desertification and land degradation control is fundamental and much rainfall still goes waste. More effective water harvesting is possible in many places. Other global issues such as bio-diversity conservation, climatic change etc. require to be tackled with land and water management in an integrated manner.

1.3 Sustainable Land Use

The concept of land degradation is inseparable from that of sustainability (WCED 1987. Dowdeswell 1995). Expressed simply, a sustainable land use is one that is able to continue without degrading the land it is using. In this case the sustainability of a particular land use depends both on the properties of the resources and the way these are managed. The feature of a resource that dictates its sustainability under a particular use is its resilience and it is important to note that the resilience of a resource will vary according to different land uses and indeed may vary from time to time, depending largely on natural seasonal and inter-annual variability and management practices and technologies.

A reasonable way to measure the resilience of a particular unit of land is to look at its ability to recover after a disturbance. Such a disturbance may be climatic, for example a drought, or human-induced, such as vegetation clearance or soil tillage. The greater the ability of an area to recover within a particular time period, the greater its resilience. In essence, land degradation is the weakening of an area's resilience. One measure of land degradation in any particular area is the cost of rehabilitation.

It is important to emphasise that sustainable land use is a function of both the natural environment and the human systems. Combating desertification, therefore, demands political, social, economic, educational approaches (Barraclough 1995) in addition to those approaches based on largely in the physical sciences that have characterized much previous work in this area. This need for an integrated approach to sustainable environmental management in dry-lands is recognized and embodied in the Convention to Combat Desertification.

1.4 Variability in Drylands

One difficulty of examining land degradation in dryland regions is their inherent variability. A definitive characteristic of drylands is their aridity, which put simply, means their lack of available moisture. An area can be said to be arid when its moisture inputs (precipitation) are exceeded by the moisture losses (evapo-transpiration) plus any changes in storage (in rivers, ground water, lakes and soil moisture). Various climatic and biological indexes have been devised to measure the aridity of drylands and these can be used to delimit dry-lands areas. However, it is important to note that although dry-lands are usually deficient in moisture on an annual basis, the moisture inputs as precipitation are notoriously variable in both time and space (Williams and Balling 1996). The more arid dry-lands areas commonly receive all their 'average' annual precipitation in just a few days. The meteorological systems responsible for rainfall in dry-lands regions are typically convective cells, formed in adverted humid air masses, which means that rainfall falls in small specific zones.

The characteristics of dry-land climate mean that these regions are highly dynamic on a timescale of weeks and months. Added to this is the variability of precipitation over longer timescale of years and decades. Droughts, in essence the absence of expected precipitation, are also characteristic of the dry-lands environment. A year, two years or several years may pass in which precipitation is well below average (Wilhite and Glantz 1985). Indigenous plants and animals have adapted their life cycles to cope with this inherent variability of the dry-lands eco-system (Polis 1991). When the duration of below "average" precipitation reaches the decadal scale, reference is often made to desiccation (Hare 1987). Beyond that, continued significant departures from a long-term average may enter the realm of climatic change.

1.5 Implications for the Study of Desertification

The implications of this environmental variability for the study of desertification are manifold. First, for the location of drylands themselves. Efforts made by UNEP by identifying and highlighting drylands are specific boundaries simply based upon the average conditions and that true ground conditions vary greatly through variable timescales (Nicholson 1978, Hulme *et. al.* 1992).

There are numerous implications for the human inhabitants of dryland regions (Mainguet 1995). The use to which land is put must be as dynamic as the environment and its resources, if those resources are to be used sustainably. Fields that produce a good crop during near-average rainfall years may have to be left fallow during a drought year, or be put to another use, if the land is not to be degraded. It is in areas where this flexible land use response does not occur that desertification takes place.

Perhaps, the most serious implication of dryland variability in the present context is for identification, monitoring and combating of desertification itself. Satellite imagery of vegetation greenness of the arid / hyper-arid zone boundaries of drylands indicate that the natural variability in climate is reflected in a green vegetarian dryland boundary that can fluctuate by up to 150 kilometer from a wet year to a preceding dry year (Tucker *et. al.* 1991). With such great natural fluctuations in the ground state of drylands, it is clearly necessary to monitor potential area of desertification over a timescale of decades before it is possible safely to state that a particular region has suffered from land degradation in the form of desertification.

Reliable identification of the locations and situations, in which land degradation takes place is essential, if variable remedies to the problems are to be reached. The actual reasons for sustainable land use taking place in a particular area may well have their roots in social and economic conditions (Blaikie 1985). Unfortunately, accurate and reliable data on the extent and severity of desertification and the rate of its progress, based on actual ground surveys, are very scarce. The existing data are often controversial and open to doubts and criticisms (Thomas and Middleton 1994). Early attempts to assess the extent of desertification on the global scale, such as the world map prepared for UNCOD (UN 1977-a) represent useful first steps towards the goal of solving the problem. But these first efforts had their own problems. Perhaps, the most important were the inconsistencies in the definitions used and the perception of desertification threatened all the world's drylands. However, when the prime motivation for studying desertification is to help relieve the problems faced by inhabitants in using drylands resources, it is clear that areas of hyper-arid desert, which by definition have very sparse biological resources, should not be included in the areas of investigation. Very few people use these regions because of their lack of resources and these areas can hardly become more desert-like.

If an exception can prove the rule, then the very largely hyper-arid country of Egypt is the major anomaly to this otherwise sound generalization, since a reliable water supply allows permanent human occupation of the Nile Valley. Hence, for the purpose of this study, the regions deemed to be susceptible to desertification are those in the arid, semi-arid and dry sub-humid zones of the world.

There may be grounds for criticizing a global approach to the problem (e.g. Warren and Agnew 1988). The complex nature of desertification means adequate assessment and consequent plans to counteract the problem can only be usefully carried out at the local scale, a central reason for the involvement of local populations and non-governmental organizations in approaches adopted by the Convention to Combat Desertification. Nonetheless, a global perception allows desertification to be evaluated relative to other global environmental issues. Furthermore, it is useful to develop a worldwide picture to identify "hotspots" at the continental, national and local scales.

Chapter 2

Status of Drylands, Desertification and Land Degradation

Precisely, desertification occurs on shallow meeting place between soil and atmosphere, where plants thrive and a balance is maintained between incoming and outgoing energy and between water received and lost. When rain falls, some of the water is taken up directly by plants, some filters into the soil, where it may remain in storage and the rest evaporates or run-off. Some soil moisture that intercepted by plants is put back into the atmosphere by the plants in transpiration. Some of the moisture may seep or aquifers, where it may remain for thousands of years or may migrate slowly from plateau to depression on back to the ocean itself.

The soil-air meeting place participates in an energy balance activated by the rays of the sun or through atmospheric heating. Some energy is reflected by the surface layer back into the atmosphere and into space. Some is held by the soil in storage, thereby warming the earth and it is this energy and that from the sun directly that is used by plants to carry out the processes of photosynthesis and growth. Some of the plants are eaten by grazers or browsers and these animals in turn may be eaten by carnivores, with all animals returning energy and moisture to the atmosphere in respiration and to the soil in the form of humus. The excretes of animals, their recomposing carcasses and the decomposition of plants supply the soil with nutrients, most densely in the topmost layers.

2.1 DESERTIFICATION, LAND DEGRADATION AND POVERTY IN DRYLANDS :

In the Sahel, like other drylands, rainfall is scanty and high variable. For longer than history remembers, nomads have pastured their herds in the arid grasslands that rim the southern edge of the Sahara Desert, such lands have seemed suitable for little else, since only nomadism can take full advantage of the fact that rainfall occurs at very low spread. The droughts of 1935-72 occurred in Sahel with an average occurrence of 284 mm per year and again in 1971 of 126 mm and worst in 1972 with only 54 mm worsened the situation in Sahel desert. The situation became catastrophic and spectacle of vast migrations and refuge camps brought the world attention. In 1974 the issue of desertification was discussed in the General Assembly of United Nations to addressal of miserable of Sahel in particular and worldwide in general. In addition to Sahel, worst hit countries were Niger. Senegal, Mali. Upper Volta and Mauritania was worst affected with drought.

Desertification is far from a novel experience for mankind. It has played a part in the decline of civilizations from the earliest historical times, because of improper drainage, salts concentrated in the irrigation lands by Sumerians and Babylons, thus destroy their agricultural productivity. Prolonged desiccation damaged the agricultural basis of the harappans, who had constructed an early civilization in what is now Pakistan. The Mediterranean littoral of Africa was more productive in Roman times than it is today. No one could say. how much land has been lost to man's use or severely degraded since the practice of agriculture began, but it is certainly a great deal, some have estimated it as high as the total amount of land now in cultivation. There is agreement that the rate of land degradation has increased significantly during last few decades and some experts have suggested that it has reached at least 50,000 square kilometers per year. This is a starting figure for a world likely to be faced with food shortages.

Man's realization that desertification is accelerating is an important factor in his increasing attention to the problem. Experience of desertification may not be new, but there is something novel in the recognition of it as a serious global problem. It is so recent, indeed, that the word "desertification" has not yet found a place in the dictionaries, with the result that there is some dispute as to how precisely it should be defined. Paving a road destroys the biological productivity of the ground covered by pavement, but few call that desertification.

Although, deserts are not without life, they can be viewed as areas with extremely limited agricultural potential. Deserts occur in a variety of types, hot and cold, stony and sandy, but all characterized by rainfall deficiencies so marked that cultivation or stock-rearing are possible only with special adaptations as by development of irrigation. Desertification as the extension or intensification of desert conditions, involves a decline in the productivity of land and it is this, which makes it fundamentally a human problem. Desertification anywhere affects the whole global community. Lowered wheat yields in one country have worldwide repercussions on the price of wheat in view of demand and supply conditions. However, the human impact of desertification is far greater on the people, who live where it is happening and who depend on arid lands for their livelihoods. This is true especially of the developing countries. There, capital is lacking to buy relief from "outside", desertification worsens conditions of poverty, brings malnutrition and disease, erodes the basis of the national economy and then brings deterioration of social services already hampered by remoteness and lack of funds. All this affects the ability of dryland communities to respond to succeeding droughts, each of which would then tend to advance further the deterioration of living standards, which represents the human aspects.

Each region of the world has its climate, an element in the global pattern of atmospheric circulation. There is also another climate, partly determined by the first, at the shallow meeting place of earth and air. Like the larger climate, this smaller on is ever changing, always balancing its water and energy budgets. Reaching the soil as rain or dew, water evaporates, run-off or sinks in to be stored, with some released by plants and animals in respiration and transpiration. Energy comes from the sun. Some is reflected back into the air, some warms the soil as stored hear. Some is taken by plants and used in the process of photosynthesis. Herbivores eat the plants and carnivores eat herbivores. Plants hold the soil in place, their waste products, as those of animals; supply the soil with nutrients, richest in the topmost layers. There are some questions that man has a profound impact on the small climate at the soil's surface whenever he makes use of land.

The drylands under threat are seen, as they are the home of one-sixth of the world's population and although they are regions of low productivity per unit of area, their total production is enormous, especially of meat, cereals, fibres and hides. Under proper management, the drylands have even greater potential for production. The problem of supplying adequate food for everyone, shelter and clothing is now recognized as urgent and a problem that increases the difficulty as the world's population continues to expand. Efforts to solve the problem with anything less than the enormous total output of the drylands have little prospect of success. The agricultural potential of the drylands is essential to the welfare of mankind and their conservation is accordingly of global concern.

1. Estimated Dryland and Population Affected :

Desertification has an impact beyond the lands immediately affected. Dust storms can move soil great distances and increased flooding may occur far downstream due to overly rapid runoff from lands denuded of trees and plants in upstream catchments undergoing desertification. If desertification is allowed to develop or proceed further on geographical scale, almost the entire population of the earth's drylands could be said to face eventual risk. They contain between 600 to 700 million people and in terms of broad areas and livelihood systems, their numbers are as given in Table 2.2.

Drylands also serve as reserves sheltering an important range of plant life, including the genetic fore-runners of many mankinds' staple grain- wheat, barley, sorghum and maize. They undoubtedly contain other potential other potential domesticates. The Green Revolution has focused new attention on the critical importance of this botanical heritage, particularly as a resource, which can be used to keep highly cultured strains, the "miracle wheat"", resistant to destruction by disease. An ecotypal reserves with a variety of interesting and useful natural settings, the drylands constitute a precious human heritage. In recent years, they have come increasingly to serve as areas to which people go and often remain in quest of health and recreation.

Table 2.2 : Estimated Dryland, Population and Livelihood Groups

(In thousands)

Region	Dryland Total Population	Livelihood Population in Drylands		
		Urban Based	Cropping Based	Animal Based
1. Mediteranean Basin	106800	42000 (39%)	60000 (57%)	4200 (4%)
2. Sub-Saharan Africa	75500	11700 (15%)	46800 (62%)	17000 (23%)
3. Asia and the Pacific	378000	106800 (28%)	260400 (69%)	10300 (3%)
4. Americas	68100	33700 (50%)	29300 (43%)	5100 (7%)
Total	628400	194200 (31%)	397100 (63%)	37100 (6%)

Note : 1. Estimates of Drylands: Meigs Classification (1953) including extremely arid, arid and semi-arid areas.

2. Population of the Region: Grouping as designed by UNEP.

3. Total World population was estimated to be 3.85 billion in 1974.

Source: UNCOD, 1977.

More than a mere threat, desertification is actively at work. A great many people live in drylands that are now undergoing the process and their livelihoods are already affected. Estimates of present losses give rise to a pessimistic outlook, suggesting that the world will lose more than one-third of its arable lands by the year 2010. Such losses will tale place, while the food requirements of the human race are rising at least as rapidly as the human population is growing. This demands an increase in food by more than one-third by the corresponding period to maintain the present dietary standards, as adequate as they may be in some parts of the world.

2. Estimates of Corrective Measures

Estimates of the present rate at which land is undergoing degradation and how much it would cost each year to rehabilitate these losses as mentioned in Table 2.3. Precise figures on losses through degradation and the cost of rehabilitation are not known, especially bearing in mind the social costs involved.

Table 2.3 : Estimates of costs and benefits of corrective measures

Particulars	Types of land			
	Irrigated	Rangeland	Rain-fed Crop	Total
1. Annual Rate of Land Degradation (000 ha)	125	3200	2500	5825
2. Estimate if not salvaged ($ per ha)	200	5	50	—
3. Value if Salvaged ($ per ha.)	1500	40	300	—
4. Gain(3) - (2) ($ per ha.)	1300	35	250	—
5. Total Million $(1) × (4)	162.5	112	625	—

Particulars	Types of land			
	Irrigated	Rangeland	Rain-fed Crop	Total
6. Estimated saving per ha. ($)	650 (250-2000)	20 (1-100)	100 (50-150)	
7. Total Cost (million $) (1 × 6)	81	64	250	—
8. Net Gain per ha. (4-6)	650	15	150	—
9. Total net benefit (million $) (1 × 8)	81	48	375	504

Notes : (a) Type of land covers arid and Semi-arid lands only, sub-humid lads omitted. Totals are estimated roughly at 4000 million hectares, of 250 millions are irrigated, 3600 million are rangeland and 170 million are rain-fed cropland.

(b) Annual rate of land degradation is based on annual rate of change of classes of land to more degraded conditions. The degree of degradation from higher to lower classes of land has been converted to more limited areas assumed to be deteriorating from land yielding highest net return, if salvaged, to land at the point of going out of production, if not salvaged. The total areas subject to desertification, in varying degrees of severity, are estimated as - by types of problem, in million hectares; water-logging (25). salinization (20), range deterioration (3200), dryland deterioration (250).

(c) Figures within parentheses give range of salvage costs. It follows from footnote (b) that cost of salvage is the maximum, equivalent to the cost of reclamation or restoration of practically completely desertified land. Because desertification is a continuous process, the move prudent course of action seems to be that corrective investment begin as soon as practicable and that it initially apply to saving lands which offer the highest returns to ensure continued maximum production.

(d) Degradation of irrigated land due to water-logging, salinization and to a lesser extent alkalinization.

Source: UNCOD, 1977

3. Poverty in Drylands

Though poverty is a relative term and it is difficult to draw a universal line for defining poverty, but the social and economic situations in which the people live and survive are not equal to all the people especially the rural based persons engaged in agricultural activities. The prime situation the people face is unequal distribution in terms of area and quality of land. Even in the areas severely affected with desertification situations, the well off people are least affected in terms of income and access to facilities. The poverty persists in view of unequal opportunities. The struggle for better quality land is visible in every society and the socially and economically powerful groups avail all the facilities like quality land, water and other inputs required for up keeping of the land. The environmental and ecological situations prevail in severely desertified areas are common to all the people residing there, but the incidence and impact of worst situations are also not equally distributed among the total society in each area. The numbers of people immediately threatened, their location and livelihood systems are shown in Table 2.4.

Table 2.4 : Estimated population and livelihood resident in areas recently undergoing severe desertification

(In thousand Nos.)

Region	Total Population	Urban Based	Cropping Based	Animal Based	Area (000 Sq. Km)
1. Mediteranean Basin	9820	2995 (31%)	5900 (60%)	925 (9%)	1320
2. Sub-Saharan Africa	16165	3072 (19%)	6014 (37%)	7079 (44%)	6850
3. Asia and the Pacific	28482	7740 (27%)	14311 (54%)	6431 (19%)	4361
4. Americas	24079	7683 (32%)	13417 (56%)	2979 (12%)	17545
Total	**78546**	**21490 (27%)**	**39642 (51%)**	**17414 (22%)**	**30076**

Source: UNCOD, 1977

Out of the 78 million people threatened, about one-third may be in a position, because of high income or other advantages, to avoid the worst consequences of desertification. This still leaves more than 50 million people, who are immediately menaced through the destruction of their livelihoods and who are faced by the grim prospect of uprooting themselves from everything familiar and of migrating to other areas frequently ill-equipped to receive them.

For the sake of elaboration of reasons of poverty as an impact of desertification and land degradation, it is necessary to classify them in to various categories as geographical and political reasons, area specific problems and peoples poverty. This is essential to elaborate the consequential impacts of desertification and land degradation.

(i) Geographical and Political Reasons

Underdevelopment and misguided government policies can combine with drought and desertification to cause hunger or even famine. The United Nations Conference on Desertification owed its genesis more to worldwide horror at the terrible effects of famine during the first period of drought in the Sahel in the early 1970s, than to concern about dryland environments alone. Famine generally occurs, when poor crop yields lead to severe shortage of food in a particular area and market mechanisms or government interventions are insufficient to balance these shortages with supplies from elsewhere in the country from overseas.

A period of prolonged, extended drought is obviously a contributing factor of poverty. Mickael Glantz (1987-a) has referred to it as "setting the stage" for famine. Amartya Sen, on the other hand, has asked "Why is it that producers of food are the first and most seriously affected by drought and famine and why do so fed town dwellers die from hunger, while rural areas are decimated by starvation and death" (Sen 1981). The implication is that famine is selective and the causes of famine are heavily influenced by social and economic factors.

Famine can arise fro misguided agricultural policies or the inadequacy of transport to ship food from one part of a country to another. Thus a country may have overall food surplus, yet some of its people can still starve. According to a report by Independent Commission on International Humanitarian issues, famine is really a "man-made disaster" and over-intensive agriculture, loss of topsoil and desertification help to establish the preconditions for its long before drought delivers the "coup de grace" (Brown 1985).

Numerous studies have linked famine with government policies that encourage the cultivation of cash crops at the expense of subsistence food crops, not necessarily with the laudable aim of preventing hunger. Government price policies, which forced small farmers to produce export crops at the expense of food crops, were blamed for the famine in Mali from 1969 to 1973 (Schmidt Wulffen 1985). These exports were needed to earn foreign currency to finance imports, create jobs in industry and increase the size of the governmental bureaucracy, but only farmers in favoured areas could buy enough food for their subsistence from the profits of cash-crop production. In northern Mali, neither cash-crop profits nor food supplies were adequate. No exchange of food was possible between different regions because of south of the country did not have a food surplus to meet the needs of the north. A similar criticism has been expressed about policies in Kenya (Hegmar *et. al.* 1982).

As Ghana and Nigeria have experienced, too great an emphasis on cash-crop can result a vicious circle, in which domestic food-crop production falls and an ever greater volume of cash crops had to be exported to pay for imports of food, which could easily be grown in the country itself (Anson-Meyer 1983). Rising food imports, whether through normal trade or as "food aid" from developed nations concerned about famine, can change consumers' taste away from locally grown staples and so depress local food production. Per capita consumption of wheat and rice in the Sahel, grew by 4.6% and 2.0% respectively between 1966-70 and 1976-80, while that of maize, millet and sorghum fell (Dregne and Miller 1985). Continuing demand for such exotic imported grain can bias a country's whole agricultural development efforts towards producing them, often by expensive irrigated cropping and this seriously hampers the rehabilitation of the subsistence farming sector. Rice production in the Sahel and other parts of West Africa has been exceeding in recent years, but not as rapidly as consumption. Only limited areas are suitable for wheat growing in the region. It is likely, therefore, that food imports will continue to rise and there is danger that increased emphasis will be placed on growing cash crops to pay for these imports. Furthermore, in times of drought, migrants are drawn to cities in search of food and employment, depriving rural areas of the farm workers needed to grow crops and further reducing agricultural production (Glantz 1987-b).

Famine may also occur as a consequence of serious political conflicts inside a country. Michel Glantz has pointed out that "Although, there were major droughts and climate-induced food shortages around toe globe in 1982-84, in various parts of Africa, India, China, Indonesia and Brazil, famines occurred only in Africa. Even a cursory review of historical record of any drought-plagued region shows that famine does not necessarily follow drought. Most recently. 31 countries in sub-Hararan Africa were declared to have drought-related food shortages. Only five of them, viz. Mozambique, Angola. Chad, Ethiopia and the Sudan, however, were plagued by famine. Each of those were also subjected to internal wars (Glantz 1987-a).

The misery of the peoples of Ethiopia and Sudan provides a good illustration of how, even in countries affected by drought, civil war can make the difference between and food shortages and famines. In Ethiopia, one million people died from starvation in 1984-85. because the government did not alert international agencies early enough to the food shortfalls likely to result from the drought (MacKenzie 1987). Although in the late 1980s, the supply of food aid was planned more in advance, with better coordination between the Ethiopian government and international agencies, the delivery of food supplies to badly affected areas in Eritrea and Tigray has been impeded by the civil wars raging there. Even the supply of food aid can be seen as a political weapon, the Eritrean rebels have organized their own relief supplies through Sudan, but have fired on UN food convoys. In one incident, twenty three trucks were destroyed. Malnutrition is, therefore, widespread and the threat of death from starvation ever present (Forse 1987, Buckoke 1988. Anon 1988-c).

Famine is therefore, as much a result of inadequate government policies and underdevelopment as it is drought. The danger is that if famine is blamed on drought, then the deeper socio-economic and agricultural changes needed to ensure food security for all would not be made, since these require long-term action and this is not attractive to politicians. Famine certainly exacerbates the land use pressures, which cause desertification, for when food is in short supply, people become desperate and over cultivate the land to get, even though they know that in long term, the land will be poorer for it.

(ii) Area Poverty

Area poverty generally refers to inadequate availability of infrastructure facilities and basic amenities. For these issues the local administration need to provide the requisite facilities as per requirement and as per availability of funds. The area poverty affects the people living in degraded areas particularly the human induced activities can be made good for retrieval. The geophysical conditions significantly affect the area resulting in low productivity, The problems faced by people in desertified areas are water-logging, salinity, alkalinity, which play significant role in decrease of the yield in view of constant degradation.

The cultivated area is the survival base of the people residing there and uneconomical cultivation due to decrease in productivity either discourages them to continue cultivation practices or shift to cash-crops which provide remunerative return. In both the situation, food production is discouraged and people have to migrate to other areas. Activities for upgrading the land are cost effective and the cultivator is unable to bear the expenses in view of inherent problems.

(iii) Individual Poverty

The most sufferers of land degradation and desertification are the poor as they possess small operational holdings and also deficient land. Land degradation further aggravates their problems in view of uneconomical holdings and uncertain monsoon conditions. The areas under desertification are generally arid and semi-arid lands where only rain-fed crops are possible, monsoon failure is general phenomenon in such areas and there the vicious circle is formed and poor people generally face various problems.

Smaller operational holding and inferior quality of land are the two significant factors for incidence of poverty. Incapability to avail high quality inputs and mechanized cultivation practices are also the reasons of low productivity from the available land resources. In dryland areas water is also an important factor, which is generally not in access to poor cultivators, as they have no voice to get benefit from the common resources of water like water harvesting units. Underground water sources are generally dried or low water charge sufficient to irrigate crops in case of monsoon failure.

In such areas, the employment opportunities are further discouraged in view of single cropped area and people are put to employment for few months in a year. They have no work relating to cultivation activities and for their survival, cutting of trees, destruction of vegetative cover and deforestation are done, which further worsen their situation.

2.2 EXTENT AND DISTRIBUTION OF DRYLANDS

Desertification occurs between soil and atmosphere, where plants thrive and a balance maintained between incoming and outgoing energy and between water received is lost. When rain falls, some of the water is taken up directly by plants, some tilters into soil, where it remain in storage and rest evaporates or run-off. Some soil moisture, intercepted by plants, is put back into the atmosphere by the plants in transpiration. Some of the moisture may seep into deeper layers to collect in underground reservoirs or aquifers, where it may remain for thousand years or may migrate slowly from plateau to depression on back to the ocean itself.

In arid situations, the cycling of water and energy takes on special characteristics because of deficient and variable rainfall and abundant solar energy fro cloudless skies. Vegetation is generally sparser than in humid areas, provides less cover to the ground surface and returns less organic matter to the topsoil. During occasional intense rainfall, run-off may occur in spate, but water at the surface tends rapidly to be lost through evaporation and in the long intervening dry spells, the soil is parched by powerful sun.

However, scanty it may be, the dryland vegetation constitutes a fundamental resource, which transforms solar energy into food and which protects as well as stabilizes the surface of the ground. This vegetation survives by adapting to water deficit in ways, which are important because they determine seasonal differences in the usefulness of dryland pastures.

Part of the plant population consists of short-lived ephemerals, which germinate and complete their life cycle rapidly after rains, remaining as seed through intervening dry periods. Such plants are commonly fleshy and palatable and are preferred by grazing animals. Other plants such as perennial grasses wither and die back to the root stock or to bulbs in dry spells and shoot anew with fresh rains. These plants form more durable pastures are attractive and palatable to stock, when green and may provide valuable hay, but are of little pastoral values, when thoroughly dried out. Nevertheless, their extensive fine root systems remain to bind the topsoil and contribute importantly to its organic content. Lastly, there are the longer-lived perennial plants, which resist water loss by such adaptations as woody stems and leathery leaves. These include the larger plants such as shrubs and trees, which remain nutritious during the dry periods, when they can provide an important food source to browsing animals, although their adaptations may reduce their palatability and attractiveness for some stock. They have the additional role of protecting the ground surface, providing shade and preserving an environment, which favours the response of important shorter-lived plants.

Under natural conditions and through appropriate strategies, the dryland eco-systems maintain a balanced exchange of water and energy, but a favourable equilibrium is rapidly disturbed, when man makes use of land, where meagre vegetation is further reduced to expose the ground surface, humus is mineralized and soil structure is lost. Rain falls directly on the soil and brakes it down. The sun bakes a thin crust, which prevents additional water from sinking in. As the water budget deteriorates in the soil beneath, the level of ground water in nearby wells may fall. The water lost to the soil store now contributes to over-rapid run-off, where the surface has been loosened or disturbed as by the trampling of animals, the topmost soil layer, with the best structure and containing the bulk of plant food. may be washed away or blown away in dust storms. The denuded soil is essentially infertile, with poor structure and water relations. All these changes constitute a shift towards a more hostile environment for plants, with the result that the vegetation responds less well to rain and produces less biomass and many plants tend to die off at a increasingly early stage of drought. Such changes are typical of desertification.

In the initial stage, desertification may merely involve a shift to a more desert-like and less productive eco-system, with water, energy and nutritional balances, less favourable to plant growth than before. Land use in arid regions poses problems, which continually menance the prevailing equilibrium. This is at least partly because of fluctuations in rainfall between drought years and good years, which are not predictable, are difficult for the land user to respond to effectively.

In dryland pastoral economies, large numbers of stock tend to build up during runs of good years, too many to be supported through inevitably ensuing drought. There is a natural reluctance to cut back on stock numbers in the first dry year and a tendency to hang on until drought is seen to be established. By the time, dryland pastures are probably being overgrazed towards a state that threatens eventual regeneration. By this time. too. prices for surplus stock have shrunk because the market is glutted and economic forces oppose destocking through sale of surplus numbers. For the same reasons, destocking may be prevented the periods critical to the regeneration of pastures following the rains that end drought.

Dryland farmers, too, show a tendency during runs of good years to extend their cropping onto over more marginal lands, into areas of higher climatic risks, pushing back the pastoralists in the process. This is particularly the case, when pressure on the land is increased through population growth or restrictive systems of land tenure or the shortsighted introduction of mechanization. The expectable but unpredictable onset of drought finds the marginal land prepared for planting, stripped of its protective natural vegetation and vulnerable to erosion. Such land enters a run of dry years without defenses and may emerge in too degraded a condition to support even livestock.

The main processes and stages of desertification are summarized here. In pastoral rangelands, there is an initial deterioration in the composition of pastures subject to excessive grazing in dry periods, particularly a reduction of proportion of edible perennial plants and an increase in the proportion of annual and inedible species. The thinning and depth of vegetation in dry seasons increases the extent of bare ground and this is followed in turn by a deterioration in the conditions at the surface of the soil that are vital to plant growth, particularly an impoverishment of plant-water relations. The response of ephemerals to rain suffers accordingly. With consequent increase in the run-off, sheet and gully erosion set in on sloping ground and the topsoil and its store are lost. All these changes mean a decrease of plant productivity and a decline in the palatability and durability of the native pastures. With continuing erosion, formerly productive lands may be lost through soil stripping and gully extension. These changes are even more drastic, where devegetation occurs in strategic areas, as on watershed uplands and the processes are advanced, where soils are exposed and disturbed in dryland cultivation.

In areas of rain-fed farming, desertification often originates on land cleared for cultivation or left fallow. Removal of the original vegetation cover exposes the soil to accelerated wind and water erosion. The heating action of rain on naked soil puddles the surface, which crusts when the sun comes out. reducing infiltration and further increasing run-off. This in-turn, leads to increased soil erosion, which ultimately, unless halted by protective measures, strips away the fertile surface soil and exposes infertile sub-soils. Gullies may form on the lower parts of slopes and impede farming operations or prevent them entirely. Sediment deposited at the foot of slopes covers plants, fills waterways and aggravates flooding in low-lying areas, which follows increased run-off from the slopes above.

Water and wind erosion work together, as redeposited silts from surfaces stripped by water erosion are particularly vulnerable to wind transport. Wind erosion starts with the movement of coarse soil particles in one part of a field, and then progresses downwind with increasing severity as bouncing soil particles knock other particles into the air in a kind of snow-balling effect. Finer materials are litted into the air and carrying away over long distance as dust; coarser sandy materials drift over the surface until they are trapped by plants in accumulation as hummocks and small dunes. Removal of the topsoil materials means the loss of the most productive and nutritious portions of the soil complex, while sterile sand accumulations cover plants and good soil. A further harmful effect of high-velocity sand drift is the destruction of young crops by the blasting impact of moving sand. Finer airborne particles may carry soil-borne diseases, irritate respiratory tracts of humans and animals, cause wear on machinery parts and reduce visibility.

The principal manifestations of desertification on irrigated lands are the sanilization and alkalinization of soils, due to inadequate leaching of salts contained in the soil or excessive irrigation water. Salinization and water logging commonly occur together. The upward movement of saline groundwater leaves salts on the surface, where water evaporates or soil is waterlogged. On soils that are not waterlogged, salinization can still occur when water containing soluble salts moves from irrigation furrows into the ridges, where crops are planted or to high spots in poorly leveled land. Under-irrigation of weakly permeable soils can also lead to salinization, if irrigation water is salty.

When left alone, dryland ecosystem disturbed by land use or stressed by drought, usually returns to their earlier degraded shape. Recovery tends to advance at a slow pace, because of more rapid recovery in years of above-average rainfall. Eventually, former water and energy balances are restored, with the recovery of the original vegetation. This is a measure of the natural resilience of the drylands.

Desertification is a global phenomenon, which afflicts drylands in more than 100 countries, of which about sixty with substantial areas of drylands are the most critically affected. Few large-scale measurements have been made, desertification is still loosely defined and too much reliant on subject assessment of experts for these estimates. The shortage of reliable data on the extent of desertification and on its social and economic impact, leads to skepticism among the scientific community about the reality and relevance of the phenomenon and is a major obstacle to securing the political and financial support of governments and international agencies for programmes, which could bring it under control. Improving the monitoring of desertification is, therefore, of the utmost importance. Consequently, the degree and extent of desertification in an area is based on the data collected through remote sensing and estimates with far from accuracy than in the past.

Estimating the actual extent of desertified land is much more difficult than producing general assessments. The ideal procedure for making estimates can involve:

(i) Defining a set of categories that may distinguish between different degrees of degradation;

(ii) Measuring the area and degree of desertification throughout each country by acombination of ground tests at sample plots and large area surveys by remote sensing techniques such as aerial photography and satellite imaging; and

(iii) Classifying the resulting data to arrive at estimates of desertified areas disaggregated by land use and degree of desertification.

Global estimates of desertification could be made either in the decentralized way, with governmental agencies in each country collecting data and forwarding them to an international agency like UNEP for collection or in a centralized manner, with the international agency making measurements itself, using its permanent staff, together with consultants and local personnel in the countries being monitored. In practices, however, estimates of the extent of desertification have largely depended upon the subjective assessments of consultants appointed by the UNEP. These assessments have been based upon a limited amount of local data, little of which have been measured and categorized in a rigorous manner.

1. Criteria for Assessment of Desertification

The extent of desertification in an area is typically categorized as slight, moderate, severe or very severe, but the criteria adopted for placing land in these categories have so far been fairly general. Slight desertification is normally disregarded, when making estimates of desertified areas, so only land in other three categories is included. Thus all estimates of desertified areas, quoted imply that the land suffers from "at least moderate desertification."

The World Map of Desertification prepared by FAO, UNESCO and the World Meteorological Organization (WMO) was one of the four maps commissioned by UNCOD. Although it later became the most widely publicized, it shows only the degree and extent of desertification hazard as moderate, high or very high by a subjective evaluation of climatic conditions, the inherent vulnerability of the land and human or animal pressure (FAO/UNESCO/WMO 1977). The first proper map of the status of desertification, another of the four UNCOD maps, was prepared by Harold Dregne. He used separate definitions of desertification status for rangeland, rain-fed cropland and irrigated cropland. He defined moderate desertification in terms of a significant increase in undesirable shrubs for rangeland and small dunes or gullies for rain-fed cropland and a reduction in the yield of irrigated crops of up to 50% owing to soil salinity. In his definition of severe desertification, undesirable shrubs dominated the flora, large gullies were present, or sheet erosion by wind and water had largely denuded the land of vegetation and irrigated crop yields were less than 50% of the potential levels owing to salinity. Very severe desertification resulted in large, shifting, barren sand dunes, large, deep and numerous gullies or the development of salt crusts on almost impermeable irrigated soils (Dregne 1977).

Dregne used a rather more specific set of criteria for his second assessment in 1983. He classed desertification as moderate if 26 to 50% of the plant community consisted of climax species, 25 to 75% of the original topsoil has been lost and soil salinity had reduced crop yields by 10 to 50%. Severe desertification required that only 10 to 25%) of the plant community consisted of climax species, soil erosion had removed all or practically all of the topsoil and soil salinity had reduced crop yields by more than 50%. On very severely desertified land, less than 10% of the plant community consisted of climax species, the land would be covered by many sand dunes or deep gullies and salt curst would have developed on irrigated cropland (Dregne 1983-a).

In his 1984 global assessment of desertification for UNEP, Jack Mabbutt classed desertification of rain-fed and irrigated croplands as moderate, if there was widespread erosion and salinization and water-logging and losses of up to 25% of crop production, severe if production losses were between 25% and 50% and very severe if they exceeded 50%. Rangelands were moderately desertified, if there was a significant reduction in vegetation cover and deterioration in species composition, a significant level of soil erosion and 25% decline in livestock carrying capacity. They would be severely and very severely desertified, if proportional reduction in carrying capacity were between 25 to 50% and greater than 50% respectively. Very severely desertified rangelands were those which could not be reclaimed economically (Mabbutt 1984).

Specially commissioned for UNCOD, identified the degree of potential desertification hazard to which dryland area was exposed. An estimated 37.6 million sq. km. Of drylands, tow-third of which were in Asia and Africa, were said to be at serious risk of desertification, as mentioned in Table 2.5. This map was not of the status of desertification and it was not known about the degree of desertification actually occurring within the vulnerable areas.

Table 2.5 : UNCOD estimates of desertification

S. No.	Continent	Area (Million Sq. Km.)	% of total
1.	Asia	13.7	36.4
2.	Africa	10.4	27.7
3.	Australia	5.7	15.2
4.	Europe	0.2	0.5
5.	North and Central America	4.3	11.4
6.	South America	3.3	8.8
	Total	**37.6**	**100.0**

Source: FAO/UNESCO/WMO

2. Estimates of Extent of Desertification

Two sets of estimates of the actual extent of desertification were published in 1983 and 1984 by Harold Dregen and Jack Mabbutt respectively. Harold Dregne estimated that 32 million sq. km. of the world's drylands were suffering from moderate or severe desertification, i.e., "at least moderate desertification." This represented almost on quarter of earth's surface and was close to the total area thought at the time of UNCOD to be at risk of, but not necessarily suffering from desertification. Of the affected area of 32.488 million sq. km, 30.486 million sq. km. were rangelands, 1.731 million sq. km. Rain-fed croplands and 0.271 million sq. km. irrigated croplands as given in Table 2.6 gives continental position of these areas.

Table 2.6 : Dregne's assessment of areas suffering at least modern desertification

(Thousand Sq. km.)

Continent	Rain-fed Cropland		Irrigated Cropland		Rangeland		Total	
	Area	%	Area	%	Area	%	Area	%
Africa	396	1.22	14	0.04	10268	31.60	10678	32.86
Asia	912	2.81	206	0.63	10889	33.51	12007	36.95
Australia	15	0.05	2	0.01	3070	9.45	3087	9.51
Europe *	42	0.13	9	0.03	155	0.48	206	0.64
North America#	247	0.76	28	0.08	2910	8.96	3185	9.80
South America	119	0.36	12	0.04	3194	15.84	3325	10.24
Total	**1731**	**5.33**	**271**	**0.83**	**30486**	**93.84**	**32488**	**100.00**

Source: Dregne (1983-a).

Jack Mabbutt's estimate was much lower. Using data obtained from replies to a special questionnaire circulated to governments of all dryland countries by UNEP and the findings of regional studies specially commissioned as part of the same programme, he estimated that 20 million sq. km. of land in the world were suffering from at least moderate desertification and of this 16.16 million sq. km. were rangeland, 3.46 million sq. km. rain-fed croplands and 0.40 million sq. km. irrigated croplands as detailed continental position in Table 2.7. Mabbutt's assessment (personal communication 1988) is that the overall area of desertified land is between 15 to 20 million sq. km.

According to Mabbutt's estimate, the world's most desertified areas were the Sudano- Sahelian region, Africa south of the Sudano-Sahelian region, North Africa and West Asia, the "Middle East," all of which had 74.4% of the total dryland areas of the world. Developing countries with 60% of their drylands desertified, were more affected than developed countries for which the corresponding proportion was only 40%. Of the 20 million sq. Km. of productive land suffering from at least moderate desertification, 4.2 million sq. km. (21%) were in Sudano Sahelian region and 3.3 million sq/ km. (16%) elsewhere in Africa; 2.5 million sq. km. (13%) in India and Pakistan and 5 million sq. Km. (25%) in other parts of Asia. Asia and Africa each account for 35-40% of all desertified rain-fed croplands and rangelands, but Asia alone had 70% of all irrigated land affected by desertification, of which 57% was in India and Pakistan (Mabbutt 1984, 1985).

Table 2.7 : MABBUTT'S assessment of areas suffering at least moderate desertification

(Thousand sq. km.)

Continent	Rain-fed cropland		Irrigated Cropland		Rangeland		Total	
	Area	%	Area	%	Area	%	Area	%
Africa	**1290**	**6.4**	**19**	**0.1**	**6100**	**30.5**	**7409**	**37.0**
Sudano-Sahel	720	3.6	8	0.1	3420	17.00	4148	20.7
South of Sudano-Sahel	420	2.1	6	Neg.	2000	10.0	2426	12.1
North Africa	150	0.7	5	Neg.	680	3.5	835	4.2
Asia	**1350**	**6.8**	**280**	**1.4**	**5850**	**29.2**	**7480**	**37.4**
West Asia	150	0.7	30	0.2	980	4.9	1160	5.8
South Asia	1050	5.3	200	1.0	1270	6.3	2520	12.6
USSR Asia	120	0.6	20	0.1	1500	7.5	1640	8.2
China and Mangolia	30	0.2	30	0.1	2100	10.5	2160	10.8
Australia	120	0.6	3	Neg.	1000	5.0	1123	5.6
Middle Europe	130	0.6	16	0.1	150	0.8	296	1.5
North America	240	1.2	40	0.2	1800	9.0	2080	10.4
South America and Mexico	330	1.7	40	0.2	1250	6.2	1620	8.1
Grand Total	**3460**	**17.3**	**398**	**2.0**	**16150**	**80.7**	**20008**	**100.0**
Alternative Total*	3350	9.6	400	1.2	31000	89.2	34750	100.0

* After adjustment to include remote, non-productive rangelands.

All percentages are of the grand total to denote the area proportionate to each category and its relationship with the total dryland areas.

Source: Mabbutt (1984).

Most of the difference between the two estimates of the total desertified area was due to the disparity in their rangeland components; 30.5 million sq. km. in Dregne's estimate but only 16.2 million sq. km. in Mabbutt's. On the other hand, it should be noted that Dregne's estimate of the area of desertified rain-fed cropland was only half of the figure, which Mabbutt gave and that for irrigated cropland was just two-thirds. Both experts agreed that Asia had the largest area of desertified irrigated cropland of any region, but Dregne considered that it also had by far the largest area of desertified rain-fed cropland.

There are two main regions why Mabbutt and Dregne produced such different estimates: a lack of reliable data and differences in overall areas of drylands within, which desertified lands were identified. Dregne's study assessment covered a smaller area than Mabbutt's and included hyper-arid, arid and semi arid lands with an overall area of 47 million sq. km. On the other hand, Mabbutt's study also covered to 60.8 million sq. km. area as detailed in Table 2.8.

Table 2.8 : Distribution of drylands by continents (Including Sub-humid Areas)

S. No.	Continent	Area (Million Sq. Km.)	% to Total
1.	Africa	21.5	35.36
2.	Asia	20.7	34.05
3.	Australia	6.5	10.69
4.	North America and Mexico-Mexico	6.1	10.03
5.	South America	4.8	7.90
6.	Europe	1.2	1.97
	Total	**60.8**	**100.00**

Source: Mabbutt (1983)

The smaller areas of desertified rain-fed and irrigated croplands in Dregne's estimates are largely result of the fact that his study did not include the sub-humid zone. The great difference between the two estimates of the area of desertified rangelands is explained in the following way. Both Mabbutt and Dregne excluded about 6 million sq. km. of slightly desertified areas in natural deserts, such as the Sahara, Arabian and Takla Makan, where the natural biological productivity was too low to support cropping or grazing and so human impact was minimal. However, Mabbutt also left over other large areas of remote, difficult or unwatered terrain, classified for convenience as rangelands but seldom, if ever used as such. He assumed that only about 30 million sq. km. of the total area of 47 million sq. km. of hyper-arid, arid and semi-arid lands were productive in an agricultural sense (Mabbutt 1984).

To achieve a greater degree of comparability with Dregne's estimate, Mabbutt, at the request of UNEP, made a second estimate, which included these remote, unproductive "rangelands". The area of desertified rangelands increased from 16.2 to 31.0 million sq. km., the area of desertified irrigated cropland remained about the same, but for some region, the area of desertified rain-fed cropland was slightly lower than the first estimate. The total area of drylands suffering from at least moderate desertification rose to 34.75 million sq. km., which was close to Dregne's estimate. On the basis of the larger total. Mabbutt estimated that about 80% of all rangelands, 60% of rain-fed croplands and 30% of irrigated lands were already at least moderately desertified. Severe desertification, in which productivity has fallen by more than a half, affected 33% of all rangelands, 30% of all rain-fed croplands and 10% of all irrigated lands, a total area of almost 1 5 million sq. km.

3. Social Impact of Desertification

Desertification has a tremendous impact on the lives of people in the drylands. Jack Mabbutt estimated that about 280 million rural people are affected by at least moderate desertification, a figure, which rises to 470 million, i.e., 10% of the world population, if urban dwellers in these areas are also considered. Most of the rural people affected by desertification live in rain-fed cropland areas in the Sudano- Sahelian region, Africa south of the Sudano-Sahelian region, the

Andean Region of South America and parts of South Asia, particularly Nepal. Farm-based populations account for 85% of the rural population in the dryland but occupy only 15% of the dryland area, so the social impact of desertification caused by rain-fed cropping is likely to be quite serious (Mabbutt 1984). Overall, fewer people are affected by the desertification of irrigated lands; the rain-fed and irrigated farming populations are similar in Asia. But this will cause the greatest losses in production and investment, particularly, in West Asia and in South Asian countries like Pakistan. Table 2.9 possesses all continental details.

Table 2.9 : Mabbutt's assessment of population affected by at least moderate desertification

(Million of People)

Continent	Rain-fed	Irrigated	Rangeland	All Moderate	%	All Severe	%
Africa	79.00	3.50	25.50	108.00	38.4	61.00	42.8
Sudano- Sahel	36.00	1.50	13.50	51.00	18.1	27.50	19.3
South of Sudano-Sahel	32.00	1.00	8.00	41.00	14.6	25.00	17.5
North Africa	11.00	1.00	4.00	16.00	5.7	8.50	6.0
Asia	56.00	50.00	17.00	123.00	43.7	55.50	38.9
West Asia	16.00	12.00	4.00	32.00	11.4	8.50	6.0
South Asia	34.50	23.00	9.00	66.50	23.6	16.00	11.2
USSR Asia	1.50	4.50	1.00	7.00	2.5	29.00	20.3
China and Mangolia	4.00	10.50	3.00	17.50	6.2	2.00	1.4
Australia	0.10	0.10	0.03	0.23	0.1	6.50	4.6
Middle Europe	13.00	1.50	2.00	16.50	5.9	0.03	Neg.
North America	2.00	1.00	1.50	4.50	1.6	6.00	4.2
South America and Mexico	22.50	2.50	4.00	29.00	10.3	13.50	9.5
Grand Total	**172.60**	**58.60**	**50.03**	**281.23**	**100.0**	**142.53**	**100.0**

Source: Mabbutt (1984)

4. Estimates of Rates of Desertification

The only available estimates of the annual rate of desertification is slightly revised version of one made by Harold Dregne for UNCOD, according to which over 202,000 sq. km. of agricultural land was said to be desertified each year to the point of yielding zero or negative net economic returns. In the revised estimates, this comprises 1,77,000 sq. Km. or rangeland, 20.000 sq. km of rain-fed cropland and 5.460 sq. km. of irrigated cropland. The potential value of agricultural production lost every year in the late 1970s was estimated at more than $ 26 billion as indicated in Table 2.10. The rate of desertification of 2,02,460 sq. km. per annum compares with an estimated rate of clearance of open woodlands in the drylands of 38,000 sq. km. per annum, of which Africa accounts for 23,000 sq. km. (Lanly, 1982).

Table 2.10 : Estimate of annual rates of desertification

S. No.	Area Category	Area (Sq. km./yr)	% to Total
1.	Rangelands	1,77,000	87.4
2.	Rain-fed Croplands	20,000	9.9
3.	Irrigated lands	5,460	2.7
	Total	**2,02,460**	**100.0**

Source: Dregne (1983-a)

No new estimate of desertification rates emerged from UNEP's special assessment in 1982-83, which included Mabbutt's 1984 study and contributions by Leonard Berry (1984-a), although qualitative indications were given of trends in the rates of desertification. Mabbutt commented that desertification was accelerating on rangelands in Africa, couth of the Sahel and in Andean South America, as well as on rain-fed croplands in tropical Africa. South Asia, South America and sub-tropical Mexico. Desertification of irrigation lands was relatively "static" everywhere. Deforestation rates were either unchanged or accelerating in most drylands, particularly in Africa, Latin America and parts of south Asia. Depletion of groundwater reserves was important in West Asia, Mediterranean Africa and the USA (Mabbutt 1984).

Leonard Berry reported that desertification continued to worsen in the nineteen countries of the Sudano-Sahelian region. He used some data from the responses of the UNEP Desertification Questionnaire, heavily supplemented with data from UN agencies and knowledgeable individuals. Each country was ranked according to whether the rate of desertification since 1977 had remained stable or increased slightly or significantly in five main categories: sand dune encroachment, rangeland deterioration, forest depletion, deterioration of irrigation systems and degradation of rain-fed cropland. Berry concluded that there was a "moderately worsening situation in 12 countries and a seriously worsening situation in 7. The picture is a somber one but realistic according to the judgement of most field observers" (Berry 1984-a).

Harold Dregne took a different view, remarking that desertification rates were particularly high in the rangelands and cultivated lands of Morocco, Egypt, Sudan, Ethiopia, Kenya, Somalia, Iraq, Iran. Pakistan and China. Rates of degradation of rangelands in the developing nations as a whole were static and either static or declining in developed countries. On the other hand, the erosion of soil by water was increasing in Ethiopia and North Africa and the wind erosion of rain-fed croplands had become more serious in Sahel, Ethiopia, India and in the USA, Australia and the USSR. Dregne agreed with Mabbutt that the rate of salinization and water-logging of irrigated lands was not increasing, although the seepage of saline groundwater to the surface of rain-fed croplands was becoming more of a problem in Australia, Canada and the USA (Dregne 1983-a. 1985).

2.3 LAND-USE IN DIFFERENT DRYLAND REGIONS OF THE WORLD

Desertification affects drylands all over the world, but tends to be concentrated in Asia and Africa, each of which accounts for 37% of the desertified land. The general review of desertification is mainly qualitative in order to highlight its regional distribution and to avoid many of the problems arising from inadequate quantitative estimates, such as those contained in the global assessments of Mabbuti (1984) and Dregne (1983-a) described in detail.

It is usually difficult to pinpoint a dominant cause of desertification in a particular region, since all four main direct causes are active throughout the drylands. However, there are regional distinctions and these are indicated where appropriate. Overgrazed rangelands account for more than 90% of the total desertified area, largely because grazing is the only feasible use for the majority of arid and semi-arid lands, although this should not obscure the importance of the other causes. Over-cultivation and poor irrigation management affect only a small percentage of the total desertified area but have a considerable social and economic impact, because the vast majority of the rural population in the drylands lives on farms.

Over-grazing is often an indirect consequence of the expansion of subsistence and cash-crop cultivation, which forces herds from traditional pastures on to drier and more marginal lands. The impact of deforestation is much higher than estimates of annual deforestation rates, although these are quite substantial, indicate that deforestation not only degrades the vegetative cover, but also makes drylands more vulnerable to desertification caused by overgrazing, over-cultivation and poor irrigation management. These estimates also understate the true magnitude of the reduction in tree cover as they ignore the removal of scattered trees, which have an important role in protecting soil from erosion.

(i) Africa

More than 80% of Africa's drylands are moderately or severely desertified. Africa accounts for over a third of all desertified land in the world and for a similar proportion of all desertified rain-fed croplands and rangelands. The region is most affected by desertification in the Sudano- Sahelian region. It consists of four main dryland zones, most of which are arid, semi-arid: the Sahel together with Mauritania and the Cape Verde Islands; the more humid Sudan savanna zone to the south of the Sahel; northeast Africa and the East African countries of Somalia, Djibouti, Kenya and Uganda. To the south of the Sudano-Sahelian region are Tanzania, Zimbabwe, Mozambique and other countries in Central and East Africa; the southern Africa, including Lesotho. Botswana and Namibia. The more developed North Africa region consists of Morocco. Algeria, Tunisia, Libya and Egypt.

Overgrazing and over-cultivation are widespread both in the Sahel region and on the northern fringe of the Sahara in Morocco. Algeria, Tunisia and Libya. Degraded lands in these two regions are subject to serious wind erosion and in extreme cases are overrun by sand dunes and drifting sand. Sand dune encroachment on rain-fed croplands is also a critical problem in coastal areas of Senegal, Somalia and other countries. Overgrazing and over-cultivation in Ethiopia, Somalia, Sudan and northern Kenya and in Botswana, Namibia and Zimbabwe, lead to problems similar to those in the Sahel, although deforested highlands in Ethiopia and Lesotho also suffer from serious water erosion. Deforestation exposes soils to erosion by wind and water and is occurring rapidly all over the continent. Grey haloes of deforested and degraded rangelands spread outwards from cities for distances of up to 100 kms or more as widespread in Africa as it is in Asia, but the salinization and water-logging of irrigated lands is prevalent in Egypt, Algeria, Tunisia and parts of Sudan, in Ethiopia and Somalia and to a lesser extent in Niger, Chad and Gambia. In Tunisia and other North African countries, the introduction of intensive mechanized cultivation techniques such as disc ploughing has led to soil erosion on fragile lands.

(ii) Asia

An estimated 36% of the Asia's drylands are desertified and they account for more than a third of all the desertified lands in the world. Asia has similar areas of desertified rangelands and rain-fed croplands to those of Africa, but contain 70% of all irrigated croplands affected by salinization and water-logging. Dry areas are mainly found in the western part of the continent in West Asia, Central Asia and south Asia. West Asia may be divided for convenience into two areas. The first includes Iraq, Syria, Jordan and other countries with semi-arid and sub-humid climates located close to the Mediterranean or in the Fertile Crescent of rivers. Tigris and Euphrates, where substantial areas of irrigated croplands are affected by salinization and water-logging

The second area is Arabia, where the extensive hyper-arid and arid lands are mainly used for grazing and rangeland degradation is the prominent form of desertification. Overgrazing is important in the extensive Central Asian drylands of China, Mongolia and the USSR, although salinization and water-logging of irrigated croplands are found there too. In South Asia, overgrazing is common on the large areas of rangelands in Iran, Afghanistan and the Indian State of Rajasthan, while salinization and water-logging are widespread in the vast irrigated croplands of the Indo-Gangetic Plain in Pakistan and India, which together account for up to 40% of all desertified irrigated croplands in the world. Water erosion of rain-fed cropland occurs all over Asia, but is a particular problem in northern Pakistan, parts of western India and on the loess plateau of China, where the Yellow River derives its name from the colour imparted to it by the huge loads of eroded soil it has carried for thousands of years.

(iii) Mediterranean Europe

Europe's semi-arid and sub-humid drylands are located around the Mediterranean in Spain, Portugal, Italy, Greece, Turkey, Cyprus and Malta. The region was the centre of western civilization for thousands of years, but large-scale deforestation, overgrazing and over-cultivation during and since that time left the vegetation heavily degraded and the soils in hilly and mountainous areas severely eroded. More recently the problem has been somewhat ameliorated by the afforestation of uplands and the control of grazing and cultivation. A fifth of the region's drylands are thought to be desertified but they account for a mere 1% of all desertified land in the world. Salinization and water-logging of irrigated cropland are serious problems in Spain, particularly in the Guadalquivir Valley in the south of the country and on the Ebro River watershed in the northeast and also in Greece and Portugal.

(iv) Australia

Australia accounts for about a tenth of all desertified land in the world. Most of the arid interior and the semi-arid or sub-humid lands in the north of the continent are only used as rangeland. Irrigated and rain-fed cropping are concentrated in moister coastal areas, although rain-fed cropping also extends to semi-arid lands in the interior. Grazing does not date back more than 125 years and as a result of this and a progressive improvement of range management, as little as a quarter of all Australian rangelands have so far been desertified. Over-grazing has degraded the vegetative cover by replacing perennial shrubs and grasses with annuals and it exposes the soil to wind and water erosion, which in the most extreme cases leads to gullying. About 30% of the rain-fed croplands are affected by soil erosion, due to over-cultivation, but another serious problem affecting rain-fed croplands in parts of West Australia and South Australia is the saline seepage, which results indirectly from the deforestation of watersheds. When tree cover is removed, the amount of water lost by evapo-transpiration into the atmosphere falls, groundwater levels rise and most groundwater flows down into the river basin, becoming saline as it dissolves salts from lower soil levels. Some of the water flows into rivers and makes them more saline, while the remainder builds up in neighbouring lowland agricultural areas, which then become water-logged and saline. Irrigated croplands in the Murray River system of Victoria and New South Wales have also been water-logged and salinized by more usual cause of poor irrigation management.

(v) North America and Mexico

North America and Mexico account for about a tenth of all desertified land worldwide. The arid zone includes northern Mexico and parts of California, New Mexico, Arizona, Utah and Nevada in the United States, but most of the key grain-growing areas in the Great Plains and Prairie regions of the USA and Canada are semi-arid and over-cultivation there has caused severe wind erosion of the soil. Although, large-scale soil conservation schemes were introduced in North America after notorious Dust Bowl episode of the 1930s, soil erosion still continues and saline seepage is also a problem. Overgrazing has led to widespread degradation of vegetation and soil on the dry rangelands of the USA and Mexico. Salinization and water-logging affect irrigated croplands in a number of valley along the Colorado and Gila Rivers, including the San Joaquin Valley in California, the Imperial and Mexicali Valleys and lower valley of the Rio Grande.

(vi) South America

Dry areas in south America are essentially confined to northeast Brazil, Argentina; areas on the Caribbean coast of Colombia and Venezuela; the Andes Mountains and the narrow western coastal strip in Chile, Peru and Ecuador centred on the Atacama Desert, one of the driest regions in the world, which receives less than 10 mm of annual rainfall. South America accounts for about tenth of the all desertified land. Deforestation and over-cultivation of rain-fed crops are causing serious soil erosion in the densely populated Andes and northern Brazil. Overgrazing has degraded vegetation and caused wind and water erosion of soil on the extensive rangelands in Argentina and in the Andes, northeast Brazil and other northern coastal areas. Irrigated croplands in western Argentina, northeast Brazil and along the coast of Peru are affected by salinization and water-logging.

2.4 EXTENT OF DESERTIFICATION AND LAND DEGRADATION IN ASIA CONTINENT

In Asia and Pacific region, land use forms threatened by the process of desertification occur not only in arid, semi-arid and sub-humid areas, but also in sub-humid zones. The forms of land use involved comprise extensive grazing of rangelands and wasteland, including nomadic pastoralism in the driest areas, rain-fed cropping and irrigated farming in semi-arid and sub-humid areas, as well as removal of vegetation for shifting cultivation . fuel, fodder and construction purposes from sub-humid and humid areas. Increased frequency of occurrence of severe droughts and devastating floods during the period of last two decades should be taken as warning to people from nature to stop its indiscriminate activities to exploit natural resources, as otherwise they had to face the problems of desertification.

The problem has now threatened much larger areas than ever before. The human and livestock population of desert areas are exerting great pressure on adjoining productive lands to meet their requirements for food and fodder. The pressure is so great that many areas in central part of India adjoining the Rajasthan Desert have become desertified within last three-four decades.(Pandeya 1982). Decrease in land productivity is an exponential function of the dynamic pathway of desertification. Viewing the problem in the context of the eco-system, once consumption increases more than primary production, the homeostatic level of the eco-system is disturbed and can no longer meet the requirement of secondary producers.

Human activities affect the landscape directly through commissioning the land for village and town expansion, agricultural purposes, fuel, urbanization, industrialization etc. and indirectly by permitting animals to overgraze, intensive agriculture and maximization of yield by adding subsidies in terms of fertilizers, pesticides and irrigation water beyond the limits of positive feedback and lowering the fertility of land by export of herbage as hay and thus the energy and minerals. Therefore, in fragile eco-system, the loss of biological productivity can easily become non-renewable, permanently reducing their capacity to support human life and may lead to the formation of deserts. The process of self-accelerating, feeding on itself and as it advances through time and space the cost of conservation and rehabilitation rise exponentially (Jain 1989).

It is interesting to note that the Asian countries are constituted in different climatic zones and accordingly possess specific problems quite different from other countries. Specific issues of desertification and land degradation in various countries are narrated on pinpointed issues.

1. Afghanistan

Afghanistan's climate and geography are the two main inter-related factors of agricultural economy, primarily dependent on irrigation as also grazing and limited dry farming are based. There is a great potential for increasing agricultural production in the country through adoption of modern technological practices and by expansion of irrigated farmlands. Barren mountains and deserts are the two main geographical features of the country, although its total geographical area of 6,57,500 sq. km has great variation of natural conditions (Hassanyar, 1977). Changes in altitude and relied are the main determinant of such diversity of natural conditions. Within the wide altitude range of 250-7450 m in Afghanistan, deserts account for about 40% of the total land area. Deserts are mainly found in the limited low land areas up to 500 m altitude and the wide areas of mountain plateaus of 500-1500 m altitude (Joyan 1968, 1988).

Irrigation is crucial and top priority factor for agricultural development of Afghanistan. From the 8 million hectares of land under cultivation in the country, 5.30 million hectare is suitable for cultivation. Of this 2.8 million hectares is presently irrigated and remaining 2.5 million hectares in unirrigated. Approximately, 55 billion cubic meter of water is available in the country annually, but only 20% is used in irrigation. For remaining areas, 30 cubic meter water is also required to irrigated remaining area fit for irrigation. Dust-E-Omakay and Dush-e-Zeray are the two small desert areas located in Kabul and Kandahar provinces, respectively.

Dust-E-Omakay is the desert area located in Zabul province about 350 km. south-west of Kabul. It lies between north latitude 31° 55' and 32° 22' and east longitude 66°33' and 67°13' (Dawlaty 1978). The area of this desert is 108810 hectares of which 1.4% is irrigated land, 19.77% is dry land, 70.55% is virgin land and 8.2% is not suitable for farming. The soils of Omakay area are divided into the capability classes II, III, IV, and V, occupying 34%, 9%, 30% and 27%

respectively (Dawlaty, 1978; WAPECA 1980). Geomorphologically, most of the land area in the reason of foot hill sloping plain and some are in the hilly plain regions. The irrigated and dryland areas in neither of the two geomorphological regions have problems of stoniness, erosion, soil texture and salinity. The virgin land areas in both the regions, particularly in hilly plains), have considerable level of the types of problems mentioned.

Vegetation in Omakay area is developed under the desert conditions. The favourable period for growth of natural vegetation prevail from February to May. The sloping plains are covered mainly with grasses, with few bushes of wild almonds. In the hilly plain, the vegetation mainly consists of grasses, but they are very scarce with predominance of small bushes. Omakay has dry continental climate with dry hot summer and moderate winter. Annual precipitation is 244, 3mm, 50% of which comes in the spring and 30-35% in the winter. Average annual evaporation is 149 mm; it reaches to the highest 399 mm in June (Joyan, 1968, WAPECA, 1980).

Dust-e-Zeray desert area is part of the Kandahar province located about 520 km south-west of Kabul. Geomorphologically, Zeray area is divided in the hilly plains slightly undulating and flat plains regions. The 70389 hectares of land in all of the three regions, are divided in to the land use classes: (1) Irrigated gray-brown, (2)Virgin lands, (3) gray brown meadows and unsuitable for farming covering 2.6%, 83.33%, 75% and 13.24% of the total respective areas. The land capability classes identified are classes II, III, IV and V occupying 6%, 15%, 62% and 17% of the total area respectively. From the land use classes, virgin lands have some problems of erosion and stoniness and very little problems of salinization. Other land use classes do not have these problems.

Natural vegetation in Zeray area is developed under short sub-humid and dry climate conditions. Grass vegetation grows from march to May and drought resisting bushes grow from may to December (Dawlaty 1978). The climate of Zeray is sharply continental. Average annual precipitation is 152 mm, 55% of which falls in spring. Average monthly evaporation is 242.5 mm with highest peak of 437 mm in July. Present land use and cropping pattern in the villages close to both the desert areas has been adopted under the constraints of soils, climate, rainfall, irrigation, local needs, market demands, infrastructural facilities, institutional supports and financial resources.

The estimate of total land area of Omakay indicates that only 1.47% of it is irrigated. The survey of land holders in the nearby villages show that 52% of these farm lands are irrigated. The main source of irrigation water fro the existing farm lands are karezes, springs and dug-wells. For efficient water transport, distribution and field application various local, traditional methods and practices are applied (UN 1985). In the farms near to Omakay, 81.4 % of irrigated lands are under crops and 18.6% grow orchards. Wheat is by far the dominant field crop, almond, grapes are the main orchard products grown there and forage and vegetables are also grown under dry farming system. Orchard plantation, except vineyards, are mostly of the mixed type (Dawlaty 1978).

The landholders survey near to Zeray area shows that 47.5% of these farm lands were irrigated and 52.5% unirrigated. In the irrigated land here, 85% are under field crops and 15% are under orchards. In addition to wheat and grapes, cotton is the main industrial crop and alfalfa is main forage crop. Arghandab River, Karezes, springs and dug wells are the main source of cultivation. Keeping livestock for draft as well as for production purposes is the complementary aspect of farming in Afghanistan.

2. Bangladesh

Bangladesh is a riverine deltaic country and is primarily dependent on the water resources fed by three major rivers namely, Brahmputra, Ganges-Jamuna and Meghna originating outside her national boundaries. Bangladesh is located between latitude 20°30' and 26°45' N and longitude of 88° and 92°45' E and has an area of about 1,44,000 sq. km. Bangladesh is one of the most densely populated areas of the World with population density in certain areas of around 772 persons per sq. km. The prominent characteristics of the topography of the country is riverine flat with the water level of the rivers mostly coinciding with land surface. Towards the east, the land is transversed by the folded mountain ranges of the Chittagong area. There are eastern bend of Himalayas, which ultimately continue in Arakan hills of Burma. The ridge and valley topography predominates. Karnafuli, Feni and Sangu are the principal rivers. Towards the North-East the Madhupur Jungle in Mymensingh district and Sylhetdistrict are the extension of Garo hills of Assam in India. The climate in tropical monsoon type with erratic distribution of rainfall throughout the year.

Bangladesh is predominantly agricultural. The basic need is to grow enough food for its people, which is increasing at a faster rate. Water is the key to agriculture for attaining self- sufficiency. With the melting of the glaciers in the Himalayas, which are the place of origin of the rivers of Bangladesh, the rivers rise flat during the summer and inundate the land by overflowing the river banks. At the end of monsoon, the water slowly subsides but causes siltation in the deltaic tract. It is a means of natural fertilization of soil. This system of "overflow Irrigation" helps the agricultural practices. Rivers are to be harnessed for regulating floods, generation of hydro-power in hilly areas like Chittagong, prevention of saline water intrusion due to tidal effects, facility of drainage, reclamation of lands, etc.

The average annual rainfall varies from 200 to 250 cm throughout the country. The underground water and the monsoon rainfall have a paradoxical situation. There is too much rainfall and river flow on one season but shortage during the rest period makes unbalanced distribution of water.

ESCAP (1981-83) has reported process of desertification in following regions:

(i) South-western region covering all of the lands south and west of the Ganges-Padma- Lower Meghna river, adjoining areas of north of Ganges in Rahshahi and Pabna districts. Geographically, this land comprises the Bagri region, the lower Ganges plain and tidal saline flood plain of the natural delta and immature delta of South Sunderban. This comprises a total area of about 53,760 sq. km. It is conjectured that the low water flow of the Ganges in dry season have caused a chain of adverse effects specially on the agricultural sector.

(ii) The salinity from the Bay of Bengal intruded into the river and the land masses in the south-Western region of the country affected the prevalent system of agriculture. Salinity showed marked increase from a normal average of 500 to 1000 micromhos per centimeter to 13,600 micromhos per centimeter in April 1976. Salinity penetrated about 132 km. further inland from the normal incursion limit of about 13 km upstream of Khulna town. General observations, however, are that the increased salinity in extreme. S.W. part is due to the saline inundation from the sea. The increased urbanization and industrialization in Khulna area has also led to the over-exploitation of water leading to the increase of salinity hazards.

(iii) The Bagri region of the Moribund delta including mature delta, saline flood plain area, has been known for its salinity in low water supply period since long. The Moribund part of delta is characterized by rivers choked with sand and unable to carry much water except when the Ganges is in high flood.

(iv) Barind tract, Rajshahi District of 1267 sq. km, to a very small extent due to uneconomic utilization of lands.

(v) Madhupur tract, Mymensing District of 12,505 sq. km, probably due to over exploitation of vegetative resources and over-grazing.

In a deltaic and moderately high rainfall area like Bangladesh, desertification problem exist only as the resources are uneconomically exploited by the inhabitants. The problems so far posed are:

(i) Salinity, due to tidal intrusion and over-exploitation of water resources, due to increased urbanism and industrialization,

(ii) Increased pressure of population on land, and

(iii) Over-grazing and deforestation in Mymensing area, particularly in Garo hills adjacent in Meghalaya plateau of India.

It is also possible that the shifting cultivation, rapid deforestation and industrialization may lead to desertification in the Karnafuli basin and in the hilly and ridge and valley terrains in Chitgong hill tracts. A portion of Barind tract in the north-western part of the country is in a state of desertification due to continuous change of the courses of rivers.

3. China

In China, desertification has affected 334,000 sq. km area, out of which, desertified lands total 176,000 sq. km. and potentially desertified lands 158,000 sq. km. About 65% of the Chinese land affected by desertification spreads over arid steppes and desert steppes; 30% arid desert zones and 5% in humid and sub-humid zone. According to the comparative analysis of aerial photos taken during the different periods from 1950s to 1970s, an area of 39,000 sq. km. has been affected by desertification within last 20 years. In rangelands, the damage to land productivity and fertility caused by desertification is estimated to an annual loss of 460 million RMB dollars. Desertification through the development of desert steppes, is believed to occupy 42.2% of Chinese desertification land, activation of fixed sand dunes 52.3% and advancement and encroachment of sand dunes at the margin of deserts is 5.5%.

These three patterns indicate that the expansion of shifting sand dunes is the major form of desertification in North China. Desertification mainly occurs in the following three types of regions:

(i) *Sandy steppe in semi-arid zone:* In the sandy steppes, the action of over-cultivation, over-grazing and over-cut for fire-wood, brings about soil erosion and development of shifting sands. It occupies 69% of desertified land area.

(ii) *Margins of deserts and oases in arid zone:* Sand dunes at the margins of deserts and oases advance and the stable dunes are activated. It occupies 26% of desertified area.

(iii) *Sandy dunes in sub-humid zone:* Sand dunes on both ancient river beds and breached alluvial fans have been caused by the changed river courses and constant over lows at the lower reaches of the rivers. This is the case with 5% of desertified land area.

The above three are almost located in the fragile ecological environment, which are characterized by the simultaneity of drought season and windy season, looseness of surfacial sandy soil and insrability of annual precipitation. In particular, in semi-arid steppe zone, the population density has increased from 10-15 persons/ sq. km. in 1949 to 40-60 persons/ sq. km. in 1980, the highest number going up to 80 persons/ sq. km. The population growth applies a greater pressure upon utilization of land resources.

4. Indonesia

In Indonesia, desertification can be termed as deforestation process on the forest lands and land degradation occurred on agricultural areas. The territorial extent is from 95 EM to 141 EM and 6 NL to 11 SL. The total land area of Indonesia is 1.95 million sq. km, consisting of approximately 17500 islands ranging from large ones to tiny atolls. Among the islands, the largest are Kalimantan- 539,500 sq. km., Sumatra- 437,600 sq. km, Irian Jaya - 422,000 sq. km., Silawesi- 189,000 sq. km., Java and Madura- 132,000 sq. km. Maluku- 75,500 sq. km. and Nusa Tenggara Islands- 73,600 sq. km. Approximately 74% of the total land area or 1.47 million sq. km. is state forest lands.

Total population of Indonesia in 1985 was about 164 million, of which about 78% or 128 million live in rural areas. About 62% of the total population lives in Java, which has only 6.6% of the total land area, where average population density is about 965 persons per sq. km. On the other hand, the population densities of the other islands are much lower, ranging from 5 to 185 persons per sq. km. Indonesia has two typical monsoon winds, the east and west monsoons. The rainy season starts from November to April and dry season from May to October. The rainfall varies from 700 mm in rain shadow areas to 7000 mm with an average of 2810 mm per year. The mean annual temperature in the coastal plains is 26 degrees centigrade and relative humidity is also high, about 70 to 90% and it is relatively stable throughout the year.

Soil is one of very important factors and soil destruction becomes a big problem nowadays. Soil classification is very wide and spread from Inceptisol and Entisol to Alfisol and Ultisol. The former soil classification of the Soil Reaearch Centre, Bogor considered that there are at least 4 groups of soil included forest land, namely;

(i) Red Yellow Podsolic, covering 80% of the total land . which spreads mostly in Somatra, Kalimantan Sulawesi and Irian Jaya.

(ii) Latosol, Andosol, Regosol and mediteran cover 12% of total lands,

(iii) Grumosol, Non Calcic Brown, covers 5% of total land,

(iv) Organosol, Podsol and alluvial- 3% of total land.

The Red yellow Podsolic spreads on the altitudes of 600 to 800 m above sea level. It is well known that the largest soil deterioration occurred in this type of soil and it may become unproductive land, enlarging very fast every year. In 1986, it was estimated that unproductive land covered 43 million hectares as shrubs of bare and grass lands. The acceleration of this unproductive land was estimated as 200-400 thousand hectare every year.

According to the temporary forest land use, the total land area of 1950 million hectares consisted as: 1430 million hectare as forest land, 55 million hectare as irrigated land, 65 million hectare dryland agriculture and 400 million hectare aquatic area. The forest land is further sub-divided in to four type of functions consisting of: Protection forest: 300 million hectare. Nature conservation and recreation forest: 190 million hectare, Production forest: 640 million hectare and Conservation forest: 300 million hectares. The total forest area consist of Beach forest, Mangrove forest, Swamp forest, Peat forest. Calcic forest and Tropical rain forest. Tropical rain forest is largest and formed as community, influenced mostly by regional climate biomass and regional biotic. Their characteristic are evergreen woody and has tow or more strata and the average height of 40-50 m. Most scientists believe that the age of tropical rain forest is 6 million years and through climatic formation, their species developed to 400 species. Their history draws good efforts of ecological equilibrium.

The degraded land is stated as critical and occurring in dryland rural areas. Critical land is usually caused by population pressure, over-grazing and over use of land for agriculture. According to World Bank Report 1987, the total critical land is increasing by about 500,000 hectare per year due to shifting cultivation and the extent of the activities occurred outside Kalimantan island ranging from 100,000 to 300,000 hectare as indicated in Table 2.11.

Table 2.11 : Total critical land by islands

(In hectares)

S. No.	Island	Outside forest land	Inside forest land	Total
1.	Sumatra	2,169888	996,056	3,165,994
2.	Java	1,018,973	128,148	1,147,121
3.	Kalimantan	294,260	901,820	1,196,080
4.	Sulawesi	914,902	958,723	1,873,625
5.	Bali	57,220	11,620	68,840
6.	West Nusa Tenggara	896,699	891,496	1,788,195
7.	Hast Nusa Tenggara	37,000	73,226	110,226
8.	Maluku	310,735	281,519	592,254
	Total	**5,949,306**	**4,315,394**	**10,264,701**

Source: Ministry of forestry, Indonesia, 1986

5. Iran

According to the most common definition, desert is a vast area in which a combination of unfavourable climatic. and geological factors plants and animals encounter severe living conditions. Survival in desert depends on tolerance and adaptability. Movement of the soil makes growth and establishment of plants impossible. On desert marginal lands, climatic conditions make possible growth and development of flora and fauna under ecological stress, yet desert margin ecosystems are quite fragile. Although, natural factors somehow increase the pace of desertification, man bears the main responsibility for intensification of this destructive process. Many vulnerable regions are even now being turned into desert. A certain view considers climate as the major contributor to desertification processes, with human factors playing a relatively minor and supporting role. But. according to Dregne (1983), desertification is impoverishment of terrestrial eco-systems under the impact of man.

The main desertification processes operating in Iran are natural causes and man-induced causes comprising of vegetation degradation, wind and water erosion, salinization, water-logging, improper land use and urbanization.

(a) Natural Causes

Iranian plateau is made of mostly of undeveloped desert soils. In central plateau, the quasi-soil layer mostly consists of the products of physical disintegration of sedimentary, volcanic and metamorphic rocks. In the most areas, sand, gravel, stone and rocks and even salt cover these lands. The northern strip of Iran due to favourable climate is more advanced as far as the soil formation and development is concerned. Average annual precipitation is about 225 mm, which is less than 1/3 world average annual precipitation of 700 mm. Uneven distribution of precipitation intensifies the dry climate prevailing over most of the country. About 1/3 of annual precipitation falls over the narrow northern strip of land, which is only 10% of the total land area. Annual precipitation in the central plateau is less than 100 mm, with quite uneven distribution in time, as it is possible that about half of this amount falls during a single night.

Maximum temperature in central plateau may reach 50°C and annual evaporation in this area may exceed 4000 mm. Severe dryness prevalent in central plateau has created frightening deserts with huge masses of shifting sand and gravel and vast salt plains.

(b) Degradation of Vegetation

Destruction of vegetation cover is a. consequence of human interference in the natural ecological balance. Man, by cutting trees and bushes and overgrazing the pastures, causes this degradation. Destruction of vegetation cover makes land naked and vulnerable to water and wind erosion. Consequently desertification is intensified. Disproportionately high ratio of livestock to the area of pastures and cutting bushes from pastures and bush-lands for fuel and industrial ad medicinal use are among the major factors in destruction of pasture and rangelands. The total area of pasture is estimated to be around 100 million hectares comprising of good pastures- 19 million hectares, medium pastures 25 million hectares and poor pastures 56 million hectares.

The pattern of grazing is uncontrolled and has no rational relationship with the carrying capacity of the rangelands. As a result, vegetation cover is seriously damaged. Moreover, pasture can not stand the severe overgrazing of livestock and consequently a gradual decline in their productivity occurs, while at the same time the number of livestock has a considerable growth rate. Yearly forage production of pastures is hardly enough to feed 16 million livestock units, while out of total livestock population of 120 million; 60 million are grazing on the pastures.

Forests have been the victims of mindless pursuit of personal profit. Forest destruction is not limited to the northern strip of Caspian coast forests, but also is the north-west and the south-west regions and desert margins, where cutting of trees and bushes are prevalent. Presently 12.74 million hectares only half of the area has remained protected with forests.

(c) Water and Wind Erosion

Soil erosion has great dimensions visible in different regions. In the mountain regions of north and north-east, mainly water erosion and in central, southern and southwestern regions, mostly wind erosion is prevalent. Muddy waters of the rivers in the high rainfall seasons carry millions of tones of topsoil, which is finally deposited as sediments behind the dams or reach lakes and seas. Studies show that around one billion tonnes of soil is annually lost by erosion (Nakhjavani 1972). If the percentage of each of main N-P-K elements in the typical soil is considered, on an annual basis, 300,000 tonnes of pure nitrogen. 10,000 tonnes of pure phosphorous and 2,000,000 tonnes of pure potassium alone is lost. The soil lost by erosion also contains living organisms, plant and animal residues and other essential elements for plant growth and development.

Wind erosion are the dominant feature of deserts. Desert winds change direction fast and keep no specific direction. About 12 million hectares of central, eastern and southern regions of Iran are covered by sand deserts. Half of the area is covered by active dunes, which inflict serious damage to the desert margin settlers (Rahbar 1986). Chaj region located at the border between Iran and Soviet Union, shifting sand is threatening 200,000 hectares of farmland (Matin and Mahboubi 1985).

(d) Salinization and Water-logging

With the exception of the northern strip of the country, most of the arable lands are more or less saline. An area of 1,180,000 hectares is saline and water logged. In the south, high underground water table and high evaporation rate, the problem of salinization is more acute and in places like Khozestan, with the construction of dams and development of irrigation network, secondary salinization has intensified.

(e) Improper Land use

Agriculture is a man made ecosystem, which is the consequence of interference in the natural ecosystem. Cultivation in-lands with 40 to 50 % slope is common in most regions, a practice intensifying the erosion and desertification process. Improper use of agricultural machinery, soil compaction, destruction of soil structure and ploughing up and down the slope are contributing factors to desertification.

6. Mongolia

Mongolia is situated in the central part of Asia, at the height of 1600 m above sea level on an average, with extremely continental climate. Semi-desert and desert areas occupy about one third of the whole territory. The Gobi desert is the far northern part of the great stony high desert of Central Asia. The yearly precipitation amount of semi desert area is 150 to 100 mm and in desert areas it goes down to 30 mm or less. The evaporation is 10-20 times more than the amount of precipitation. For keeping 10 mm precipitation in the soil, 1000 mm precipitation is required in a year. Thunder storms occur only in July and August and winter and spring are very dry. During winter, the temperature goes down to -35 °C, sometimes even less and in spring blows the strong wind. In summer, air temperature rises up to 32-38 °C. Amplitude of the absolute temperature in a year reaches 100 °C.

In western desert, the wind is very active during the spring from April to May, but during the summer from June to August, the speed decreases. During the autumn, it riscs again but not so intensive as in spring. The maximum number of days with strong wind in May at one station Khovd in western Mangolia is 15 days and for a year 47 days. The wind speed reaches 24 m/s once a year. 28 m/s once in 5 years, 32 m/s once in 32 years and 34 m/s once in 34 years.

Desertification is a problem involving the sensitive and inconsistent i.e., unsustainable natural environment of arid area and irrational use and exploitation of this environment for providing means for living. Over-exploitation of arid areas resulted in degradation of three natural elements: vegetation, soil and water. Investigation of desertification problems started with comparison of old maps with satellite pictures since 1970s are evident that sand boundaries have expanded by several kilometers in some areas.

7. Nepal

Nepal, a Himalayan country lying between the latitude of 26°22' to 30°27' north and longitude 80°04' to 88°12' east is bordered by China in the north and India in the east, west and south. The country comprises of an area of 147,181 sq. km. From north to south, Nepal is divided in to five physiographic regions, namely the High Himalayas, the High Mountains- Transitional Zone, the Middle Mountains, the Siwaliks and the Terai covering 22.7. 20.1, 30.1, 12.8 and 14.3 percent of the total area of country respectively.

About 80% of the total rainfall occurs in four months from July to September. Due to its geographical diversity, the climate varies from sub-tropical mainly Terai, Siwalik and lower parts of the Middle Mountains to temperate mainly in the upper parts of the Middle Mountains and to alpine mainly in transition zone and the High Himalayas.

Administratively, the region is divided into five development regions, fourteen zones and seventy vide districts. The districts are further divided into Panchayats, the smallest administrative unit. Nepal is drained by three rivers, the Koshi, Gandaki and Karnali and their tributaries. All the rivers of the country ultimately flow into the Bay of Bengal. The population of the country was 15 million in 1981 with decadal growth rate of 2.66 percent.

The major land use namely agriculture, forestry, grazing and shrub land cover about 26.8, 38.1, 11.9 and 4.7 percent of the country respectively. Perpetual snow, rock and sand, gravel and boulders cover about 3.5, 13.3 and 1.4 percent of country respectively. About 26,834 sq. km. area with snow, rock, sand, gravel and boulders is devoid of vegetation and incapable of supporting any considerable population, which are similar to the characteristics of desert.

Nepal is losing an estimated 240 million cubic meters of soil annually, which is equal to 1.63 mm of soil removed from surface of 147,181 sq. km. area. In the course of building up an inland delta, Koshi river has shifted over 110 km. from east to west during the period from 1731 to 1963. Besides destroying about 7800 sq. km. of land in Bihar, India and 1300 sq. km. in Nepal by sand deposits, it has wiped out towns and villages in the course of its shifting (Gole and Chitale, 1966). The satellite imageries revealed that 40,000 sq. km. island is taking shape in the Bay of Bengal. The island is being formed by the deposition of silt transferred by the Himalayan rivers (Sterling 1976) of which Nepal is the major contributor. The river beds are estimated to rise 10-30 centimeters per year (Nepal 1975).

8. Thailand

Thailand with 51.36 million hectares of land area, lies in the tropical zone between latitude 8° and 20°N. It has been covered with various kind of rich forest, an important natural resource for developing country. Thailand forests, covering 27.2 million hectares i.e., 53 percent of country at time, have vastly depleted for cash crops like cassava, sugar cane and jute. The more serious overuse of natural resources have caused severe environmental impacts. Forest cover has remained only 9.24 million hectares i.e., 18 percent. Destruction of forests has generally affected both eco-systems and rural village lives. One ecological impact is the nationwide drought. The Farm Research Institute had reported that the rainfall recorded 8 to 14 percent lesser than last three decades. The most seriously affected region is the North-east, which also faces salinization over increasing area. Rural people become impoverished because of unequal use of natural resources. Due to poverty about 1.2 million families were pushed to encroach on national reserve forests to establish farms.

9. Vietnam

Vietnam is situated in eastern part of Asia, is now affected by land degradation. Total area of the country is 33.1 million sq. km. and its location is 6°5' - 23°22 N and longitude 102°07 to 117°20'E. South of the 16th parallel latitude, the climate is typically tropical, north of the 16th latitude it is tropical and influenced by north-east wind. Mean annual temperature in the low land is 25-27 °C in the south and 22-23°C in the north. In high mountain areas, the climate is mild ranging between 15-19 °C.

Rainfall in most of the country is 1700-2000 mm/ year and in some parts it is over 3000 mm/year. Rainfall is usually concentrated in two or three months causing floods and soil erosion. Specially, the climate of Thuan Hai region in south central Vietnam is semi arid with annual rainfall of 600-700 mm/ year. About 3/4 part of the country is mountainous and hilly with elevation mainly from 50-700 m. The peak of the highest mountain is 3314 m. Most of the soils are yellowish red feralit with rather steep slopes, easily eroded. Alluvial plains are in 3 million hectares area.

The population is over 64 million, mostly farmers in alluvial plains. The ethnic minorities, which amount about 2 million, live in mountainous regions with slash-and-burn cultivation system. The land area per capita is very small- 0.1 hectare in agriculture, 0.14 hectare in forest.

Vietnam lies in the humid tropics and there exist no typical deserts and arid zones like Africa, Middle East and Australia. The concept of desertification has been broadened to mean "the diminution or destruction of the biological potential of the land, which can lead ultimately to desert like conditions." With this definition, there are 13.8 million hectares of forest land already fell into desertification. 9.7 million hectare of forest have turned into grass savannas, seriously eroded and lateritized and biotic potential of the soil is diminishing day by day. The causes for such a situation are- shifting cultivation in mountainous areas, exceeded timber and fire-wood exploitation. Annual forest fire and prolonged wars in more than 30 years:

2.5 STATUS OF DRYLANDS AND DESERTIFICATION AND LAND DEGRADATION IN INDIA

India is endowed with wide variety of climate, ecological regions, land and water resources. However, with barely 2.4 % of the total land area of the world, the country has to support 16.7% of the total human population and about 18% of the total livestock population of the world. This has put enormous pressure on natural resources. Ecosystems are highly complex systems relating to a number of factors, both biotic and abiotic, governing them. Natural eco-systems by and large have a high resilience for stability and regeneration. However, continued interference and relentless pressures on utilization of resources leads to an upset of this balance. If these issues are not effectively and adequately addressed in a holistic manner, they can lead to major environmental problems like depletion of vegetation cover, increase in soil erosion, decline in water table and loss of biodiversity all of which directly impact the very survival.

Desertification is a complex phenomena resulting from factors of physical, biological, socio-economic, cultural and political nature. Desertification is not confined to the desert areas or to the arid, semi-aird and dry sub-humid regions. Land degradation has a direct impact on land and other natural resources, which result in reduction in reduced agricultural productivity, loss of bio-diversity and vegetation cover, decline in groundwater and availability of water in the affected regions. All these lead to a decline in the quality of life, eventually affecting the socio-economic status of the region. A major part of the country is degraded in some form or other. In some regions, problems of land degradation have reached serious proportions threatening the very existence of the people.

1. General Topography

India's mainland comprises of four geographical areas: the northern Mountains, , the vast Indo-Gangetic plains, the Southern Deccan Peninsula bounded by the Western and Eastern Ghats and the coastal plains and islands. Total geographical area of the country is 32.87.263 sq. km. and total population as per 2001 census is 1027 millions with a density of 324 persons per sq. km.

(i) *Northern Mountains:* Corresponding with the Himalayan Zone, along with country's northern boundaries including Jammu and Kashmir, Himachal Pradesh, north-west Uttar Pradesh, Sikkim, part of Assam and the North-Eastern states of Arunachal Pradesh, Nagaland, Manipur, Mizoram, Tripura and Meghalaya. The Himalayas comprise of mountain ranges, which form an indomitable physical barrier as the world's biggest and largest mountain range. The Himalayas also contain the cold arid deserts and fertile valleys.

(ii) *The Great Plains:* Also known as the Indo-Gangetic plain is formed by the basin of three distinct river systems- the Indus, the Ganga and Brahmaputra. The plains extend from Rajasthan in the west to Brahmaputra valley in the East. This region covers the entire states of Punjab, Haryana, Union Territory of Chandigarh, Delhi, major parts of

Uttar Pradesh, Uttaranchal, Bihar, West Bengal and parts of Assam. These plains comprise one of the world's greatest stretches of flat and deep alluvium and are among the most densely populated areas of the world. Desert region, which contains the Great Thar Desert, extends from the edge of Rann of Kutchh to larger parts of Rajasthan and lower regions of Punjab and Haryana.

(iii) *The Deccan Penninsula:* This zone covers the whole of south India covering states of Tamil Nadu, Karnataka. Orrisa, Puriliya of West Bengal. The Indo-Gangetic plains and the peninsular plateau by mountain and hill ranges known as Aravalli, Vindhya, Satpura, Ajanta and Maikala ranges.

(iv) *The Coastal Plains and Inlands:* The peninsula on either side by the Eastern Ghats and the Western Ghats. On the other side of the Ghats, outward to the sea lies a coastal strip. The western coastal plains lie between the Western Ghats and the Arabian Sea in the west, whereas the Eastern Coastal Plains face the Bay of Bengal in the East. This is also a region with very high-density population.

(a) Climate

India is mainly a tropical country but due to great altitudinal variations, almost all climatic conditions from hot desert to cold desert exist. There are four seasons: (i) Winter from December to February, (ii) Summer from March to June, (iii) South-west monsoon season from June to September and (iv) post monsoon season during October-November. During the post monsoon season, commonly known as winter monsoon, rains begin over north India and pass over the Bay of Bengal before reaching the Andaman's and the south-east coast. However, the south-west or the summer monsoon is the main source of rainfall in the country providing 80% of the precipitation.

(b) Soils

India has a wide range of soils, each type being particular of a specific locality. Alluvial and black cotton soils are the two most important soil groups for agricultural production. Alluvial soils cover about 78 million hectare i.e., 24% of the total land and occur in the great Indo-Gangelic Plains, in the valleys of Narmada and Tapti in Madhya Pradesh and the Cauvery in Tamil Nadu. These soils are considered very good for production of wheat, rice, other cereals and pulses, oilseeds, sugarcane etc. The black cotton soils cover about 51.8 million hectare and are found in the States of Maharashtra, Gujarat, Madhya Pradesh, Karnataka. Andhra Pradesh, Tamil Nadu, Uttar Pradesh and Rajasthan. These are also considered good for cultivation of cotton, cereals, pulses, oilseeds, citrus fruits, vegetables etc. In addition. Red soils have been estimated to occur in 51.8 million hectare area and primarily found in Tamil Nadu. Karnataka, Kerala, Maharashtra. Andhra Pradesh. Madhya Pradesh, Bihar and West Bengal. These are most suited for rice, millet, tobacco and vegetable cultivation. Laterite and lateritic soils occur in 12.6 million hectare. These are not considered good for agriculture. The area of desert soils is about 37 million hectare and not deemed fit for agriculture.

(c) water

Rainfall annual occurrence is about 4000 cubic km. of water through precipitation. About 80% of the annual rainfall is mainly from the South-West monsoon season of June to September, followed by the North-West monsoon. in November-December. Most of rainfall is therefore few months in a year and almost the remaining period remains dry. The rainfall is highly variable both spatially and in quantity among the 35 meteorological sub-divisions, the country has been divided into. It varies from as low as 100 mm in western Rajasthan to as high as 9000 mm in Meghalaya in the north-east India. Monsoon rainfall is usually torrential in intensity. This results in tremendous run-off resulting in soil erosion (Sharma and Paul 1999).

Water Resources: It is estimated that out of total precipitation of around 400 million hectares metres in the country, the surface water availability is 187 million hectares. Of this, only 50% is put. to beneficial use due to topographical and other constraints. Usable water resources of about 1100 cubic km., surface sources e.g., rivers account for 60% and ground water sources are 40%.

Surface Water: The country has been divided into 20 river basins, comprising of 12 major river basins, each having a catchment area exceeding 20,000 sq. km. and eight composite river basins (Sharma and Paul 1999). In addition, other water resources include reservoirs, tanks, ponds and lakes, which cover 7 million hectares of surface area. There are 14 major river systems. The rivers may be classified as : (i) Himalayan Rivers, (ii) Peninsular rivers, (iii) Coastal rivers and (iv) rivers of inland drainage basin. Due to uneven precipitation, the availability- of renewable fresh-water varies enormously in different river basins. The Himalayan rivers are short in length with limited catchment areas. The rivers and tributaries of the peninsular and coastal rivers are intermittent and non-perennial in nature. The streams of the inland drainage basins of western Rajasthan are few and with little water holding capacity. Floods and drought effect vast areas. A third of the country is drought-prone. Floods affect an average area of around 9 million hectares per year and area susceptible to floods is 40 million hectares.

Groundwater potential varies in different regions. Due to heavy extraction and limited recharge, the groundwater is depleting very fast, particularly marked in most of the dryland regions of the states of Andhra Pradesh, Karnataka, Rajasthan, Madhya Pradesh. Orissa and Maharashtra.

(d) Forest

Forests show greatest variation and range depending upon rainfall, soil topography and climatic factors. The forests range from tropical rain-forests to dry thorn forests and mountain-temperate forests. There are four major types and 16 detailed forest types and more than half of the forest area is tropical-moist and dry-deciduous types as given in Table 2.12.

Table 2.12 : Major forest types found in india

(Area in Million Hectares)

Forest Type	Sub-Type	Area	%	Occurrence in States/ Union Territories
Tropical	Tropical Wet Evergreen Forests	4.5	5.8	Arunachal Pradesh. Assam, Karnataka, Kerala, Mizoram, Manipur, Nagaland, Tamil Nadu, Sikkim, Andaman and Nicobar islands, Goa.
	Semi-Ever-Green Forest	1.9	2.5	Assam, Karnataka, Kerala, Maharashtra, Nagaland, Orissa, Tamil Nadu, Andaman and Nicobar Islands, Goa.
	Moist Deciduous Forest	23.3	30.3	Andhra Pradesh, Assam, Bihar, Gujarat, Karnataka, Kerala, Madhya Pradesh, Maharashtra, Manipur, Meghalaya, Mizoram, Tripura, Nagaland, Orissa, Tamil Nadu, Uttar Pradesh, West Bengal, Andaman and Nicobar Islands, Goa, Dadra and Nagar Haveli.
	Littoral Swamp Forest	0.7	0.9	Andhra Pradesh, Gujarat, Maharashtra, Orissa. Tamil Nadu, West Bengal, Andaman and Nicobar Islands.
Semi-tropical	Dry Deciduous Forest	29.4	38.2	Andhra Pradesh, Bihar, Gujarat, Haryana, Himachal Pradesh, Karnataka, Kerala, Madhya Pradesh, Maharashtra, J and K, Orissa, Punjab. Rajasthan Tamil Nadu, Uttar Pradesh.
	Thorn Forest	5.2	6.7	Andhra Pradesh, Gujarat, Haryana, Himachal Pradesh, Madhya Pradesh, Karnataka, Maharashtra, Punjab, Rajasthan, Tamil Nadu, Uttar Pradesh.
	Dry Evergreen-Forest	0.1	0.1	Andhra Pradesh, Tamil Nadu.
	Sub-tropical Broad leaved Hill forest	0.3	0.4	Assam, Meghalaya.
	Sub-tropical Pine Forest	3.7	5:0	Arunachal Pradesh. Himachal Pradesh, J and K, Manipur, Meghalaya, Nagaland, Sikkim, Uttar Pradesh, Haryana, Punjab.
	Sub-tropical dry Evergreen forest	0.2	0.2	Himachal Pradesh, J and K
J & K, HP	Montane Wet Temperate Forest	1.6 .	2.0	Arunachal Pradesh, Karnataka, Manipur, Nagaland, Sikkim, Tamil Nadu.
	Himalayan Moist Temperate Forest	2.6	3.4	Himachal Pradesh, J and K, Uttar Pradesh.
	Himalayan Dry Temperate Forest	0.2	0.2	J and K, Himachal Pradesh.
Alpine	Sub-Alpine Moist Alpine Dry Alpine Shrub		4.3	J and K, Uttar Pradesh.

Source: Forest Survey of India, 1995.

(e) Bio-diversity

The country is rich in bio-diversity in view of its diverse physiography and climatic conditions. The country falls in the confluence of three major bio-geographic realms- the Indo- Malayan, Eurasian and Afro-Tropical. The country is divided into ten bio-geographic zones: (i) Trans- Himalayas, (ii) Himalayas, (iii) Indian Deserts, (iv) Semi-Arid areas, (v) Western Ghats, (vi) Deccan Penninsula, (vii) Gangetic Plain, (viii) North-East India, (ix) Islands and (x) Coasts. India is one of the 12 mega biodiversity countries of the world. Out of the 18 unique bio-diversity Hotspots", which contain about 20% of the world's flora, two, namely- north-eastern Himalayas and the Western Ghats are located in India. In order to protect and conserve the bio-diversity, a number of "protected areas", in the from of National parks and sanctuaries have been set up. The bio-geographic locations of these are given in Table 2.13.

Table 2.13 : Bio-geographic zones of protected areas

S. No.	Bio-geographic Zone	Number of National Parks and Sanctuaries	Area (Sq. Km.)	%
1.	Trans-Himalayas	7	17002.50	11.11
2.	Himalayas	67	20890.52	13.65
3.	Deserts	6	16076.08	10.51
4.	Semi Arid	91	15302.60	10.00
5.	Western Ghats	56	13425.01	8.78
6.	Deccan Peninsula	137	51151.74	33.43
7.	Gangetic Plain	36	7648.92	5.00
8.	Coasts	26	5619.58	3.67
9.	North-East	27	4354.12	2.85
10.	Islands	103	1529.27	1.00
	Total	**556**	**153000.34**	**100.00**

Source: Wild Life India, 2000

(f) Agriculture

Agriculture contributes to 29.93% of the India's Gross Domestic Product. The overall growth of the Indian economy has depended much on the performance of agriculture. With a share of 2.7% of the world agricultural production, India is ranked sixth in the world. Major part of agriculture, is rain-fed, extended to 87 million hectare and constituting nearly 61% of the net cultivated area. The wide variation in rainfall and potential evapo-transpiration decides the actual land use and vegetation cover available. A large percentage area of cultivation growing 90% coarse cereals, 76% oilseeds, 65% cotton and 50% rice are rain-fed. Agriculture also is the single largest source of employment, even though its contribution to the national economy has been shrinking over the years. In 1990-91, the total main work force of 285.42 million, 64.90% constituted the agriculture workforce.

Irrigation potential is 139.5 million hectares comprising of 58.5 million hectare from major and medium schemes, 15 million hectare from minor irrigation and 66 million from ground water exploitation. But most irrigation projects are operating at a low efficiency in the range of 30-40%, thereby losing 60-70% of the irrigation water during conveyance. Even after achieving full irrigation potential, nearly 50% of the total cultivated are will remain rain-fed.

2. Bio-Climatic Regions

The agro-ecological regions fall into 6 major climatic regions are (i) Arid, (ii) Semi- arid, (iii) Dry sub-humid, (iv) Moist Sub-humid, (v) Humid and (vi) Per-humid. Based on the classification laid down by the National Bureau of Soil Science and Land Use Planning (NBS and LUP) in the publication "Agro-Ecological Sub-regions of India for Planning and Development" (Velayutham 1999) and the Method of the dryland classification of bio-climatic zonation of the country, main areas/ districts in the concerned regions falling in the various bio-climatic regions. Thornthwaite system of classification is used for agricultural planning in the country considering the annual rainfall, moisture index and other aspects such as the growing period of crops, given in Table 2.14.

Table 2.14 : Classification of dryland regions using thornthwaite classification

S. No.	Region	Annual Av Rainfall (in mm)	Moisture Index	Growing Period (in days)	Total Land Area	
					In m. ha.	In %
1.	Glaciers and Others	–	–	–	5.2	1.5
2.	Hyper Arid	<100	<83.3	0-50	22.9	7.0
3.	Typic Arid	100-500	-66.7 to -83.2	60-90	22.7	7.0
4.	Semi-Arid (Dry)	500-750	-50 to-66.6	90-120	51.2	15.6
5.	Semi-Arid (Moist)	750-850	-49.9 to -33.4	120-150	72.2	22.0
6.	Sub-humid (Dry)	850-1000	-33.3 to 0	150-180	54.1	16.5
7.	Sub-humid (Moist)	1000-1500	0 to 20	180-270	39.8	12.1
8.	Dry/Moist Sub-humid Transition	1000-1500	0 to 20	210-270	21.0	6.4
9.	Humid		21 to 99.9	210-330	16.6	5.1
10.	Per-humid	>250.0	>100	>300	20.5	6.3
11.	Transition Humid/ Per-humid	>2500	>100	>300	1.8	0.5
	Total				**328.0**	**100**

The area of bio-climatic regions at SI. No. 1-6. which fall under the dryland regions as per Thornthwaite Classification =228.3 million hectares.

Source: National Bureau of Soil Survey and Landuse Planning, 2001

3. Type and Extent of Land Degradation

As per assessment made by the Ministry of Agriculture in 1985, the total geographical area of about 329.72 million hectares, approximately 173.64 million hectare are subject to moderate to severe form of degradation, which constitute 52.8% of the total geographical area detailed out in Table 2.15.

Table 2.15 : Estimation of land degradation status in india

(In Million Hectares)

S. No.	Type of Land Degradation	Area	%
1.	Water Erosion	107.12	61.69
2.	Wind Erosion	17.79	10.25
3.	Ravines	3.97	2.29
4.	Salt Affected	7.61	4.38
5.	Water-logging	8.52	4.91
6.	Mines and Quarry Wastes	0.04	Neg.
7.	Degraded Land due to Shifting Cultivation	4.91	2.83
8.	Degraded Forest Lands	19.49	11.22
9.	Special Problems	2.73	1.57
10.	Coastal Sandy Areas	1.46	0.84
	Total	**173.64**	**100.00**

Source: Ministry of Agriculture, 1985.

Central Research institute for Dryland Agriculture (CRIDA) has attempted to identify the areas falling within the dryland regions as per Thornthwaite Method of classification and the extent of degradation occurring in these regions. About 130 million hectare area in three climatic regions- arid, semi-arid and dry sub-humid, are affected by varying degrees and types of degradation.

4. Climate and Desertification

Rainfall distribution varies from place to place. In general, about 80% of annual rainfall is received during the South-west monsoon season from June to September. Rainfall exceeds 1000 mm in areas of longitude 78*E . It exceeds 2500 mm almost the entire West Coast and Western Ghats and over the most of Assam and adjoining Sub-Himalayas and West Bengal. It decreases from east to west and also from south to north. The peninsula has an elongated area of low rainfall of less than 600 mm.

In arid zone areas, water deficit prevails in all the months of the year. Rajasthan, Saurashtra and Kutch come under the hot arid area. A small area in the Deccan peninsula latitude 15* is also arid. Semi-arid areas with water surplus during some months and deficient in come under semi-arid areas. Parts of Haryana, Punjab, West Uttar Pradesh and entire peninsular parts of Western Ghats are semi-arid. Dry Sub-humid zone covers parts of Northern Plains, Central Highlands, eastern plateau, parts of Eastern Ghats and plains and parts of western Himalayas fall within this region.

There are 556 meteorological surface observatories, engaged in collection of information on weather and climate like rainfall, temperature, wind pressure etc. In addition, there are more than 9000 reporting and non-reporting rain gauges, 211 agromet observatories, 216 evaporation stations. 43 soil moisture stations, 41 evavo-transpiration stations. 83 dew fall stations, 21 snow gauge stations, 46 radiation measurement stations, 10 back ground pollution stations and 7 ozone observation stations established in various locations for carrying out specific monitoring. In addition, there are 65 upper air observing stations for recording upper air wind speed and direction, temperature, humidity. 20 satellite picture receiving stations and 10 cyclone detection radars. In north-west, 7 pilot balloon-cum-micro-meteorological stations to provide low-level wind data for anti-locust operations.

5. Major Factors Causing Desertification

Desertification is land degradation in drylands due to a number of factors including climatic variations and human activities. Man-made causes include expansion of agriculture and unsustainable agricultural practiceslike over-cultivation, nutrient inputs, poor irrigation practices, deforestation and over-grazing. Unsustainable resource management practices are often induced by population pressure, social conflicts and disruption of social systems, inappropriate government policies and poverty. People affected by desertification often need to draw on their limited assets in order to survive, which accentuates their poverty. This constitutes a vicious cycle linking deteriorating natural resources to deteriorating livelihoods as people need to encroach further on fragile soils, sparse vegetation and limited water resources to meet their basic needs for food, shelter and livelihood. Many of the complex casual relationships are not fully understood. It is often difficult to separate the causes from the effects. Major causes of desertification are:

S. No.	Major Factors	Causes of Desertification
1.	Unsustainable Agricultural Practices	• Extensive and frequent cropping of agricultural areas • Excessive use of fertilizers • Shifting cultivation without allowing adequate period of recovery.
2.	Unsustainable Water Management	• Poor and inefficient irrigation practices. • Over abstraction of ground water, particularly in the coastal regions resulting in saline intrusion into aquifers.
3.	Conversion of land for other purposes	• Prime forest into agricultural use. • Agricultural land for other uses. • Encroachment of cities and towns into agricultural land.
4.	Deforestation	• Unsustainable forest management practices. • Forest land clearances for agriculture including shifting cultivation • Changes of land use for projects, energy roadways etc. • Over-grazing, excessive fuel wood collection. • Uncontrolled logging andillegal felling of trees • Forest fires.
5.	Industrial, mining and other activities	• Without applying satisfactory measures for prevention of land degradation and land rehabilitation.
6.	Demographic pressures	• Human and livestock.
7.	Frequent droughts and failure of monsoon	• Their impact with global climatic phenomena.

6. Impacts of Desertification

Different processes involved in land degradation include- wind erosion, water erosion, salinity and alkalinity, and water logging. These cause severe impact on desertification.

(i) Impact of Human Population Pressure on Forest Resources

Population pressures per unit area of forest is one of the highest in the world. In 1991, the national average density of population per sq. km. of forest land was 1320 and ranged from 2860 in the north-western states to 191 in the north-eastern region. Consumption of wood is considerably higher. In 1990, the excess removal of fuel wood was 250 million cubic meters, which is more than 5 times of the world wide impacts.

(ii) Impact on Bio-diversity

Owing to degradation of forests and natural habitats for expansion of agriculture, river valley projects, industrial and urban developments, the bio-diversity are under threat for survival itself. The environment impacts are loss of ecological stability, loss of bio-diversity, reduction in carbon sink capability' and their affects on climate, floods, droughts, damages of watersheds, soil erosion, silting up of reservoirs, desertification and drought and changes in hydrological regime. The social and economic impacts are loss of employment and food and livelihood security.

(iii) Impact of Livestock Population Pressure

Number of livestock grazing on forest land was 35 million in 1957-58, which was increased to 60 million in 1973-74 and 90 million in 1995. The combined availability of fodder from all sources from all sources was 434 million tones against the estimate requirement of 882 million tones. The deficit is met froni grazing in forest land causing adverse impact.

(iv) Impact of Depletion of Vegetative Cover

The gradual change in eco-system through loss and replacement of one species with another affects the natural vegetative cover and average rainfall decline more than 8 to 12% (Singh *et. al.* 2000)

(v) Impact of Wind Erosion

Sand dunes and other sandy land forms in the Thar desert are most vulnerable to wind erosion. Sandy land forms in the western parts of the desert are more unstable due to unstable and deficient rainfall and increasing gradient in wind velocity from east to west direction. The threshold velocity for initiating wind erosion has been estimated to be around 10 km/ hour (Prasad and Biswas 1999). However the wind velocity is as much as 30 km./hour are common, leading to loss of topsoil and terrain deformation in the affected regions.

(vi) Impacts of Water Erosion

The national average of rate of soil erosion, based on existing soil loss data has been 16.35 tonnes/ hectare/year, yielding a total soil loss of 5.3 billion tones annually (Prasad and Biswas 1999). About 64% of the eroded soil comes from the Shivaliks, the western Ghats and the north-eastern States. Of the eroded soil, nearly 29% of the total eroded soil are permanently lost to sea, 10% end up in reservoirs and about 61% transferred from one place to another.

(vii) Impact of Soil Erosion on Soil Fertility

Soil erosion is directly linked to deterioration of soil health, which in turn affects crop productivity and sustainability (Prasad and Biswas 1999). According to Tata Energy Research Institute (TERI), New Delhi, the economic losses caused by upper crop yields and reduced reservoir capacity has been estimated to be in the range of 89-232 billion rupees, as a result of loss of 11-26% of agricultural inputs.

(viii) Impacts of Inefficient Water Management

The impact has been observed at all levels like cities, province and villages leading to drought like situations. In places of acute water scarcity, long hours are spent for collection of water, which affects the quality of life and a direct loss to the economy. The availability of renewable fresh water resources per capita has fallen from 6000 per cubic meters in 1947 to about 2300, cubic meters in 1997 (TERI report). According to a World Bank study, India, losses 1.2 - 6.0 million tonnes of food grain production every year due to water logging (Prasad and Biswas, 1999).

(ix) Impact of Desertification on Migration and Urbanization

The urbanization in developing countries result pressures from declining quality of life in rural areas. The urban population was 217 million which was 25.73% of total population of 1991, which was increased to. 285 millions i.e. 27.78% of total population of 2001. The desertification has well-established relationship with migration from rural areas. Forced migration to urban areas due to desertification leads to an inability to adjust to exigencies of urban living and hence creation of slums and associated social stress.

Chapter 3

Integrated Indicators of Desertification and Land Degradation

Desertification is a slow and silent process of land degradation and a phenomenon-involving climate, soils, flora, fauna and man. It may be regarded as a form of deterioration of ecosystems. due to change in the pressure of human use systems. The significance of the process of deterioration derives from its magnitude in amount of land and numbers of people affected, the rate at which it occurs and its implications for the future well-being of mankind.

Desertification arises from fragility of dryland ecosystems, under excessive pressure of human use or change in land use causing loss in productivity and ability to recover Stability and resilience are major factors in the viability of all systems- social and cultural as well as physical and biological. Human use systems exhibit considerable coping ability or resilience in the face of desertification, continuously accelerating pressure to disrupt social and cultural systems that support a particular human use system. Desertification can develop from natural causes alone and in any climatic zone of the world and human use system in irrational manner converts the fertile lands into arid, semi-arid and sub-humid lands. Desertification is the causative system of breakdown of natural and environmental systems, while human use systems diminish the productivity of the soil resulting in totally loss to the human community.

There is an utmost necessity to diagnose the desertification through the whole range of physical, biological and social processes relevant to the ecosystems and human use systems of drylands. Attention to the social processes is particularly important, as people are largely responsible for desertification. Human factors causing desertification may not be in step with the physical consequences as productivity measured by income or total volume of yield may actually increase as physical symptoms of desertification become apparent. This harmony may be caused by such things as intensification of land use or the discovery of an alternative use of the degraded resources.

Despite the prodigious efforts made in recent years by the scientific community, interested governments and international organizations, both the control desertification and to raise consciousness of it. There still remains considerable doubt about its nature, extent and its relationship with other phenomena. Although knowledge, technology and expertise are available for the treatment of each of the acknowledged causes of desertification, treatment is rarely successfully carried out. Failure can be due to a number of factors, the most obvious being tardy diagnosis and piecemeal treatment of problems with adequate integration of all the factors involved.

The main objective of the Nairobi Seminar on Desertification (SANSOD) was to define a set of indicators of desertification, which could be used to improve information systems for scientists, planners and policy makers. In order to reach of consensus on a set of indicators, it was deemed vitally important to coordinate the basic definition of desertification, the indicators of the process, the methods to obtain indicator data to suffice the requirement.

The first definition quoted from U.N. Plan of Action to Combat Desertification (A/CONG. 74/36): "Desertification is the diminution or destruction of the biological potential of the land and can lead ultimately to desert-like conditions. It is an aspect of the widespread deterioration of eco-systems and has diminished or destroyed the biological potential, i.e., plant and animal production, for multiple use purposes at a time, when increased productivity is needed to support growing population in quest of development. Important factors in contemporary society- the struggle for development and the efforts to increase food production and to adapt and apply modern technologies, set against a background of population growth and demographic change-interlock in a network of cause and effect. Progress in development, planned population growth and improvements in all types of biological production and relevant technologies must, therefore, be integrated. The deterioration of productive eco-system is an obvious and serious threat to human progress. In general, the quest for ever greater productivity has intensified exploitation and has carried disturbance by man into less productive and more fragile lands. Over-exploitation gives rise to degradation of vegetation, soil and water, the three elements, which serve as the natural foundation fur human existence. In exceptionally fragile eco-systems, such as those on the desert margins, the loss of biological productivity through the degradation of plant, animal, soil and water resources can easily become irreversible and permanently reduce their capacity to support human life. Desertification is a self-accelerating process, feeding on itself and as it advances, rehabilitation costs rise exponentially. Action to combat desertification is required urgently before the costs of rehabilitation rise beyond practical possibility or before the opportunity to act is lost forever."

The physical and biological factors in this definition are plant production and productivity, animal production, soil degradation and water degradation. The social factors are struggle for development, population growth and demographic change, efforts to increase food production and modern unadapted technologies.

The second definition is from Harold Dregne (1976:12); "Desertification is the process of impoverishment of arid, semiarid and some subhumid ecosystems by the combined impacts of man's activities and drought. It is the process of change in these eco-systems that can be measured by reduced productivity of desirable plants, alterations in the biomass and the diversity of the micro and macro fauna and flora, accelerated soil deterioration and increased hazards for human occupancy."

In this definition, the physical and biological factors are: drought, plant productivity, biomass, fauna distribution and type, flora distribution and type and soil deterioration. The social factors are man's activities and increased hazards for human occupancy.

SANSOD addressed itself principally to the first recommendation in the Plan of Action to Combat Desertification, which calls for assessment of desertification through the application of comparable indices. The indices, as such, are not spelt out in the United Nations document and the goal was to secure as much stress as possible on the most useful set of indices or indicators. The recommendation of SANSOD was, " it is recommended that, where it is not yet done, desertification and the degradation processes leading to it be assessed and evaluated as the affect people, the physical elements of the environment like, earth, water and air; and the plant and animal products of the land and water in territories experiencing or likely to experience it, using in the first instance existing data and carrying out new surveys, whenever required, with the aim of defining priorities for national and local programmes and projects in those areas, where population or resources are directly threatened. It is also recommended that, in the assessment of desertification, comparable indices should, whenever practicable, be applied in all regions exposed to the threat of desertification" (NOCOD, 1974:9).

A word is necessary here on the nature of the indicators. The word "indicator" has been introduced into several scientific disciplines as a technical term. As such, it generally denotes a statistic or the presence of a phenomenon judged to carry a specific information value. The use of the term "indicator" is well established in disciplines such as chemistry, geology and ecology. It signifies the condition of the environment, the comprehensive and rather general definition of Webster's new Collegiate Dictionary (1965) is quoted here: "An organism or ecological community so strictly associated with particular environmental conditions that its presence is indicative of the existence of these conditions."

In short, an indicator must be diagnostic of an interrelated set of phenomena. However, it is important to note that this definition is not applicable to physical or social indicators. This definition is especially appropriate in the organic domain. The use of the indicator concept in the social sciences-has been defined by land n the Science article on "Social Indicators" (Sheldon and Parke 1975: 697) The term social indicators refer to social statistics that are components in a social system model including socio-psychological, economic, demographic and ecological or of some particular segment or process thereof. The important point is that the criterion for classifying a social statistic as an indicator in its informative value which derives from its empirically verified nexus in a conceptualization of social process.

The use of the term developed by the seminar was more general and involved both qualitative and quantitative information, since for maximum utility, the indicators must be applicable for diagnostic purposes in areas for which quantitative data are not available. It is also stressed that these indicators were not specifying all of the variables involved in desertification and that there is not a one-to-one correlation between variable and indicator. Indicators are key to the' recognition of the process. Many of these are physical phenomena, which can be measured or seen, other are measured through statistical trends.

Application of the indicators may take different forms according to the nature of the area and the problem. It is envisaged that in areas considered to be at risk, the task could be assigned for assessing the vulnerability, determination of risk or diagnosing the desertification problem by measuring indicators. In some cases, quantitative information could be available for a large proportion of the indicators, in other cases, even qualitative information for relatively few of them could serve the purpose. In either case, application of the indicators should facilitate the best possible for immediate decision-making. In this connection following distinctions are proposed:

1. Persons with short-term training can collect data.
2. Compilation, interpretation and analysis should be done only trained persons interms of the sensitivity of indicators.
3. Indicator methodology requires a time dimension and in turn, requires a temporally defined database. Desertification is also perceived as having at least two phases. During the first phase, characteristics of desertification may be apparent but are mild and potentially reversible. During the second phase, the processes become increasingly difficult to reverse. Establishing numerical and qualitative standards for both of these phases is among the methodological goals. This must be kept in view that desertification is a process and phase definition is to some extent arbitrary.
4. Some observations need to be scheduled empirically, depending on the occurrence of the phenomena e.g., dust storms. Other repetitive measurements can be scheduled in advance.
5. Indicators must be directly measurable, not referred from other evidence.
6. The unit of geographic reference for each indicator must be specified. Indicators of soils, vegetation and various social processes, relate to different sized areas and different time frame.
7. The phenomena causing desertification for which indicators have been defined do not occur within the same areal frame, the method by which

they are observed are applicable in areal units of varying size, one square meter to trans-continental and the suspices under which methods must be employed are most commonly national or international in scope- the distribution of the phenomena, the methods and suspices all create a configuration, which alters the use of indicators.

The appropriate scale at which monitoring of physical indicators is to be done is a major issue. Macro-scale and micro-scale as well as varying mixes at differing scale, each had their supporters. Among the specific devices advocated to cope with the problem, the most notable were the sampling of indicators along a gradient from hyper-arid to sub-humid and the employment of a complex of functionally integrated sites linking, slope and valley systems. The relatively latest technology of remote sensing allows efficient and regular scanning of vast areas by computer in order to pick up certain diagnostic features like changes in land use or in reflectance, which may then be investigated selectively and intensively on the ground, it may also be used to aid in the construction of area frames.

One further complication is the areal relationship between cause and effect. Although, most biological indicators relate directly to the area of desertification, some physical indicators and many social indicators are indirect and relate to situations outside the areas of desertification. In many cases, the primary causes of desertification are to found outside the region undergoing the process and derive from social processes that develop in centers of economic and political activity located in more densely populated and highly productive areas. At the national and regional levels, a constant watch should be maintained on social and economic development; planned and unplanned that may adversely affect the pressure of human use on resources in areas vulnerable to desertification. Development projects should be inventoried for their anti-desertification features.

An important objective is the development of a methodology, which uses controlled comparison. Comparison is done to determine the changes already occurred. The task is to determine the changes occurred after initial measurement. The basis of comparison is to match inhabited and uninhabited areas of comparable soils, vegetation and climate. Therefore, response to climatic variability can be controlled for different types of land use. The use of controlled comparison would reduce the number of sites, which require monitoring without compromising accuracy.

The definition of "uninhabited" and "inhabited" are relative and depend upon two important qualifications. First, the distance to the nearest customary habitation and the second, the length of time an area has been "uninhabited" or "inhabited". Most habitats that are capable of supporting human population are inhabited or remain uninhabited because they are under a form of administrative issues that forbids human habitation. The relationship of man's activity in causing desertification is so important, it is equally important to make the attempt to locate areas which are uninhabited, whether through a system reserves or naturally so.

3.1 HYDROLOGICAL INDICATORS

A knowledge of hydrological processes is fundamental to understand the causes of desertification and in devising means of improving the sustainability of land in arid and semi arid areas. Degradation of land and vegetation is widespread in dryland areas and constitutes a major environmental problem. Although, the boundaries around dryland areas are difficult to define and are neither static nor abrupt, it is estimated that dryland comprise between one-third and a half of the Earth's total land surface. It is also estimated, albeit crudely, that areas affected by degradation, at least moderately, may be as much as 75 percent of the total drylands of the world (Ogallo 1994). Although, there is considerable dispute over the true extent of dryland degradation (Thomas and Middleton 1994), it is agreed that degradation is very widespread and that it affects profoundly the lives of millions of people. Table 3.1 gives extent of dryland areas of the world:

Table 3.1 : Definition and extent of dryland areas

Zone	Definition	% earth's surface
1. Hyper-arid	P/PET < 0.05	7.5
2. Arid	0.05 P/PET < 0.20	12.1
3. Semi-arid	0.20 P/PET < 0.50	17.7
4. Dry sub-humid	0.50 P/PET < 0.65	9.9

Note : P/PET Ratio of annual precipitation to annual Thornthwaite potential Evaporation

Source: UNEP (1992)

It was only some 10,000 years ago, in the early Holocene period that man started to practice agriculture. Instead of being solely an acquirer of food resources, man became a producer of them. Natural and cultural constraints on population were relaxed and seasonally nomadic composites were replaced by sedentary villages (Lawton and Wilke 1979). Agriculture enabled a single nuclear family to be supported by a smaller area of land and sedentary village made possible increased food storage and development of specialized crafts. Although the causes or motives for the development of agriculture are complex and variable fro region to region, it is evident that many agriculturally dependent economies developed in drier regions of the world e.g., Near East, southern Mexican Highlands.

Desertification is currently perceived as the degradation of drylands, which occurs as a subtle, dispersed and continuous process mainly far away from the desert fringes, with outright conversion of once fertile land into desert occurring only in extreme cases (Wallace 1994). Current definitions also recognize that climatic and human factors may contribute to dryland degradation. Each of them has a different causal mechanism, however, needs to be addressed in a different way. Agriculture in dryland regions is fraught with difficulty once population pressure reaches a critical level. At this level, it becomes less likely that areas will be left fallow, nutrients are depleted, soils degrade and yield decline (Figure 3.1 and 3.2). There is serious risk that farmers become struck in a poverty trap. There is also a serious risk that farmers are forced to cultivate and over-graze marginal lands e.g. steep hill slopes, which are even more prone to soil erosion. The net result is reduction in vegetation cover, land degradation and a steady decline in carrying capacity.

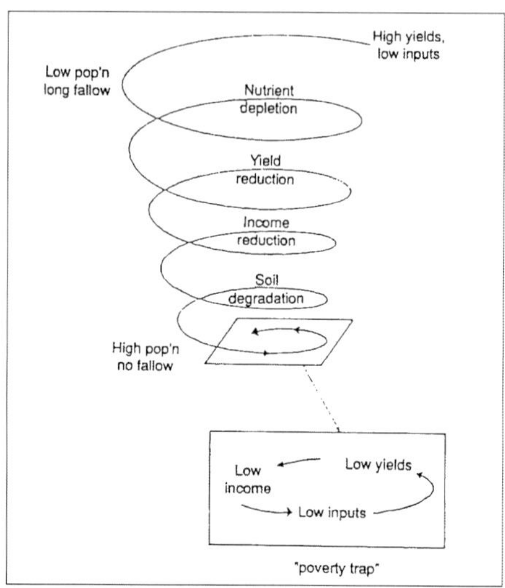

Figure 3.1 : *Diagram showing that increasing populations and continuous, cropping without inputs can lead to land degradation (after McCown and Jones, 1990).*

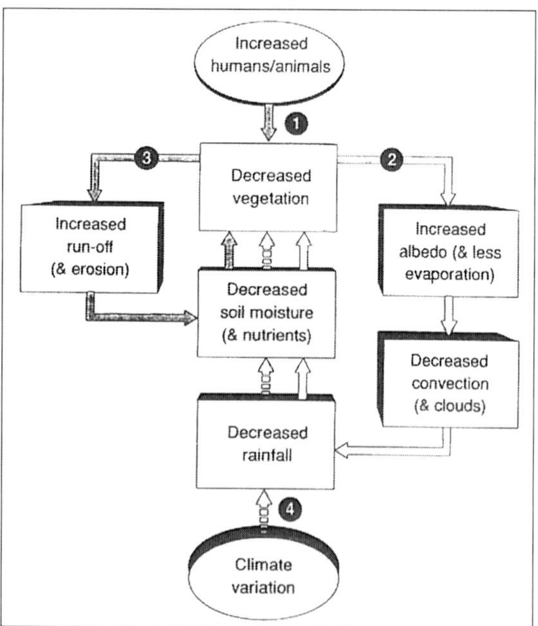

Figure 3.2 : *Framework for the possible mechanisms of dryland degradation: (1) Anthropogenic; (2) Land surface-atmosphere feedback; (3) Hydrological feedback; (4) Climate change.*

Another possible framework for explaining the processes of land degradation is illustrated in Figure 3.1 where any one or combination of four mechanisms can cause a decrease in vegetation. Direct anthropogenic pressures resulting from poor or inappropriate land and resource management and leading to loss of vegetation as a consequence of such factors as over-grazing, soil erosion or deforestation (Mechanism-1 of Figure 3.2). Loss of vegetation cover may trigger feedback mechanisms, which can propagate further land degradation via the land surface atmosphere feedback (Mechanism-2). This occurs when a decrease in evaporation and an increase in the amount of radiation reflected back to atmosphere (albedo) reduce cloud formation and rainfall, causing a negative feedback, which may further reduce vegetation. A third possible mechanism contributing to dryland degradation is hydrological (Mechanism-3). This can occur, when a decrease in ground cover associated with poor land management results in increased and more rapid run-off, decreased soil moisture storage and reduced groundwater recharge, In this situation, less of the rain that does fall on degraded land is available for plant growth, the risk of severe soil erosion is increased and less ground water will be available either for deep-rooted trees or as a resource that can be used by rural communities. The hydrological feedback can also exist in the absence of any climatic change (Mechanism-4). Here, external influences stemming from sea-surface temperature anomalies, humid tropical deforestation and / or CO_2- induced climate change are thought to be associated with drought and degradation in arid zones such as the West Africa Sahel. Another example of possible linkage between sea surface temperatures and rainfall is the connection between rainfall and crop yields in southern Africa and sea temperatures in eastern equatorial Pacific Ocean (Cane *et. al.* 1994).

In recent years, there has been growing interest is using participatory extension and rural development techniques to promote integrated resource management at scales varying from the farm to catchments. Results from a number of projects in Africa and worldwide, give some cause for optimism that these techniques can be successful in combating desertification and in improving the quality of life of rural communities (Anon 1990, Bottrall 1992, Hennessy 1993, Blackmore 1994).

As many of the problems affecting land and water resources are inter-related and cannot be solved individually, strategies of integrated resource management are necessary complex. Strategies must include consideration of crop production, agriculture, livestock management and forestry as well as socio-economics, health, gender issues and education. Studies must also reconcile, as far as possible, the often conflicting requirements and aspirations of the many institutions that have interest in the resources of catchments i.e., land, water, people etc. This is particularly important in developing countries where, in general, water related legislation has not kept pace with requirements and where effective management mechanisms are lacking. It should also be noted that competition for water resources in particularly acute in many developing countries (Engelman and LeRoy 1993).

Although donor agencies are showing considerable interest in Integrated Catchment Resource Management (ICRM) as a means of improving the sustainability of land and water resources, it is clear that there is little consensus about the sphere and contents of ICRM. ICRM is an improvement in catchment planning, whereby relevant institutions work constructively and in collaboration. Auerbach (1994) argued that programmes of combating desertification must reconcile technology centred and people centred approaches within differing timeframe. (Figure 3.3).

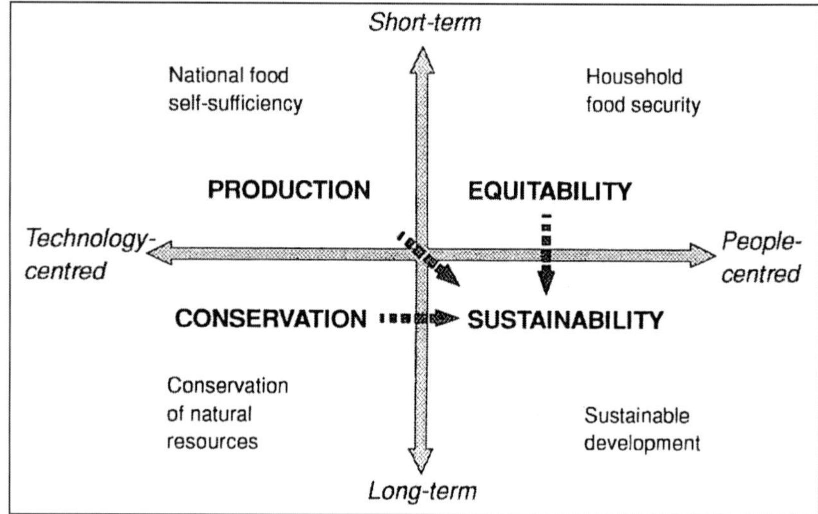

***Figure 3.3** : Model for sustainable agricultural development (after Auerbach, 1994).*

It is clear that hydrological mechanisms are central to many of the issues of desertification. Hydrologists have an increasingly significant role to play both in its study and in the resultant recommendations for remedial measures. There are four key areas where future hydrological work should focus.

1. In the studies of fundamental hydrological processes associated with desertification. These studies should concentrate on separating and evaluating the degradative influence of anthropogenic factors e.g., overgrazing, cropping of marginal lands, which are being studied in some detail and the feed back mechanism, which these factors may trigger e.g. hydrological and land-surface -atmosphere feedback.
2. Research is required to develop better qualitative indicators of degradation especially those, which can be dictated using remote sensing techniques. Progress of defining the current extent and rate of increase of dryland degradation through these indicators.
3. Focus on growth of water efficiency in arid zones need to be explored. This is essential for plant and crop survival in arid environment and to ensure sustainable balance between natural habitat preservation and optimal productivity given water and nutrient availability.

4. Field studies of IRCM are required for hydrological, agronomic and socio-economic potential of a holistic approach in halting and reversing the desertification.

The scale of the problem may be vast but the challenge and opportunity for the application and dedication of hydrological expertise is clear. It will develop sustainable management systems for desertification with greater focus on hydrological processes.

3.2 PHYSICAL INDICATORS

Physical indicators consists of following aspects of desertification as suggested in Nairobi Conference of on Desertification:

1. *Soil:* consisting of effective soil depth, soil organic matter, crusts, dust, dust-storms and sand-storms, salinization and alkalization.
2. *Water:* consisting of depth of ground water and quality, standing open water and status of drainage system.
3. Relative reflectance.

Selection of diagnostic physical indicators of desertification is intended to assist scientists, administrators, managers and planners in determining the impact of a particular use of land is lessening or increasing desertification. The indicators are adaptable for use in individual land holdings, districts and nations. Most of them can be measured by simple field techniques that require little training. A few require laboratory analyses of field samples. Trained persons must do interpretation. The physical indicators appear in Table 3.2.

Table 3.2 : Physical indicators of desertification

Indicator Classes	Sources of observation	Frequency of observation	Background data variables	Testing method	Observation area
A. *Soil Indicators*					
1. Effective soil depth (changes indicate soil removal and sedimentation	Ground and remote*	Monthly, Annually	Rainfall Slope of land Soil texture Soil structure Wind speed	Probe	Watershed or drainage catchment
2. Soil organic matter	Ground (permits estimate of organic matter by inference from colour	Monthly, Annually	—	Require chemical analyses	Soil type
3. Crusts	Ground and remote	Monthly, Annually	—	observation	Soil type
4. Dust, dust-storm and sand-storms	Ground and remote	Daily	Wind speed Temperature soil moisture geomorpho-logy	Air density measurements observation	Dust and dust-storms are local and regional or global; sandstorms are local
5. Saninization and alkalizations	Ground	Monthly, Annually		Require chemical analyses observation of plant growth	Soil/hydro-logical setting (irrigated lands)
B. *Water Indicators*					
1. Water ground depth and quality	Ground	Monthly, Annually		Need chemical analysis for quality Probe	Regional and local (irrigated lands)
2. Standing open water	Ground and remote	Monthly, Annually	Area depth and turbidity	Observation	Local
3. Status of Drainage system	Ground and remote	Monthly, Annually	Water flow and sediment load	Observation	Watershed (Basin)

Indicator Classes	Sources of observation	Frequency of observation	Background data variables	Testing method	Observation area
C. ***Relative Reflectance***					
	Measured with satellite air photo and ground instruments	Seasonally	Plant cover, soil surface, soil moisture and soil colour. Certain observations must be taken every time. Topoadaphic setting. Antecedant rainfall **	Observation. Inapplicable wth overcast skies or very turbid air. Requires satellite technology for large area coverage	Regional for the most part. Some local and global.

* Remote means airborne or satellite platforms for measurement.

** Rainfall not a primary indicator

Source: UN Plan of Action to Combat Desertification (UNCOD 1977)

Rainfall is also a prime climatic variable, whose variability is very high in both space and time. However it is not a good indicator of desertification because short-term statistics of rainfall is not reliable indicators of desertification although long-term statistics like 5 and 10 year mean annual rainfall may correlate well with changes in desertification.

The number of dust and sand storms occurring seasonally or annually, the presence or absence of soil crusts changes in the character of runoff patterns in watercourses the amount of soil organic matter and the area and turbidity of water standing in ponds or lakes are indicators of changes in soil and landscape characteristics. They are most widely applicable to grazing lands and cultivated drylands, but are also vulnerable for indicating land changes under other uses. Determination of the quantity of soil organic matter is dome by laboratory analysis; the others are measured by simple quantitative and qualitative field observations.

Two important factors that are not directly measured are soil fertility and soil structure. Changes in soil fertility are referred from erosion and organic matter measurements. Structural changes are inferred from data on soil erosion and deposition. soil organic matter and soil crusting. Since change is the factor to be evaluated in estimating management practices affecting desertification.

Biological and social indicators also provide information on land conditions and sued in conjunction with social indicators. The presence or absence and the density of particular plant species can be useful indicators of soil fertility, soil structure, salinization, water-logging, erosion and soil moisture.

1. SOILS

Soils are the main element affected through desertification. Soil deterioration causes through various natural and human induced factors. Indicators of soil effects are narrated below:

(a) Depth of Soil above Root-Inhibiting Layers

The indicator of soil depth is in a class of closely related soil variables comprising the depth and the processes of water and air erosion, sedimentation and exposure of hard-pans. This class of variables is considered together and the inter-relationships understood carefully before using any one variable as an indicator. Effective soil depth is an indicator of water and wind erosion on all lands, whatever their use. Desertification brings about general land degradation and loss, of vegetation cover, soil erosion is a phenomenon associated with the desertification process. Erosion and its corollary, sedimentation, both impact on land use and it is therefore, important to have an effective system to monitor soil depth. The root-inhibiting layers to which soil depth is measured consist of bedrock, hardpans or gravelly sub-strata.

The decrease of effective soil depth by erosion of the soil causes a decrease of the water available to the plants and micro-organisms. The soil organic material is decreased at the same lime. The amount of soil available for the root system of plants decreases to the vanishing point as bedrock or hardpans are exposed. Conditions become less favourable for the absorption of run-off. These changes can be manifested by changes in the species of plants growing in the area: the physical indicator of soil depth and the biological indicator of key plant species are, at times, closely linked.

An increase by sedimentation is favourable only if it occurs in small amounts; otherwise, especially in an arid environment, it produces a prolonged, if not irreversible, degradation of the eco-system. In am case, either by wind or water erosion, change in soil depth corresponds to a degradation of structure. The following are specific indicators:

(i) With wind erosion, at the first time, the surface horizons become poorer in silt, clay and organic matter and them thin strata of coarse sand, more or less layered, become visible. In the second stage, the classic forms of nebkas, barkans, dunes or residual regs appear, as well as other large scale phenomena, which become more and more irreversible. In certain cases, whirlwinds and winds of localized dust, which are also a function of the climate, appear Dust devils, however, are not always due to desertification.

(ii) With water erosion, the first stage is represented by sheet erosion and the appearance of light-colored sub-surface horizons and small gullies. At the second stage, erosion produces deep gullies which, if numerous, lead to the appearance of badlands and practically irreversible land degradation.

These indicators appear locally and often regionally. They are observed on the ground and often by air and tele-defection and they are interpreted locally or regionally. This latter task, which is very complicated, draws on geomorphology and hydro-geology and soil science. Method for measuring the depth of the soil over a root-inhibiting layer. Sampling points must be located at different locations of landscape e.g., high places, low places, slopes etc.

The use of remote sensing is based on observations that the characteristics of soil erosion and sedimentation are clear in remotely sensed imagery. Reflectance changes as a result of an exposed lower soil horizon and deposition of alleviated material. In many semi-arid areas the surface or "A" horizon is characterized by relatively high organic content and low surface reflectance, or "B" horizon. This reflectance change can be readily identified and monitored by employing remote sensing techniques.

(b) Amount of Organic Matter in soil

The amount of organic matter in soil is an indicator of changes in soil and landscape characteristics. The loss of organic matter because of improper soil management causes a degradation of soil structure and a loss of water and nutrients available to plants and micro-organisms. The soil becomes less resistant to variations of rainfall and the effectiveness of fertilizers decrease. The loss of organic matter is usually reversible in the first stage. As the loss continues, it facilitates accelerated erosion by wind and water. The surface of the soil often becomes crusted, especially in a semi-arid to sub-humid climate.

This indicator may also be considered to be a biological indicator, because organic matter is the living part of the soil. It is also linked to the biological/agricultural series for its relationship to plant nutrients, specifically to a "decrease in nutrients for plants and micro-organisms, principally in their assimilable form. This is due to the decrease in the richest soil elements, organic matter and clay. Here again, changes in the species of plants present can signal the physical change.

Method: The appearance of the phenomenon, its observation and its interpretation are essentially local. The direct determination of this indicator can be done only by sampling on the ground and laboratory analysis. Aerial and satellite observation may give indirect estimate of changes in soil organic matter content (Bremner 1965).

(c) Presence of soil Crusts

This indicator is characteristically present in range-lands and areas under dryland cultivation. The presence of soil scurst indicate changes in both the soil and landscape characteristics. Crusts are hard packed soil surfaces Extreme erosion can cause the degradation of soil structure to proceed far enough to seal the surface, leading to a great reduction in water penetration and the amount of water available to plants and an increase of run-off.

It may be noted, however, that soil crusts can form in areas controlled as reserves, where recent use of land for grazing, cultivation or other use is forbidden, so the use of this indicator is not confined to human land use systems. Some evidence for this was reported from work in the Karakoum Desert in the Soviet Union, where a long-term reserve system was found to have unfavourable changes, The soil surface became hard and crusted even in sand deserts; it. prevents the normal regrowth of vegetation. Instead lichens became widespread.

However, in terms of long-range soil conditions, Shields (1953, 1964) noted that the crusts characteristic of abandoned and degraded arid lands, formed by binding strands of blue-green algae, are also the first stage in recovery of the soil. By checking wind and water erosion and at the same time, increasing soil nitrogen, these algal crusts pave the way for higher plant succession.

Method: Soil crusts caused by land degradation can be identified by he hardness and platy structure of thin surface layers.

(d) Dust, Dust Storms and Sand Storms

Studies suggest that dust storms are a significant indicator of desertification. It is most important that the occurrence, incidence and severity of dust storms be carefully correlated with desertification phenomena. Cause and effect and presumably the dust storms themselves, need to be tied closely to other desertification processes.

Dust, dust storms and sand storms comprise an indicator class application to many but not all regions undergoing desertification. Sand and dust refer to different soil fractions. One important variable associated with this indicator class is opacity or visibility, both being measures of the ability to see objects in such a storm. This physical indicator class is strongly dependent on rainfall and the soil moisture present before the onset of a wind storm. Dust storms, in particular, indicate desertification at their source, which may well be different from the place, where they are observed, since the storms can travel thousands of miles.

Method: measure the visibility, from one point, of a landscape feature such as a distant building or a hill. Dust storms are also' nominated by Berry and Ford (1977: 21-24), their summary on data sources and methods is quoted- "Longitudinal analysis showing periodicity or changes in intensity in dust storms have rarely been attempted. Fortunately, the United States National Climatic Centre in Asheville, North Carolina, has an abundance of unanalyzed data Satellite data from meteorological programs and LANDSAT serve to provide a now overview of the phenomenon. The North Carolina data is gathered from whether stations all over the world, which account for every occurrence, when visibility at the weather station was less than 6 miles. Many of the explanations for low visibility include dust storms, thus developing an excellent record of incidence, timing and place of dust storms, which could easily be correlated with precipitation records for same time period. Baseline data on dust storms could assist in using current and future dust incidence as an indicator."

Frequency, intensity and direction of dust storms is inversely related to the quality of vegetation. The experience of the Great Plains in the United States in the 1930s indicates that dust storms create a multiplier effect. Increased dust destroys additional ground cover, which causes more severe dust storms, etc. Careful recording and monitoring of these storms can be of great value in pinpointing, where problem areas are getting severe and where action plans should be implemented. Cost of Monitoring dust storms could be moderate LANDSAT and meteorological satellites already are in place. Monitoring for storms, such as that of the National Environmental Satellite Service is already underway. The addition of dust storms as a new variable to monitor should not add greatly to existing cost. More specific methods are proposed by Rapp, LeHouerou and Lundholm (1976: 235).

(e) Salinization and Alkalinization

Both soil and ground water are subject to salinization and alkalization, particularly where irrigation has been used on arid and sub-humid land. The presence of this indicator is so frequent due to irrigation and it affects ground water as well as soil. It is clearly related to other indicators like depth to ground water and quality of water. These are used to identify the magnitude of and change in salinization and water logging of irrigated lands. Depth of ground water is a simple measurement; the other measurements require laboratory analyses of soil and water samples.

Canal irrigation schemes, have often been introduced without due regard for drainage and related problems. Peterson (1970:24) describes the consequences: "to avoid water-logging, more lands are left fallow, but the salt rises in the then-unleached soils under the hot desert sun. Caught between the twin hazards of water-logging and salinity, the farmers press for more intensive irrigation that the land can bear and this leads to the cumulative aggravation of salinity and sodicity that are associated with calcium precipitation. This is the story of ancient Mesopotamia and the modern Indus. More level the land and tighter the soil, the greater the hazard; initiation of perennial irrigation, in contrast to seasonal and eventually full control of the river for irrigation, each in turn aggravates the threat."

The use of poor quality ground water from tube wells also induces salinity and has adverse influence on soil productivity. A high evaporation rate in an arid climate can cause the build up of soil minerals. Even when water used for irrigation is not of particularly poor quality, if the evapo-transpiration rate is very high, the water evaporates, rather than draining off or leaching through the soil, leaving minerals the water contained and causing a gradual build up of salts or alkali. Evapo-transpiration is a poorly understood process, the primary evidence of a high rate is salinization or alkalization of the soil.

As the soil processes, salinization and alkalization are clearly related. These physical indicators are linked closely to biological indicators, especially in the early stages of desertification. Salinization is the accumulation of excess exchangeable sodium, often leading to an increase in the pH of the soil. Both processes are unfavourable for plants and micro-organisms and changes in the composition of the plant community may signal the physical changes in the soil. Degradation of the soil structure accompanies these changes, a decrease in soil permeability, porosity, depth of penetration of water and amount of available water. At an advanced stage, saline or alkali deposits are visible as white, gray or black patches. Saline soils may have a layered or pseudo-sandy, powdery surface soil structure.

Method: Salinization and alkalization are generally local phenomena. In their early stages, identification call for laboratory analyses of field, samples; at an advanced stage non-specialists can easily perceive the visible signs in the field. The United States Department of Agriculture Handbook contains the detailed methodology, which is used for determining salinity (1954). Furthermore, as Berry and Ford suggest (1977: 26), it should be possible to devise a low-cost remote sensing monitoring operation using LANDSAT imagery.

2. WATER INDICATORS

Water is one of the most important feature of vegetation and drylands face the problems of water scarcity. Most of the land degradations in drylands are the impact of less water or uncertain availability of rain water. The desertification in view of water indicators are narrated below:

(a) Depth of Ground Water and Quality of Water

This indicator is especially relevant to irrigated lands. Variation in the levels of aquifers, wells and water tables are important factors. Ground water depth is very important to the potential use of wells to tap underground aquifers. If there is a drop in the level in non-irrigated low-lying areas, it indicates an increase in desertified conditions for plants, animals and man. If there is an increase in he ground water level, on the other hand, in irrigated areas, it indicates the approach of water-logged conditions. Water-logging is a classic problem in irrigated areas with poor drainage. The critical level for ground water thus varies in different situations and its influence on land productivity is subject to local interpretation. Another key factor is. ground water quality. Chemical analysis in needed to determine water quality and the expense of this analysis limits the use of this indicator.

Method: This indicator is measured by sampling on a regular basis of the ground water in aquifers and wells and chemical analysis of the quality of the samples.

(b) Area Covered and Turbidity of Surface Water

Standing or open water is an integrated indicator of several variables in the water balance of an area. Rainfall is the key incoming variable, while temperature, runoff and percolation through the soil are key outgoing variables. Standing water smoothes out the high variability of rainfall. The amount of standing water may indicate increased or decreased runoff. Turbidity increases the degree of salinity of standing water, if measured a short time after runoff occurs. Sediments, borne by the water, is removed to lower levels, where in small amounts it may enhance the soils, but in large amounts, it may "clog" permeable soil with clayey or silty compacted deposits, thus effect of the phenomenon is not obvious.

Method: Surface water and its turbidity is observable by remote sensing and aerial photography.

(c) Change in Water Flow and Sedimentation Load in Watercourses

The status of a drainage system is an indicator class somewhat less than most of the other physical indicator classes. It refers to the magnitude and continuity of water flows in an entire drainage basis, including the sediment load. For a specific drainage basin, the size and variability in position of a single water channel might be identified to be used as a simpler indicator of this class. Caution is needed in interpreting this indicator. For instance, reduction in the volume of water in wadis and irregular flow may indicate desertification in the receiving area or may be due to desertification of upstream areas and to poor water distribution. Also, water flow or runoff may increase due to reduced soil absorption or sparse vegetation associated with desertification. Frequently rainfall occurs in heavy downpours even though the annual total is low, thus increasing the likelihood of high runoff.

Method: This indicator is monitored at local level.

3. RELATIVE REFLECTANCE INDICATOR

The last of the physical indicators is relative reflectance. With refraction and absorption, reflectance is the basic property of sunlight striking the earth's surface. It is measured by sensors mounted in orbiting satellites, aircraft and ground based instruments. Reflectance, per se, is not an indicator of desertification, but relative reflectance, i.e., changes in reflectance in one site or a comparison of reflectance in more than one, can provide a basic data source for correlation with other variables. Reflectance depends on several background variables and therefore, must be interpreted with care. Its dependence on other variables is not yet clear enough for it to be used uncritically except in a few cases, where it has been studied carefully. Relative reflectance is included, in part, because synoptic or repetitive data coverage is available and is likely to continue to be so in identified parts of the world and because further integration of research results is very important.

Method: Data are gathered with remote sensing instruments and analyzed by photo interpretation techniques and machine processing.

3.3 ECOLOGICAL AND ENVIRONMENTAL INDICATORS

The changes in land use and infrastructure development in arid, semi arid and hub-humid regions significantly affect on ecological and environmental balance of the area. The impact are varied and have mixed reaction on the area and the people surviving in the area in short and long term basis. In one view of the causes of desertification, resource abuse occurs as part of human activities for food production. Therefore, the effects visualized in plants, human beings, wild and domestic animals, vertebrate and invertebrate to yield or expressed in terms of vegetation.

In a number of places, former rangelands have been or are being converted into areas for cultivation and such dryland cultivation forms another important spatial zone for examination of desertification processes. In such fields, the state of soil and the amount of moisture in the soil are especially important; to assess the impact on the environment and ecology of the area.

The usual assumption is that monitoring actively cultivated fields means gathering data on planting decisions. Determining the pressure of key species is relevant only in construction of cropping inventory or in identifying weed problems. The rate of regeneration of vegetation and the succession of species that invade formerly cultivated fields and fields in fallow cycle are key factors for monitoring purposes. Sobrinho and many others consider fallow fields that do not regenerate vegetation at an anticipated rate to be the foci at the very start of desertification.

Changes in wasteland, land unsuitable for either forage or cropping, may occur as a result of human use, indirectly or for reasons not related to man's use, Wasteland include disturbed lands or formerly cultivated lands or grazed lands, which are no longer suitable for either activity. Maintaining unused natural areas is also important to provide a control area to compare with land under some form of active use. Some land is kept in reserve in nearly every country and these reserves, of course serve some purpose.

Residential areas and changes in the amount of land devoted to residential areas are also of special importance for evaluation. The space in the immediate vicinity of houses can be kept fertile with mulching, waste water, special cultivation, weeding etc., or the reverse may be taking place and either trend in important.

Although no single indicator is intended to indicate desertification by itself, a combination of changes in indicators signify the trends. These indicators are not given in order of importance, but are measuring rod to witness the impact of desertification. Table 3.6 gives details indicators relevant for adjudging the ecological and environment from desertification.

Table 3.3 : Ecological and Environment Indicators

Indicator Class	Source of Data	Frequency of Observation	Background Data Variables	Limitations	Remarks	Sample Units in Area of Observation
1. Vegetable						
(a) Cover	% of bare ground / ground cover in dry seasons: aerial photos and observation, remote sensing	Annually - dry season Periodic- rainy season	% of ligneous/ herbaceous plants. % annuals/ perennials	Plant types must be distinguished from amount of cover.	Water-holes may be more overgrazed than range in general- sampling important	Ecological unit: sample stratified according to vegetation class/ land use - Cultivated - Grazing land - Wasteland - Fallow - Forest - Human settlement - Preservation plot
(b) Above ground biomass	Chlorophyll contents: remote sensing Transects and quadrates: ground sample and laboratory analysis Preservation plots Amount of humus in soil, litter	Monthly/ Seasonally 3-10 years intervals Long-term measure Short-term index	Net assimilation Climate Soil type Erosion Background biomass % of ligneous/ herbaceous plants	Effects of root pathogens. Effects of fire, especially on litter Effects of grazing intensity	Sampling of transects and quadrates can be done by trained local workers. Analysis of samples and other measures requires professional training	As (a) above Comparison of preservation plots with other classes crucial.

Indicator Class	Source of Data	Frequency of Observation	Background Data Variables	Limitations	Remarks	Sample Units in Area of Observation
(c) Key species	Interviews with local experts Local historical records, manuals Ground observation of transects	List compilation: 6 months-1 year Observation: seasonally	Range condition and trend: based on relation of soil condition and key species Herd composition the effects pf selective grazing of different animal spp.	List of key species are site- specific Reaction of plants to human activity requires trained ecologist	Important to monitor distribution and frequency of species and changes or trends. With regard to physical adaption, plants are classified as: Heterotrophs xerotrophs	Watershed. Also as (a) above.
2. Animal						
(a) Key species of wild-life: Frequency and distribution	Arthropods: -Tenebrionids -Solifuges -Acopions Small mammals and birds (Systematic reconnaissance flights- large mammals)	Seasonally Seasonally Seasonally	Ratio of black or desert colour/ other colour fauna. Relation and competition with domestic herds	Species appear in environments where they compete most favourably, not where they are best adapted. Large mammals too mobile to be primary indicators	Site-specific inventories must be established. Measurable by non-scientists	Ecological unit or watershed

Indicator Class	Source of Data	Frequency of Observation	Background Data Variables	Limitations	Remarks	Sample Units in Area of Observation
(b) Population of domestic animals	Numbers and distribution of large and small stock: -livestock management -studies of pastoralism; -Ground enumeration	Seasonally	Carrying capacity of range: physical and vegetation indicators determine amount and location of available water and forage	Observation-unit must correspond to area being used as single socio-economic unit.	Carrying capacity in estimated in animal units/ month.	Ecological units or watershed.
(c) Herd composition	Composition of herds by species, age, sex, livestock censuses by ground enumeration, aerial photos	Seasonally	Physical and vegetation indicators determine diverse types of forage available Variation in rainfall and spacing of water points. Needs of different species	Trends more significant than composition at one time	Changes in ratios of sheep: goat or cattle: camel may be indicative of desertification	Ecological unit or watershed

Indicator Class	Source of Data	Frequency of Observation	Background Data Variables	Limitations	Remarks	Sample Units in Area of Observation
(d) Production	Live weight of animals at market: abattoir statistics	Annual statistical review Annual interview	Distance to grazing; local interview data Trade and purchase of supplementary cereals : local interview data	Black market sales often unrecorded	—	Market area/ ethnolinguistic area
3. Yield						
	Grams of dry weight /square meter / year	Annual	Short-term effect of fertilizers	—	Must be measured by trained scientist	Ecological unit: sample stratified by vegetation class, as per 1(a)

Source : Prett and Gwynne, 1977.

1. Vegetation

Plants that are indicators of increasing desertification are not necessarily the dominant species in a community. It is for this reason that observations are based on key species.

(a) Cover

All the plants need to be considered. The frequency or density of the cover during the dry season is of primary concern, although the trend is the response of vegetation in the rainy season and in both the density and trend, following rainy seasons of different total accumulations, is important. Although only one measure may be necessary, bare ground or amount of cover, it is best to tabulate both so that confusion is minimized in any compilation of observations. Due to strong influence of the rainy season, the time at which an observation is taken in anyone' place is very important.

The word "Canopy Cover" is also used instead of ground cover. The amount of vegetation cover may be quite different from the quality of that cover. Bushes and weeds may provide cover, yet key species indicators of desertification should be used together. The amount of cover is closely related to indicators associated with the use of patterns of a particular area. Plant cover can be strongly influenced, in range areas, by the location of water. A tendency to overgraze around water can itself become a focus of desertification. Changes in species composition and vigour with distance from water can also be a measure of increasing likelihood of desertification. Relative location of of sampling area, therefore, becomes a critical matter if an adequate assessment is to be made in varying types of range management.

Method: Aerial photography can be used to measure the amount of bare ground. Satellite imagery may also show the amount of bare ground.

(b) Above Ground Biomass

These indicators mostly consist vegetation i.e., grass, forbs, trees and shrubs. Strictly speaking, it is not phytomass, i.e., mass consisting of plants or the standing crop of plants, because precise measurement which excludes nematodes, insect deposits and so forth, is highly technical and difficult to achieve under field conditions. Biomass does not include animals. Above ground biomass is the preferred measure, because trained technicians can use it. However, Pandeya, among others, stressed that below ground biomass, included in photomass, is an important measure of aridity. The ratio of the above-ground to below-ground biomass can be radically different in semi-arid and arid areas from areas where moisture is higher and more reliable. Furthermore, the appearance of "open, evenly spaced shoots is a result of extensive root development and competition. What is therefore apparently a structurally open community on the surface may be biologically closed beneath the surface (Tivy 1977).

Root pathogens can be of considerable significance in reducing the viability of plants or killing them outright. They should be considered; especially if the measured biomass is lower that one would expect based on rainfall, grazing pressure, habitat type etc. Pathogens are difficult to observe and assess, but if below-ground phytomass is taken into consideration, the below-ground root pathogens should also be considered. A number of factors are known to interact, influencing the rate at which is produced, as well as affecting the basic mechanisms of photosynthesis. The rate of carbon assimilation in semi-arid and arid areas is most obviously linked to the availability of water , but temperature, climatic variables, abiotic or soil variables, presence, of minerals, humidity and direct and diffuse solar radiation are all relevant. The amount of litter is also a factor closely related to biomass, the greater the amount of litter, the better the condition. Decrease in the amount of litter is evidence of downward trend.

Method: Measurement of annual and perennial distribution and production should be made under controlled conditions (preservation plots or fenced exclosures), as well as in open areas. The distinction between annuals and perennials is basic especially if the amount of perennial cover present during the dry season is used as a key indicator.

(c) Yield

The definition of yield, primary productivity and net productivity were made available by Terhune. These definitions are as follows:

(i) Primary Productivity is the rate at which organic material is created by photosynthesis. Primary productivity is limited by such factors as availability of light, soil nutrients and water. Water is of course, critical in arid and semi-arid environments. Not only water is sued during photosynthesis, in the chemical reaction, when carbon dioxide is fixed into sugar that serves as the basis of other organic compounds, but water is necessarily lost from moist cell surfaces in leaves and green stems, because surface pores must be kept open so that carbon dioxide from the air can be observed.

(ii) Net Primary Productivity is the amount of CO2 incorporated into organic material in plants in excess of the organic material, used by the plant for its own metabolic needs. For desert scrubland throughout the world this ranges from 10 to 250 grams dry weight/square meter/ year, with a world average of 70 g/m2/year.

(iii) Yield (harvest. That fraction of net productivity that it is feasible to remove without destroying the basis of productivity. The word "productivity" is often used to mean "yield". Yield is always lower than net productivity. Yields are at most 30% of net productivity in the case of cereals and wood.

Distinction between productivity as a scientific term and yield, essentially as an economic terms and productivity is used rather widely and carries more than one specific meaning. A downward trend in the yield of species used for firewood and for grazing is one important indicator of desertification.

Method: Yield is measured in grams of dry weight per square meter. The techniques used are the areas of trained scientists.

(d) Key Species: Distribution and Frequency

Changes in plant density and increase in the relative number of weedy species, of xerophytic species and/ or of unpalatable plant species are characteristic of a downward trend. For these reasons, the possibility of isolating and identifying the presence or absence of certain species, of determining their distribution, especially in relation to other features and of measuring their frequency over time all suggest empirically useful means of monitoring desertification. Again, the amount of plant cover, the standing crop and the occurrence of certain species are all closely related and presumably these three indicators would be used together. However, the area unit of observation may differ for these indicators, since the measurement of biomass is a general procedure, whereas presence or absence of a species probably will be site-specific, even though the conceptual framework is also general.

Rumburg (Personal Communication 1978) emphasizes the importance of sampling in General Method for Vegetation Indicators, a method applicable to all indicators covered. Accordingly, "Sampling is crucial to the prediction of range trends. Procedures for sampling of vegetation depend on the size and density. Mixed stands of trees, shrubs and herbs require different methods within the same sample site. Cover can be estimated by ocular or measuring techniques that are relatively simple for larger plants. Herbaceous plants are probably are best inventories by frequency techniques if properly employed. In all cases, selection of the specific method, plot size and location require extensive training and experience in sampling vegetation. Only qualified individuals should locate, select and design sampling procedure, and then they should validate the precision and efficiency with trial runs. Thereafter, technicians familiar with the plants and measuring procedures can collect the data.

2. Animal Indicators

Faunal indicators include key species of wildlife as well as several indicators pertaining to aspects of the domestic animal population, herd composition and production.

(a) Key Species : Frequency and Distribution

Changes in the distribution of desert fauna can, presumably, be indicative of desertification. However, according to Cloudsley- Thompson, it is essential to recognize the limitations on the usefulness of animal data. It may be suggestive of desert conditions, faunal data is to be used mainly to supplement the information from other sources.

Animal species are not necessarily found in the environments to which they are apparently best adapted, but rather in the environments in which they compete most favourably with other species Birds, vertebrates and some species of small animals, particularly rodents, can be identified as key indicators of desertification: As with plants, the inventory of key species must be established independently.

In addition to examining the frequency and distribution of various species, another type of frequency analysis may also be revealing Desert animals are characterically either pale or black. Changes in the proportions of species showing other types of coloration could be indicative of changes in aridity. In the Sahel, sand grouse, wheatears and rodents are associated with the shift or transition from perennial to annual grasses (Swift, personal communication). Rodents, especially, can be used as indicator species in long-term research studies (Henry Setzer, personal communication 1974).

Among invertebrate taxa, key species, those best adapted to desert conditions, include tenebrionid beetles and most scorpions. It is quite possible that the relative abundance of these taxa, compared with that of arthropod taxa characteristic of mesic biomes, could provide a quantitative measure of aridity. In extreme desert. regions, difficulties arise in obtaining sufficient numbers of specimens to provide statistically significant data. Finally, ectoparasites of mammals especially tick, fleas and mites that serve as human and domestic animal disease vectors, are relevant. The procedure adopted by Setzar mentions that biological samples can be processed, identification made and the results integrated with existing data sources. It can be useful to:

(i) Provide each country with synoptic collections of small mammals and ectoparasites for use in local identifications of ecologically and medically important organisms;

(ii) Develop keys and distribution maps of various groups and species as well as summaries of their economic importance;

(iii) Provide ectoparasite identifications and distribution as potential indicators of vectors of diseases to man and domestic livestock; and

(iv) Integrate data with those for other indicators.

Indicators species should be separately identified for rangelands, dryland farming areas and irrigated land.

(b) Population of Domestic Animals

By "population" is meant the numbers, frequency and distribution of large and small stocks. Large stocks include cattle, camels, donkeys and horses and small stocks are sheep and goat. This indicator is closely related to herd composition and production. The almost universal use of concept of fangeland carrying capacity makes the actual numbers in relation to unit of grazing land of special importance. It is also important to stress the relationship of number and distribution of animals to vegetation types and to soils and season of use. These considerations form the basis of rangeland management. The carrying capacity of renewable resources related to desertification, because it emphasize the importance of:

(i) One concept of carrying capacity reflects that level at which population of native organisms can be maintained on a given area of land throughout a climatic cycle. In its purest form, this concept states that the carrying capacity is limited by that harshest period during the climatic cycle.

(ii) Carrying capacity is normally considered to be site-specific. In other words, the basic biological carrying capacity of a particular soil and vegetation type is often determined and reported in animal units, number of people, etc., that a particular geographical area can support. However, from human standpoint, it is often impractical to limit carrying capacity concepts to a single site or even a series of closely interrelated sites. The carrying capacity of a particular range site for cattle may be quite low, if cattle are maintained on it year round, but if these livestock are removed for short periods of. time to graze crop aftermath or other fodder sources, thus allowing a deferment of he rangeland to take place, the carrying capacity for site may be raised as long as it is managed as part of a large system (Peterson 1977).

(iii) The concept of carrying capacity in renewable resource management is used to indicate the level of sustained use at which the resources will not be permanently damaged or its productive capacity lowered While the concept may be extended to all renewable resources, it is classically applied to grazing, basic vegetative productivity vis-à-vis animal, units. It could be applied to the tillage of rainfed croplands vis-à-vis erosion or to air pollution associated with urbanization. Sustained yield is maintained by a regulatory process that contains the use of the resource within its carrying capacity (Peterson 1977).

The number of animals, the total area and the degree of control in adjusting, each are the principal factors in range management. Carrying capacity is a relative figure depending on management, of which there are two radically different types, indigenous management and commercial management. Under indigenous management, a variety of mechanisms, rights and constraints can operate and are subject to change. Where the total range area is fixed and known or change or subject to calculation, the principal more of adjustment is change in the number of animals in he number of animals. In countries, where the transhumance is practiced, the movement of animals is the principal mode of adjustment. A further set of variables is rainfall and other climatic factors. If fluctuations are averaged over a large area, carrying capacity is likely to be higher than the capacity of smaller sections of the area.

If data on animal population are to have any predictive value, the area unit within which population is calculated should correspond fairly well with an area that is socio-economically coherent. To define it too narrowly, it is to underestimate the economic productivity of a region and the strategies for adjusting its use to its potential.

Method: Animal population data recorded by commercial ranches are converted to units, rather that being expressed in number. An animal unit (AU) is widely accepted as a mature cow with calf, or their equivalent. Horses, sheep and goats commonly are converted to animal units at the rates of 1.25, 0.2 and 0.17 respectively. Young animals between weaning and maturity vary from 0.6 to 0.9 of their adult female equivalent animal units. These conversion factors are indicators of equivalent amounts of forage needed by different kinds and classes of domestic animals with vide difference in food habits. Animal unit equivalent refers to forage needed by AU for one year.

An animal unit month'(AUM) is the amount of forage required by an animal unit for one month of grazing. Related terms for specific instances include band day, sheep day, cow day, cow month and many others that refer to different kinds of animals and time periods. Animal unit equivalent symbolize mixed herds as groups of average animals, whether they are different kind of mature animals that change little in feed requirements from day to day or growing animals that gradually increase their daily intake. Coordination of varying AUM requirements and forage increments in day to day livestock management still is a matter of judgment by the manager. Animal unit month of grazing or a variant such as steer months, are widely sued as leasing units and as a basis for pasture rental. Grazing fees on public lands and on many private pastures are attached to animal units on a yearly or monthly basis (Heady 1975).

Stocking rate is the actual number of animals or animal units on a specific area for a specified period of time, usually for a grazing season. In areas where bookkeeping is not suitable or available, methods have been devised for enumerating animal population or units (Heady 1975, Gwynne and Croze 1975-a and b).

(c) Animal Herd Composition

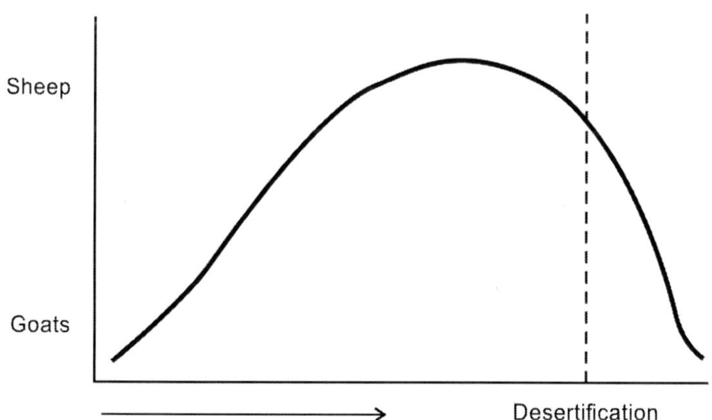

Figure 3.4 : *The Sheep: Goat ratio trend and desertification*

Given the global frame of reference, animal herd composition has three different meanings: (i) single species composition by age and sex, (ii) mix of domestic species, also factored by age and sex, and (iii) mix of domestic and wildlife species also (if possible) by age and sex. Some of the techniques developed by demographers for human populations are being used for animal populations. J. Swift has constructed life tables of cattle for Mali and have constructed life tables for Fulani and Tuareg herds in Mali and Niger including relationship between herds and households (Dahl and Hjort, 1976).

In small stock, the sheep: goat ratio can shift and can be indicative of a trend towards desertification. Sheep are market animal, so an increase may indicate enhanced pressure on the range. Since sheep are more sensitive to drought, a change in the proportion of sheep and goats, with an increase in the number of goats, is an index of desertification, but changes in tire land and plants may have resulted from earlier increase in the number of sheep. In large stock, the cattle: camel ratio may shift in the same way. That is, cattle are the market animals; an increase in favour of camel as the more drought resistant animal may be indicative to desertification, but the earlier increase in the number of cattle may have provided the predisposing conditions. (Swift, Personal communication, 1978). This position has been depicted in Figure 3.4 and 3.5 respectively.

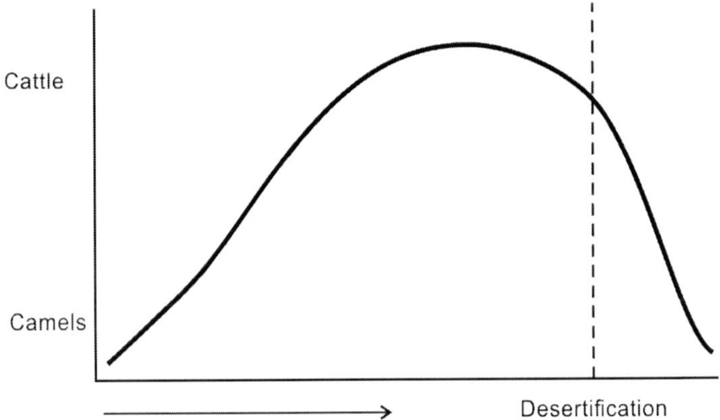

Figure 3.5 *: The Cattle: Camel ratio trend and desertification*

The concept of niche is widely used by range ecologists in analyzing a mix of animal species. Heady (1975) has the mechanisms operative in species mix determined by Lamprey, as well as his summary of Tanzanian data. The fact that several kinds of animals occupy the same niche and are in direct competition with each other. More likely, they occupy separate but overlapping parts of the. total area. Lamprey (1963), after lengthy study of numerous grazing and browsing animals in north central Tanzania, suggested that separation in accomplished in six different ways:

(i) Species concentrate in different parts of the vegetational mosaic.

(ii) Species select different types of food.

(iii) Species separate topographically on a seasonal basis.

(iv) Species select the same area but at different seasons.

(v) Species select different feeding levels in he vegetation.

(vi) Species separate on a vegetational basis according to season of food stress.

First two of these principles are illustrated in Table 3.4, which shows that occupied habitat is not the same as food selection. The third principle, topographic separation, is accomplished in the dry season, when impala and Grant's gazelle use upland areas because they can live without free drinking water, while other species move to lower areas near water. Elephants illustrate the fourth principle by their ability to range more that 40 kilometers from water in the dry season placing them on forages that are used by other species only in the wet season. Black rhino, dikdik and giraffe illustrate the fifth principle because they browse, but they feed at different levels in the woody vegetation. The sixth type of separation is illustrated by zebra and wildebeest, which live closer together and near water in the dry season; but separate' to different parts of the grassland in the wet season. These separations illustrate types of horizontal and vertical stratification among animals.

Table 3.4 : Habitat and Food preference in tanzanian animals

<table>
<tr><th colspan="3">Habitat Preference</th><th colspan="3">% Selection of Food Plants</th></tr>
<tr><td>Grassland</td><td>Open Woodland</td><td>Dense Woodland</td><td>Grasses and Sedges</td><td>Herbs</td><td>Shrubs and Trees</td></tr>
<tr><td colspan="3">|- Grants Gazelle-|</td><td>90.67</td><td>4.00</td><td>5.33</td></tr>
<tr><td colspan="3">|-Wildebeest-------|</td><td>94.48</td><td>3.45</td><td>2.07</td></tr>
<tr><td colspan="3">|- Zebra-------------|</td><td>92.57</td><td>5.53</td><td>1.90</td></tr>
<tr><td colspan="3">|- Hartebeest---------|</td><td>96.25</td><td>2.50</td><td>1.25</td></tr>
<tr><td colspan="3">|- Eland----------------|</td><td>70.18</td><td>8.77</td><td>21.05</td></tr>
<tr><td colspan="3">|- Cape Buffalo-------- |</td><td>93.64</td><td>1.21</td><td>5.15</td></tr>
<tr><td colspan="3">|- Giraffe-----------------|</td><td>0.83</td><td>1.10</td><td>98.07</td></tr>
<tr><td colspan="3">|- Impala------------------|</td><td>92.62</td><td>1:23</td><td>6.15</td></tr>
<tr><td colspan="3">|- Wart Hog---------------|</td><td>88.32</td><td>9.28</td><td>2.40</td></tr>
<tr><td colspan="3">|- Rhinoceros-------------|</td><td>38.00</td><td>9.33</td><td>52.67</td></tr>
<tr><td colspan="3">|- Elephant----------------------|</td><td>12.4.7</td><td>1.29</td><td>86.24</td></tr>
<tr><td colspan="3">|- Waterbuck-----------|</td><td>95.14</td><td>0.81</td><td>4.05</td></tr>
<tr><td colspan="3">|- Dikdik-----------------|</td><td>17.95</td><td>2.56</td><td>79.49</td></tr>
<tr><td colspan="3">|-Lesser Kudu---------|</td><td>66.67</td><td>4.75</td><td>28.58</td></tr>
<tr><td colspan="3">|-Cheetah------|</td><td></td><td></td><td></td></tr>
<tr><td colspan="3">|- Hunting Dog--|</td><td></td><td></td><td></td></tr>
<tr><td colspan="3">|- Lion----------------------|</td><td></td><td></td><td></td></tr>
<tr><td colspan="3">|- Leopard----------------------|</td><td></td><td></td><td></td></tr>
</table>

Source: Heady 1975:141

Changes in herd composition indicative of increased stress among species which are less drought-resistant would, in turn, be indicative of desertification.

Method: Herd composition, like domestic animal population, can be determined by ground count or with aerial sampling in low-flying light aircraft (Gwynne and Croze 1997- a and b; Pratt and Gwynne 1977).

(d) Production

Production in pastoral economies particularly, includes both animals that reach the markets and animals raised for subsistence. In addition, animal products-milk, blood, wool and hides, form an important part of the off-take and are generally subsistence commodities, although surpluses are sold. As far as desert expansion is concerned, as yet not quantifiable crucial indicator would be the biomass of domestic animals above a threshold, determined by rainfall, evaporation, vegetation and other non-zoological indicators, according to Cloudsley- Thompson. It has been suggested that trends in carcass weights on marketed animals provide an important indicator that correlates with the total animal production and for which statistics are more readily obtained.

Method: Berry and Ford (1977) identify animal productivity as one of their regional level indicators. The method they suggest for the Sahel is the following:

"Abattoir Data- The Sahel produces large quantities of meat. Region-wide information on (a) weight, (b) number, and (c) sex of stock sold to abattoirs at key market towns would provide succinct indicators of changes in livestock numbers, off-take and range productivity, all of which are indicators of the condition of grasslands. Mechanisms are already in place in towns such as Mopti, Bamako, Gao. Niamey, Timbuctu, Kano, Abidjan, Accra and Dakar. Retrieval of this information would not require new organizations nor should it involve major expense. The regional monitoring centre should be able to establish base-line data on normal abattoir activities and thereby have a base against which current and future data can be compared."

The Ecological and Environmental indicators for flora and fauna include indicators which, applied to rangelands. determine trend in range condition. Range trend and range condition are thus composite indicators. Range trend can be included directly in the list of indicators, since 90% of the area is subject to present or potential desertification in rangeland (Dillard Gates, personal communication), their concern with this measurement is understandable, Furthermore, at present half of world's cattle, a third of the sheep and two-thirds of all goats are living in semi-arid or arid areas (Schechter 1977). In practice, most monitoring of desertification deals with rangeland.

Range condition is the amount of forage that a range can be expected to produce under good management, given the variables of climate and soil (Terhune 1978 and Humphrey 1962). The trend in range condition, either being downgraded or improving from conditions that result from management, implies a temporal dimension or a means of making a predictive qualitative judgment.

One of the better indicators of trends, in addition to condition of the soils, is the presence of plants of greater or lesser forage value, classed as increasers, decreasers or invaders. One area of specific difference between the conventional guides to range evaluation and this set of indicators is the emphasis placed on key species rather than on an inventory or appraisal of all the plants in the plant community.

3.4 SOCIAL AND ECONOMIC INDICATORS

Social and economic indicators are applicable to people living in areas undergoing desertification. The changes in infant and adult mortality, increased patterns of migration and levels of poverty are associated with the onset of desertification. Socio-economic indicators are used to analyze the behaviour of the people whose actions lead to desertification. The desertification can be localized spatially and human population can be involved with desertification may interact in significant ways with the efforts made by national and international organization for combating it. The people residing in dryland areas face various problems and indicators are developed to assess their problems. For systematic approach, these indicators have been classified in following categories:

1. Types of and changes in land and water use comprising of irrigation, dryland cultivation, pastoralism, mining, firewood, tourism and water.
2. Settlement Pattern consisting of new settlement, expansion of settlement and sedentarization, diversification of settlement and settlement abandonment.
3. Human Biological Parameters, which include population structure and rates, measures of nutritional status and public health indices.
4. Social and Economic Processes like, conflict, migration, redistribution patterns, marginalization and cash versus subsistence.

1. Types of and Changes in Land and Water use

Land and water are the most crucial factors for survival of human beings, flora and fauna. In the areas prone to desertification, various aspects of social and economic nature are affected in view of hardships. These aspects are narrated hereunder.

(a) Irrigation

The social and economic indicator aspect of irrigation has to do with management with the delivery of the optimal amount of water at the proper times and with allocation of water (Hunt 1977). Since water is the scarce commodity in irrigation System, the concentration so much on the amount of water available and its allocation that the balance between delivering to much and too little at a time is lost. According to Hunt (1977), two main types of social organizations are linked to irrigation systems: the local community. Where water users are also neighbours and kin and where water manager knows to other members of the community and relates to them in a variety of ways: and secondly, a more comprehensive and impersonal system, where the manager is a specialist not otherwise related to users and allocation of water is managed by an administrative system divorced from any other social functions. Most changes in type of social system regulating water use are from first type of system to the second. The switch towards the second type of social system from the first type of community organization is associated with greatly increased hazard of salimzation and alkalization (McGuire Gibson, personal communication) and in general a system of social organization in transition in the quality of maintenance.

The key problem in most irrigation systems is that water management is separate from control over the land proper and effective salinization control requires both. Stress in these systems, occurs with falling production and land ruined, by salinization and alkalization, is manifested in an increase in water rights court cases, in violence and in canal maintenance reduction.

Method: Social factors to monitor are : separate or water management from land management, changes in type of social organization controlling the irrigation system; reduced maintenance efforts and an increase in formal and informal disputes over water rights.

(b) Dryland Agriculture: Changes and Types of Dryland Agriculture usage

The type of agriculture practiced in dryland areas and especially changes in it are both important. Much scientific research provides support that conversion of marginal drylands from range use in agricultural use is one of the principal cause of desertification. An obvious consequence of such change in the destruction or the degrdation of perennial cover, especially the amount of dry season cover and changes in the regeneration rate of vegetation, which are major biological/ agricultural indicators.

Schechter's (1977) estimate of differential production of food calories between range use and cultivation is essential to understand this phenomenon: where there is pressure to increase production, the motivation to convert land from range to agriculture grows very strong. "Since the turn of century, large areas of former pasture land on the fringe of deserts have been cultivated to supply the needs rapidly growing population. In the more arid areas, where rainfall is inadequate for cultivation of grain or where the type of terrain makes cultivation impractical, the pasturing of livestock remains the most efficient way of producing food. Above a minimum of about 150 mm of winter and 250 mm of summer rain, however, cultrvated agriculture is considered more productive with variations depending upon other ecological factors, the cultivator as a primary user of vegetation produces food calories more efficiently than the pastoralist, who is a secondary user of vegetation. The same unit of land, where rainfall is sufficient, can support three times the population, when converted from a milk economy to grain production. Hence, the continuous tremendous pressure for the expansion of cultivated agriculture, mostly at the expense of the best existing pasture lands, which have been pushed back to the least hospitable zones."

In the face of such pressure, the incentive is strong to choose short-term high yield techniques over farming strategies that ensure sustained yield over the long term. One consequence is the introduction of mechanized deep plowing techniques, which destroy perennial and dry season cover. Given the high variability of rainfall in semi arid regions, fields are placed under cultivation in places where rainfall may be Sufficient in "good" years but adequate over the long run. In the "bad" years, the denuded soil is rapidly eroded.

Gonzales (1978) has noted the importance of the conversion of "good" rangeland to plow agriculture as an aspect of land tenure reform, with degradation of the range. This observation has also been made else-where by R. Peterson (1970), together with its sequel of fall in productivity.

The reason for cultivating former rangelands and especially for their mechanized cultivation, may well have to do with exogenous factors, such as investment rates and the distribution of the investment (Spooner 1978). If the input is local, made by individuals or groups, who expect to remain on the land as long-term users, they are likely to opt for farm practices that are conservative of the land resources. If investment is primarily from the outside and commercial, it is more likely to encourage abusive techniques, Bennett's (1978) observations in the North American wheat belt illustrate the two possible outcomes.

"In the Canadian section of the grain belt, the smaller operators could be divided into two classes. One group followed extremely conservative, satisfying programs, planting on a 1:1 fallowing system, with the type of crop never varying despite predictions of boom prices for alternative crops Most of them used correct minimal tillage but were often handicapped in adhering to strict-standards because of their lack of proper machines. The other group of small-scale operators displayed a different regime; attempting to take advantage of every change in the market, they used their land to the fullest, varied their crops, rented additional land and bought expensive machinery on credit. The condition of their fields was poor and their neighbours usually complained about soil blowing, when wind came from their direction."

"Other samples of operators were studied in. Montana, in order to determine the effect on production practices of more generous subsidies provided by U. S. Government, through parity and other policies. The results indicated that over time, about a decade, many more of the smaller operators than in Canada were inclined to follow the abusive patterns described previously. When asked about this, most of them observed that they might as well- after all, the Government keeps us going anyway. That is, they reacted to the income subsidies for marginal operators in this high-uncertainty agriculture by overusing their resources."

"However, the larger enterprise, the less inclined the operator was to do this. Subsidies for the larger operators encouraged them to retrench and to conserve their resources. The level of income satisfaction could be reached in these cases by existing production returns plus the extra funds supplied by subsidy programs"

Method: The criteria for site selection subsume some of the methodological considerations. One of the more important is comparison of cultivation technologies and practices in various types of situations and in comparison with fallow and uncultivated plots. Such comparison can use data from field work and from aerial and satellite data sources.

(c) Pastoralism

Rangelands, potentially or actually in the process of desertification, account for as much as ninety percent of all such land, the social indicators for use in rangeland pastoral systems are most important. The amount of land involved is of paramount importance; no more than twenty two percent of the people living in such areas are involved in animal husbandry (Kates, Johnson and Haring 1977). The related social indicators are changes in type of land use and in herd composition. The rate at which this occurs is important in monitoring desertification, since shrinkage in the. range or grazing area, due to the encroachment of agriculture, means that move animals are pastured per unit of available land.

Nomadic pastoralists, often in contrast to commercial livestock ranching, have tended to make "ecologically astute" use of their grazing and animal resources, combining mobility, flexibility and diversity. Traditional pastoralism, as well as modern range systems, has proved responsive to both short and long - term trends in market demand and directed planning. Pushed too far, either system can result in degradational land use changes.

Method: The methods used to monitor desertification in pastoral areas are closely related to those used to monitor dryland farming and irrigation, since changes in resource use are criterial in all competing forms of land use. Swift (Personal Communication 1978) has developed a productivity index for his work among pastoralists. Milk and meat production are measured in megajoules (MJ) and related to land unit and labour unit necessary for their production. Such a productivity index can also be tied to the productivity and yield measurements used by the biologists. Swift has also devised a "minimum consumption bundle" which, for the Somali consists of the quantity of sorghum and cloth needed to support a family per year.

(d) Mining

In the United States, according to Pimentel, Terhune et. al (1976),. strip mining disturbs 150,000 acres of land per year directly and affects from three to five times as much land in addition. It also causes stream pollution. Reclamation of mined lands in the United States formed the subject of AAAs panel papers and estimates given for reclamation potential (N.Gonzales, ed. 1977). The decisions to change land use from range or agriculture to mining are complex. Mined lands are included here because unreclaimed lands remove such land from food production and are subject to desertification processes.

Method: Changes in reflectance characteristics associated with mining make a real estimate possible with remote sensing data sources. Licenses or financing to engage in strip mining also involve the preparation of Environmental Impact Assessment statements, which can be examined.

(e) Firewood

This indicator covers the use of land and its vegetation for firewood and charcoal production and changes in that use as a function of settlement change. Firewood and charcoal are used for domestic fuels by a third of the world's population, the per caput requirement is one ton a year (Schechter 1977). Calculations made show a net deficit between demand and supply. Women customarily collect firewood in the vicinity of residential areas, most commonly. Change is also occasioned by converting new areas of residential use or by growth from village to town or town to city as with Dodoma, in Tanzania, with attendant new or increased firewood demands.

Method: In populated areas, the firewood collection zone can be discerned and measured by changes in relative reflectivity, itself an important measurement of desertification, as seen in aerial and satellite data. When combined with field observations, such linkages may be attributed to firewood collection. However, a number of other uses, such as grazing or cultivation, may have comparable results in changing the reflectivity. In areas, where there are active programmes of Afforestation, the change in relative reflectivity may to in the other direction and become a measure of success of anti-desertification programme.

(f) Tourism

Tourism in the American southwest (Andrews 1974) and elsewhere (Drouhin 1970), in its commercial aspects, is a totally new use of arid lands, even though travelers have been attracted to desert landscapes for a very long time. Tourism is based on a judicious mixture of high technology (for water, food and accommodation) and the appeal of the generally sparse population and other salutary features of desert life. Nevertheless, the change from one type of land use or reserve to another can be a cause or a result of the effects of desertification. Tourism is no exception. The dangers are negative effects on the vegetation, soil structure changes from off-road vehicles, local lowering of ground water supplies to meet tourist hotel needs and removal of commercially attractive plants or destruction of those that hamper tourist activities.

(g) Water

Changes of water use through water holes, dug wells etc. and impact of tourism on plant cover and soil erosion can be monitored with the biological indicators relating to vegetation and the physical indicators of soil and water.

2. Settlement Patterns

Changes in settlement patterns, especially in rural populations and in relation to energy sources can either reflect increasing desertification or the success of anti-desertification programmes. Natural population movements can be a response to improving conditions of Short or long duration; they can also act as early signals of desertification occurring as a result of changes in population.

(a) New Settlements

New settlements took form of urban centers or of villages. New towns and cities started forming through commercial development and planning, their habitants are primarily migrants into the area from distance. In 1974, AAAS and ICASALS joined forces to publish a book entitled "Urbanization in the Arid Lands". New towns in American south-west were settled with their open yards, spread out town layout, grass lawns, water borne sewage systems and houses of northern architectural design were modeled on towns in the wetter and cooler parts of the country. Desertification accompanies urbanization of this type, a desertification seen in lowering of ground water table, vegetation changes in surrounding areas, air pollution and an increase in demand for desert products.

The Turan case study (UNCOD 1977-a) points to other indirect effects of urbanization such as the demand for meat in the growing city of Tehran, where a rising standard of living combined with rising population unfavourably affected the dry pastoral hinterlands. The consequences of new settlement need not be destructive. Some new settlements in Kirayat area of Israel (Abt 1978), appear to be increasing agricultural productivity, based on capital intensive agriculture, the desertification potential is nor yet known.

On smaller scale, new villages are formed to accommodate increasing population encourages conversion of pastoral land in to cultivated lands where productivity is comparatively lesser.

(b) Expansion of Settlement and Sedentarization

Sedentarization refers to pastoral nomads, who are encouraged, forced or spontaneously changing from tent pastoralism to a sedentary life style. Desertification is seen as a possible, appropriate or even desirable alternative, the need to settle nomadic people need to be seen as possible future effects on. desertification. Such possibilities have been discussed by Kate's, Johnson and haring (1997) as consequential impact of pastoral people who have to move to other places in view of displacement from such areas.

(c) Diversification of Settlements

Under the economic transition, which characterizes so many formerly closed or semi-closed societies, settlements in low-density areas are not only being formed or expanded, but are also becoming more diversified. Verification of negative impacts of such diversification can only come through field surveys.

(d) Settlement Abandonment

A final evidence of severe desertification is the abandonment of settlements. The coastal Peruvian town and canal system was abandoned for two reasons: plate tectonics, which started dune formation and a major flood caused by torrential rains in an area with virtually no rainfall for very long time. Desertification followed these natural events. In 1973, in the Sahel, villages on the northern perimeter were apparently abandoned during the field season of 1973, because there was no evidence of cultivations nor of vegetation grazing lands in their vicinity (Reining 1973).

Method for Settlements: On any substantial scale, to approach the size of the area in which settlement changes take place, the most feasible method for determining changes in the location, both formation and abandonment and size of settlements is with the aid of remote sensing data sources combined with sampling. Small-nucleated settlements can be ascertained quite reliably because each settlement is in high contrast with its immediate environment (Reining 1973, 1975). Dispersed settlements and the relative density of vegetation can be compared with uninhabited areas, which are otherwise comparable, as far as vegetation and soils are concerned. Again, the approach involves comparison and contrast.

3. Human / Biological Indicators

Population growth is one or even the main reason for desertification. Many of the social indicators take in to account changes in land use, changes in settlement pattern, pressure on water, soils, vegetative resources are inexorable result of population growth.

(a) Population Structure and Rate

A baseline estimate of the population subject to desertification, with breakdowns by degree of severity and by type of livelihood is provided by Kates, Johnson and Haring (1977). Table 3.5 and 3.6 give requisite details:

Table 3.5 : Estimates of dryland populations by regions and livelihood groups (in thousands)

Region	Drylands Total Population	Livelihood Population in Drylands		
		Urban Based	Agriculture Based	Animal Based
Mediterranean Basin	106800	42000 39%	60000 57%	4200 4%
Sub-Saharan Africa	75500	11700 15%	46800 62%	17000 23%
Asia and the Pacific	378000	106800 28%	260400 69%	10300 3%
Americas	68100	33700 50%	29300 43%	5100 7%
Total	**628400**	**194200 31%**	**297100 63%**	**37100 6%**

Note : 1. Meigs classification of drylands (1953) including extremely arid, arid and semi arid areas

2. Groupings as designed by UNEP

3. Total world population was estimated to be 3.86 billion in 1974.

Source: Kates, Johnson and Haring (1977).

Table 3.6 : Estimates of population and livelihood resident in areas recently undergoing severe desertification

(in thousands)

Regions	Total Population	Urban Based	Agriculture Based	Animal Based	Area (Sq.km)
Mediterranean Basin	9820	2995	5900	925	1320000
Sub-Saharan Africa	16165	3072	6014	7079	6850000
Asia and the Pacific	28482	7740	14311	6431	4361000
Americas	24079	7683	13417	2979	17545000
Total	**79546**	**21490**	**39642**	**17414**	**30076000**

Source: Dregne

The mobility of population in arid, semi-arid and hyper-arid areas through trans-humance, seasonal, periodic and permanent migration takes rates and population, which are made for number of persons per defined area, both difficult to secure and of less than normal reliability. Douglas Johnson suggests using these rates, in particular: infant and old age mortality as they are most vulnerable to nutritional deficiencies and female fertility, an important demographic indicator, because a drop in the rate of live birth is a sensitive measure of stress.

(b) Measures of Nutritional Status

Two measures of nutritional status were employed in the Sahel during drought period. Both are reported by Caldwell (1975) - "A British team comprising of Seamon, Holt, Rivers and Murlis, 1973, worked in the most northerly and poorest province of Upper Volta at the height of the drought in July 1973, the hungry period, before the new food from the rains could be harvested. They used three tests of malnutrition, clinical assessment from a range of Symptoms; the ratio of upper arm circumference to. height and the ratio of weight to height. Teams sent out by the Atlanta Center for Disease Control employing the Stuart-Meredith Standard, examined 3500 children."

The ratio of weight to height, named the Stuart-Meredith Standard, was used in the Sahel by the Center for Disease Control Physicians, who designed a sampling scheme and measured the same children twice- first, during the growing season and secondly, a month after harvest. A failure to gain during this period is considered indicative of severe under-nourishment. Upper arm circumstance, weight and height give greater reliability without much additional cost in effort and provide some indication of body style, as well.

(c) Public Health

Disease incidence, especially of infectious diseases, causes much of the illness and death in developing countries, where desertification is a problem. The array of causes of mortality in developed countries is known to be quite different, with the diseases of old age and environmentally related causes important. Nevertheless, some of the infectious diseases reported in public health statistics seem to have a lower incidence in semi-arid environments of the same approximate level of development, possibly due to lower population densities and reduced water sources, where disease organisms can breed (J. Johnson, Personal Communication). In the time of crisis, however, when refugees are forced to flee their homelands and live in crowed conditions, they prove to be highly vulnerable to infectious diseases such as measles, typhus, tuberculosis or polio. Increased levels of certain ‘ water-borne diseases, such as schistosomiasis and oncherceriasis, are associated with irrigation projects using water from high dams.

Method: Data collection of medical and public health provide a database for analysis and interpretation.

4. SOCIAL PROCESS PARAMETERS

Key areas in social and economic organizations are cause or effects to desertification. There are certain composite signs or indicators, which are roots or the signs of social and economic distress or vulnerability These are discussed in following paragraphs.

(a) Conflict

Conflict arises from so many different causes like court cases, feuds, witchcraft accusation or armed conflict are attributed to desertification and are obvious form of behaviour treated as early warning signals.

(b) Migration

Migration of individuals or of families also occurs for many different reasons. The word “transhumance” is widely used for seasonal migration and unless the patterns change in important ways, should not be included as indicative of desertification. But periodic migration and marked changes in other migration trends can be very important. Analysis of migration trends need to be studied in terms of comparative annual trends, factored for age and sex, major demographic process and on the basis of local, regional or national levels.

(c) Redistribution Patterns

Numerous social mechanisms exist to redistribute wealth. Some are ceremonial and others have a functional foundation. Reduction in productivity can lead to changes in these redistribution mechanisms. Failure of these devices e.g., bride-price, stock friends, feasts, festivals etc. can lead to collapse of the indigenous social systems, a sort of “social desertification” the abandonment of potential potentially productive economic modes and a loss of overall production irrespective of its impact, which in some cases maybe positive on physical or biotic environment (d. Johnson).

(d) Marginalization

The process of marginalization often accompanies the development of mechanization, industrialization and motorized communication. There is a tendency for the investment process involved in development to reinforce itself and lead to an increasing gap between rich and poor areas. The agricultural and range lands vulnerable to desertification end to be both poor and sparsely populated, although these factors can be independent to each other.

The lack of responsibility towards resource productivity can also lead to unwitting discrimination against the people living in poor areas by people in he centers of power. In making decisions about future development, managers tend to ignore the traditional technologies, which often have interesting and useful adaptive components and thus lose the benefits of local enterprise. The people in marginal areas, by some token, often develop an inferiority complex, which makes it more difficult for them to gain access to services, which may have to come from outside the vulnerable areas. Marginalization decreases the sense of responsibility towards the resources of poorer areas.

On a larger scale, many of the countries most severely threatened by desertification are also among the poorest economically of the world's nations and people. Shackled with a colonial heritage that encouraged export production of cash crops, the consequences of this for proper land use can be calamitous on a national level.

(e) Cash Versus Subsistence

Sudden changes in the relation between cash and subsistence cropping may well promote desertification, although the pathways along which this develops are often quite variable. Expansion of cash cropping may force subsistence production to concentrate energy, labour and available' moisture on cash crops can leave large, unvegetated areas exposed to soil erosion In modern societies, with well-deyeloped census systems, monitoring the cash income of farmers is also useful. Smce it measures the reality of resources available to meet felt-needs and financial obligations, a fluctuation in cash income indicates the ability of farmers to hold on to their farms and may. suggest the degree of risk and damage to the environment they are willing to accept in order to do so.

Chapter 4

Processes and Causes of Desertification and Land Degradation

4.1 CLIMATIC SURFACE AND DESIGNATION OF ARIDITY ZONES

At the global or regional scale, aridity zones have been defined by using a variety of approaches including drainage characteristics (De Martonne and Aufrere 1928),vegetation types (Shantz 1956) and by climatic characteristics (e.g. Pick 1894).Climatic approaches have the widest relevance because climatic influences and many other environmental attributes imparts important limitations on human activities. Aridity can be simply defined as an overall moisture deficit in average climatic conditions (Agnew and Anderson 1992). Various atmospheric and environmental factors contribute to the occurrence of aridity (Williams and Balling 1996, Thomas 1997-a), and these may vary in their occurrence and distribution through time (Hulme 1992, Balling 1994-a). It is therefore, important to outline the manner by which aridity zones have been defined and determined, since these form the basis of designation of regions in which the desertification has been examined.

Aridity zones are based on an evaluation of the relationship between key climatic variables, creating an index of moisture deficit, termed the Aridity Index. To calculate the index values, it is necessary to compare incoming moisture totals with potential outgoing moisture (Kemp 1994). Although, this is determined by the ratio of mean annual precipitation to mean annual atmospheric evaporative demand, usually expressed as potential evapo-transpiration, there are number of ways, in which this index could be derived from climatic data (Meigs 1953, UN 1977-a, Hulme 1996-a, Le Houerou 1996). This makes it important to explain the way, in which the Aridity Index has been determined, as well as outlining, how it differs from other methods, for example used for the FAO/ UNESCO/ WMO map of world arid regions (UNESCO 1977), produced for the 1977 survey of desertification (UN 1977-a).

(a) Data used in the Determination of Climatic Surfaces and Aridity Zones

The map of climate surface, including aridity zones, were derived by the Climatic Research Unit (CRU), University of East Angola, UK from their data sets of worldwide monthly mean precipitation and temperature values (Hulme and Marsh 1990). To produce maps of climatic surfaces, data can either be interpreted on the basis of the total available data set, or for specifically designated time periods. When whole data sets are used, mean variable values are calculated for data from individual weather stations, regardless of the length of time for which data exist. This approach maximizes the data potential, since it allows the full spatial and temporal ranges of data to contribute to the analysis. However, if this approach is taken, other problems arise, particularly when comparisons are made between data from stations with different record lengths. This approach can lead to too much confidence being placed in values derived from data runs of only a few years, and is potentially a major difficulty for drylands, where marked climatic inter-annual variability is a normal feature (Le Houerou 1979, Agnew 1990).

An alternative approach to determining the extent and distribution of aridity zones uses data for specified time periods or 'timebands' (Hulme and Marsh 1990). A timeband approach is used in this study and offers a number of advantages over the use of temporally unrestricted data sets. First, it allows greater confidence to be placed in the compatibility of data from different locations. Second, it creates the possibility of producing sequential climate surface maps for a series of designated timebands, allowing changes in the locations and extent of aridity zones to be identified and even predicted for the future (Hulme 1996-a). In this study stations with data for the timeband 1951-80 were used to produce surfaces of mean annual global precipitation (P) (Figure 4.1), temperature (T) as inputs to the calculation of the Aridity Index. Initially, 2769 station means were used as source of P (Figure 4.2) and 1834 for T (Figure 4.3).

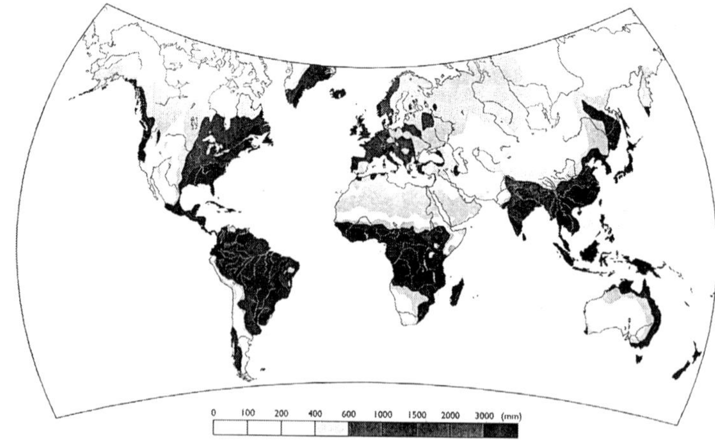

Approximate equatorial scale 1:195 million *Source:* CRU/UEA, UNEP/GRID

Figure 4.1 : *Mean annual precepitation 1951-80*

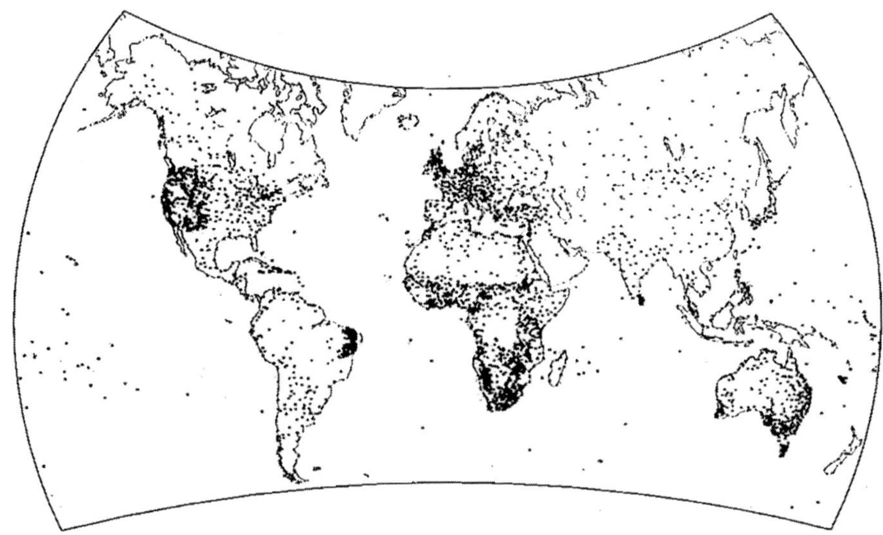

Approximate equatorial scale 1:317 million *Source:* CRU/UEA, UNEP/GRID

***Figure 4.2** : Precepitation stations (number of stations : 3758 1951-80)*

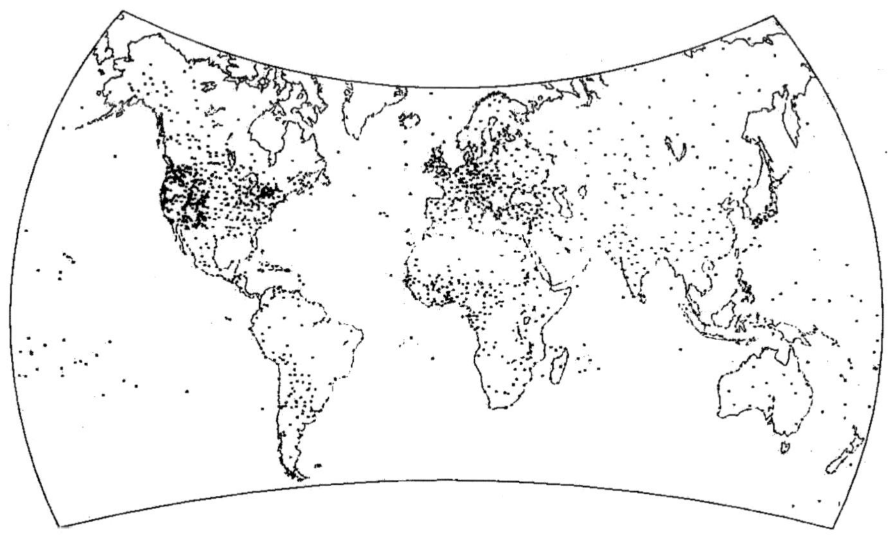

Approximate equatorial scale 1:317 million *Source:* CRU/UEA, UNEP/GRID

***Figure 4.3** : Temperature stations (number of stations : 1834 1951-80)*

In view of lack of robustness of P data for extrapolating point data for area coverage data from a further 989 stations were added for areas, where the initial coverage was sparse, involving some relaxation in selection criteria. This enabled a grid of 0.5 (degree latitude and longitude), cells to be generated for the production of climate surfaces. While this might be considered to be a coarse grid and although data are available for a higher resolution grid for some sub-areas, it generates the best overall global coverage that is currently available and is directly comparable with the grid used in the current generation of General Calculation Models (Hulme *et. al.* 1992).

(b) Determining Evapo-transpiration Values

The calculation of Aridity Index requires determination of the moisture loss or potential evapo-transpiration (PET) values from the climatic surface data. PET can be determined in three ways, each having its own limitations and advantages. First, it can be established through direct measurement using lysimeters, evaporation pans or armometers (Jackson 1989, Ward and Robinson 1990). At the global scale, such an approach is impractical while the availability of existing data is poor and in any case, the equipment used in reported studies is not standardized and, therefore, not always directly comparable. Second, PET can be calculated using empirical formulae. The method of Penman (1948) is commonly cited, but its calculation requires a large body of directly measured meteorological data, including solar radiation, wind velocity, relative humidity and temperature. Its application at the global scale is, therefore, again constrained by data availability.

A more practical approach utilizes knowledge and understanding of the empirical relationship between measured PET and values of more readily gained environmental variables (Thornthwaite 1948). This approach allows PET to be calculated from just two parameters: mean monthly temperature data and the average number of daylight hours by month. The Thornthwaite method has a practibility relevant to the scale of this study and was used by Meigs (1953) in the production of a map of world aridity for UNESCO. Though, more sophisticated methods exist for calculation of evapo-transpiration rates for different crops (Wright 1982), or for calculating the soil moisture deficit in relation to precipitation and evapo-transpiration (Palutikof *et. al.* 1982, Palutikof 1986), such approaches were not used here as once more they require the input of data that are not widely available.

A global Thornthwaite PET surface calculated from the temperature surface data for a 0.5 degree resolution grid. The Thornthwaite method is known systematically to underestimate PET for dry conditions and to overestimate values for moisture and cold environments (Mather and Ambroziak 1986, Hulme and marsh 1990, McKenney and Rosenberg 1993). Consequently an empirical adjustment factor was applied to the data (Hulme and Marsh 1990) to bring the values closely in line with those of the Penman method. The global annual PET surface is shown on Figure 4.4.

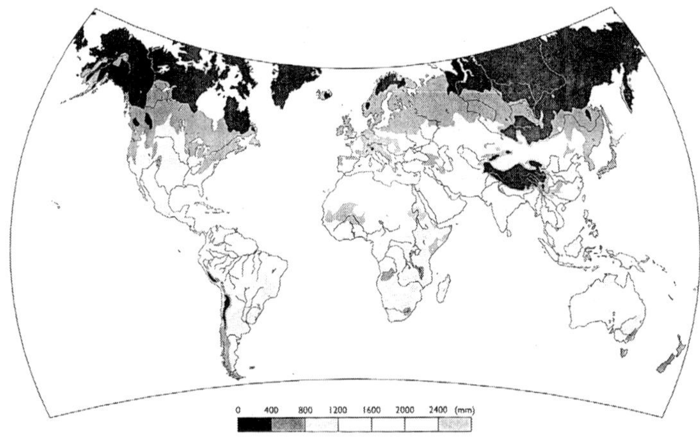

Approximate equatorial scale 1:195 million *Source:* CRU/UEA, UNEP/GRID

***Figure 4.4** : Mean annual poetential evapotranspiration, 1951-80*

(c) The Aridity index

The Aridity Index (AI) was derived from the climatic surface maps and calculated on monthly basis, as the ratio P/ PET. The monthly values were then averaged to provide both a map of mean annual potential moisture availability for the period 1951-80 and a global climatic classification based on potential moisture availability. AI values of <1.0 indicate an annual moisture deficit.

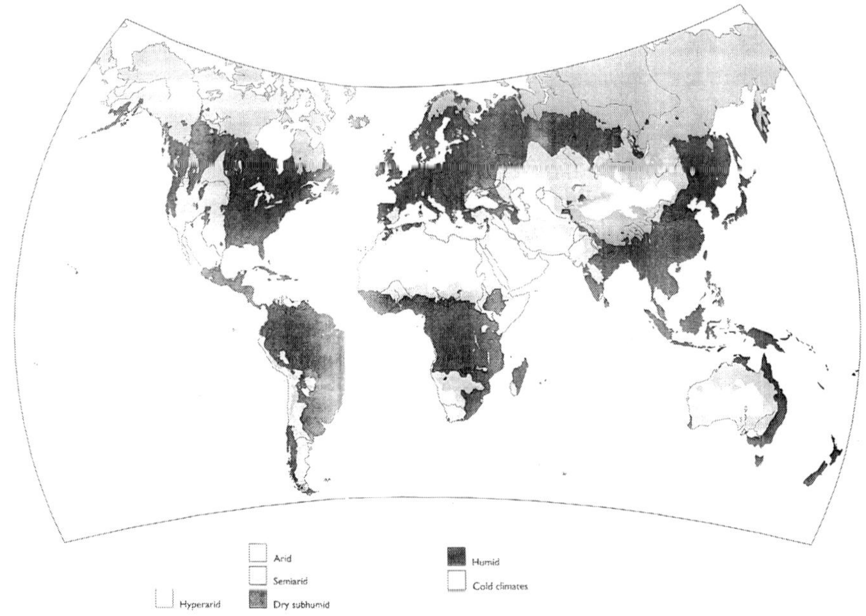

Approximate equatorial scale 1:104 million *Source:* CRU/UEA, UNEP/GRID

***Figure 4.5** : Aridity zones*

Aridity Index values were classified so that climatic zones could be delimited (Figure 4.5). The classification of AI values was modified in a number of ways from that used in UNESCO (1977). The AI boundary between arid and hyper-arid zones was increased from the value of 0.03 used in UNESCO (1977) to 0.05 (Hulme and Marsh 1990), as part of the adjustment for the Thornthwaite method underestimation of PET in very dry environments. The boundary between dry sub-humid and humid was also changed because of overestimation of values where conditions are moister. A separate designation of cold and mountain climates (not shown on Figure 4.5) was introduced because cold environments present a different range of problems than those in warm dryland environments. As such, these areas are excluded from the category of susceptible drylands in this degradation study, as the climate is too cold to allow a crop-growing season.

(d) Types of Aridity

Aridity can be defined in several ways, but most simply it represents a lack of moisture in average climatic conditions, caused by one of four climatic situations (Suzuki 1981, Thomas 1997-a), which may interact in case of individual areas. These are:

(i) Atmospheric stability, where, particularly in parts of the tropics and sub-tropics, anticyclonic subsidence results in zones of stable, moisture deficit, air;

(ii) Continentality, where distance from oceans prevents the penetration of moisture-bearing winds into continental interiors;

(iii) Topography, where mountain barriers create rain shadow zones; and

(iv) Cold oceanic currents, which contribute to coastal desert zones by reducing sea-surface evaporation.

Varies in intensity from place to place, aridity due to different levels of moisture deficit. The nomenclature used in this study reflects the terminology widely used in scientific literature since Thornthwait (1948) and Meigs (1953). The extent and distribution of the different aridity zones determined in the study for the period 1951-80 are given in Table 4.1 and Figure 4.6.

Table 4.1 : Aridity zones by Continental regions

(in Million Hectares)

Aridity Zone	Africa	Asia	Australia	Europe	North America	South America	Total
Cold	0.0	1082.5	0.0	27.9	616.9	37.7	1765.0
Humid	1007.6	1224.3	218.9	622.9	838.5	1188.1	5100.4
Dry Sub-humid	268.7	352.7	51.3	183.5	231.5	207.0	1294.7
Semi Arid	513.8	693.4	309.0	105.2	419.4	264.5	2305.3
Arid	503.5	625.7	303.0	11.0	81.5	44.5	1569.2
Hyper-arid	672.0	277.3	0.0	0.0	3.1	25.7	978.1
Total	**2965.6**	**4256.0**	**882.2**	**950.5**	**2190.9**	**1767.5**	**13012.7**

Source: World Atlas of Desertification Volume II.

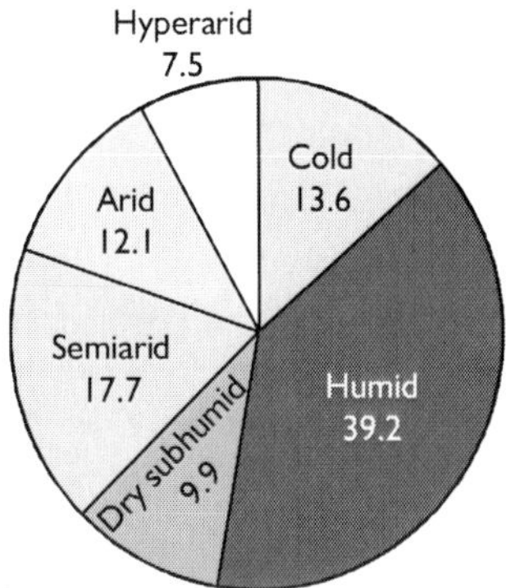

Figure 4.6 : *Global Land Area by Aridity Zone (%)*

It is inevitable that at the global scale, each zone embraces a range of environmental characteristics. However, there are broad climatic and environmental attributes that can be used to provide a useful summary of each zone (Grove 1977):

Dry Sub-Humid Areas: (0.50 P/PET < 0.65) have highly seasonal rainfall regimes with less than 25% inter-annual rainfall variability and agriculture is widely practised. UNESCO (1977) found such areas very susceptible to degradation, probably enhanced by the seasonability of rainfall, drought periods and the increasing intensity of human use. It is, therefore, appropriate to retain areas designated as dry hub-humid in this study and to regard them as part of the overall global arid realm. To this effect, dry sub-humid areas are included in the definition of desertification.

Semi-Arid Areas: (0.20 P/PET < 0.50) have distinctly highly seasonal rainfall regimes and mean annual values up to 800 mm in winter regimes. Inter-annual variability is nonetheless high (25-50%) so despite the apparent suitability for grazing of semi-arid grasslands, this and sedentary agricultural activities are susceptible to seasonal and inter-annual moisture deficiency.

Arid Areas: (0.05 P/PET < 0.20) have mean annual precipitation values up to approximately 200 mm and inter-annual variability in the 50-100 % range. Pastoralism is possible but without mobility or the use of groundwater resources, it is highly susceptible to climate variability.

Hyper-Arid Environments (P/PET < 0.05) have highly variable rainfall both inter-annually (up to 100%) and on a monthly basis such that there is no seasonal rainfall regime. In virtually all cases where data are available, yearlong periods without rainfall have been recorded. These areas are the true deserts and as such offer very limited opportunities for human activities.

(e) Problems of Delineating Dryland Boundaries

Dryland boundaries are neither static nor abrupt (Tucker *et. al.* 1985 a, b and c, 1991, Hellden 1991). This is not surprising given the high inter-annual variability in mean rainfall and the occurrence of drought, which may last for periods of several years at a time. Attempts to locate these climatically derived boundaries on the ground or identify them in terms of features such as soil type or natural vegetation are likely to fail (Thomas and Middleton 1994). Physical changes are likely to be gradual and they will be modified by human induced processes such as grazing, deforestation and burning. Identification of climatic changes from shifts in boundaries, therefore, needs to proceed with caution given the dynamism that is inherent in dryland climatic regimes and the fact that drought can cause spectacular but temporary changes in natural vegetation (Tucker *et. al.* 1991).

It is also recognized that individual aridity zones do not represent homogeneous climates, either in the long term or during a particular timeband (Babaev *et. al.* 1993). As already noted, specific P/PET ratios can be derived from a wide range of values for individual meteorological parameters.

(f) Comparison with the 1977 UNESCO Map of Aridity

It is useful to identify how the map of world aridity zones differs from the widely used UNESCO (1977) map, which according to Heathcote (1983) is very similar to the map of Meigs (1953). Objective comparisons between UNESCO (1977) and Map No. 4.5 cannot be made, however, because of their different scientific bases, UNESCO (1977) only exists in paper form and was not derived from a geo-referenced data set that could be used as a basis for comparison (Hulme *et. al.* 1992). The UNESCO map was, as far as, it is known, derived from a timeless database, details of which are unclear, and certainly does not identify the range of time periods covered by different station data sets.

For areas, where the spatial coverage of meteorological stations was limited, local experts were consulted to suggest or adjust the boundaries between aridity zones. Finally, the calculation of PET values in the UNESCO study employed the Penman (1948) method, requiring a wider range of meteorological data inputs. Although the PET values in the study calculated using the Thornthwaite (1948) method, were empirically adjusted to better fit Penman values, there may still be statistical divergence in some areas of the Aridity Index values (Hulme and Marsh 1990).

Overall, it is believed that the utility of the map prepared by CRU/ UEA, UNEP/GRID and adopted in this study is greater that UNESCO (1977). The use of timeband climatological data for the calculation of the Aridity Index will permit future comparisons with maps for other designated time periods, as Hulme (1992) has done for African precipitation surfaces and Hulme *et. al.* (1992) for aridity surfaces. As the 1961-90 surface was unavailable with global coverage, the period 1951-80 has been taken. Additionally, the use of an empirically based method to calculate PET may be more pragmatic than using the Penman method at the global scale, not just because of the problems of generating the necessary data inputs but because it utilizes fewer meteorological data and, therefore, reduces errors inherited from primary data collection.

4.2 CLIMATIC VARIABILITY AND CHANGE IN DRYLANDS

The study of desertification and land degradation is complicated by the natural environmental variability that occurs in dryland areas (Thomas and Middleton 1994). Accurate identification of the causes of desertification at any particular location and thus suitable strategies for its treatment, can only be made by paying close attention both to the human use and possible mismanagement of natural resources and also to the way in which dryland ecosystems and their resources respond to natural climatic variations. This is especially important given the definition of desertification used here, which is that adopted in Agenda 21 and embodied in the Convention to Combat Desertification (CCD) (UN 1994). This incorporates the three separate elements of short term drought, long term climatic fluctuations and land degradation caused by human actions (Cardy 1993). While this study concentrates on human induced soil degradation, it is important to understand the dynamic nature of some of the natural elements in the dryland equation. This is particularly vital given that climatic variability not only impacts directly on ecosystem but indirectly through influencing human actions and activities, including land use opportunities and migration (Olsson 1983, 1993, Cardy 1993, Norse 1994). As Agnew and Warren (1996) note, variability is not simply a series of changes but a series of impacts on communities. Because aridity zones are climatically defined, changes in climate that alter the extent of aridity zones will implicitly impact on the scale and extent of desertification problems (Hulme et al. 1993).

The inherent natural dynamism of dryland ecosystems is very largely governed by the variability in climatic parameters that characterize such regions (Agnew and Anderson 1992). Chief amongst these is precipitation, the input of moisture into drylands. While many dryland areas receive important inputs of moisture fro dew (Evernari 1985), and some other s from fog, for example in western Namibia (Seely 1978), rainfall is still the key source of moisture in most of the world's dryland regions, however sporadic and unreliable its occurrence may be. It is not sufficient to consider rainfall inputs alone, however, since the effectiveness of rainfall, which is the amount available for plant growth or other uses, is also dependent on outputs. The main output route, evapotranspiration (ET), is influenced by parameters such as plant cover and type, wind speed and perhaps most importantly, temperature.

It is also valuable to consider the timescales over which climatic fluctuations in drylands occur, since their recognition and identification influences the strategies for survival that societies have to adopt (Warren and Khogali 1992, Thomas and Middleton 1994, Williams and Balling 1996). In this section, therefore, the relationships between climate change and desertification will be considered and the scale of climate change and variability will be explained. Following this, a closer examination of the variability off rainfall and temperature in the world's drylands, serves as a contextual background to the considerations of dryland degradation that follow.

(a) Interactions between Climate Change and Desertification and Land Degradation

In their book, Interactions between Climate and Desertification, produced by WMO and UNEP, Williams and Balling (1996) noted that desertification can affect climate as well as climate effecting desertification. Climatic change induced at the global scale by human actions, notably the emission of greenhouse gases, will impact on the extent of drylands and the nature of climatic regimes within them. Specifically in drylands, human actions can alter the nature of the land surface, which in turn, may modify the energy balance in the lower atmosphere and the occurrence of rainfall.

Climatic changes, regardless of their cause, can impact on desertification processes and human activities that lead to land degradation. Any changes of climate in the future are likely to be relevant to issues of desertification and land degradation, particularly if the balance between moisture gain and moisture loss through evapotranspiration is altered. Such changes will alter the extent and distribution of drylands and may intensify or even perhaps reduce, problems of moisture availability and drought occurrence, by changes in climatic variability. Even if absolute changes in moisture availability are limited, changes in the seasonality of climate can have significant impacts on the distribution of vegetation communities, on soil moisture availability and on the potential for specific human, especially agricultural activities. Desertification may even be triggered or exacerbated in productive areas by pressures brought to bear in other locations. The occurrence of a prolonged drought, or an increase in aridity in one area, will inevitably increase the pressures to cultivate and produce more food in other areas.

(b) Climatic Changes

There is considerable well established evidence that demonstrates the occurrence of marked global climatic changes at the millennia and greater time scale (Williams *et. al.* 1994, Kadomura 1994), and it is well attested that such changes have affected arid zones, leading to both their expansion and contraction in the past (Thomas 1997-b). These long term changes, driven principally by external forcing mechanisms, particularly changes in earth-sun relationships, have had important effects on development of landscape, soils and animal and plant distribution in drylands. These changes must, however, be contrasted with climate changes and variations that occur within the scale of human lifespans and which have a more direct and immediate effect on society-environment relationships.

While it is clear that many factors control and influence climate at global, regional and local scales, with many of these factors, currently only poorly understood, it is widely agreed that human actions themselves have an important impact on climate (Williams and Balling 1996). What remain uncertain are the likely magnitudes of human-induced climate changes, and their impacts on areas susceptible on desertification.

At global scale, it is widely accepted that the build up of radiatively active gases, the so called 'greenhouse gases' in the atmosphere, particularly carbon dioxide (CO_2), methane (CH_4) and chlorofluorocarbons (CFSs), is contributing on global warming (IPCC 1990). Dryland areas are globally important stores of stores of carbon (Adams *et. al.* 1990). and land degradation in these areas is contributing 5 to 10 %of total greenhouse gas accumulation (Williams and Balling (1996). It is, therefore, important to increase the understanding of the patterns of carbon storage in drylands and to aim to adapt and improve land management systems to reduce the flux of carbon to the atmosphere.

At the regional scales, it has long been suggested that human-induced soil and vegetation change in drylands may modify rainfall amounts (Otterman 1974, Charney 1975). In this 'biophysical feedback model it is hypothesized that vegetation reduction both lowers the contributions plants make to atmospheric moisture by evapotranspiration, and increases surface albedo and thereby atmospheric stability through reduced lower atmosphere warming. While/there have been numerous studies to investigate this phenomenon and the principles involved (Charney *et. al.* 1975, Walker and Rowntree 1977, Sud and Molod 1988, Le Houerou 1993), the possible contribution it makes to climate changes in drylands remain debated (Schlesinger *et. al.* 1990, Williams and Balling 1996). Overall climate change is clearly an important issue related in a number of ways and at a range of temporal and spatial scales to desertification. The complex interactions that are likely to exist between its different components and how these relate to desertification processes (Figure 4.7) are clearly vital areas for further investigation.

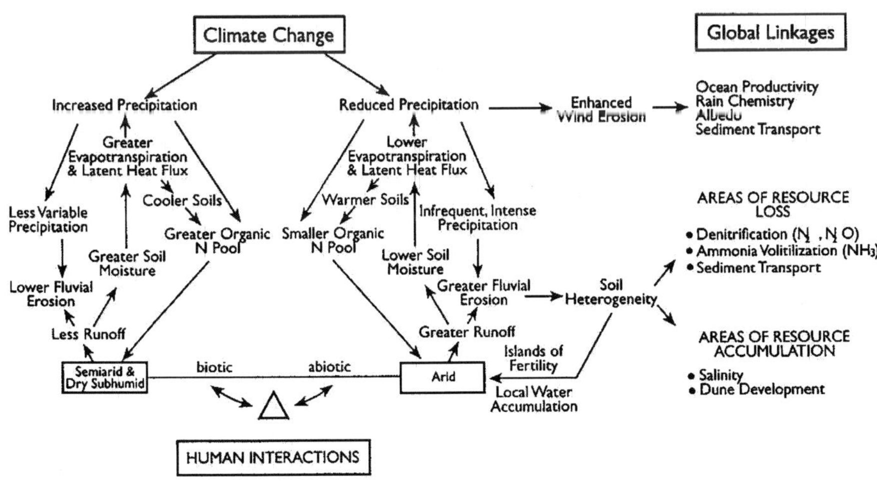

Source: based on Schlesinger *et. al.* 1990) (Atlas Vol. II page 9)

***Figure 4.7** : Possible Linkages between Climate Change, Environmental Change and Human Activities During Desertification*

(c) Climatic Variability

Climatic variability refers to the year-to-year variability of individual climatic parameters around longer-term mean values, and is inherent in dryland areas (Le Houerou 1996). Climatic variability can occur whether longer-term climate is stable or is changing, for example, year to year rainfall values may fluctuate around an overall trend towards wetter or drier conditions, Balling (1994-b) has shown that for the dryland areas of North America. Between 1890 and 1990, there were annual variations in temperature anomalies around an overall warming trend of 0.86 degree Celsius (Figure 4.8)

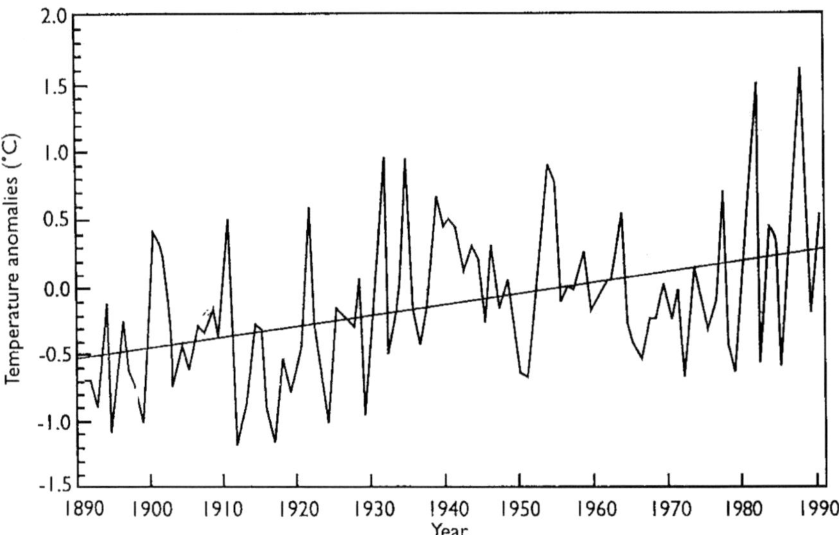

Figure 4.8 : *Temperature Anomalies for the Drylands of Mexico, The United States and Canada for the Period 1891-1990. The linear trend line reveals astatistically significant warming (after Balling 1994-b)*

Climatic variability may be linked to variations in climate forcing mechanisms. Cycles of change in solar output, identified by sun spot changes (Meadows 1975), operate at scales of 11, 60, 80 and 180 years, and are believed to impact on climatic variability (Seitz *et. al.* 1989, Warren *et. al.* 1996). Links between sea-surface temperature and atmospheric circulation changes, particularly the 'El Nino Southern oscillation' (ENSO) event (Lockwood 1984), are thought to have particularly significant impacts on climate in low latitude regions. Though, a complete understanding of the specific impact of such links on climate remains to be ascertained (Hidore and Oliver 1993), the occurrence of marked droughts and periods of significantly enhanced rainfall in some arid areas have been related to ENSO cycles, which have 3-7 year periodicities (Warren *et. al.* 1996).

Rainfall variability is an important aspect of dryland climates, and is described under each aridity zone. While aridity results from low average moisture availability, human activities can be compromised further by the uncertainty that exists about rainfall occurring in any particular location in any one season or year. Despite the identification of quasi-cyclical trends in rainfall variability in some dryland regions (Tyson 1986), the unpredictability of year-to-year (or inter-annual) variations of rainfall in drylands makes planning for agricultural activities difficult. Interannual rainfall variations have great relevance to issue of land degradation, and the failure of rains in one location can increase the pressure on neighbouring lands that have been better watered. Significantly, periods of better rainfall may contribute to desertification as well, by increasing the pressure of cultivation, which then creates difficulties during subsequent drier periods. Gonzalez Loyarte (1996) has noted this effect in semiarid and dry subhumid areas of Argentina.

Marked rainfall deficiency below average conditions in any one-year is commonly referred to as drought, but this is a very imprecise term (Kemp 1994) that has been defined in many different ways (WMO 1975). It is particularly important to distinguish between agricultural droughts, which is expressed in terms of the moisture requirements of crops and meteorological drought, which refers to a moisture deficit in relation to mean precipitation. Wilhite and Glantz (1985) catalogue a range of definitions that have been used in different studies and to some extent the appropriateness of any definition relates to the purpose for which it is intended. A further complication is that it is becoming increasingly common distinguish drought, considered to be a period of below average rainfall lasting for a year or so, from a period desiccation, which is a decadal scale reduction in moisture availability (Hare 1987).

The nature of climatc systems that bring rainfall to drylands, particularly the occurrence of convective rainfall cells and the seasonal movements of the Inter-Tropical Convergence Zone (ITCZ) in low latitudes, contributes to rainfall uncertainty (Hastenrath and Heller 1977). As well as temporal variability, these systems contribute to spatial variability in rainfall. In Niger, for examples, Agnew (1992) has shown that during the 10-year period commencing in 1967, some areas of he country experienced 10 drought years, while others less than 200 km away had 2 or less. This variability at the country level demonstrates just how difficult it is to produce reliable global summaries of drought. Hulme (1992) has also shown that is also possible to identify variability within the seasonality of rainfall in some African drylands.

(d) Examples of Rainfall Variability in Dry-land Areas

The nature, characteristics and variability of rainfall anomalies in susceptible drylands can be investigated by recourse to four examples, based on data supplied by CRU. Other analyses and comparisons of rainfall trends in dryland areas include those of Hulme (1996-a) for nine different global dryland areas, and Nicholson (1989) and Hulme (1992) for Africa. Figure 4.5 to 4.8 show time series graphs of annual rainfall for dryland regions in three different continents: the Sahel region of Africa, which extends from the Atlantic coast to 35 degree east; the north-eastern region of Brazil, from the equator to 10 degree south; northern China, from 100 degree east to the China Sea and from 35 degree north to the borders of Mongolia and Russia; and the whole of Pakistan. The areas are shown on Figure 4.9.

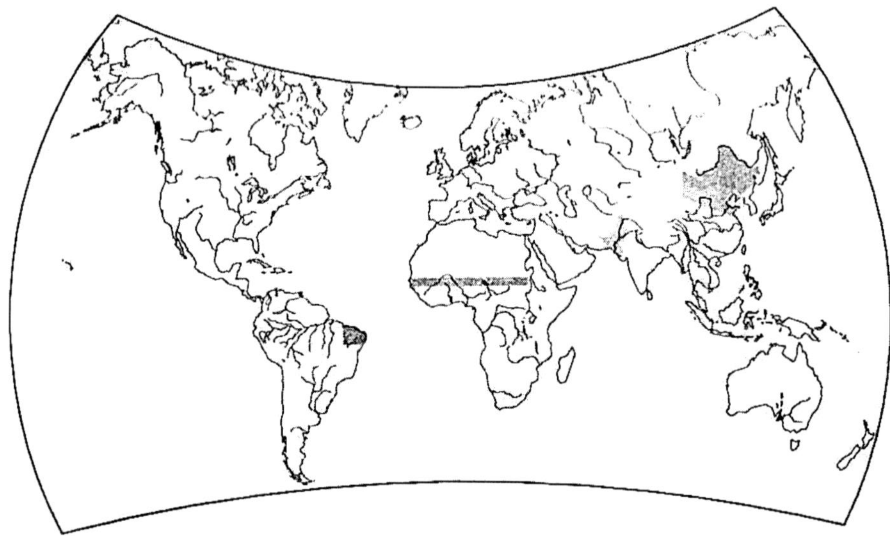

Approximate equatorial scale 1:350 million

Source: CRU/UEA

Figure 4.9 : *Location of drylands regions referred to in Figures 4.10-4.13*

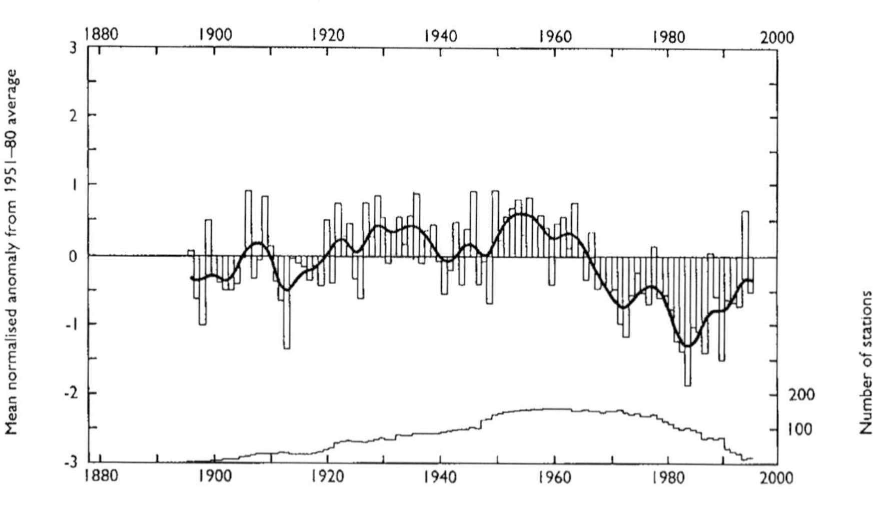

Source: CRU/UEA

Figure 4.10 : *Sahel annual rainfall, 1896-1995 (1951-80 main = 524 mm)*

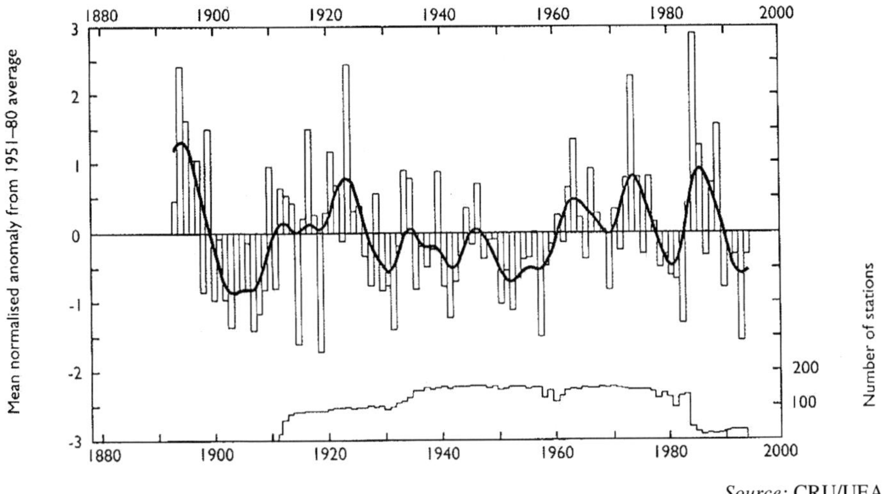

Source: CRU/UEA

Figure 4.11 : *North-eastern Brazil annual rainfall, 1893-1995 (1951-80 mean = 829 mm)*

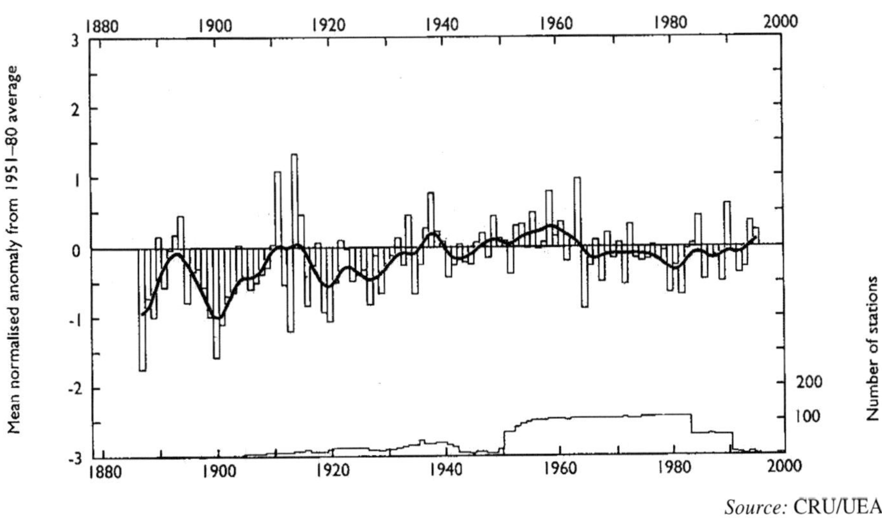

Source: CRU/UEA

Figure 4.12 : *Northren China annual rainfall, 1887-95 (1951-80 mean = 491 mm)*

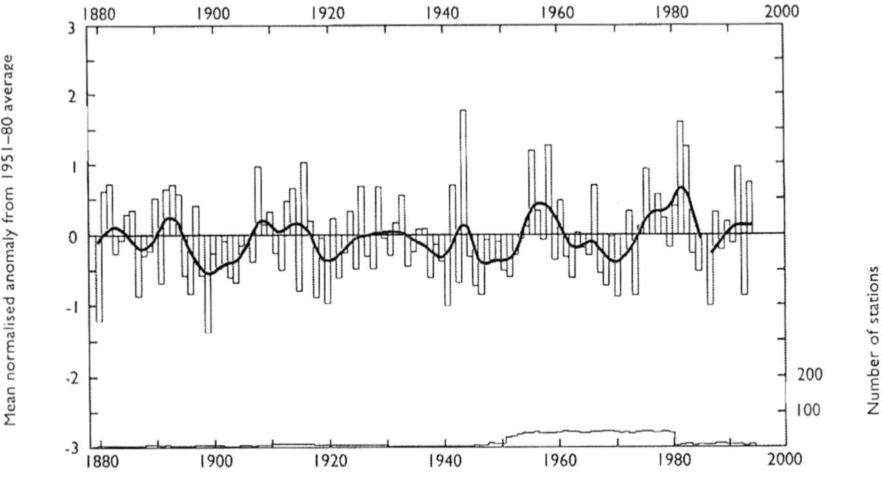

Source: CRU/UEA

Figure 4.13 : *Pakistan annual rainfall, 1880-1994 (1950-80 mean = 435 mm)*

To construct each graph, data from each area have been handled as narrated. For each station that contributes to a graph, the annual rainfall series has been normalized by subtracting the long-term mean from each value. The difference was then divided by the long term standard deviations used were derived from the 1951-80 data, which corresponds to the time series used for calculating Aridity Index. Normalising the data series in this way enhances comparability, since mean values become close to zero and standard deviation close to one, for each data set. The spatial mean rainfall anomalies have been calculated by averaging the values for all stations in each area with data.

While the number of stations with data throughout each period in each of the three areas is not constant, the time series are the best available, as they represent a spatial average for each area, they reliably reflect the annual rainfall anomalies in each area in question. The same approach has been used by Hulme 1992) in an analysis of rainfall variations in different areas of Africa. The smooth curves fitted in each of Figures 4.5-4.8 are 10-point Gaussian filters fitted through the data, which suppress variations of less than a decade. Thus the curves give a reliable indication of trends in rainfall anomalies.

All four graphs (Figures 4.5-4.8) demonstrate how dryland regions experience prolonged periods of above and below mean rainfall excess and deficit (drought). In the Sahel, (Figure 4.5), 1898 and 1913 were particularly marked drought years. The 1930s and 1950s were prolonged periods of above average-rainfall across the whole region, during which livestock populations markedly increased (Warren and Khagoli 1992). The rainfall deficit periods around 1900, preceding the early 1920s and 1940-42 have been suggested by lamb and Peppier (1991) to relate to marked ENSO activity, but overall the Sahel displays a poorer relationship between ENSO events and drought occurrence than many dryland areas (Williams and Balling 1996).

Extended drought conditions, contributing to a period of desiccation, have occurred in Sahel since the late 1960s. Across the region as a whole, only 1978, 1988 and 1994 experienced above average conditions during this period. If the tri-decades 1931-60 and 1961-90 are compared, the rainfall decline has been of the magnitude of 20-40% (Hulme and Kelly 1993) and exceeds the duration of events that might normally be expected to be associated with ENSO events (Le Houerou 1996). There is perhaps some evidence that the 1990s have not been quite as dry as the 1980s, but there is no clear sign of a return to the wetter decades of the mid century. There is debate about whether the recent desiccation period represents the onset of a marked climatic change.

Examination of historical and sedimentary records, however shows that this recent episode of desiccation is not unique, since equivalent or even longer periods of moisture deficit have occurred in the Sahel on several occasions during the last 500 years (Nicholson 1978) and even the last millennium (Maley 1989). Despite doubts regarding the longer term significance of the current long period of rainfall deficiency in the Sahel, and despite sub-regional variations within the overall picture of Sahelian rainfall deficiency (Agrew 1992), it is probably sensible to regard the rainfall conditions of the last 30 years as a more appropriate basis for current planning decisions than a longer-term mean value.

Figure 4.9 shows that in the north-eastern region of Brazil dramatic variations in rainfall anomalies occur over short periods. For example, 1917 was an extremely wet year, but 1915 and 1919 experienced severe and extreme droughts, and in mid 1980s, years of high positive rainfall anomalies were closely juxtaposed to a severe deficit in 1983. As well as short-term variations, Figure 4.9 also shows that NE Brazil experienced marked runs of years with rainfall above and below mean values.

Care should, however, be exercised in extrapolating region-wide impact of these trends on human activities, since the region is also noted for the very high spatial variability of its rainfall (Harzallah *et. al.* 1996). This is because a wide range of climatic systems impact on the region (Hastenrath and Heller 1977, Chu 1983). In the drier parts of the region there is nonetheless a very strong correlation between rainfall departures from the mean and sea-surface temperature anomalies (SSTA) linked to ENSO (Rao *et. al.* 1986. Ropelewski and Halpert 1987), with droughts commonly occurring in El Nino years (Kemp 1994). The years of extreme rainfall and drought commented on above particularly well associated with SSTA (Harzallah *et. al.* 1996, Roucou *et. al.* 1996), suggesting that improvements in understanding the causes and periodicities of SSTA could provide an important predictor of future droughts and floods in this region.

Comparison of the curves in Figures 4.5 and 4.6 indicates that certain times of extreme rainfall anomaly in NE Brazil correspond inversely to those in the Sahel. This inverse relationship may be not be coincidental since teleconnections, the tendency for variations in climate in one part of the world to be linked or to provide changes in other areas, may be an important factor in explaining aspects of climatic variations in dryland areas (Glantz *et. al.* 1991). One of the mechanisms that links NE Brazilian and Sahelian rainfall, sometimes inversely, seems to be founding the thermohaline circulation of the Atlantic Ocean.

The rainfall graph for northern China (Figure 4.14) also shows high year on year variations, such as those from 1913 to 1914 and 1965 to 1966, as well as prolonged periods of above and below rainfall, notably the period of desiccation spanning the turn of the 19th to 20th century and throughout the 1920s. There is some evidence to link drought in northern China with ENSO events (Williams and Balling 1996), and the availability of old historical records for the desert areas of China does show that runs of drought years have been a feature of the region's climate during the last 300 years (Yan *et. al.* 1992).

The Pakistan rainfall time series (Figure 4.15) is based on data from rather fewer stations than the other regions. A maximum of about 50 stations contribute data compared to more than 100 for the other areas. Prior to 1940, the series is based on only 5 stations. With this note of caution borne in mind, the series nevertheless gives little indication of any long-term trends in rainfall in Pakistan as a whole. Again variability can be high from year to year; for example in 1941-44 and 1992-94. The 1900s and 1945-55 were long dry periods, while 1975-85 was a relatively wet period.

It is particularly the years of rainfall deficiency that have relevance to the study of desertification and dryland development. The four case studies presented here demonstrate that such years are common both as shorter-term periods of drought and more extended periods of desiccation. It must not be forgotten that the overall rainfall deficiency, the timing of rainfall events is highly important to human activities. The focus here on meteorological drought, but the severity of the impact of drought to some extent depends on the types of land use being practiced and variety of crops being grown in an area. A certain level of rainfall deficiency at a particular time of year may mean the critical loss of sorghum crop for example, but be more advantageous for yields of millet in a neighbourhood field. Understanding the causes of drought and linkages with controlling factors is undoubtedly important for future attempts to forecast its occurrence, but the ability to predict may differ markedly from area to area depending on how strong linkages with casual mechanisms and teleconnections can be established (Nicholls and Katz 1991).

(e) Changes in Climatic Surfaces 1930-59 to 1960-89

The variability of dryland climates makes it likely that the extent of different dryland zones will vary over time. Figures 4.14, 4.15 and 4.16 give a spatial perspective on this variability, showing changes in the global climatic surfaces for precipitation, temperature and the Aridity Index between two- 30 year periods, 1930-59 and 1969-89. The surfaces were constructed in the same manner with the values of change presented in Map No 4.9 being one approach to identifying the fluctuations in the area of drylands that have occurred (Hulme *et. al.* 1992).

The rainfall graphs for the Sahel and NE Brazil (Figure 4.10 and 4.11) are confirmed in Figure 4.14, which shows these regions respectively to have been drier and wetter in the 1960-89 period. Most of the Sahel west of Ethiopia has been considerably drier, with an annual precipitation decline between 10 and 30%. In southern Somalia by contrast, where the principal controls on climate are different, annual precipitation has increased by over 30% in recent decades. Similar increases have occurred in NE Brazil and in the Middle Asian region east of the Caspian Sea. Lesser but nonetheless significant increases have occurred in western Australia and parts of the south-west USA.

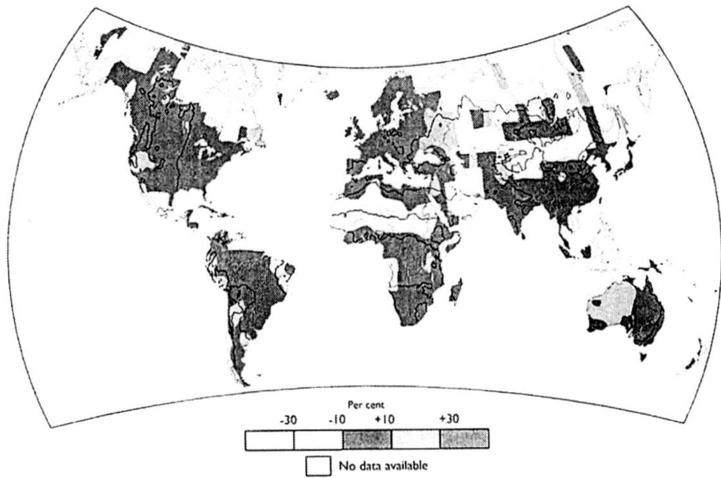

Approximate equatorial scale 1:350 million *Source:* CRU/UEA

Figure 4.14 : *Mean annual precipitation change from 1930-59 to 1960-89*

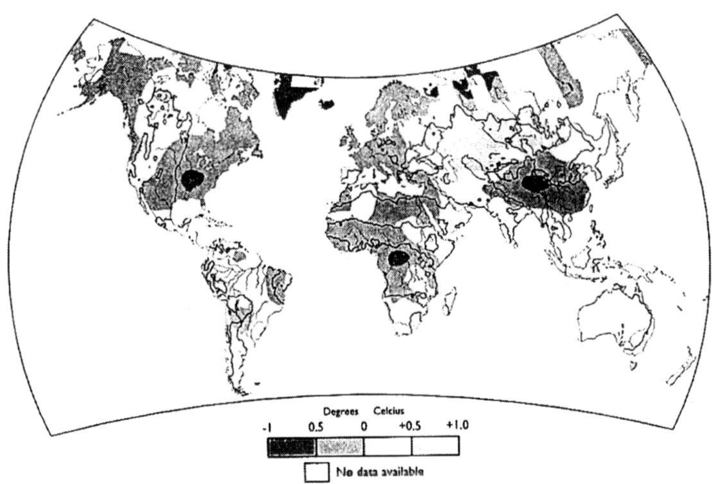

Approximate equatorial scale 1:350 million *Source:* CRU/UEA

Figure 4.15 : *Mean annual tempreature change from 1930-59 to 1960-89*

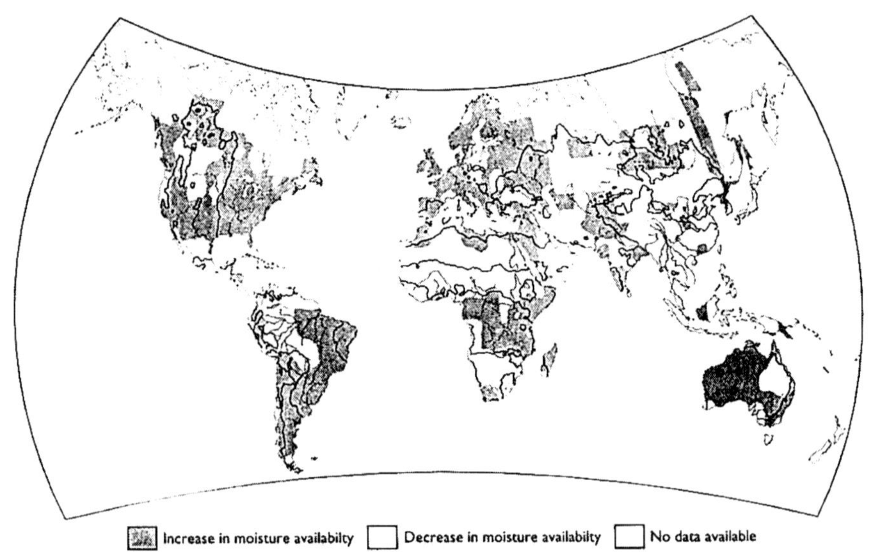

Approximate equatorial scale 1:350 million *Source:* CRU/UEA

Figure 4.16 : *Mean annual Aridity Index change from 1930-59 to 1960-89*

Mean annual temperature changes between the two 30-year periods are shown in Figure 4.15. Increases of up to 0.05 degree C have occurred in the dryland areas of southern Africa, parts of the Sahara and the eastern Sahel, SE Europe, southern Asia including India, northern China and Australia. In contrast, a reduction in mean annual temperature occurred in the Tibetan and Kunlun Mountain drylands.

Figure 4.16 shows changes in Aridity Index values between the two periods and as such represents the composite impact of precipitation and temperature changes. Parts of northern Mexico and NE Brazil have experienced enhanced moisture availability in the later period. Most striking, however, is the decrease in moisture availability in Sahel fro Ethiopia westwards and in parts of Maghreb region. Southern African drylands also experienced declining moisture availability, for although precipitation was little changed, temperatures decreased. Hulme *et. al.* (1992) carried out a similar analysis of Aridity Index changes for nearly identical 30-year periods. They have estimated that in Africa from 1931-60 and 1961-90, 63% of the continent experienced a decline in Aridity Index values (that is, areas getting more arid), with around 7% of the continent's land area being reclassified into a drier aridity zone. Changes were concentrated north of equator and resulted in the extent of hyperarid areas increasing by 50 million hectares and the overall extent of humid areas declining by 26 million hectares (Hulme *et. al.* 1992).

(f) Global Warming

Human actions are now widely believed to be contributing to climate change at the global level. Indeed, Hulme (1996-b) indicates that there is little doubt that the world is warming under this influence (Figure 4.16). Doubts do exist though about the nature, timing and distribution of the regional consequences of global warming. This situation arises due to the complexity of influences on and linkages between, the climates of individual regions as well as the combination of climatic variables that can be influenced by change. For example, it is not just overall annual temperature and precipitation values that may alter, but other more complicated variables such as the seasonal distribution of changes, the timing of the onset of rainfall periods and the intensity of rainfall events.

Various routes are being followed to try to establish the levels of changes in temperature and precipitation that might be expected to occur during the coming years. The main method being used is numerical modeling using General Circulation Models (GCMs), where various scenarios of the impact of enhanced atmospheric greenhouse gas concentrations on climate can be examined (IPCC 1990, 1995), and predictions based on short-term observational data and longer-term proxy records (Williams and Balling 1996).

Most outputs from models predict that during the coming years, year-round temperatures in drylands will increase during all seasons, but there is considerable variance between models concerning future precipitation trends in dryland areas (Williams and Balling 1996). It can, however, be instructive to view possible trends in the context of observed 20th century climatic patterns of drylands. Using a gridded database similar to that used in constructing the climatic surfaces, Hulme (1996-a) has shown that in all the world's dryland areas, there have been overall warming trends.

In most dryland areas, the annual and decadal levels, warmer years have been associated with drier conditions. The only exceptions to this are the drylands of Patagonia and in Middle Asia, where warmer conditions have been associated with enhanced rainfall. The problems faced by climatic modelers in predicting future trends are well demonstrated by the fact that the warmer drier relationship has only been statistically significant in the Maghreb countries, the Sahel and southern Africa (Hulme 1996-b). Nevertheless, the warning for inhabitants of drylands is clear; global warming is likely to further reduce the already limited availability of moisture. This serious implication, along with other more detailed and localized effects, must be taken into account in all efforts to solve desertification problems in the 21st century, and emphasizes the need for stronger action to be taken.

4.3 SOIL DEGRADATION/ LAND DEGRADATION

Soil is an integral part of most terrestrial ecosystems and serves a fundamental function in supporting human communities, primarily through its importance as a medium in which crops and other plants grow. Soil degradation/ land degradation, therefore, is an environmental issue of crucial concern to all societies (Barrow 1991, Blaikie and Brookfield 1987). Its occurrence is generally agreed to be widespread and its effects threaten the sustainability of human life (UNEP 1982, WCED 1987). Identifying the geographical distribution of soil degradation/ land degradation and quantifying the areas affected is. therefore, a task of great importance. The GLASOD project was formulated as the first step towards an appraisal of soil degradation/ land degradation worldwide, with the immediate objective of strengthening the awareness of decision makers and policy makers on the dangers resulting from inappropriate land and soil management to the global well being and leading to a basis for the establishment of priorities for action programmes (Oldeman 1988; 1).

The most direct away to create awareness is a visual representation severity on the global scale (Figure 4.17) embodies both the degree of degradation and its spatial occurrence within individual mapping units. It, therefore, provides both a comparative framework for the overall degradation problem throughout the world's drylands and an overview before the specific types of degradation are considered in the maps.

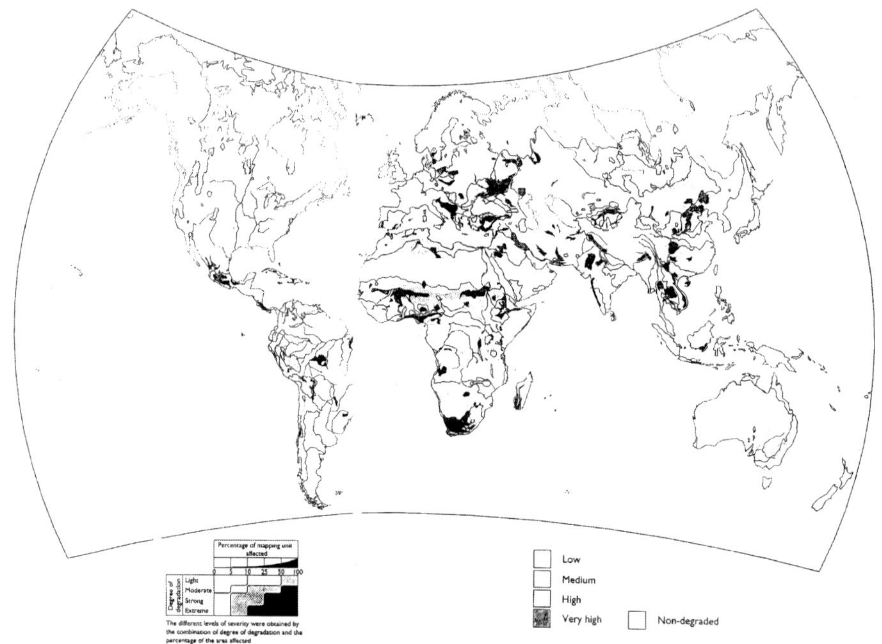

Approximate equatorial scale 1:104 million *Source:* UNEP/ISRIC

***Figure 4.17** : Soil degradation severity*

(a) Methodology

GLASOD defines soil degradation/ land degradation as 'human induced phenomena, which lower the current and / or future capacity of soil to support human life.' Two principal categories of human-induced soil degradation/ land degradation processes are recognized as mentioned in Table 4.2. The first deals with degradation by displacement of soil material, principally by water erosion and wind erosion. The second deals with internal soil deterioration by physical and chemical processes. Soil erosion processes are recognized to cause degradation both on-site and off-site, while degradation by deterioration only includes in situ effects on land that has been abandoned or forced into less-intensive uses. This latter category does not include cyclic fluctuations of land conditions for relatively stable agricultural systems, in which the soil is actively managed to maintain its productivity, nor does it include gradual changes in the chemical composition as a result of soil-forming processes. Particular areas may of course suffer degradation from more than one type of process. Soil compacted by the use of heavy agricultural machinery, for example is often characterized by enhanced runoff rates and increased soil erosion by water.

Table 4.2 : Summary of Glasod Soil Degradation Types

(Oldeman 1988)

Degradation Type	Description
I. SOIL DISPLACEMENT:	
A. WATER EROSION	
(a) On-site	
(i) Loss of topsoil	Uniform loss by surface wash and sheet erosion.
(ii) Terrain deformation	Irregular displacement characterized by major rills, gullies or mass movement.
(b) Off-site	
(i) Sedimentation	Of harbours, lakes, reservoirs and burial of coral, shellfish beds and seaweed.
(ii) Flooding	Including by riverbeds filling, riverbank erosion, excessive siltation of basin land.
B. WIND EROSION	
(a) On-site	
(i) Loss of topsoil	Uniform displacement by deflation
(ii) Terrain deformation	Uneven displacement characterized by major deflation hollows, hummocks or dunes
(b) Off-site	
(i) Over blowing	Encroachment on structures such as roads and buildings, sand blasting of vegetation.
II. INTERNAL SOIL DETERIORATION	
A. CHEMICAL DETERIORATION	
(i) Loss of nutrients	Often leading to reduced production (e.g. accelerated acidification of soils in the humid tropics)
(ii) Pollution and acidification from bio-industrial sources	Excessive addition of chemicals (e.g. organic manure, acid rain)
(iii) Salinization	Caused by activities such as irrigation
(iv) Discontinuation of flood- induced fertility	Resulting from any conservation method that controls flooding and leads to discontinuation of natural replenishment of nutrients by flooding.
(v) Gleysation	As a result of water logging.
(vi) Other chemical problems	Such as catclay formation upon drainage of some coastal swamps; negative chemical changes and development of toxicities in paddy fields.

Degradation Type	Description
B. PHYSICAL DETERIORATION	
(i) Sealing and crusting of topsoil	
(ii) Compaction	By heavy machinery on a soil with weak structural stability or on soils in which humus is depleted.
(iii) Deterioration of soil Structure	Due to dispersion of soil material by Na (and Mg) salts in the sub-soil (sodication)
(iv) Water logging	Human-induced soil hydromorphism; flooding and submergence (excluding paddy fields)
(v) Aridification	Human induced changes of the soil moisture regime towards an aridic regime (e.g. by lowering of the local base groundwater level by changing grassland to wheat cultivation) but excluding deep ground water depletion
(vi) Subsidence of organic soils	By drainage oxidation.

Source: World Atlas on Desertification II Edition (Oldeman 1988)

Although, soil degradation/ land degradation is affected by a range of human-induced and natural processes (Figure 4.10), it is important to stress that it is by definition a social problem. While purely environmental processes such as leaching and erosion occur with or without human interference, the description of these processes as 'degradation' implies social criteria relating land to its current or possible future uses (Blaikie 1985, Blaikie and Brookfield 1987). The GLASOD database and maps indicate areas of soil degradation caused by human misuse, such as soil erosion enhanced above natural rates by the clearance of vegetation for use as fuel. The causes of degradation are noted and these causes all represent forms of human mismanagement. Areas where the sustainability of human land uses is threatened by natural processes, such as the blowing of salt from a salt flat on to nearby cropland, are not identified on the maps.

In the field, however, the distinction between such 'active' and 'passive' human involvement is not always clear-cut. This problem is often particularly acute in dryland areas where the environment is naturally high dynamic. During a drought period, for example. vegetation cover is naturally less extensive than during non-drought periods. The distinction between enhanced soil erosion driven by climate and that due to human activities will inevitably be subjective to some degree. Nonetheless, this distinction is important because it would affect the strategies adopted to combat desertification.

If degradation is a function of natural processes, which determine the previously sustainable land use, the solution may be to change the land use, particularly in societies, which are unable to implement expensive soil conservation or rehabilitation schemes. If, by contrast, degradation results from poor land management form of management may be the solution. It is appropriate in either case to focus on the human side of the desertification equation, because that is the side that can be altered successfully, identifying whether human activities are involved actively or passively in a vital first step.

The GLASOD project was executed by more than 250 soil scientists from all over the world with 20 regional coordinators. The first step was to divide the global land surface into physiographic mapping units. These units were delineated so that each showed a certain degree of homogeneity of topography, geology, soils, climate, vegetation and land use. The next step was to evaluate the degree, relative extent, the rate in the recent past and causative factors for each type of human-induced soil degradation shown in table no. 4.12 at occurred in the physiographic unit.

Table 4.3 : Proportions of Stable, Soil-Degraded and non-used land by regions for all Climate Zones (%)

Region	Stable Land	Degraded land	Non-used Land *	Total
1. Africa	39	17	44	100
2. Asia	48	18	34	100
3. Australia	61	12	27	100
4. Europe	61	23	16	100
5. North America	54	7	39	100
6. South America	72	14	14	100
World	**52**	**15**	**33**	**100**

* Non-used land is land not used for agricultural land, permanent pasture or covered by forest or woodland.

Source: GLASOD

In the absence of scientific measurements and monitoring of soil degradation/ land degradation for most of the world's land surface, the data that did exist for each unit were supplemented by expert opinions from the participants, who were selected for their intimate knowledge of local conditions. Using a standardized set of guidelines to ensure uniformity in reporting, soil degradation was assessed over the recent past, ideally averaged over the previous 5 or 10 years, but in some cases the time period extended over recent decades.

The involvement of a large number of scientists with personal experience of regional conditions is an important strength of GLASOD. The complex interplay of soil degradation/ land degradation processes and their spatial variability needs to be assessed according to local conditions and forms of land use. This point is well illustrated in the case of off-site effects of soil erosion. Soil is often lost from steep highland slopes and deposited in lower, move level sites. If the lower site contains a reservoir, the accumulation of sediment is a problem; it causes a loss of storage capacity requiring expensive remedial efforts, as described by Tolouie *et. al.* (1993) in an example of north-western Iran.

In other situations, however, a similar transport of soil from upper slopes may be regarded as beneficial to lower areas. In southern Zimbabwe, Scoones (1992) reports that the transfer of sediment in this way can actually give significant advantage for production because the lower valley areas are better locations for water retention and, therefore, cultivation, livestock production and human occupation (Ingram 1991). The same situation has been described in the pastoral lands of central Australia (Pickup and Chewings 1988).

This spatial variability in the operation of degradation processes also illustrates an important point concerning the scale of analysis. In the Zimbabwe example, Scoones (1992) noted that despite indications of increased degradation in his study areas, livestock production had not declined, although this could have been because negative effects on production lagged the initial degradation. The scale of the GLASOD mapping polygons of large areas on the ground are coloured according to an average severity of degradation. Actual areas of degraded soil within each polygon are shown in the accompanying tables, but is should be noted that within individual polygons, there may still be areas that benefit from degradation processes elsewhere. The global maps are designed to give an overall, if exaggerated, impression of the scale of the soil degradation problem worldwide. Once key problem areas are identified on the global scale, however, more detailed appraisals of local situations are needed before appropriate solutions to degradation problems can be developed.

Areas identified as non-degraded in the GLASOD survey are designed wither 'stable' or 'non-used' by human populations. Stable areas are either those regions, where human intervention is minimal due to very low population densities, or regions where soil improvement or protection programmes have been successfully implemented. Non-used areas are those not used for agricultural purposes, permanent pasture or covered by forest or woodland. This category includes land with little value for agricultural use, such as the large tracts of hyperarid desert in the central Sahara and parts of Arabia, but land in this category may also be potentially productive, but has been put to other land use e.g. build on areas and roads.

Severity is calculated according to a combination of the degree of soil degradation and the percentage of the area affected. The degree to which soil is degraded has been estimated in relation to changes in agricultural suitability, in relation to reduced productivity and in some cases in relation to its biotic functions. The degree of degradation is classified according to the following scheme:

(a) *None:* There is no sign of present degradation from water or wind erosion, from chemical or physical deterioration; all original biotic functions are intact. Suchland is considered stable.

(b) *Light:* The terrain is suitable for use in local farming systems, but with some what reduced agricultural productivity. Restoration to full productivity is possible by modifications of the management system. The original biotic functions are still largely intact.

(c) *Moderate:* The terrain is still suitable for use in local farming systems, but with greatly reduced agricultural productivity. Major improvements are required tore store productivity, such as draining for water-logged or salinized land or counter banks for eroded land. The original biotic functions are partially destroyed.

(d) *Strong:* The terrain is not reclaimable at the farm level. Major engineering worksare required for terrain restoration. The original biotic functions are largely destroyed.

(e) *Extreme:* The terrain is not reclaimable and impossible to restore. The original biotic functions are completely destroyed.

The importance of the land use in question should be noted here. While a moderately degraded soil for example may require major structural alterations to restore productivity under one particular land use, for a less demanding land use the alterations required may be less intensive. The relative extent of degradation was expressed using five categories:

(i) Infrequent : Up to 5 % of the unit is affected,

(ii) Common : 6-10 % of the unit is affected

(iii) Frequent : 11-25 % of the unit is affected.

(iv) Very Frequent : 26-50 % of the unit is affected.

(v) Dominant : More than 50% of the unit is affected.

With four categories for the degree of degradation and five categories for the extent of degradation, 20 combinations are possible. These combinations have been grouped into four severity classes as illustrated in Figure 4.17 and each severity class is given different shading.

(b) Interpreting the Maps and Tables

The maps indicate the areas in which degradation occurs and severity of degradation according to the accompanying key. The tables, however, provide a more detailed indication of the actual areas within units affected. It is important to note that while maps indicate degradation severity, which is a combination of the degree of degradation and its extent, the tables show the information according to the actual affected areas by degree of degradation.

Figure 4.17 illustrates the overall global assessment of human induced soil degradation/ land degradation. Comparison of Figure 4.17 and Table No. 4.3 illustrates the point that the maps are designed to give an overall indication of where degraded areas are located, but because the methodology dictates that each mapping polygon is coloured according to the degradation severity, the map appears to exaggerate the overall area affected. Hence, as Table 4.3 shows, 15% of the global land area shown in Figure 4.17 suffers soil degradation/ land degradation, while 52% is non-degraded stable land. At the continental level, Europe is shown to suffer the most widespread soil degradation/ land degradation problem, affecting 23% of the continent's land area. By contrast, just 7% of North America is so affected. Table 4.3 also highlights the wide range in the relative proportions of the continents classified as non-used i.e., not used for agriculture, permanent pasture or covered by forest or woodland. While

Europe and South America both have less than 20% of their land areas in this category, more than 40% in Africa is not currently used for agricultural or other biomass uses.

Table 4.4 gives the soil degradation/ land degradation areas inside and outside the susceptible drylands. Worldwide, just over one billion hectares or 20% of the global extent of susceptible dryland soils is currently being degraded by human activity. A similar absolute area of soils outside the susceptible drylands is being degraded, but this represents approximately 13% of the total non-susceptible dryland area. Figure 4.18 and Table 4.4 indicate that in the light, moderate and extreme categories more land area is susceptible dryland regions is being degraded that in non-dryland regions categorized as 'others' in Table 4.4. When total used land, disregarded as non-used areas, inside and outside the susceptible drylands is considered, not less that 29% of the world's used susceptible dryland soils are degraded, in contrast to 15% of the global area used outside the susceptible drylands.

Table 4.4 : Soil Degradation Degree by Region: Inside the Drylands (Susceptible) and Outside the Drylands (others) Areas

(Million Hectares)

Region	**S O**	**Light**	**Mode-rate**	**Strong**	**Extreme**	**Total Degraded**	**Total Non-degraed**	**Total**
Africa	S	118.0	127.2	70.7	3.5	319.4	966.6	1286.0
	O	55.7	64.6	52.8	1.7	174.8	1504.8	1679.6
Asia	S	156.7	170.1	43.0	0.5	370.3	1301.5	1671.8
	O	137.8	174.2	64.6	0.0	376.6	2207.6	2584.2
Australasia	S	83.6	2.4	1.1	0.4	87.5	575.8	663.3
	O	13.0	1.6	0.8	0.0	15.4	203.5	218.9
Europe	S	13.8	80.7	1.8	3.1	99.4	200.3	299.7
	O	46.7	63.8	8.9	0.0	119.4	531.4	650.8
North America	S	13.4	58.8	7.3	0.0	79.5	652.9	732.4
	O	5.5	53.7	19.5	0.0	78.7	1379.8	1458.5
South America	S	41.8	31.1	6.2	0.0	79.1	436.9	516.0
	O	63.0	82.4	18.9	0.0	164.3	1087.2	1251.5
World	S	427.3	470.3	130.1	7.5	1035.2	4134.0	5169.2
	O	321.7	440.3	165.5	1.7	929.2	6914.3	7843.5
Total		**749.0**	**910.6**	**295.6**	**9.2**	**1964.4**	**11048.3**	**13012.7**

S = Susceptible O = Others.

Source: GLASOD

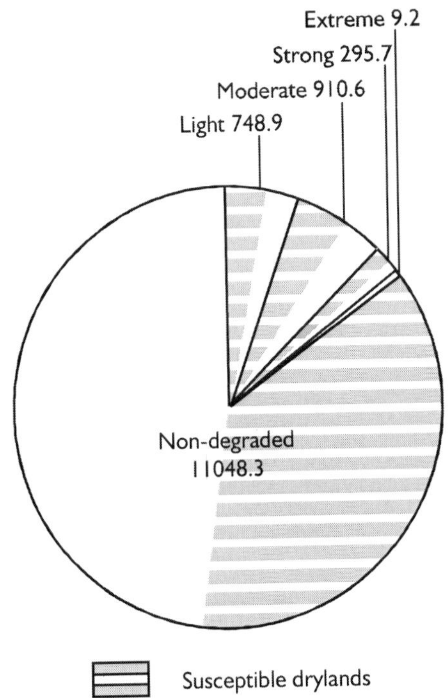

Figure 4.18 : *Global land area by degree of soil degradation (Million hectares)*

The continental breakdown of the data reveals that a greater proportion of Europe's susceptible dryland soils suffer from degradation (99.4 million hectares; 32% of the total area) than in any other continent. Most of these degraded soils occur in southern Portugal, Spain, Corsica, Sicily, southern Italy, Greece, Ukraine and Russia, although the large majority of these areas are degraded to a light or moderate degree. In North America, by contrast, 11% of the susceptible dryland soils are degraded, again largely to a light or moderate degree. A light and moderate degree of degradation also predominates in South America, where 15% of susceptible dryland soils are degraded. Australia's 13% affected area is very largely degraded to a light degree. Asia, with 22% of its susceptible dryland soils degraded, has far greater areas affected to a strong and extreme degree, but the dryland soil degradation/ land degradation situation is most critical in Africa. With 25% of its susceptible drylands affected by soil degradation/ land degradation, Africa has the highest proportion of its susceptible dryland soils affected to a strong and extreme degree.

4.4 SOIL DEGRADATION/ LAND DEGRADATION IN DRYLANDS

The study concentrates on arid, semi-arid and dry sub-humid areas, the susceptible drylands, within which almost 20% of the total global land area is recorded as experiencing soil degradation/ land degradation. Hyper-arid areas are excluded from the main consideration; they generally have very strong desert characteristics and nutrient-deficient soils, giving limited potential for degradation by many processes and they offer only limited opportunities for degradation-induced human land uses. A notable exception to this rule, the Nile Valley, is discussed in more detail in, Figure 4.19, therefore, concentrates on areas where natural dryland soil properties are susceptible to degradation and where land use activities may cause desertification.

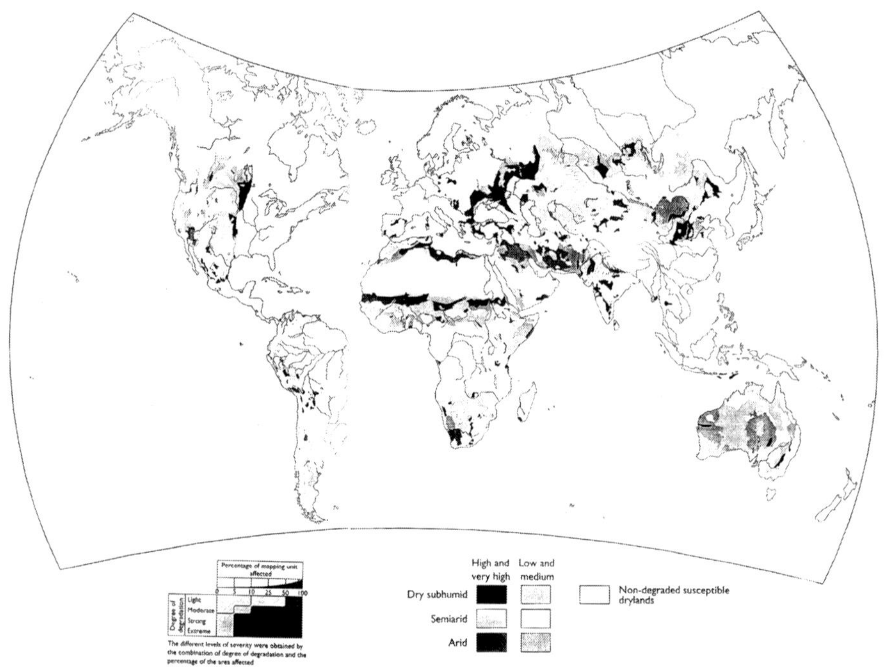

Approximate equatorial scale 1:104 million

Source: UNEP/ISRIC, CRU/UEA

Figure 4.19 : *Soil Degradation severity in susceptible Drylands*

(a) Soil Degradation in Drylands

The soils commonly found in dryland regions are distinctive in a number of ways (Dunkerley and Brown 1997). They have low organic matter contents, due to the generally low plant biomass and are, therefore, dominantly mineral soils of an immature or skeletal type. Many, but not all types tend to exhibit a low clay content compared with soils from more humid areas. Low precipitation levels also mean that dryland soils are little affected by leaching, so that soluble salts tend to accumulate at a certain level in the soil profile dependent upon the local water table or the depth of moisture infiltration. A classification of dryland soil types, their characteristics and their relative preponderance is shown in Table No. 4.5.

Table 4.5 : Predominant Soils of Dryland Regions

(After FAO 1993-a, Lal 1990 and Dregne 1976).

USDA Soil order	FAO major soil groups	Description	Moisture storage capacity	Inherent soil fertility	Proporti on of world land area (%)
Enti sols	Aerno-sols Lithi-sols Fluvi-sols	Sandy to loamy soils of recent origin formed from wind blown sands on the margins of deserts or lacustrine or marine deposits. Used for nomadic grazing, dry farming or irrigated agriculture. Mostly on flat land, loose and easily worked and easily eroded by wind. They may receive fresh sediments and hence be regularly rejuvenated.	Low, seasonal water-logging risk	Low to moderate	13.1
Aridi sols	Xero sols, Yermo sols	Formed in environments with a pronounced dry season. Medium to fine textured soils that may contain high soluble salts, calcium carbonate or gypsum, forming surface or sub-surface crusts or calcite pendants. Loose structure and stony in some paces. Easily erodible by wind and water.	Low, moderate to high	Moderate to high	11.3
Molli sols	Chernozems Kastanozems Phaeonems Rendzinas	Semi-arid steppe soils, mostly pcrous, well-aerated and stable structures with high organic matter contents. Used for pastures or dry farming. Can be easily leached of their nutrients, but less erodible.	Moderate to very high	High	3.7
Alfi sols	Luvi sols	Soils common in flat or gently sloping land with well defined clayey horizons. Favourable physical characteristics, suitable for a wide range of agricultural use. Usually well-drained and are highly susceptible to water erosion	Moderate to high	Moderate	2.1
Verti sols	Verti sols	Develop in sediments or weathered rock that are fine in texture and form smectite clay, which swells when wet and cracks when dry. These soils have low infiltration rates, high susceptibility to runoff and erosion and are quickly affected during drought periods. They have great agricultural potentials for a variety of crops, but special management practices are required.	Moderate to high	High	1.3
Total					**31.5**

An important characteristic that makes dryland soils especially vulnerable to degradation is the slowness of their recovery from a disturbance. Since water is often not available, or only in very limited amounts in a few places, new soil is formed slowly, although actual data on dryland soil formation rates are relatively hard to come by in the literature, not least because it is a slow process. For the loess soils of the US Great Plains, Kirkby (1980) suggests a soil formation rate of about 0.2 mm y-1, but the figure drops to 0.02 mm y-1 in the arid south-west. In semi-arid Kenya, Lal (1984) reports a rate of 1 mm y-1 but Dunne *et. al.* (1978) estimate that rates may fall below 0.01 mm y-1.

The general deficiency in moisture means that salts, once accumulated, tend to remain in situ. Moisture deficiency also discourages recolonization by plants that have been removed or damaged and the build-up of organic matter is slow since it is usually consumed rapidly under warm conditions. Reestablishment rates for soil bacterial and fungal populations have been little studied, but the nitrogen fixation capability of dryland soil is thought to require at least 50 years to recover from a serious disturbance (Belnap 1995). Although dryland soils tend to be characterized by low nutrient contents (Buckley *et. al.* 1987). and hence have a low level to recover after degradation, the ability of soils in dry areas to recover from negative changes in the soil/land surface is generally lower that in humid areas. In other words, dryland soils have a low resilience.

(b) Susceptibility to Degradation

In drylands, susceptibility to degradation by specific processes varies spatially and is affected by a range of environmental characteristics. Drylands are more susceptible to wind erosion than any other biome because soils tend to be dry, poorly cemented and sparsely covered by protective vegetation. Susceptibility to wind erosion once any protective cover has been disturbed, is affected by the particle size distribution or the soil. Soils with high clay content will naturally offer greater resilience to deflation than silty or sandy soils because of their increased particle cohesion.

With mixed grain sizes, selective winnowing of the deflatable component can dramatically reduce soil fertility and resilience (Chepil and Woodruff 1963). Extensive areas of ancient wind-lain sediments in drylands are particularly vulnerable to deflation following damage to the protective vegetation cover. Such reactivation of quaternary Aeolian sediments has been noted in many drylands, including the African Sahel (Thiemeyer 1992), the Great Plains of the USA (McCauley *et. al.* 1981), north-west India (Kumar *et. al.* 1983) and China (Zhu Zenda and Liu Shu 1983).

Compared to other world climatic zones; dryland areas have long been recognized as particularly susceptible to water erosion (Figure 4.20) although the precipitation levels are relatively low and do not support a round-year cover of protective vegetation, they are often highly erosive when rain does fall (Langbein and Schumm 1958). This generic relationship highlights the importance of soil degradation in drylands. In areas that receive an effective mean annual precipitation above 300 mm, increased precipitation encourages greater vegetation growth, which protects the soil from water erosion, while below the 300 mm-mark, erosion declines as precipitation levels decline. Topography affects susceptibility to water erosion. High relative relief encourages runoff even when only moderate of ground surface have lost the protection of a vegetation cover, Water erosion is rarely spatially continuous and often manifests itself through rilling and gullying (Moss *et. al.* 1982). The soils with most potential for loss of nutrients are those that naturally contain higher nutrient concentrations in the first place. These are the very soils, which offer some of the greatest potential for agricultural development.

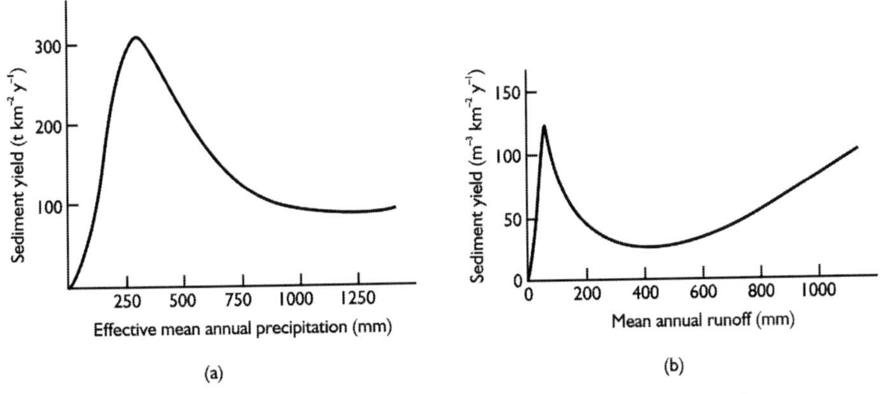

Figure 4.20 : *Proposed Relationship between Sediment Yield and (a) Effective Annual Precipitation (Langbein and Schumm 1958); (b) Mean Annual Runoff (Douglas 1967).*

Like wind erosion, some forms of in-situ degradation are most commonly encouraged in drylands. Salinization is an obvious example. Salt affected soils occur mostly in dryland regions, where evapotranspiration exceeds rainfall and hence leaching and transportation of salts to the oceans is not as effective as in more humid regions (Rhoades 1990). Salinization occurs locally where water is available, either from groundwater sources or from perennial rivers with headwaters in more humid areas, as in the case of parts of the Colorado River basin of the USA and the Indus valley in Pakistan. While salt-affected soils are found extensively under natural conditions, salinity problems of great concern to agriculturalists arise when previously productive soil becomes salinized as a result of irrigation (Szabolcs 1987) or the replacement of natural vegetation by certain crops (Berg *et. al.* 1910, so called secondary salinization (Thomas and Middleton 1993). Salinity problems on poorly managed irrigation schemes are also often associated with physical deterioration of dryland soils through waterlogging. This combination of degradation processes, so ironic in drylands in that they are intimately linked to excessive application of irrigation water combined with poor drainage, has been widely reported from all over the dryland realm (Abrol *et. al.* 1988, Szabolcs 1987).

Overall, susceptibility to degradation, in total or with regard to specific processes, is by no means even throughout the drylands. The occurrence of degradation reflects a range of natural contemporary and antecedent conditions and processes and the wide variety of land uses practiced by human occupants of such regions.

(c) Spatial Patterns

Some 1035 million hectares of the world's susceptible drylands are affected by soil degradation, of which about 87% are in the light and moderate categories. The distribution by continent and by aridity zone is given in Table 4.6. High and very high severity degradation can be seen to be a phenomenon that is particularly significant is Africa, whether considered in relative or absolute terms. Soil degradation affects some 320 million hectares in Africa out of a total of 1286 million hectares or about a quarter of the susceptible drylands.

Table 4.6 : Soil Degradation Degree by Susceptible Dryland Areas

(Million Hectares)

Region	Aridity Zone	Light and Moderate	Strong and Extreme	Total
Africa	Dry subhumid	25.2	12.1	37.3
	Semiarid	69.9	39.6	109.5
	Arid	150.2	22.3	172.5
Asia	Dry subhumid	70.6	7.7	78.3
	Semiarid	124.2	17.2	141.4
	Arid	131.9	18.8	150.7
Australasia	Dry subhumid	4.2	0.6	4.8
	Semiarid	32.9	1.0	33.9
	Arid	48.9	0.0	48.9
Europe	Dry subhumid	59.0	2.3	61.3
	Semiarid	30.8	2.6	33.4
	Arid	4.8	0.0	4.8
North America	Dry subhumid	15.0	3.2	18.2
	Semiarid	50.9	2.3	53.2
	Arid	6.3	1.6	7.9
South America	Dry subhumid	21.4	2.3	23.7
	Semiarid	43.9	4.0	47.9
	Arid	7.5	0.0	7.5
Total		**897.6**	**137.6**	**1035.2**

Source: GLASOD

Regardless of how much of the areas are affected, the map indicates that the distribution of severest degradation in Africa, Asia and North America occurs on the wetter margins of drylands. Scrutiny of Table 4.6, however, suggests a rather different picture. Degradation in Africa and Asia is largely concentrated in semiarid and arid areas; actual areas affected increase from dry subhumid zones, though the semiarid zones, to peak in their extent in arid areas, although Figure 4.19 indicates that degradation appears to be minimal in the northern Sahel. A similar pattern is identifiable in Australasia, albeit that the level of overall severity is consistently lower than in Africa and Asia. In the Americas, by contrast, greater areas of land are degraded in the semiarid zone that in the dry subhumid and arid zones combined. The situation is different again in Europe, where degradation in the dry subhumid zone greatly exceeds that in the semiarid and arid parts of the continent.

(d) The Relative Importance of Different Degradation Types

At the world scale, the GLASOD database indicates that dryland degradation is dominated by soil erosion as indicated in Table 4.7. Water erosion accounts 45% of the degraded area and wind erosion 42%. Chemical deterioration accounts for 10% and physical deterioration just 3%. This situation, however, varies according to aridity zone. Given likely variations in vegetation cover, it is not surprising that 60% of the soil degradation in arid zones is by wind erosion , a figure which falls to 21 % in dry sub-humid areas. The reverse trend is found for water erosion, to which the wetter dryland areas are inevitably going to be more susceptible. Water erosion is dominant form of degradation in dry subhumid and semiarid zones, accounting for 63% and 51% of the degradation in these zones respectively. In arid areas, water erosion accounts for 29% of the degradation. These and other trends are explored in greater detail in the analyses of different degradation types at the global scale.

Table 4.7 : Soil Degradation Type by Susceptible Dryland Climate Zones

(Million Hectares)

Climatic Zone	Water Erosion	Wind Erosion	Chemical Deterioration	Physical Deterioration	Total
1. Dry Subhumid	141.0	46.8	22.5	13.3	223.6
2. Semiarid	213.0	150.3	40.9	15.1	419.3
3. Arid	113.4	235.3	37.3	6.3	392.3
Total	**467.4**	**432.4**	**100.7**	**34.7**	**1035.2**

Source: GLASOD

The relative importance of water erosion for all continents except Africa is shown by data in Table No. 4.8. In Asia, Europe, North America and South America water erosion is responsible for degrading slightly more land than wind erosion, while in Australasia, water erosion is by far the most dominant form, affecting 80% of the susceptible drylands to wind erosion's 18%. In Africa, however, more land is degraded by wind erosion than by water erosion. Only in Asia is the proportion of dryland affected by chemical deterioration greater than the global average. Here, the figure is 14%. The global average proportion of land affected by physical deterioration is exceeded only n Europe, where 9% of the susceptible drylands are affected.

Table 4.8 : Soil Degradation Type by Region in Susceptible Dryland Areas
(Million Hectares)

Region	Type of Soil Degradation				
	Water Erosion	**Wind Erosion**	**Chemical Deterioration**	**Physical Deterioration**	**Total**
1. Africa	119.1	159.9	26.5	13.9	319.4
2. Asia	157.5	153.2	50.2	9.6	370.5
3. Australasia	69.6	16.0	0.6	1.2	87.4
4. Europe	48.1	38.6	4.1	8.6	99.4
5. North America	38.4	37.8	2.2	1.0	79.4
6. South America	34.7	26.9	17.0	0.4	79.0
Total	**467.4**	**432.4**	**100.7**	**34.7**	**1035.2**

Source: GLASOD

For all types of degradation except chemical deterioration, the largest areas as susceptible dryland affected are degraded to a moderate degree as given in Table 4.9. For wind erosion, 95% of all degradation is in the light and moderate categories. The proportion of land degraded by water erosion to these degrees is 82%, while for physical deterioration, it is 74%. Of all types of degradation, the greatest proportion of land affected to a strong degree is in situ processes; 24% of the susceptible drylands degraded by chemical deterioration, and 26 % of the susceptible drylands degraded by physical deterioration. No soils affected by physical deterioration are degraded to an extreme degree.

These global patterns are investigated further on specific degradation types, followed by analysis of the human activities causing degradation and an appraisal of the relationships between soil degradation and vegetation.

Table 4.9 : Soil Degradation Type by Degree in Susceptible Dryland Areas
(Million Hectares)

Region	Type of Soil Degradation				
	Water Erosion	**Wind Erosion**	**Chemical Deterio-ration**	**Physical Deterio-ration**	**Total**
1. Light	175.1	197.2	44.3	10.8	427.4
2. Moderate	208.5	215.4	31.4	15.0	470.3
3. Strong	79.0	18.0	24.2	8.9	130.1
4. Extreme	4.8	1.8	0.8	0.0	7.4
Total	**467.4**	**432.4**	**100.7**	**34.7**	**1035.2**

Source: GLASOD

4.5 SOIL EROSION

Soil erosion is a natural phenomenon occurring over much of the earth's land surface, but its extent and intensity have been greatly increased by human activities over many centuries (Brown and Wolf 1984). While wind erosion might intuitively be expected to be the most serious hazard in drylands, because of the low levels of rainfall and relatively sparse vegetation covers, these factors do not discount the significance of erosion by water. Indeed, the occurrence of many rainfall events in association with high intensity convective storms can give rise to very high levels of surface runoff and, therefore, erosion (Yassoglou 1995).

Soil erosion has a dual significance when land degradation is considered, not only does it affect the productivity of the land in areas experiencing the problem, but it can also have negative off-site impacts in areas that receive eroded soil. For example, contributing to reservoir sedimentation (Barrow 1987) and the occurrence of dust storms that may travel hundreds of kilometers from source areas (Worster 1979). Negative off-site impacts can also threaten the productivity of agricultural systems, for example, Mohammed *et. al.* (1995) report how wind blown sand threatening the north-western area of Gezira irrigation schemes in the Sudan. Positive off-site impacts can sometimes accrue too; when erosion leads to the movement of soil from steep slopes that are difficult to cultivate to flatter, more easily accessible, valley bottoms.

Human induced soil erosion is not a new phenomenon in drylands. It has been documented in the Mediterranean basin from as long as 400 BC (Rubio 1995), while sedimentary evidence points to cultivation-induced erosion occurring in Mexico before the Spanish conquest (O'Hara *et. al.* 1993). However it has been during the twentieth century, with the expansion of agriculture to lands that are marginal for crop production and the intensification of agricultural methods, that soil erosion has become an issue of considerable global concern. The 1930s 'Dust Bowl' on the Great Plains of the USA was particularly important in generating a realization of how susceptible dryland environments can be vulnerable to soil erosion by both wind and water (Bennett 1938). Concern about the susceptibility of drylands to soil erosion and an awareness of a need to better understand the physical process and social pressures leading to soil loss, have been a characteristic of scientific and anti-degradation initiatives over several decades (Kellogg 1956, UNOSTD 1987, Fantechi *et. al.* 1995).

Soil erosion embraces a complex set. of processes and interacting factors. Whether soil is eroded or conserved depends on the balance achieved between erosivity and erodibility (Morgan 1995). Erosivity refers to the forces that can liberate particles from the main soil mass and is controlled primarily by natural criteria such as wind strength and rainfall intensity during a storm (Table 4.10) Erodibility is the susceptibility of a soil to the loss of material, which is influenced by both physical soil characteristics and land use and management techniques. Some surface characteristics may enhance one erosion type but inhibit another. A crusted soil may be less susceptible to wind erosion but more susceptible to erosion by running water, while the timing of human actions that enhance erodibility relative to erosivity events is also important in terms of whether actual erosion is enhanced or not (Kar 1996).

Table 4.10 : Summary of Factors Affecting Soil Erosivity and Erodibility

EROSIVITY	
OF WIND	**OF WATER**
* Frequency of strong winds	* Frequency of rainfall events
* Duration of windy events	* Storm duration
* Wind velocity	* Raindrop impact velocity
* Wind turbulence	* Raindrop size
ERODIBILITY	
* Particle size and shape	* Slope angle (water erosion)
* Organic material (binding) content	* Vegetation cover density
* Clay content	* Plant height and shape
* Cohesiveness	* Leaf/ stem fineness
* Infiltration capacity	* Presence of surface crusts
* Moisture content	* Plant orientation (wind erosion)
* Porosity and permeability	

Source: GLASOD

By reducing natural plant densities, replacing a perennial plant cover with a seasonal crop, ploughing steep slopes and causing surface compaction by livestock trampling or the use of mechanized farming implements, human actions can readily enhance the potential for erosion (Thomas and Middleton 1994, Belnap 1995). Ploughing during the dry season in anticipation of later rains can lead to high water erosion rates during the first rain storms that occur, but rain fail or are late, the soft ploughed surface is left vulnerable to wind erosion (Dresch 1986, Kar 1996). Particularly in semiarid lands, there is the potential for linkages between water and wind erosion, with rain drop impacts loosening particles that can subsequently be blown away.

The GLASOD survey included an assessment of both water and wind erosion and the human factors that may contribute to enhanced erosion. Box No. 1 outlines the methodology for the assessment of water erosion and Box No. 2 that for wind erosion. The respective methodologies allowed all the key in-site components of soil erosion that lower productive potential to be assessed: top soil erosion by sheetwash and rilling (Savat and De Ploey 1982); gullying (Oostwoud Wildenes and Bryan 1994); pipe development Ybryan and Harvey 1985); Aeolian deflation (Watson 1990) and the development of deflation hollows and hummocky dunes (Zhao 1988).

BOX 1 : Glasod Methodology for Assessing Degree of Water Erosion

The definitions and criteria used for the designation and identification of water erosion were as follows (after Oldeman 1988). The categories incorporate an assessment of both the loss of top soil by erosion and terrain deformation through rilling, gullying and, where appropriate, various forms of mass movement. The definitions are as follows:

Slight: For soils with a rooting depth exceeding 50 cm., part of the top soil has been removed. Shallow rills with a spacing of 20-50 m may be present. Thin soils have rills at least 50 m apart. In areas under pastoralism the perennial plant cover or the original or optimal plant cover extends over at least 70% of the surface.*

Moderate: All the top soil will have been removed from deep soils. Rills may be present, less than 20 m apart. Gully development will have occurred at a spacing of 20-50 m. Thin soils will have lost part of the top soil and are likely to have rills with a 20-50 m spacing. Perennial/optimal vegetation cover in pastoral areas reduced to 30-70%. *

Strong: All the top soil and part of the sub-soil will have been removed from areas of deep soils, with moderately deep gullies less than 20 m apart. All the top soil will have been removed from areas of thin soils, exposing bedrock, weathered bedrock, or a hardpan. In pastoral areas the perennial/ original/ optimal vegetation cover will be less than 30%. *

Extreme: The general criterion, that the land is unreclaimable and impossible to restore, applies. Figure 4.21 shows land units in which human-induced water erosion occurs. The map representation does not necessarily mean that water erosion is the only form of degradation occurring in a particular unit, nor that water erosion is the dominant degradation process in a unit.

* Known maximum coverage of perennials under good management as practiced during some time in the past.

BOX 2 : Glasod Methodology for Assessing Wind Erosion

The nature of wind erosion necessitates the recognition of three effects (Oldeman *et. al.* 1990). These are the loss of topsoil, defined as a loss by which a uniform part of the topsoil is removed, terrain deformation, in which there is an irregular or uneven displacement of soil material resulting in major hollows, hummocks or dunes and the overblowing the sediments affecting physical structures such as roads and buildings.

Taking these three effects into account, wind erosion was considered in the GLASOD survey using the following criteria:

Slight: In deep soils; topsoil partly removed and / or dew (10-40% of area) shallow (0-5 cm) hollow; in shallow soils; very few (<10%) shallow hollows, in pastoral country; groundcover of perennials of the original/optimal vegetation > 70%.*

Moderate: In deep soils; all topsoil removed or with common (40-70% Of area) shallow (0-5 cm) hollows or few (10-40%) moderately deep (5-15 cm) hollows; in shallow soils; partly removed or few (10-40%) shallow (0-5 cm) hollows; in pastoral areas; groundcover of perennials of the original / optimal vegetation 30-70%.*

Severe: In deep soils; all topsoil and part of subsoil removed or with many (>70% of area) shallow (0-5 cm) or common (40-70%) moderately deep (5-15 cm) or few (10-40%) deep (15 cm) hollows / blowouts; in shallow soils; all topsoil removed bedrock or hardpan is exposed. In pastoral country; groundcover of perennials of the original / optimal vegetation is < 30%.*

Figure 4.22 shows land units in which human-induced wind erosion occurs. The map representation does not necessarily mean that wind erosion is the only form of degradation occurring in a particular unit, nor that wind erosion is the dominant degradation process in a unit.

* Known maximum coverage of perennial under good management as practiced during some time in the past.

A. WATER EROSION

(a) The Nature of Dryland Water Erosion

A soil is commonly unable to absorb all the energy to absorb all the energy and mass provided by rain drops hitting the ground surface (Yariv 1976, Stocking 1977). A plant cover will act as an interceptor, reducing direct drop impact on the soil surface and reducing the energy available for erosion. Reductions in vegetation cover or changes in vegetation type, brought about by land use decisions and/ or moisture-deficient periods are therefore likely to have a significant impact on erosion rates in drylands (Abrahams *et. al.* 1995). Bare or partially exposed soils are vulnerable to the dislodgement of individual particles by drop impact, and any subsequent generation of overland flow, which transports the particles down slope; but runoff and sediment yields fall dramatically as plant cover increases (Francis and Thornes 1990). The ability of a soil to absorb raindrop energy is therefore dependent on antecedent environmental conditions and the nature of the rainfall event.

In drylands, the sporadic nature of rainfall events make the study of water erosion processes difficult, while the range of variables that interact to influence the actual occurrence of erosion further complicates elucidation of the problem (Thornes *et. al.* 1996). To this end, soil erosion models, for example those developed by the MEDALUS research programme, are making significant contributions to enhanced understanding (Kirkby *et. al.* 1990, Thornes 1995). The ability of a soil to resist erosion is dependent upon the range of particle sizes, the chemistry and organic content as well as the nature and extent of the vegetation cover (Imeson 1995). It is therefore, possible to categorize the soils of drylands according to their sensitivity to erosion based on the occurrence of key criteria mentioned in Table No. 4.11. The depth of a soil also influence, its ability to absorb and store rainfall (Yassoglou 1995), while overland flow is more likely to be generated on steeper slopes. Dryland rainfall events are often noted for their intensity, which may exceed soil infiltration rates, which can itself attain higher velocities than in non-dryland areas.

(b) Processes of Water Erosion

Water erosion results not only in the loss of top-soil and in off-site effects, but also in a reduction of water and nutrient storage capacity. The processes of water erosion that occur on slopes are as follows:

(i) *Splash Erosion:* The impact of raindrops may liberate particles from the soil surface. On slopes, this contributes to the general down-slope movement of particles, or can loosen particles, which are then transported down-slope by flow or wash processes.

(ii) *Sheetwash or Interill Erosion:* It occurs as a continuous film of water, when the ground surface is smooth or as a myriad of small interconnected rivulets on rougher surfaces. Sheetwash is effective in eroding particles loosened by both drop impact and the progressive increase of soil water content that occurs during rainfall event (Lal 1990). Sheetwash normally affects the upper parts of hillslopes and where it operates across an entire slope can lead to the reduction in fertility in upslope locations and a corresponding increase at the slope foot (Scoones 1992). Plant productivity can mirror this pattern. Sheetwash can be difficult to monitor and may, therefore, be underestimated.

(iii) *Rilling:* Rilling results from the concentration of overland flow (Scoging 1989). The depth of water in rills is greater and more turbulent than in sheetwash, giving the potential for larger particles to be entrained. Rills develop into networks that can, over time, extend laterally and up slope. However, they can be removed by ploughing and need not be an obstacle to agriculture, though they will reappear unless remedial action is taken to deal with their cause.

(iv) *Gullying*: Gullying can result from the widening and deepening of rills or by a change in surface conditions on a slpoe leading to a sudden increase in runoff (Oostwoud Wijdens and Gerits 1994). A Gully can be defined as having a steep head and sides, wider than 0.3 m and deeper than 0.6 m. Development can be rapid and not only do gulliesmast as effective conduits for the removeal of soil from fields, but they obstruct movement and inhibit the use of mechanized farming methods.

(v) *Piping:* It is erosion through the development of subsurface tunnels. This can occur naturally, particularly in dispersive soils detailed in Table No. 4.23 or those subject to marked action by burrowing animals (Yair and Rutin 1981),but it is enhanced by a reduction in surface vegetation and a loss of internal binding by roots.

Table 4.11 : Five categories of soils sensitive or resistant to water erosion.

Modified in imeson (1995)

Characteristics	Susceptible to (+); resistant to (-)
1. Soil Sensitive to loss of organic matter because of low clay content.	+ Sheetwash erosion
2. Soils with moderately high clay contents that shrink and swell.	+ Rill and gully erosion - Sheetwash erosion
3. Sandy soil with high infiltration rates'; sensitive to crusting	If crusted + sheetwash erosion (Also + wind erosion)
4. Soils with high content of water soluble salts which enhance dispersive potential	+ Gully erosion + Piping
5. Soils with high stone and rock fragment content	- All erosion types through enhanced infiltration or + Erosion through funneling of surface water.

(c) Human Activities and Water Erosion

Water erosion is readily accelerated to many human actions in dryland ecosystems, but notably through the intensification and expansion of agriculture and a range of activities, including fuel wood collection and high intensity pastoralism, that can reduce the vegetation cover. The move towards greater farming intensities can lead to the abandonment and even removal of traditional soil conservation methods such as terracing to facilitate mechanized cultivation (Thomas and Middleton 1994). On the one hand On the one hand, mechanization can lead to surface compaction that enhances runoff, but on the other mechanized deep ploughing can loosen soils and enhance their susceptibility to runoff erosion (Dresh 1986).

The pressures that cause the intensification or expansion of agriculture commonly accrue from the need to feed growing population. However, more people working on the land can also reduce erosion risks by water and enhance productivity, if particular attention is paid to soil conservation techniques. Such techniques include the construction or restoration of terracing on the steep slopes, or the construction of bunds that trap water and sediment that moves downslope. Bunds can also be part of runoff-trapping water harvesting schemes that irrigate crops as wess as reduce erosion (Rapp and Hasteen-Dehlin 1990, Atampugre 1993), and may therefore serve a dual benefit in susceptible drylands.

(d) Dryland Water Erosion Assessment

GLASOD data given in Table No. 4.13 indicate that human-induced water erosion effects about 140 million hectares in dry subhumid areas, 213 million hectares in semiarid areas and 113 million hectares in arid areas. This translates to 9% of the global susceptible drylands. It even effects 11 million hectares in the hyperarid zone. Translated to percentages of the total land area by aridity zones, this indicates that 11% of the subhumid zone, 9% of the semiarid zone and 7% of the arid zone are subject to water erosion degradation, with an additional 1% of the hyperarid areas affected. Water erosion affects 48% of the total degraded land area in susceptible drylands. It is thus a substantial problem, though one that varies significantly both in terms of distribution (Table No. 4.12) and degree of severity.

Table 4.12 : Degree of water erosion in susceptible dryland areas

(Million hectares)

Degree	Africa	Asia	Australasia	Europe	North America	South America	Total
1. Light	28.5	49.6	67.5	6.4	10.3	12.8	175.1
2. Moderate	36.6	91.2	2.1	38.0	23.9	16.7	208.5
3. Strong	51.5	16.7	0.0	1.4	4.2	5.2	79.0
4. Extreme	2.5	0.0	0.0	2.3	0.0	0.0	4.8
Total	**119.1**	**157.5**	**69.6**	**48.1**	**38.4**	**34.7**	**467.4**

Source: GLASOD

Table 4.13 : Distribution of susceptible dryland water erosion by climate zone in each continent

(Million Hectares)

Region	Dry Subhumid	Semiarid	Arid	Total
1. Africa	25.1	59.2	34.8	119.1
2. Asia	54.9	69.9	32.7	157.5
3. Australasia	4.1	26.3	39.3	69.7
4. Europe	34.7	12.8	0.6	48.1
5. North America	10.7	24.4	3.3	38.4
6. South America	11.5	20.6	2.5	34.6
Total	**141.0**	**213.2**	**113.2**	**467.4**

Source: GLASOD

Only European and African susceptible drylands contain areas that have experienced extreme water erosion Figure 4.21. affecting more than 2 million hectares in each case. In Europe, water erosion is a long-standing problem in the dry subhumid and semiarid Mediterranean lands (Grove 1996), with Yassoglou (1987) noting that about 40% of the land area of Spain and Greece has suffered serious reductions in productivity as a result. Murcia, in south-east Spain, has experienced particularly severe water erosion problems (Albaladejo 1990) with one or two erosion events occurring in most years (Lopez Bermudez and

Albaladejo 1999). Erosion severity is very high under both traditional vine and almond stands and more intensive land use.

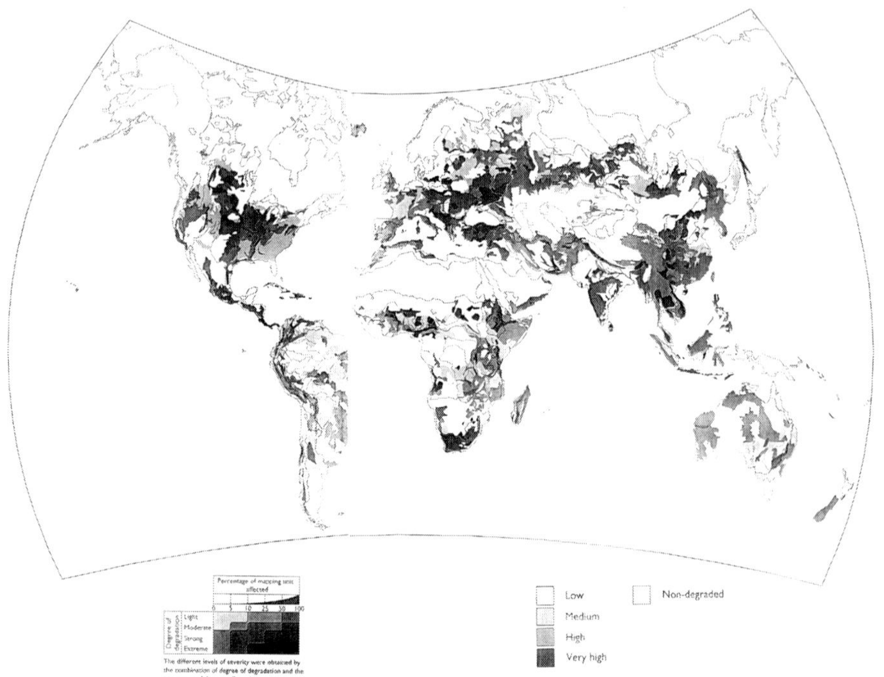

Approximate equatorial scale 1:104 million

Source: UNEP/ISRIC

Figure 4.21 : *Water Erosion Severity*

When overall severity is considered, in addition to parts of southern Europe, water erosion is a major degradation problem in parts of Africa's Sahel belt, Yatenga Province in Burkina Faso, where gully erosion has been especially severe (Atampugre 1993). In northern Ethiopia up to one billion tones of top soil are lost each year from steep slopes (Brown and Wolf 1986). This problem also occurs in some of the more densely populated dry subhumid areas of Kenya. In the GLASOD report of Hakkeling (1989), the recent restoration of terracing is reported to be leading to reduced water erosion problems in some Kenyan areas possessing steep slopes. This is a good example of positive actions that can be taken by local communities to lessen desertification problems.

In hilly northern Somalia, conservation measures started in the 1970s have been hampered or have fallen into disrepair due to the civil war in the 1990s. However, the region is sparsely populated and it is only around settlements that erosion problems attain moderate proportions (Hakkeling 1989). In dryland Africa, south of the equator, major water erosion problem areas occur notably in intensively settled former homelands in South Africa (Dardis *et. al.* 1988), but also in the more sparsely settled Karoo, where steepwash is significant (Rowntree 1988).

In some cases, erosion can be attributed to the intensification and inappropriate use of agricultural practices on steep slopes, as in the Loess Plateau area of China. Light to moderate degree water erosion leading to the loss of topsoil affects over 50% of the dry subhumid areas to the west and north of Beijing, according to GLASOD reports. While the clearance of natural vegetation for agriculture is the prime reason for this, the recommendation of terracing assists in the amelioration of the problem. In other cases, severity can be related to specific changes in human activities. In Yemen high water erosion, severity is related to the recent breakdown of traditional soil management terraces and their associated cultivation systems, to significant erosion damage.

B. WIND EROSION

(a) Nature of Wind Erosion

When and where wind erosion occurs, depends upon the mutual interaction between erosivity and erodibility factors (Bullard *et. al.* 1997). Before wind erosion occurs, it is necessary for wind speed to exceed a threshold velocity, related to the particle size and coherence characteristics of the soil (Gillette *et. al.* 1980), with atmospheric turbulence increasingly recognized as important in the process of Aeolian entrainment (Wiggs 1997). Key aspects of the soil surface with regard to wind erosion include surface roughness, the presence of stones and boulders and perhaps most importantly, the percentage cover, height, density, structure and orientation of vegetation (Ash and Wasson 1983). In some marginal drylands, for example in Rajasthan, India, ancient dune sands laid down under even drier climatic conditions in the past, are particularly susceptible to erosion, when they are utilized for agriculture (Kar and Joshi 1992).

(b) Processes of Wind Erosion

Wind erosion of soils has many environmental impacts, perhaps most importantly for the farmer, but also for a range of other human activities. The problems caused by wind erosion occur both on-site and off-site and can be looked at conveniently according to the three fundamental phases of aeolian activity; deflation, transport and deposition.

Soil deflation, which preferentially removes sand and silt-sized particles, occurs through three main processes. The finest, silt-sized, particles may be removed from the soil surface in suspension, which may transport silt grains high into the atmosphere leading to long distance transport. Fine and medium-sized sand-grains are commonly moved by saltation, whereby particles are raised from the ground surface and transported forward in a hopping motion. The coarsest sand grains mainly move by creep, which is a forward rolling motion of grains. In practice, during a sediment-moving event, all three processes often operatc simultaneously.

Since two of the most important influences on soil structure are the ratio of sand, silt and clay and the presence of binding agents such as those produced by decomposing organic matter, removal of silt and organics has adverse effects on structure. These particles also exert important influences upon the soil's ability to retain moisture, hence their removal reduces a soil's moisture-retention capacity (Fryrear 1981). The fine silt and clay particles also have the maximum concentration of nutrients attached to them, so that their removal reduces fertility.

Abrasion by moving soil particles is probably the most serious problem associated with the transport phase of wind erosion. Soil clods may be blasted by bouncing particles, impoverishing soil structure and rendering it more erodible and crops are abraded and in extreme cases cut from their roots (Fryrear *et. al.* 1973). The abrasion of plants enhances evapotranspiration losses and may in fact be a more significant factor in causing falling yields than the loss of the soil itself (Thomas and Middleton 1994). Structures such as walls, telephone poles and fences can also be affected by wind erosion and it may be necessary to undertake protection or stabilization measures. Large volumes of eroded material in dust clouds present visibility problems to road and air transport and can adversely affect radio and satellite communications (Middleton 1997). The inhalation of fine particles can cause respiratory problems for humans and animals and some particles may be disease-carrying pathogens (Ervin and Lee 1994).

Material deposited by wind can bury and kill plants, fill ditches and block roads, railways, runways and pipelines (Watson 1990). Salts transferred by siltation events can increase the salinity of groundwater and be highly destructive of buildings. There may, however, be positive aspects to the deposition of wind-eroded soils. Their high nutrient content provides additional fertility to marine and terrestrial ecosystems, the latter both in terms of soils and as direct nutrient inputs through leaves to certain plant types such as rice, wheat and grasses.

(c) Human Activities and Wind Erosion

Human activities most commonly responsible for inducing wind erosion are those that change or remove vegetation cover and those that destabilize natural soil surfaces. There is a variety o causes for such environmental damage. Vegetation is cleared for agriculture and building and used for fuel and fodder. It is also modified by cropping practices. Stable soil surfaces may be disturbed by ploughing, by the trampling of animals, by off-road vehicle use, construction, mining or military maneuvers. In Kuwait, Saudi Arabia and Iraq, the breaking up of desert surfaces by tank movements during the Gulf war has led to both the activating of underlying sands and an increase of dust storms (El-Baz 1994).

Dry soils with low organic content are most susceptible to wind erosion and if human activities reduce surface vegetation cover in such areas, then the potential for wind erosion is increased. Attempts to conduct agriculture have expanded towards the dry margins of susceptible drylands and such soils are often found close to the limits of the hyperarid deserts, then wind erosion may increase. The effects of wind erosion may be limited, where only small fields are cultivated, but the creation of large cultivatable areas suitable for mechanized techniques creates large potentially deflectable surfaces.

Livestock production is a common and appropriate activity in many dryland areas unsuitable for agriculture, which include extensive areas of infertile sandy soils in northern and southern Africa, Australia and the Middle East. Excessive grazing pressure may lower vegetation covers, increasing susceptibility to deflation. Processes of bush encroachment may, however, occur when this happens and while generally regarded as negative in a productivity sense, may enhance ground cover and reduce the likelihood of wind erosion.

(d) Dryland Wind Erosion Assessment

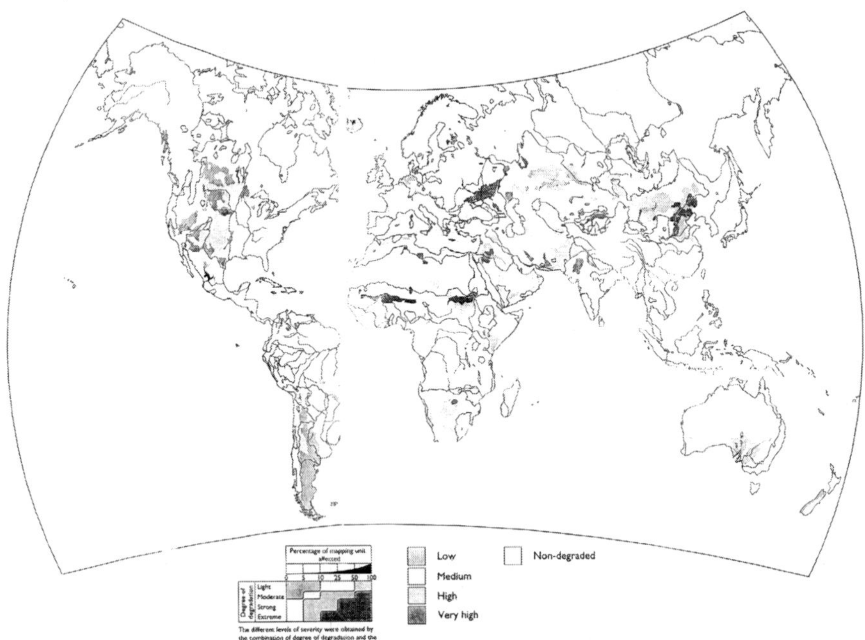

Approximate equatorial scale 1:104 million *Source:* UNEP/ISRIC

***Figure 4.22** : Wind Erosion Severity*

Figure 4.22 clearly indicates that human-induced wind erosion is very largely confined to the world's drylands. Some regions immediately become apparent as suffering from a significant wind erosion problem. These are the Sahel (Middleton 1985, Nickling and Wolfe 1994); the Meghreb; a broad belt of land stretching from the northern shores of the Black Sea to the north of the Caspian Sea, where nutrient depletion by wind is a particular problem (Petrova 1982); Mesopotamia and parts of the Levant; and the Thar Desert of India (Kaushalya 1992). The importance of treating the map as a general overview is appreciated, when looking at the large zone of high wind erosion severity on the Arabian Peninsula. This large area lies in the hyperarid heartland of Saudi Arabia, where human population densities are very low and the problem is confined to highly localized situations.

Overall, wind erosion problems in susceptible drylands increase as conditions get drier (Table 4.13 given earlier). For individual continents, however, where the relative occurrence of different climatic zones varies, this relationship does not always hold. In European drylands, for example, the dominance of dry subhumid and semiarid conditions relative to arid areas means that the total extent of wind erosion-affected lands in the latter climate zone is relatively small. On the global scale, human-induced wind erosion affects 432.4 million hectares or 8% of the total susceptible dryland area (Table 4.14 and Figure 4.23). in terms of the total degraded area of susceptible drylands, wind erosion is an important degradation process affecting 42% of the area.

Table 4.14 : Degree of wind erosion in susceptible dryland areas

(Million Hectares)

Region	Degree of Wind Erosion				
	Light	**Moderate**	**Strong**	**Extreme**	**Total**
Africa	78.1	74.2	6.6	1.0	159.9
Asia	80.5	62.9	9.7	0.1	153.2
Australasia	15.9	0.0	0.1	0.0	16.0
Europe	1.3	36.6	0.0	0.7	38.6
North America	2.6	33.6	1.6	0.0	37.8
South America	18.8	8.1	0.0	0.0	26.9
Total	**197.2**	**215.4**	**18.0**	**1.8**	**432.4**

Source: GLASOD

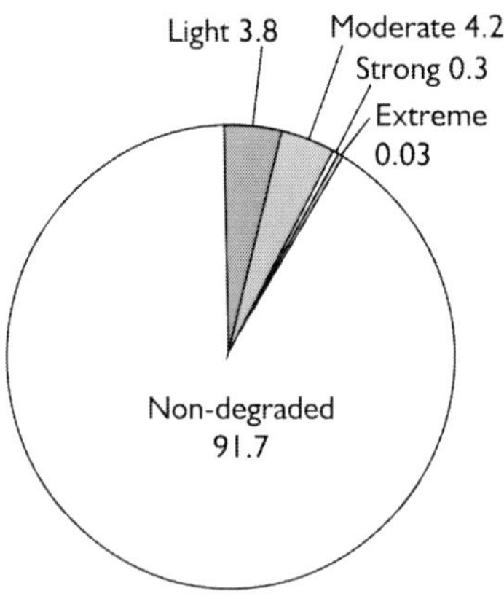

Source: GLASOD

Figure 4.23 : *Global susceptible drylands degraded by wind erosion (%).*

In absolute terms, wind erosion problems in susceptible drylands are most extensive in Africa and Asia, both showing areas in excess of 150 million hectares affected. In the Sahel, wind erosion may well be the dominant geomorphic process (Nickling and Wolfe 1994), and while human activities have undoubtedly contributed to the problem (Mainguet 1985), the occurrence of prolonged drought has exacerbated the potential for agriculture and livestock to have erosional impacts, making it difficult to differentiate human and natural causes (Suliman 1988). In China problems associated with drifting sand encroaching on cultivated land are well documented (Zhu Zhenda and Liu Shu 1983).

The GLASOD reports show that in some areas, for example, the Hetao Plain, which has a mean annual rainfall of only 150-300 mm, it is the over cultivated fine soils of alluvial origin that are most susceptible to wind erosion. There are nonetheless some significant success stories in China of the stabilization and rehabilitation of wind-eroded land (Figure 4.30) (Mitchell and Fullen 1994). In Inner Mongolia, for example, it has been suggested that the area of shifting sand resulting from human activities has been reduced by 90% between 1984 and 1994 (Wang Tao and Zhu Zhenda 1996).

The 16 million hectares affected by human induced wind erosion in Australia are located in three zones, in Western Australia, South Australia and Northern Territory. In South America, the 27 million hectares affected are largely to be found in the Argentinian regions of the Pampas and Patagonia, but problems also exist in other areas. GLASOD reports note the deflation of top soil occurs in the cold semi-desert areas of the Bolivian altiplano. Although, the windiness of the low rainfall area (c.200 mm p.a.) makes it susceptible to natural wind erosion, grazing pressures and the clearance of shrubs for firewood are enhancing the problem above background levels. In semiarid north-east Brazil, the potential for human-induced erosion has been debated (Fearnside 1979, Le Prun 1981). The survey carried out by GLASOD reveals that where clearance of trees and shrubs has taken place to facilitate the production of annual crops, particularly maize, beans and cassava, the occurrence of both wind and water erosion has increased, especially in areas of soils derived from crystalline basement rocks. In the Magdallanes region of southern Chile, intensive sheep grazing has contributed to the expansion of wind erosion within an area covering 2 million hectares.

The ploughing up of grassland for grain cultivation has created massive wind erosion problems in the Great Plains of North America, notably in the 1930s, and in the wheat belt of southern Russia and Kazakhstan in the 1950s and these continued activities are probably largely responsible for the human induced wind erosion in these areas today (Lockeretz 1978, Ervin and Lee 1994). Similar activities go a long way to explain the wind erosion areas of the Canadian Prairies (Wheaton and Chakravarti 1990). The situation in Africa is more complex. Excessive livestock grazing pressures, excessive cultivation of dryland soils and cutting wood for fuel are commonly quoted as causes of land degradation in the Sahel. In Hausaland Nigeria and in Mali, Mainguet (1991) reports situations, where the wholesale removal of topsoil has resulted from wind erosion, resulting from agriculture. Despite the problems caused by wind erosion, Dregne (1992) has noted that at the regional scale, the impact of soil productivity may be relatively limited, because Aeolian erosion is frequently very localized and the off-site costs and impacts, particularly of dust, may far exceed on-site impacts (Piper 1989).

4.6 SOIL DETERIORATION

Dryland soil degradation by processes of chemical and physical deterioration has been shown to occur on a relative minor scale compared to the effects of soil erosion. This said, however, these forms of degradation can hit specific localities, particularly hard. Such localities may also be in hyperarid regions, such as in desert oases and in the Nile Valley. Many of the notable examples in this respect are due to the fact that a number of the wide range of different processes embraced within the chemical and physical deterioration categories can be traced to a single form of human activity; irrigation.

Poorly managed dryland irrigation schemes can result in soil salinization, gleysation, water-logging and sodication. These problems reduce the crop yields from irrigated lands and ultimately can result in their abandonment. They are difficulties, which have been faced since the beginning of irrigated agriculture on the floodplains of Tigris and Euphrates in ancient Mesopotamia more than 6000 years ago (Jacobsen and Adams 1958), but the scale of the problem has increased significantly in recent decades due to large-scale efforts to bring additional area under irrigation.

Other forms of in-situ degradation also have a long history. Trampling by livestock, causing compaction, sealing and crusting, must have been a problem in localized areas such as water-holes for as long as people have herded animals in drylands. As with other types of in-situ degradation, such as the gradual loss of soil nutrients due to shortening of cropland fallows periods, rising human and animal population on desert margins have increased the scale and intensity of these human-induced dryland degradation processes in recent times (Warren *et. al.* 1996).

(A) Processes of Chemical Deterioration

GLASOD methodology recognizes several processes on the chemical deterioration category; nutrient and organic matter loss, salinization, gleysation, acidification and pollution. The study here concentrates on those mechanisms particularly important in drylands; nutrient loss and salinization and the methodologies used to determine the occurrence and degree of degradation by these processes are shown in Box No. 3 and 4 respectively.

BOX 3 : Glasod Methodology for Assessing Degree of Nutrient Depletion

The criteria used to assess the degree of degradàtion due to nutrient depletion were the organic matter contents; the parent material and climatic conditions. Nutrients decline by leaching or due to extraction by. plant roots without adequate replacement was identified by a decline of organic matter, P, and/or cation exchange capacity (Ca, Mg, K). The definitions adopted were as follows (Oldeman 1988):

Slight: Cleared and cultivated grassland or savanna on inherently poor soils in tropical regions. Formerly forested areas cleared and cultivated in tropical regions on soils with relatively rich parent materials.

Moderate: Formerly forested areas cleared and cultivated on soils with moderately rich parent materials, where subsequent annual cropping is not being sustained by adequate fertilization.

Severe: Formerly forested areas cleared and cultivated on soils with inherently poor parent materials, with a low caution exchange capacity, where all above ground biomass is removed during clearing and subsequent crop growth is poor or non-existent and cannot be improved by the addition of N fertilizer alone.

Extreme: Formerly forested areas cleared by removal of all above ground biomass, on soils with inherently poor parent materials, where no crop growth occurs and forest regeneration is not possible.

BOX 4 : Glasod Methodology for Assessing Degree of Salinization

The degree of salinization has been taken as the relative change in soil salinity status over the last 50 years (Oldeman 1988). Salinity can be measured in terms of electrical conductivity (EC) with changeable sodium percentage (ESP) and pH also delimited. Soils have been characterized as follows:

Non-saline	EC < 5 mS/cm.	ESP < 15%, pH < 8.5
Slightly saline	EC 5-8 mS/cm.	ESP < 15%, pH < 8.5
Moderately saline	EC 9-16 mS/cm.	ESP < 15%, pH < 8.5
Severely saline	EC > 16 mS/ cm.	ESP < 15%, pH < 8.5

Human - induced salinization has been identified by changes in the soil salinity status:

Slight: Salinity increase of one class;

Moderate: Salinity increase of two classes;

Severe: Salinity increase of three classes.

(i) Nutrient Depletion

Soil nutrient depletion is a component of chemical degradation relevant to many dryland locations. Nutrients are essential to healthy plant growth, so that their loss from a soil often results in reduced crop yields. Nutrients become depleted when their export from a field exceeds their input and can, therefore, be caused either by an increase of exports or a decrease in inputs. Export processes include extraction by crops, losses due to leaching and erosion and to volatilization and denitrification, while common inputs are applications of fertilizers and manure, the restitution of crop residues, nitrogen fixation, atmospheric deposition of rain and dust and enrichment by weathering of soil minerals. The depletion of nutrients is often intimately linked to a decline in soil organic matter. Soil organic matter is important not only because it is prime source of the nutrients needed by plants, but also because of its positive effect on the water retention capacity of a soil, soil structure, biological activity and cation exchange capacity (Allison 1973).

In practice, a number of human actions tend to promote loss of nutrients. Perhaps the most commonly quoted are over-cultivation and insufficient application of replacement nutrients. Shortening or even the total abandonment of fallowing reduces of opportunity for land to recover from producing a crop, as Khogali (1991) outlines for part the Kordofan region of Sudan. Crops that are particularly demanding of nutrients can lead to severe nutrient depletion, often resulting in acidification of the soil. In Senegal, soils of the groundnut fields immediately east of Dakar have been seriously acidified with nearly 3,00,000 hectares affected in 1983 according to Paye (1990).

Nutrient depletion can also occur through the clearance of natural vegetation in savanna areas of poor soils, where much of the nutrient store is held in the savanna vegetation rather than the soil (Proctor 1989). It can also result, where the replenishment of soil fertility by annual flooding is prevented by dam construction and the control of river flood regimes.

Nutrient depletion can also be related to soil erosion, where increased runoff removes nutrients in solution or enhanced deflation removes them in suspension. Conversely, nutrient depletion caused by a reduction in organic matter content may eventually result in enhanced erosion; less organic matter tends to decrease the stability of soil aggregates, rendering them more prone to destruction and entrainment.

(ii) Salinization

High evaporation rates in drylands contribute to the surface accumulation of salts that would be leached out of the system in wetter environments. Salinization, therefore, refers to the surface or near-surface accumulation of salts, mainly chlorides, sulphates and carbonates of sodium, calcium and magnesium. Salt accumulation reduces soil pore space and the ability to hold soil, air and nutrients. High salt concentrations are toxic to many plants especially during the seedling stage. Plants can suffer from salt burn and salt stress, causing inability to take up the moisture necessary for growth. Salinization problem also embraces alkalization, the excess accumulation of sodium, which rises in association with the dissolved salt load of irrigation water (Rhoades 1990).

Though there are extensive areas of natural salt-affected soils, salinization is a particular problem when human activities cause soil salinity to rise significantly; so-called secondary salinization. Drylands can be susceptible to this problem for several reasons; the predisposing climate; the need to irrigate soils to generate or increase crop yields; and the occurrence of suitable terrain for irrigation in flat, low-lying situations, the localities with a natural predilection to salinity. The areas of irrigated land worldwide has increased from around 8 million hectares in 1980 to about 250 million in the early 1990s (FAO 199-b), much of it in drylands, creating an enormous potential for the problem.

Human-induced salinization and alkalization affect nearly 50% of all the irrigated land in arid and semiarid regions (Abrol *et. al.* 1998). these occur in one or more of the following situations according to Szabolcs (1976):

1. Accumulation of salts from poor quality water.
2. Increase in the level of groundwater:
   (a) the salt content of the groundwater accumulates in deeper soil layers;
   (b) the rising groundwater transports salts from the deeper soil layers to the surface or surface layers;
   (c) the rising water table limits natural drainage hinders the leaching of salts.
3. Lack or low effectiveness of drainage systems in irrigated soils.

It is a paradox of water-deficient drylands that one of the major factors contributing to irrigation-induced salinization is waterlogging due to poor drainage. If water is applied in amounts exceeding those used by plants, the local water-table may be raised. This in turn mobilizes stored soil salts, which then accumulate in the root zone or at the surface through capillary rise or 'evaporative pumping.' Even without waterlogging, evaporation will concentrate the salts carried in irrigation water and left in the soil after water use by plants. In many places, this is enhanced, where groundwater is used for irrigation as it is often more saline than water from surface sources.

The effects of salinity on crop yields depend upon crop type, since different crops are susceptible to different concentrations and forms of salinity (Maas 1986). Crops like barley, cotton and sugar beet are fairly salt-tolerant; cowpea. rice and alfalfa much less so.

Salinization can also occur without irrigation. If natural vegetation is replaced by a crop that uses less water, root zone waterlogging and subsequent evaporative pumping can ensue, a process known as saline seep. The problem has been reported from US Great plains (Berg *et. al.* 1991, Halvorson and Black 1974) and southern Australia (Bettenay 1986).

(iii) Chemical Deterioration in Drylands

The GLASOD study indicates that in the susceptible drylands, 23 million hectares in the dry subhumid aridity zone are affected by chemical deterioration, 41 million hectares in semiarid areas and 37 millions hectares in arid areas, with an additional 11 million hectares in the hyperaridity zone. In percentage terms, 22% of semiarid areas, 2% of dry subhumid and arid areas and 1% in the hyperarid zone are affected. The map shows all land units in which human-induced chemical deterioration occurs. It does not necessarily mean that chemical deterioration is the only form of degradation occurring in a particular unit, nor that chemical deterioration is the dominant degradation process in a unit. In total, some 10% of the soil degradation experienced in the susceptible drylands can be accounted for by chemical processes, of which undoubtedly, the most dominant are nutrient depletion and salinization.

Figure 4.24 indicates that very high severity chemical deterioration is the drylands occurs as pockets in Senegal and the Sudan of Africa. In Iraq and Syria in the Middle East and in the Middle Asian States. Aral Sea Basin and the Euphrates Valley are particularly affected. In Syria, where it has been estimated that about 45 % of the irrigated area is affected to various degrees of soil salinization (Ilaiwi *et. al.* 1992), irrigation on the Euphrates dates back more than 6000 years. However, salinity has only become a significant problem since the late 1940s, when the introduction of diesel pumps allowed a significant expansion of irrigated lands.

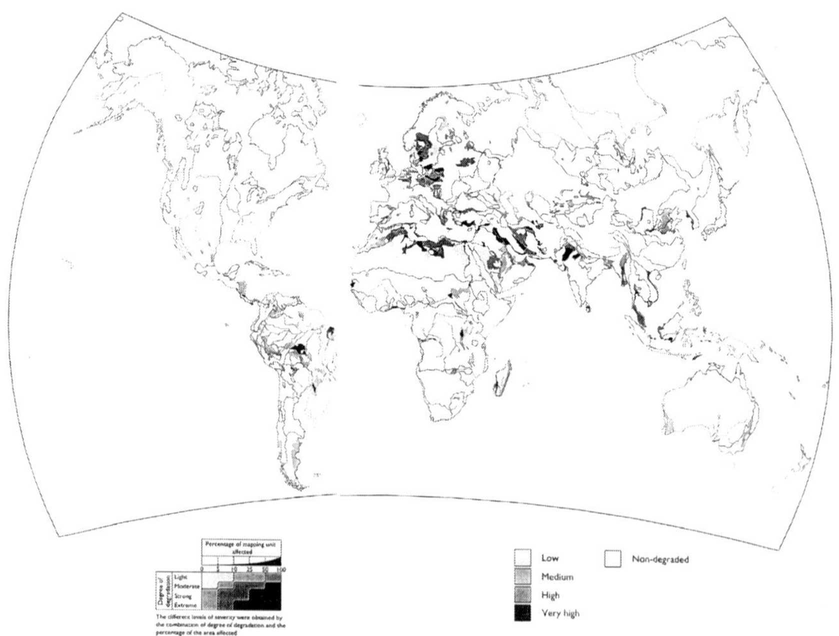

Approximate equatorial scale 1:104 million

Source: UNEP/ISRIC

***Figure 4.24** : Chemical Deterioration Severity*

Misuse of irrigation water and a compete absence of drainage resulted in large areas being abandoned in the 1960s due to excessive salt accumulations. Downstream in Iraq, over 60% of the land irrigated under the 1953 Mussayeb Project was affected by salinization by 1970, and subsequent reclamation attempts have also produced mixed results. (Iraq 1980). Salinization has been a persistent problem in Uzbekistan since irrigation commenced in 1902. The problem is particularly severe on the Golodnaya Steppe, near Tashkent, where over 80% of the irrigated area is now saline, while the irrigation schemes at delta of the Amu Darya, where it enters the Aral Sea, are similarly affected (Smith 1992).

Chemical deterioration reaches high severity status in several areas. Irrigation supports about 78% of Pakistan's cropland, but the Indus Basin is now badly affected by salinization (Ghassemi *et. al.* 1995). The Nile Valley in Egypt experiences persistent chemical degradation but the problem is not just caused by salinization. The contribution of the Aswan High Dam, which has no provision for the passing of sediment, has deprived locations downstream of nutrient replenishment from river silt (Abu-Zeid 1989). Salinization is an increasing problem in numerous oases throughout North Africa as motorized pumps have replaced traditional methods of water use. Elsewhere, in Africa's cultivated drylands, as fallow periods are reduced, nutrient depletion is a more serious threat to soils (Thomas and Middleton 1993).

Chemical deterioration is not restricted to developing countries. The lower Colorado River, south-western USA, supports several irrigation schemes. The Wellton-Mohawk project in Arizona, completed in 1952, has suffered particularly from waterlogging and high salinity and though a drainage scheme has been implemented, the return of saline wastewater to the Colorado generated its own set of problems for 5.00.000 people in Mexico, who also rely on the river's waters. Agreement on a solution to salinity continues to be a serious issue (Holburt 1982).

The severity of soil degradation through chemical deterioration also reaches a medium degree in southern areas of Western Australia, where the replacement of native eucalyptus forest by shallow-rooted grasses and crops has lead to an increase in groundwater recharge rates and an increase in salinity, a process that has reversed in some parts by a range of land management techniques including reforestation (Bari and Schofield 1992). In the east, the Murray River catchment has also undergone clearance of natural perennial vegetation and replacement with cereal drops and grazing land. As in Western Australia, these land uses do not use all available moisture, leading to surface waterlogging and salt accumulation, as well as to substantial changes in water flows to the lower reaches of the Murray, resulting in large scale dieback of eucalyptus trees (Grieve 1987, Taylor *et. al.* 1996).

(B) Processes of Physical Deterioration

Six forms of in-situ soil degradation by physical processes are recognized in the GLASOD methodology. These are the sealing and crusting of topsoil, the compaction of topsoil, deterioration of soil structure due to dispersion of soil material by salts in the subsoil (sodication), waterlogging, aridification and subsidence of organic soils as depicted on Figure 4.25.

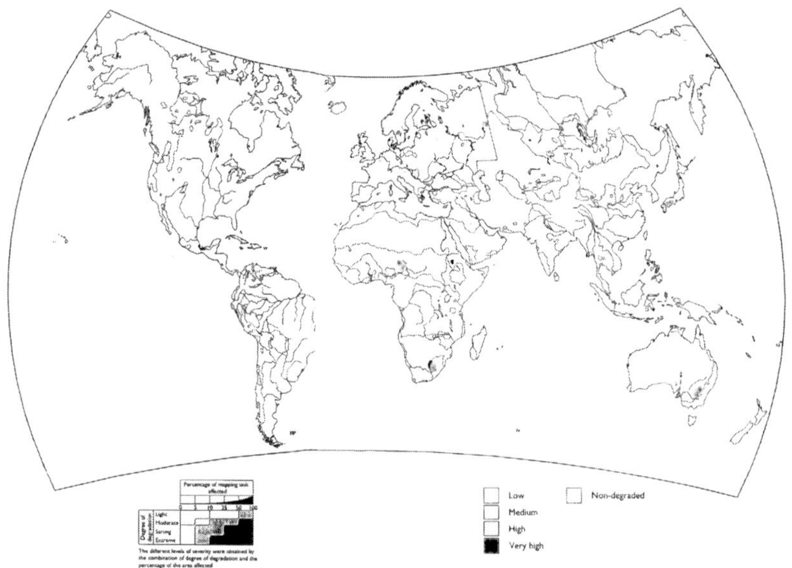

Approximate equatorial scale 1:104 million *Source:* UNEP/ISRIC

Figure 4.25 : *Physical Deterioration Severity*

(i) Compaction, Sealing and Crusting

The compaction, sealing and crusting of soil surface occurs under virtually all climatic and physical conditions. The most common causes of compaction are the use of heavy machinery and trampling by livestock on soils with a low structural stability. Livestock trampling can also lead to crusting, although the most common cause of crusting and sealing is due to the clogging of soil pores by fine-grained silt and clay particles dispersed by raindrop impact (Farres 1978). While the term sealing is commonly used to refer to reorganization of the surface soil layer during a rainstorm, crusting is the hardening of this surface seal as the soil dries out.

These effects commonly occur in areas where vegetation does not adequately protect the soil surface from the impact of raindrops. The sparse vegetation cover may of course be a natural occurrence, but equally it could result from human clearance, in which case it is included in the GLASOD classification. Soils with a low humus content, poorly sorted sand fractions and appreciable silt contents are particularly vulnerable. Crusting may also be a product of salinization.

The effects of sealing, compaction and crusting are numerous. They hinder the tillage of arable soils and impede or delay the emergence of seedlings and the penetration of roots. The diversity and abundance of soil biota may be adversely affected and nutrient cycles can be altered by the lowered nitrogen and carbon inputs and the slowed decomposition of soil organic matter, resulting in a decrease of nutrient levels in associated vascular plants (Belnap 1995).

Soil water infiltration capacity is also diminished, affecting soil moisture properties and causing increased surface runoff and often higher erosion. Field measurements in Israel, for example, show that on sandy soil, crusting can reduce the infiltration capacity from 100 mm h-1 to 8 mm h-1 and on a loess soil from 45 mm h-1 to 5 mm h-1 (Mortin *et. al.* 1981) (Here H raised to power -1 to be depicted). Although the overall effects of rock fragments are complex and scale-dependent (Poesen and Bunte 1996), it is interesting to note that clearing of surface gravel deposits to expose finer soil beneath to rain drop impact, consequent sealing and crusting and enhanced runoff, has been deliberately employed as a dryland farming technique since ancient times.

The increased runoff is channeled to cropland. This practice has been carried out to great effect in parts of the Middle East for millennia ; Lavee *et. al.* (1997) have shown that selective removal of stones from surfaces in the Negev can enhance runoff by 250% during small rainfall events. Today a greater range of compaction techniques using heavy machinery and water repellent materials is available for the same agricultural end, as well as for livestock, industrial and urban reservoirs. This human modification of land surface is often referred to as 'water-harvesting'.

(ii) Sodication

Sodication is a physical consequence of salinization, largely dealt with in GLASOD as a chemical degradation process. Sodication occurs when saline water in a soil, often from irrigation in drylands, is concentrated by evapotranspiration, leaving sodium ions dominant in the soil solution because calcium and magnesium components tend to precipitate as carbonates. The sodium ions tend to be absorbed by aggregates of very fine clay particles or 'colloids', which consequently become broken down or 'deflocculated'. The result is a structureless soil, which is unfavourable to root development, almost impermeable to water and highly erodible, particularly by piping (Rosewell 1970).

(iii) Waterlogging

Waterlogging includes flooding, by river water and submergence by rainwater, when caused by human intervention in natural drainage systems. However, it usually results from poorly managed irrigation systems, where water is applied in excess of the needs of the crops and the soil infiltration rate (Table 4.15). Waterlogging leads to the severe loss of soil air content. In other conditions, plants are stressed due to a shortage of oxygen for metabolism by the roots and the micro-organisms responsible for biodegradation of organic material are inhibited or killed. Organic matter accumulates as the mineralization and humification of plant debris is retarded, freeing acids that can release toxic substances with consequent effects on plant growth (Crawford 1989). Waterlogging also causes problems of sodification salinization.

Table 4.15 : Soil infiltration rates and their links with irrigation, waterlogging and crusting
(Including information from Gairon and Hadas 1973 and Morin *et. al.* 1981)

Infiltration Class	Infiltration rate mm h^{-1}	Suitable irrigation type	Impact on water logging and crusting
Very Low	<5	Flood irrigation	Excessive irrigation commonly causes ponding and waterlogging that may in turn lead to salinization.
Low	5-15	Low-intensity sprinklers possible	Excessive irrigation may cause ponding and waterlogging, leading to salinization. Crusting may result on soils with naturally low infiltration rates if irrigation too intensive. Low infiltration may, however, be an outcome of crusting or compaction caused by agricultural activities.
Medium	15-25	Flood irrigation only on small ponds	Careful irrigation regimes may optimize agricultural potentials and minimize degradation.
High	25-50	Sprinklers possible, flood irrigation impossible	Rapid infiltration can lead to waterlogging if impermeable sub-surface horizons present.
Very High	>50	All irrigation difficult	Compaction by livestock, mechanized agriculture, can reduce infiltration rate on sandy soils from 100 to 8 mm h-1 and on silty soils from 45 to 5 mm h-1.

Source: GLASOD and Gairon and Hadas 1973 and Morin *et. al.* 1981

(iv) Aridification

Aridification of soils is the human-induced change of the soil moisture regime towards a more water-deficient soil system. It may be caused by the lowering of the local ground-water levels, other than by deep groundwater extraction. Such depletion may occur in areas where river or lake water is used for domestic or agricultural purposes. Aridification may also occur in areas where natural vegetation is replaced by a crop that needs greater moisture for successful growth. The replacement of dryland grass with wheat is an example here. Successful mitigation of aridification problems in north-western China.

(v) Subsidence of Organic Soils

The subsidence of organic soils, due to excessive drainage and/ or oxidation, is only included in areas, where the agricultural potential of the land is adversely affected. This occurs when peaty materials become susceptible to oxidation after drainage has lowered the water table, leaving the peat susceptible to oxidation and deflation, hence lowering the land in a similar manner to the way in which clay soils shrink when desiccated (Doornkamp 1993). Peat is often highly productive in the early stages of post drainage, so that its loss affects fertility. Conversely, in many cases, drainage may lead to subsidence but also an increase in agricultural potential of land, in which case it is not shown on the map.

(vi) Dryland Physical Deterioration

In terms of the area of the global susceptible drylands affected, physical degradation is the least important of the four degradation types identified in the GLASOD methodology. Physical degradation has affected 335 million hectares or less than 1% of the global susceptible dryland area and is mostly confined to croplands. Most of the physical deterioration occurs in the dry subhumid (13 million hectares or 1% of the area) and semiarid zones (15 million hectares or 0.7%). Arid areas experience relatively little degradation by physical processes (7 million hectares or 0.4%), but most of this degradation occurs in grazing areas.

In Africa, most of the 14 million hectares area affected are degraded to a moderate or strong degree. Soil crusting and compaction are the most widespread types of degradation, affecting the wheat and fruit growing belts in the Western Cape and along the Orange River in South Africa and in the central parts of Sahelian latitudes around Lake Chad and in the western parts of the Sahel, largely in western Mali and on the Senegal/ Mauritania border (Hoogmoed 1986).

Asian drylands affected by physical degradation are almost exclusively degraded to a light or moderate degree. An area of high severity physical degradation due to compaction and crusting occurs in Lower Mesopotamia, while medium and low sererities are found in Afghanistan parts of Middle Asia, areas of Siberia west of Lake Baikal and on the floodplains of the lower Volga River.

The 9 million hectares of susceptible drylands affected in Europe are all degraded to a light or moderate degree primarily by compaction and crusting. These areas are located in southern parts of Portugal, Spain and to the west of the Black Sea. In the Lower Alentejo region of Portugal, a long history of wheat farming has contributed to the region's extensive lithosols (soil with hard rock occurring at less than 10 cm depth). Enhanced sealing and crusting is just one of numerous and often contradictory, effects that rock fragments can exert on soil (Poesen and Bunte 1996), and the extensive use of mechanized farming in this part of Portugal has also contributed to the degradation of structure (Roxo *et. al.* 1996). Moderate degrees of compaction and crusting affects soils throughout the irrigated croplands that cover about 50% of the dry subhumid zone of southern Romania. The GLASOD survey also reported that associated problems of water erosion were widespread in this area.

Although, the area of susceptible dryland affected in North America is small (1 million hectares), these regions are degraded to a very high severity by waterlogging. They are located in the dryland regions of Mexico, in Sinaloa State on the Pacific coast and in Tamaulipsa State of the Gulf of Mexico.

Low severity physical degradation characterizes the affected areas in Bolivia in South America. GLASOD survey reports indicated that parts of these high-altitude dry alpine grasslands, which are extensively grazed, suffer from subsidence of organic soils, as well as compaction and crusting and allied loss of topsoil by wind erosion. High severity compaction and crusting associated with agricultural activity occurs in western Venezuela.

Oila compaction and crusting is widespread in the prime wheat belts north of Murray River in New South Wales Australia. Many years of mechanical rillage on the soils in this area has also created impervious 'plow-pans' just below the surface of some soils. These plow-pans are impenetrable to both crop roots and moisture and crops are increasingly vulnerable to drought stress as a result (Harte 1984).

4.7 CAUSES OF SOIL DEGRADATION /LAND DEGRADATION

'Human-induced soil degradation' implies by definition a social problem. Many of the types of soil degradation outlined occur with or without human interference, but their inclusion in this atlas is dependent upon circumstances, in which natural processes interact with human activities and in most cases, this means they have been initiated or accelerated by human mismanagement in a particular area. More fundamental social and economic forces can also be identified as compelling or encouraging people to cause degradation, while in some cases degradation, itself may combine with such forces to cause social changes such as migration. Here, the focus is on the actual land uses. This focus is crucial because identification of the nature of the land use that has led to soil degradation allows appropriate ameliorative management responses to be planned. Such responses must aim to make land use sustainable. This may mean either a complete change of land use or a modification of the existing one.

(a) Nature and Effects of Soil/ Land Degradation Causes

Investigators deputed by GLASOD had identified one or two of the five soil degradation causes for each land unit:

(1) Deforestation and removal of the natural vegetation;

(2) Overgrazing;

(3) Agricultural activities;

(4) Over-exploitation of vegetation for domestic use; and

(5) Bio-industrial activities.

Highlighting more than one such cause in important at this scale of analysis, since single causes are unlikely to be the only ones operating in a particular mapping polygon, as other large-scale studies of vegetation change have found (Ringrose and Matheson 1992). In addition, some causes are often the precursors to others, such as the removal of natural vegetation to make way for cropland and attributing particular types of degradation to one or the other may not be straightforward.

Deforestation and removal of the natural vegetation cover is defined as the total removal of natural vegetation from stretches of land. In most cases, this natural vegetation is forest or woodland and the degradation that often results is a function of the ability of trees to protect soils from erosion. Their root systems and the organic matter they supply help to stabilize soil, while water uptake through the roots and interception via the canopy help to reduce the frequency and intensity of raindrop impact and local runoff. Vegetation is cleared to make way for a new land use, which. may still depend upon a vegetation cover in the case of agricultural uses like cropping or cattle raising or new commercial forestry plantation. Conversely, some clearance is undertaken to remove vegetation permanently, such as for road construction, urban and industrial development.

There is a wide range of consequences stemming from such action. In areas, where the vegetation removal is effectively permanent, erosion can be very high during the period between vegetation clearance and construction, but after construction the new land use effectively terminates the resource potential of the soil. Nonetheless, the new land use may affect surrounding soils. Impermeable land surfaces of urban areas, create greater runoff than from natural soils, and this runoff, channeled through storm drains, can be highly erosive, when it reaches uncovered soils or natural water courses (Cooke *et. al.* 1982)

The methods used for land clearance affect both the amount of soil loss and the subsequent yields of any crops planted. Mechanical clearance can significantly increase the soil bulk density and infiltration rates, when compared to slash-and-burn, as Lal and Cummings (1979) show for an alfisol in Nigeria. The bulk density of 0.9 Mg. M-l on uncleared land was increased to 1.12 mg: M-l after slash-and-burn clearance and 1.25 mg. M-l after mechanical methods were used. The response of crop yields to different land clearance methods varies considerably, however, depending on the degree of soil disturbance and the subsequent cultivation method. Another advantage of the use of the fire during clearance is that it releases phosphorus and other nutrients for use by crops, but studies on the effects of burning on semiarid gorse scrubland in south-eastern Spain indicate that such advantages are short-lived, if a new vegetation cover does not become established before an erosive event (Curreira and Neill 1995). When the first significant rains fell on the study plots, the loss of nutrients by water erosion was on order of magnitude higher in the burned areas compared to undisturbed control plots.

Once a new cover is established, however, different forms of degradation may occur as the soil establishes a new equilibrium with its vegetative overlay.

Changes in vegetation quality and coverage may mean increased erosion. Different plants require different types and amount of nutrients and consequently their decomposing leaf litter similarly return different types and quantities of materials back to the soil. Any resulting degradation on cropland may then be classified under agricultural activities.

Overgrazing includes both the actual removal of biomass by grazing animals and other effects of livestock like trampling and consequent compaction. Heavy grazing and other forms of degradation due to livestock occur in areas either where the wrong sort of animals are being used. A common consequence of heavy grazing pressure is a decrease in the vegetative cover, leading to increased erosion by water or wind. These effects may be concentrated in certain zones like around water holes (Hanan *et. al.* 1991), or they may extend over large areas (Takeuchi *et. al.* 1995).

Another widespread effect of intensive grazing is the encroachment of unpalatable or noxious shrubs into grazing lands, but such encroachments certainly influence grazing potential it is not identified in GLASOD as degradation unless the soil itself is affected. Typical effects of long-term grazing of semi-arid grasslands in this respect are an increase in the spatial and temporal heterogeneity of water, nitrogen and other soil resources, which promotes the invasion by desert shrubs, in turn leading to a further localization of soil resources under shrub canopies in a process of positive feedback. In the barren area between shrubs, soil fertility is decreased by erosion and gaseous emissions (Schlesinger *et. al.* 1990). Increased runoff and erosion result in stripping of the soil. A horizon, the formation of desert pavement in intershrub areas and the development of rills (Abrahams *et. al.* 1995).

It is also recognized at this point that some notes of cautions concerning the deleterious effects of rangeland overgrazing have been sounded recently. While halos of bare, compacted soil are undoubtedly characteristic of many rangeland watering holes, the degradation represented a lack of vegetation and trampled soils is offset to some extent by increased soil nutrients from livestock faces and urine (Barker *et. al.* 1990. Perkins and Thomas 1993-a and b). Ideas on the effects of grazing on rangeland vegetation are also currently being substantially revised in a way that amounts to a paradigm shift (Smith 1988, Ellis and Swift 1988). The 'old paradigm', based on Clements' (1916) model of vegetation successions and ecological stability and the consequent concept of a fixed carrying capacity, has been superseded in the eyes of many rangeland ecologists. Current thinking depicts semiarid ecosystems as seldom, if ever, reaching equilibrium. Rather, they are in a state of more-or-less constant flux, driven but disturbances such as drought, fire and insect attack. The implications for pastoral management are now being evaluated (Scoones and Graham 1994).

Agricultural activities that lead to degradation encompass all aspects of inappropriate agricultural management, including the introduction of techniques without adequate evaluation of their likely effects. It includes a wide variety of mismanagement types, which result in various forms of degradation. Among these are the improper uses of irrigation water and poor drainage leading to problems like salinization (Rhoades 1990), the absence of anti-erosion measures (Morgan 1995), the shortening of fallow periods in shift cultivation resulting in soil exhaustion, the insufficient or excessive use of fertilizers and the improper use of heavy machinery.

The ways in which routine agricultural soil management practices affect soil degradation are many and varied. The net effect on wind erosion of surface ridges produced by tillage on a field, depends on the height and lateral frequency of the ridges, their shape, orientation to erosive winds and their proportion of erodible to non-erodible grains. Likewise, the crop itself influences the amount of soil lost to wind erosion by virtue of its quantity i.e., proportion of ground surface covered and quantity i.e., height, density and flexibility. These properties vary with vegetation type and for a given type according to the season (Middleton 1990). The extension of cropland into desert margins has been a trend noted in many parts of the world in recent decades and many of these newly cultivated lands have suffered from widespread degradation as well as inherent threats of drought (Glantz 1994).

Over-exploitation of vegetation for domestic use encompasses the use of vegetation for such purpose as fuel wood, fencing and construction. In contrast to the above mentioned deforestation causes over-exploitation for domestic purposes does not usually lead to complete removal of vegetation. The use of vegetation is, however, at a degree that is beyond the natural capability of vegetation to renew itself and thus results in a degraded vegetation cover. This exposes soils to increased erosion and in the longer term, will rob the soil of inputs of nutrients and organic matter from decomposing leaf litter in a similar way to those described for deforestation above.

Bio-industrial activities are those that lead to all forms of pollution outlined in the chemical deterioration section. These include the accumulation of urban and industrial wastes, the excessive use of pesticides, oil spills and acidification by airborne pollutants. Such causes are associated with intensive agricultural, industrial and urban land use and in practice this cause shows very little impact on the world's drylands, according to the GLASOD survey.

(b) Causes of Soil Degradation in Drylands

Overgrazing has been long regarded as a scapegoat for many dryland degradation problems, but in recent years, studies by anthropologists (McCabe 1990) and ecologist (Behnke and Scoones 1993) have led to are-evaluation of the effects of grazing pressure on dryland eco-systems. According to GLASOD survey, overgrazing is the most important cause of degradation in the dryland areas of Australia, Africa, Europe and Asia, where it affects 90%, 58%, 42% and 32% of the total degraded dryland area respectively as given in Table No. 4.16 and Figure 4.26. It is a problem in all dryland areas of Australia and Africa. In Australia, the loss of perennial grasses has been widespread from grasslands, savannas and open woodlands, often with replacement by unpalatable shrubs and heavy grazing pressures continue to reduce ground cover, especially during droughts, laying landscapes bare to wind and water erosion (Ludwig and Tongway 1995).

Approximate equatorial scale 1:317 million

Source: UNEP/ISRIC

Figure 4.26 : *Areas affected by overgrazing*

Africa is the continent most commonly associated with dryland degradation by intensive grazing and examples of the deleterious effects of excessive grazing pressures have been cited from north of the Sahara (Sghaier and Seiwert 1993), the Sahel region (Aweto and Adejumobi 1991) and in the continent's southern drylands (Kerley *et. al.* 1995). Prevailing drought in Sahelian latitudes since the late 1960s saw a large concentration of livestock in moister southern parts of the region as herders competed with other land users for sparse vegetative resources. In some cases, however, the effects of overgrazing are being re-examined in the light of the paradigm shift, as Dean *et. al.* (1995) advocate for the semiarid Karoo of south Africa.

A theme common to many examples of intensive grazing in Africa is the sedentarisation of previous nomacid people, whose livestock are consequently concentrated in much smaller ranges, as Sghaier and Seiwert (1993) report from the margins of the Sahara in Tunisia. Both sedentarisation and the encroachment of cropland on the traditional grazing areas, pushing herders into more marginal rangeland are also common processes throughout the Middle East (Johnson 1993). Some of the effects in the Elburz mountains in Iran are documented by Klein and Lacoste (1994). In Saudi Arabia, where sedentarisation has been encouraged by government policy since the 1950s, the nomadic proportion of the country's population has declined from 70% in the 1930s to just 3% in 1990 (Child and Grainger 1990).

Table 4.16 : Main causes of soil degradation in susceptible drylands and other areas (Million hectares)

Region	Aridity Zone	Overgrazing	Deforestation	Agricultural activities	Over-exploitation
Africa	S	184.6	18.6	62.2	54.0
	O	58.5	48.2	59.2	8.7
Asia	S	118.8	111.5	96.7	42.3
	O	78.5	186.3	107.6	3.8
Australasia	S	78.5	4.2	4.8	0.0
	O	4.0	8.1	3.2	0.0
Europe	S	41.3	38.9	18.3	0.0
	O	8.7	44.9	45.6	0.5
N. America	S	27.7	4.3	41.4	6.1
	O	10.2	13.6	49.1	5.4
S. America	S	26.2	32.2	11.6	9.1
	O	41.7	67.8	51.9	2.9
TOTAL		**678.7**	**578.6**	**551.6**	**132.8**

Region	Aridity Zone	Bio-Industrial activities	Total degraded	Nondegraded	Total
Africa	S	0.0	319.4	966.6	1286.0
	O	0.2	174.8	1504.9	1679.7
Asia	S	1.0	370.3	1301.5	1671.8
	O	0.4	376.6	2207.5	2584.1
Australasia	S	0.0	87.5	575.8	663.3
	O	0.1	15.4	203.5	218.9
Europe	S	0.9	99.4	200.2	299.6
	O	19.7	119.4	531.4	650.8
N. America	S	0.0	79.5	652.9	732.4
	O	0.4	78.7	1379.8	1458.5
S. America	S	0.0	79.1	436.9	516.0
	O	0.0	164.3	1087.3	1251.6
TOTAL		**22.7**	**1964.4**	**11048.3**	**13012.7**

S= Susceptible O= Others

Source: GLASOD

Sheep and goat herding remain widespread, but 60% of the stock owners are now settled, semi-settled or short-range nomads and even truly nomadic herders now tend to stay in one locality for longer. Traditionally, during times of drought, when vegetation became scarce, stock numbers decreased so that environmental degradation was checked, but sedentarisation, combined with modern aids to herders like boreholes, water trucks and supplementary forage, has undermined this traditional self-regulation. Consequent overstocking of rangelands was thought to have severely degraded 85% of the land in the 1970s (Child and Grainger 1990).

Elsewhere in Asia, degradation has also been caused b the intensification of grazing by growing numbers of livestock, as Kharin and Tsolmon (1994) report from Mongolia. Takeuchi *et. al.* (1995) have documented extensive degradation in neighbouring Chinese Inner Mongolia, where Quaternary sand deposits have been disturbed on a large scale following a substantial increase in livestock in the 1950s and 1960s, The area of remobilized sand dune fields has expanded by 2.3 times in the semiarid Kerqin Sandy Lands in the Last 5.0 years.

In South America, where overgrazing is the second most widespread cause of degradation, much of Argentina is affected. Widespread soil erosion in Potagonia, is attributed to the heavy grazing pressures associated with sheep, monoculture in the region (Iglesias 1992). In Europe, the overgrazing problem is largely confined to the Ukraine and southern Russia north of the Caucasus Mountains. Ray *et. al.* 91993) report that the extinction of some perennial grass species on heavily grazed parts of the Ukrainia stepps are associated with physical changes in the soils. In the Russian Republic of Kalmukia, Zonn (1995) documents the widespread intensification of grazing pressures, mainly by sheep, during the period of Soviet rule.

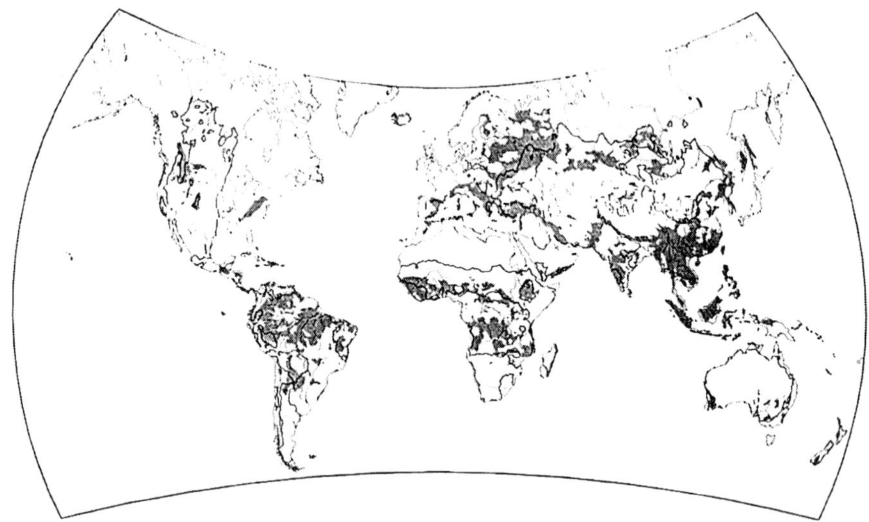

Approximate equatorial scale 1:317 million — *Source:* UNEP/ISRIC

Figure 4.27 : *Area affected by deforastation*

Deforestation and removal of the natural vegetation cover (Figure 4.27) is the primary cause of degradation in South America, affecting 41% of the 79 million hectares of drylands damaged. These are largely concentrated in north-east Brazil, on the Caribbean coasts of Venezuela and Colombia and in northern Argentina. GLASOD surveys of north-east Brazil found water and wind erosion problems of often shallow soils of the crystalline basement complex where deforestation of deciduous scrub or 'caatingas'. had taken place for annual crops and 'improved' pastures. The semiarid Chacoan forests of northern Argentina have been exploited for more than 100 years, initially because the wood made excellent railway sleepers, but more recent clearance has been for agricultural expansion, prompted by high grain prices and positive rainfall balances since the 1970s (Margarita and Loyarte 1996).

In European and Asian drylands, deforestation closely follows overgrazing as the second most important cause of degradation, affecting 39% and 30% of all degraded dryland area respectively. European areas affected include southern Spain, Sicily and southern Greece. In Asia the main deforestation problem areas are located in a broad belt from Turkey along the Zagros Mountains and into Pakistan. Deforestation is the least extensive cause of soil degradation in African drylands and is very largely confined to dry subhumid areas as indicated in Figure 4.27. In the savanna areas of northern Ghana, woodland clearance to expand the grazing area and for charcoal-making was the most common cause of water erosion during the intensive rainy season according to GLASOD reports.

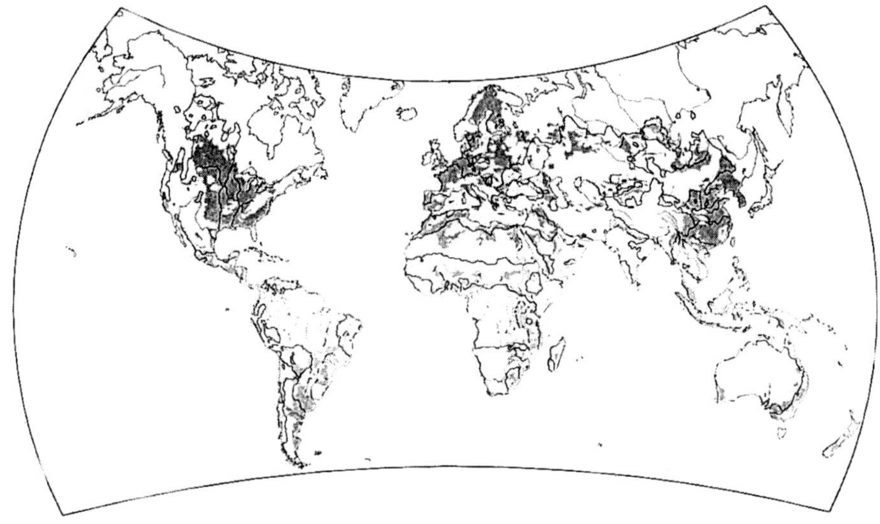

Approximate equatorial scale 1:317 million *Source:* UNEP/ISRIC

***Figure 4.28** : Area affected by agriculture activities*

Agricultural activities. (Figure 4.28) represent the most important cause of dryland degradation in North America, where they account for 52% of the degraded dryland area. These degraded zones are located in northern Mexico, the Great Plains of the USA and the Canadian Prairies. Although the effects of soil degradation in the Great Plains and" the Prairies can be masked by improved crop strains and plentiful addition's of fertilizers, Fryrear (1981) has noted substantial long-term declines in yields of sorghum and kafir in the Texas Panhandle over 30-40 years, a trend attributable at least in part of the effects of wind erosion. Indeed, despite the effects of the 1930s Dust Bowl, poor agricultural practices continue to play a role in exacerbating dust storms during periods of drought (Ervin and Lee 1994).

Agricultural activities are also important in Asia (26% of the degraded dryland area), Africa (19%), Europe (18%) and South America (15%). Among the Asian areas, worst affected are Mesopotamia, where degradation affects poorly managed irrigation schemes and grassland areas converted to cereal cultivation, as in Syria, where extensive wind erosion has resulted. Parts of Arabian Peninsula used fro irrigation crop cultivation; some areas in Siberia and Central Asia, Mongolia and northern China are also affected. In Africa, areas north of the Sahara are worst affected, where the desire to expand the cropland area has pushed the tractor and multidisc plough deeper into the desert-fringe steppe (Dresch 1986). This example of misapplication of temperate European practices in a dryland environment is mirrored elsewhere on the continent, where colonial administrators discouraged local techniques for land husbandry. The introduction of the plough to southern Zimbabwe more than 70 years ago encouraged the almost total disregard for indigenous techniques for soil and water conservation. The value of these indigenous farming practices in only now being recognized (Hagmann and Murwira 1996).

The most serious agricultural causes of soil degradation in Mediterranean Europe are linked to rural depopulation and significant changes in traditional cultivation practices since the late 1950s. Extensive grazing and wheat cultivation has been largely replaced by intensive agriculture based on tree crops, horticulture and irrigation. The abandonment of traditional terrace systems have affected runoff rates, erosion and downstream flooding. Declining demand for many traditional dry products like almonds and figs has also resulted in abandonment of traditional tree crops in some areas. Overgrazing and fire combine to degrade the soils of these formerly cultivated areas (Margaris *et. al.* 1996). In South America, the degradation problems in the highly populated , predominantly agricultural semiarid areas of north-east Brazil are outlined by Magalhaes and Magee (1994), while Molina (1993) reports that about a quarter of Argentina's agricultural land suffers from some degree of erosion.

The over exploitation of vegetation for domestic use (Figure 4.29) only accounts for more than a tenth of degraded drylands in Africa (17%), South America (12%) and Asia (11%). In Africa this is a problem predominantly located in the Sudano-Sahelian region, where much of the resultant degradation is concentrated around urban areas that have expanded very rapidly since the late 1960s, as rural migrants fled the effects of drought. Denuded areas susceptible to enhanced erosion have been reported from numerous Sahelian cities (Michel and Louembe 1992). In the case of Niger, Spaeth (1996) suggests that at current rates of clear-cutting, for agricultural land as well as for fuelwood, the entire country's forest biome will have disappeared by the end of this decade.

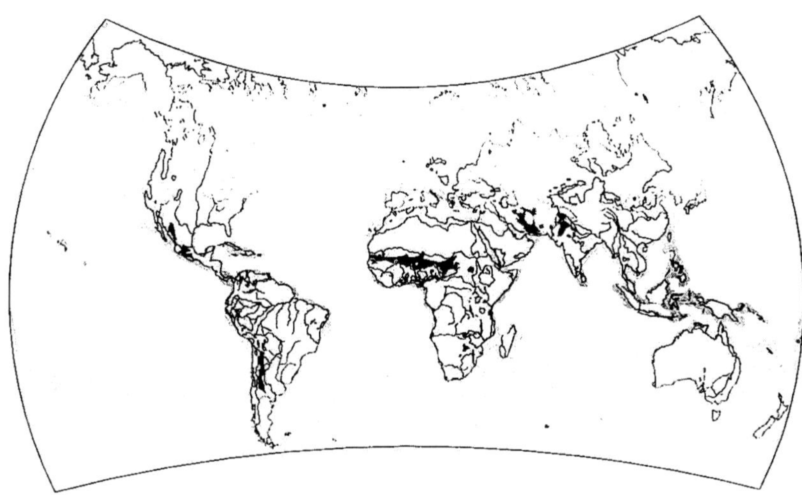

Approximate equatorial scale 1:317 million *Source:* UNEP/ISRIC

Figure 4.29 : *Area affected by overexploitation of vegetation for domestic use*

Despite evidence of declining woodland resources in the hinterlands of some Sahelian cities, pattern of depletion are not always straightforward. In the Kano region of northern Nigeria, Nichol (1989) found that tree density had increased in recent times in the immediate vicinity of the city as the transport of fuelwood by donkey had been displaced by the long-distance trucks. I decline in tree density was recorded in the zone 70-250 kms. from Kano. Although, there is evidence of denudation and consequent soil degradation in some parts of Sahel, the fuelwood crisis that was predicted for the region in the 1970s and 1980s has largely failed to materialize. Poor survey data, the importance of replacement fuels such as crop residues and dung and the effect of increasing woodfuel prices as local supplies dwindle are all factors that make the situation more complex than was proposed (Leach and Mearns 1988 RPTES 1996). Distinguishing between tree felling for fuel and to clear land for agriculture is another difficulty that obscures the picture, but more recent surveys of fuelwood stocks and demand in the Sahel are much less pessimistic than those of previous decades (Millington *et. al.* 1994).

In South America, the over-exploitation of vegetation for domestic use is largely confined to dryland areas of north-western Argentina and southern Bolivia, where scrubis collected for firewood. In Asia, it is mainly a cause of degradation in Iran, Pakistan and north-western India. In Pakistan, the few remaining stands of tropical thorn forests which once covered the Punjab Plains, largely cleared for irrigated agriculture, are under continuing pressure for their long-standing use as a source of fuelwood (Khan 1994).Fuelwood collection around many urban centers in India has significantly reduced forest areas in their hinterlands in recent decades, as the urban poor have been forced to turn increasingly to fuelwood by rising prices of kerosene, coal and charcoal. One study of major Indian cities using satellite imagery (Bowonder *et. al.* 1988) has shown that more than half of the closed forest cover within a100-km radius around many dryland cities was lost in the 10 years as given in Table No. 4.17.

Table 4.17 : Loss of closed forest cover around selected dryland urban centres in India (After Bowondor *et. al.* 1988)

Urban Centre	State	Closed Forest Cover (in Sq. kms)		Loss (%)
		1972-75	1980-82	1972-82
Ajmer	Rajasthan	259	124	52
Jaipur	Rajasthan	1534	786	49
Amritsar	Punjab	208	111	47
Bhavnagar	Gujarat	112	9	92
Hyderabad	Andhra Pradesh	40	26	35
Bhopal	Madhya Pradesh	3031	1417	53
Gwalior	Madhya Pradesh	1353	515	62
Indore	Madhya Pradesh	3770	1070	71

Source: Bowondor

In Rajasthan, the use of cattle dung to supplement fuelwood during the summer dry season and in drought periods has been decreasing due to low fodder availability and high livestock mortality (Kumar and Bhandari 1993-a and b).

Bio-industrial activities account for very small areas of dryland degradation in Asia, around Tehran in northern Iran due to industrial pollution and in the Aral Sea basin due to the excessive application of pesticides (Glazovsky 1995).

4.8 SOIL DEGRADATION AND VEGETATION

The GLASOD survey focused on degradation within the soil system. No consideration of land degradation and desertification is complete without vegetation; however, the links between desertification, soil degradation and vegetation are complex and multifarious.

It has been visualized that plant cover is the core issue in desertification (Thornes 1995) and a range of factors point to both its environmental and social significance in dryland areas (Box 5). However, vegetation is. a complex factor to incorporate in assessments of the extent of desertification for a number of reasons.

(1) It is now known that, dryland vegetation communities are hot naturally stable, but vary in response to fluctuations in controlling variables, particularly soil moisture, rainfall and natural fires Consequently, dryland vegetation communities can be described as 'disequilibrium systems' (Smith 1988, Ellis and Swift 1998, Ellis *et. al.* 1991), with the characteristics of a particular systems at any time being driven by preceding events.

(2) The widespread and. frequent occurrence of drought means that the natural status of many vegetation communities in the world's drylands is likely to be highly dynamic in time.

(3) Hinders assessment that try to distinguish changes in vegetation systems that area consequence of drought and from which natural recovery is likely (Kassas 1995), from those that are caused by degrading human actions. This point is made more apparent, when it is realized that the ecology of dryland vegetation communities is relatively under-researched relative to those in more temperate environments (Warren 1995).

BOX 5 : Environmental and Social Significance of Dryland Vegetation

- Vegetation is a major pathway for the loss of water from drylands, through evapotranspiration.
- Plant cover is a major factor influencing the erodibility of dryland soils both by wind and water.
- Vegetation is a key source of soil nutrients and therefore, changes in vegetation cover can significantly contribute to the chemical deterioration of soils.
- Natural vegetation supports pastoralism, which is a major land use in dryland areas.
- Vegetation is an important source of' domestic fuel in many developing world drylands.
- Vegetation communities and plant cover densities are susceptible to changes caused by natural (drought) and human factors.
- Changes in vegetation communities may be important indicator of pending change within the soil systems.

Current understandings therefore, indicates that vegetation and soils in drylands display differing resilience to disturbance. Although, vegetation communities can be readily disturbed and changed, recovery rates are relatively fast (Westoby *et. al.* 1989), when compared to soil degradation, through shower in drylands than in other environments. An effectively permanent undesirable change to vegetation communities usually occurs, when either the soil resource itself or the mechanisms employed by plants to allow recovery have been severely degraded. In other situations, where rangeland degradation seems to have occurred, it often appears to be the land use system that maintains the vegetation in a degraded or changed state, rather than a change of soil properties that would make the vegetation degraded irreversible (Dougill *et. al.* 1997).

In many locations, rigorous evidence that would allow an assessment of whether effectively permanent vegetation change has occurred is sorely lacking (Dean *et. al.* 1995). There is, therefore, an obvious need for long-term monitoring programmes (Dean *et. al.* 1996) and the development of methodologies, which permit the significance of vegetation changes to be assessed (Pickup *et. al.* 1994).

An additional problem arises as changes in vegetation that are 'effective permanent' can be interpreted in a number of ways, in terms relevant to a particular production strategy, to the average productivity of natural vegetation changes or in terms of longer time scale. Where significant vegetation changes are occurring, their relationship can also be complex. Bush encroachment, which is the replacement of grasses by perennial herbaceous plants, may in some livestock systems lower production (Ringrose *et. al.* 1990) and therefore, be regarded as significant degradation, but in others it may be regarded as offering opportunities, as the browse resources that are created can be a useful grazing substitute at times of moisture stress, particularly- in mixed livestock systems (Behnke and Scoones 1993, Cox and Dougill 1995).

When bush encroachment reaches very dense state, as often occurs in distinct annular zones in African grazing systems (Parkins and Thomas 1993-b), it also increases ground cover and can therefore, inhibit soil erosion. However, if it results in a reduction in biodiversity, it can lower the overall biological resource base. It is issues such as these contribute to the debates and controversies that surround current understanding and interpretations of the significance of dryland vegetation changes (Adams 1996, Cox and Dougill 1996). Soil and water degradation are of great importance; however, they are not necessarily coincident as indicators of land degradation.

Strategies displayed by arid land plants for survival at times of moisture deficiency include reducing production and lying dormant during dry periods (Tolsma *et. al.* 1987), and generating large seed banks during productive periods (Skarpe 1991). In fact, many dryland vegetation communities respond to rainfall, when it occurs but otherwise display very limited or no productivity: It may thus be difficult to distinguish, whether a particular vegetation state at a particular time is a consequence of natural short-term climate variability, livestock or human impacts or a longer-term trend induced by soil degradation. These factors contribute to the absence of an adequate data base of vegetation degradation, but assessments of the extent of rangeland vegetation change have been made at the global scale.

Kassas *et. al.* (1991) estimated that, in addition to 1035 million hectares of susceptible drylands that the GLASOD survey identified as having degraded soils, a further 2559 million hectares or 50% of the total susceptible dryland area, was subject to degradation of vegetation. There is, however, according to Warren *et. al.* (1996), a formidable body of evidence pointing to the occurrence of excessive grazing pressures that threaten sustainability in some semiarid areas. Evidence that support this contention comes from recent studies in parts of Middle Asia, in the Kyrgyz Republic, where there is marked over-utilization of lowland ecosystems by sheep herders (Schillhorn Van Veen 1995); China including Inner Mongolia, where in some areas up to 25% of vegetation is heavily degraded with a further 55% experiencing light and moderate degradation (Humphrey and Sneath 1996); and areas of the Sahel belt of Africa (Le Houerou 1990).

Two useful relationships between soil degradation and vegetation can be determined. First, vegetation cover may help to indicate vulnerability to soil degradation. A limited plant cover or low biomass, whether natural or the result of human disturbance, can often indicate a greater susceptibility to degradation by soil displacement (Thornes 1995). Second, vegetation cover or community type may indicate that soil degradation has occurred or is taking place: The latter is difficult to determine meaningfully except through detailed ground survey (Dean *et. al.* 1996) at a greater resolution than that used for studies of natural climatic variability, but useful information relating to the former relationship can be collected at global (Justice *et. al.* 1985) and regional (Ringrose *et. al.* 1990) scales.

(A) Characterizing Vegetation

Vegetation can be characterized in many ways, by plant community composition. Production is another useful way of characterizing vegetation in a form that allows large scale interregional comparisons to be made and can also be incorporated in studies that attempt to identify the causes of such changes. Studies that have evaluated the causes of changes in vegetation production in parts of the Sahel have found that those related to livestock impacts in the Ferto areas of Senegal (Hanan *et. al.* 1991) and the Gourma area of Mali (Hiernaux 1996).

Primary vegetation production can be defined in two ways:

(a) Gross Primary Production equals the energy fixed by plants during photosynthesis.

(b) Net Primary Production equals the energy fixed by plants in photosynthesis minus the energy lost in transpiration. Net primary production can be measured on the ground by biomass, the volume of plant matter per unit area. Although, this is a static measure and is less environmentally significant in degraded studies than the maximum potential production for a given environment, it can be approximately evaluated from remotely sensed data allowing global patterns to be determined. Thomas and Middleton (1994) evaluate the contributions of different remote sensing sources to desertification studies, which include the use of Landsat Multi Spectral Scanner (MSS) data (Tsoar el. al. 1991, Matheson and Ringrose 1994),Landsat Thematic Mapper TMdata (olsson 1984) and high resolution SPOT data(Guyot 1990) to evaluate regional vegetation change. National Oceanic and Atmospheric Administration (NOAA) satellite data has the greatest utility because of its wide spatial coverage and high temporal resolution.

(B) A Global Vegetation Assessment: NDVI and GVI

Within ecosystems, it is possible to identify changes in the patterns of vegetation and the spatial patterns of change through time, by using satellite data sources. The Normalized Difference Vegetation Index (NDVI) is a qualitative index of vegetation phenology. The Global Vegetation Index (GVI). supplies by NOAA, is a weekly NDVI value, available with global coverage, derived from surface reflectance data generated fro NOAA. Advanced Very High Resolution Radiometer (AVHEE) satellite waveband information, GVI values have a 16 km x 16 km spatial resolution and are supplied in the form of the highest value occurring each week (Dregne and Tucker 1988). A simple mean of daily values cannot be used because of the problems that values for days with cloud cover, would introduce. GVI values are not-specified but they generally to the capacity of plant canopy to evapotranspiration rates. The index can, therefore, by used as an indicator of vegetation biomass (Tucker 1980, Justice *et. al.* 1985), though several studies suggest that the correspondence of the index with actual values is weak under some environmental conditions (Tucker *et. al.* 1985 a, b, c).

In the context of this study, global GVI surface has been produced taking the mean of individual monthly maxima for the timeband 1983 to 1990. The continuous spectrum of GVI values is represented on the map as a five-grade scale from low to high indicating biomass differences. It was observed that as with the map of aridity zones, GVI values are not static, as they will respond not just to human disturbance but to natural meteorological variability, particularly seasonal variability in drylands.

(C) Interpreting the Soil Degradation and Vegetation map

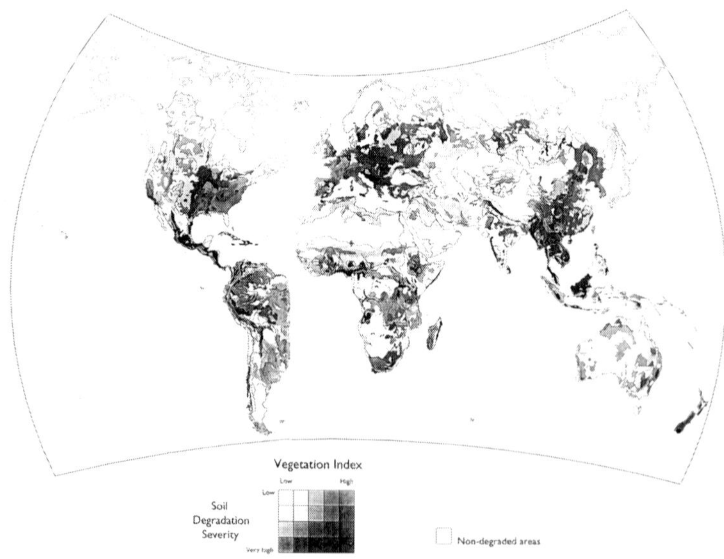

Approximate equatorial scale 1:104 million

Source: UNEP/ISRIC, NOAA

***Figure 4.30** :Soil degradation severity and vegetation*

Figure 4.30 gives an integrated assessment of overall soil degradation from the GLASOD survey and vegetation production from the NDVI. The four degrees of degradation severity have been combined with the five grades of the vegetation index to produce a 20-colour grid showing general relationship between the degree of degradation of susceptible areas and vegetation production.

The meaning of the relationship between the two variables requires careful interpretation. No casual links can be inferred from this map between vegetation production and soil degradation, nor one between the severity of degradation and types of vegetation community. An indication of low vegetation production does not mean that it has been directly or indirectly influenced by human actions. The map gives no indication of natural or managed vegetation communities or vegetation quality. It might generally be assumed that a high vegetation index value indicates a more luxuriant and beneficial vegetation, but in semiarid rangeland areas, it could represent the replacement of palatable grassland with scrub vegetation through the process of bush encroachment which may in fact offer protection to the ground surface against further soil erosion. High values can, therefore, also include disturbed vegetation communities.

Best use can be made by taking the map to highlight areas within which susceptibility to soil or vegetation degradation may exist or have been realized. For example, mapping units possessing a colour from the top left part of the grid key may well contain areas susceptible to wind and water erosion. This is because, degradation severity during GLASOD survey period was low, but with a low vegetation index also indicated, ground cover in probably restricted in at least part of the mapping unit, creating soils vulnerable to wind or water erosion. The most productive and least degraded environments are represented by colours from the top right of the key. With high GVI values, such areas also possessed the least susceptibility to soil erosion during the study period.

Conversely, parts of the map represented by colours from the bottom left of the key are most likely to contain ecosystems displaying the greatest overall response to damage by human actions. Soil degradation is highly, creating situations unfavourable to biotic processes, with plant communities appearing to have realized this negative potential of having low biomass. Caution needs to be exercised, when interpreting the spatial extent of such conditions though, because of the range of scenarios tat any degradation severity can represent and the size of individual mapping polhygons.

(D) Spatial Patterns Within Drylands

Figure 4.30 allows several general interpretations to be made cautiously. High severity-low biomass areas are found throughout the susceptible drylands of North Africa, parts of southern Africa, Arabia, Iran, Pakistan, the southern steppes of Middle Asia and in Spain. These are the drylands containing areas where the greatest overall environmental degradation has occurred. Although, many mapping polygons in Australia and North America, with the exception of the lower Colorado River basin and the Alberta badlands contain areas that have experienced limited soil degradation, low to medium biomass values suggest that these are locations where potential for degradation by erosion exists, especially if land uses were to intensify (Pickup and Chewing 1988). In the semiarid Patagonian grasslands of Argentina, local factors such as aspect may also be important in determining the critical relationship between moisture availability, biomass and degradation (Coronato and Bertiller 1996). Finally, the drylands in the rain-shadow area of south-eastern Peru possess unusual situation of soils that have been highly degraded but which also appear to have high GVI values. The ability of the map to highlight such features and situations, which are difficult to explain points to their usefulness in indicating areas where the status and origins of degradation and relationships between soil degradation and vegetation status, warrant more detailed investigation.

Chapter 5

Consequences of Desertification and Land Degradation

Desertification has too far long been the poor relative of environmental issues. It is mainly experienced in rural areas of the world and in many cases in developing countries. In both cases, it has been less visible to politicians and policy makers than problems such as urban air pollution and disposal of hazardous wastes. Yet, as the world's population expands, desertification will become of increasing concern. Measures to combat its advances have until now been hampered by lack of political commitment and lack of funds. But it is a serious problem and well deserving the priority status. Desertification is accelerating and the costs of combating it, rises with each passing year.

Desertification is destroying the food-producing capacity of vast tracts of drylands in every continent of the world. Many millions of people living in almost 100 countries including the Unites States, The Soviet Union. Africa, Asia, Central and South America and Australia, suffer its effects. Of the world's drylands used for agriculture, about 70% are affected by some degree by various forms of land degradation. One-sixth of the world's population is threatened by the effects if desertification. Their numbers are rising alarmingly. In 1977, 57 million people were suffering the direct effect of land, which had degraded to the extent, where it no longer produced enough food to sustain them. By 1984, their number had risen to 135 million.

Every year nearly six million hectares of previously productive land in arid, semi-arid and dry sub-humid areas loses its capacity to produce food. In terms of income lost, this destruction of productive capacity costs the world in 1990 prices about US$ 42.3 billion each year. By 2010 this figure is estimated to cross US$ 100 billion per year.

Clearly, the situation is serious, with rapidly expanding global population. The world can ill afford to lose any of its productive land, let along such large areas. Desertification is a global problem, demanding urgent global action and this action must involve more than a campaign against the processes of desertification. It must be taken as an essential part of broad process of development and the provision of basic human needs.

Although desertification is a serious problem of land degradation in arid, semi-arid and dry sub-humid areas resulting in many adverse human impacts: yet there is no adequate global database available for assessment of vegetation degradation, overgrazing and land use.

There are three major causes of soil degradation; overgrazing, deforestation and unsustainable agricultural practices. Overgrazing and deforestation destroy the protective layer of plants, which cover drylands, making it possible for wind and water erosion to strip away the fertile top layers of soil. Unsustainable agricultural practices sap the soil of nutrients overload it with salts, dry it out. Compact or seal its surface and cause water-logging, which removes air from the soil and allows toxic substances to accumulate.

More than 6.1 billion hectares, over one third of the earth's land area is dryland. Nearly one billion hectares of this area are naturally hyper-arid deserts, with very low biological productivity. The remaining 5.1 billion hectares are made up of arid, semi-arid and sub-humid areas, part of which have become desert since the dawn of civilization, while other parts of these areas are still being degraded by human action today. These lands are the habitat and the source of livelihood for about one quarter of world's population. These are areas characterized by the persistent natural menace of recurrent drought, a natural hazard, accentuated by unbalanced management of natural resources. Particularly severe drought years in Sahelian region of Africa from 1968 to 1973 and their tragic effects on the people of the region drew worldwide attention to the problems of human survival and development in drylands, particularly on desert margins.

The concept of degradation is inseparable from sustainability. Expressed simply, a sustainable land use is one that is able to continue without degrading the land it is using. In this case, sustainability of a particular land use depends both on the properties of the resource and the way it is managed. The feature of a resource that determine its sustainability under a particular use is its resilience and it is important that the resilience of a resource system varies according to different land uses and indeed may vary from time to time, depending largely on seasonal and inter-annual variability and management practices and technologies.

A reliable way to measure the resilience of a particular unit of land is to look at its ability to recover after a disturbance. Such a disturbance may be climatic like drought, human induced like vegetation clearance or soil tillage. The greater the disturbance the area can recover from, the greater its resilience. In essence, land degradation is the weakening of an area's resilience and one measure of land degradation is the cost of rehabilitation.

One serious difficulty of examining land degradation in dryland regions is their inherent variability. A definitive characteristic of drylands is their aridity, which put simply, means their lack of available moisture. An area can be said to be arid, when its moisture inputs that is precipitation are exceeded by the moisture losses by evapo-transpiration plus any changes in storage in rivers, groundwater, lakes and soil moisture. Various climatic and biological indexes have been used to measure the aridity of drylands and these can be used to delimit dryland areas. However, it is important that drylands are usually deficient in moisture on an annual basis, the moisture inputs as precipitation are notoriously variable in both time and space. It is not unheard of, for the more extreme dryland areas to receive all their "average" annual precipitation in just one rainfall event and commonly such regions receive their average annual total in just few days. The meteorological systems that bring rainfall to dryland regions are typically convective cells, which means that spatially rainfall often falls in small specific zones.

These characteristics of dryland climate, to which indigenous plants and animals adapted, mean that these regions are highly dynamic on a timescale of weeks and months. Added to this variability of precipitation of longer timescales of years and decades. Droughts in essence, the absence of expected precipitation, are characteristic of the dryland environment. A year, two years or several years may pass in which, precipitation is well below "average". Again indigenous plants and animals have adapted to cope with this inherent variability of dryland ecosystem.

The implications of this environmental variability of desertification are manifold. Firstly, it has implications for the location of drylands themselves. Although for the purposes of assessment, dryland areas are given specific boundaries, it is important to note that these boundaries are simply based on average conditions where as actual ground conditions vary greatly through variable timescales.

There are many implications of human inhabitants of dryland regions. The use to which land is put is as dynamic as the environment it self. Fields that produce good crop of millet during average rainfall years, may have to be left fallow during a drought year or put to another use. if the land is not to be degraded. It is in areas, where this flexible land use response does not occur that desertification or land degradation takes place.

The most serious implication of dryland variability in the present context is for the identification, monitoring and combating of desertification itself. Satellite imagery of vegetation greenness on the fringes of drylands indicate that the natural variability in climate is reflected in a green vegetation dryland boundary that can fluctuate by up to 200 km from a dry year to a following wet year. With such great natural fluctuations in the ground state of drylands, it is clearly necessary to monitor potential areas of desertification and land degradation over a timescale of decades before it is possible to safely state that a particular region has suffered from land degradation in the form of desertification.

Reliable identification of the locations and situations in which such land degradation takes place is essential if variable remedies to the problem are to be reached. The actual regions for unsustainable land use taking place in a particular region may well have their roots in social and economic conditions. But before these underlying causes can be tackled, it is necessary to locate and quantify the nature of problem. The most important is the perception that desertification threatens all the world's drylands. However, the prime motivation for studying desertification is to help and relieve the problems faced by inhabitants in using dryland resources, it is clear that hyper-arid desert, which by definition have very sparse biological resources, should not be included in the areas as these regions lack the resources and these regions can hardly become more desert-like. Hence, for the desertification, the regions deemed to be susceptible to desertification are those arid, semi-arid and dry sub-humid zones and these regions are referred to as the "susceptible drylands."

The complex nature of desertification means that adequate assessment and consequent plans to counteract the problem can only be usefully carried out on the local scale. Nonetheless, it is useful to attain a worldwide appreciation of he phenomenon in order to estimate its magnitude worldwide and to identify more specific problem areas at the national and local scales.

5.1 GEOGRAPHICAL REGIONS

Aridity Index (AI) calculated as the ratio P/PET, was derived from the climatic surface maps. Values were averaged to provide a map of annual potential moisture availability for the period 1950-1980 and a global climatic classification based on potential moisture availability, with AI values < 1.0 indicating an annual moisture deficit. The zonation of the classification in shown in the legend on Figure 5.1.

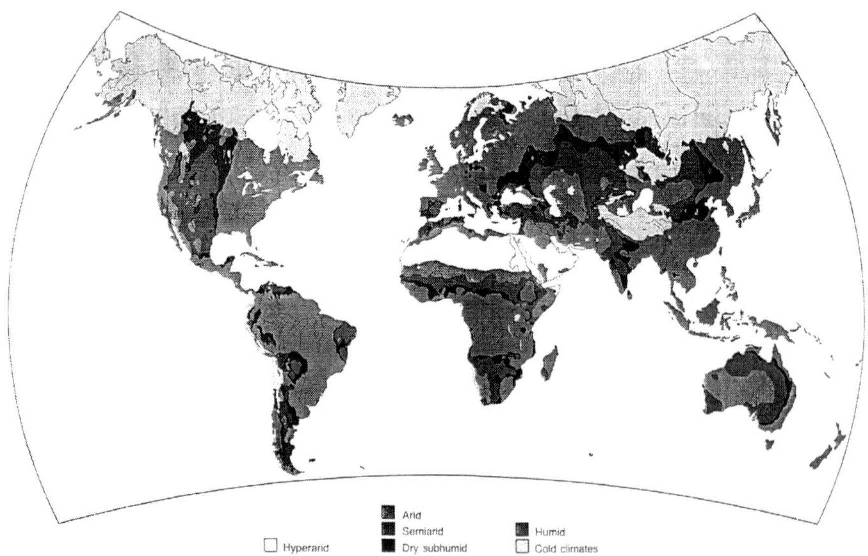

Approximate equatorial scale 1:115 million *Source:* CRU/UEA, UNEP/GRID

***Figure 5.1** : Aridity zones*

The P/PET boundary between arid and hyper-arid zones were increased from the 0.03 used in 1977 UNESCO map of the world distribution of arid regions to 0.05, to compensate for the Thornthwaite method underestimation of PET in very dry environments. The boundary between dry sub-humid and humid zones was introduced in order to exclude moisture areas from consideration under desertification. The separate designation of cold mountain and tundra climates based on a temperature criterion was introduced because these areas may suffer from an annual moisture deficit, such cold environments present a different range of problems than those in warm dryland environments. As such these areas are excluded from further consideration of desertification issues. Aridity zones by continents are given in Table No. 5.1 representing the classification in percentage terms of respective continent and global position to understand the problem in significant terms.

Table 5.1 : Aridity Zones by Continents

(In percentage)

Aridity Zone	Africa	Asia	Australia	Europe	North America	South America	Total
Cold	0.00	25.43	0.00	2.93	28.16	2.13	13.56
Humid	33.98	28.77	24.81	65.53	38.27	67.22	39.20
Dry sub-humid	9.06	8.29	5.81	19.31	10.57	11.71	9.95
Semi Arid	17.32	16.29	35.03	11.07	19.14	14.97	17.71
Arid	16.98	14.70	34.35	1.16	3.72	2.52	12.06
Hyper-arid	22.66	6.52	0.00	0.00	0.14	1.45	7.52
Total	**100.00**	**100.00**	**100.00**	**100.00**	**100.00**	**100.00**	**100.00**

Source: World Atlas of Desertification, Volume I

1. Dryland types

Drylands are defined in several ways, but most simply aridity represents lack of moisture in average climatic conditions, caused by one of four climatic situations, which may interact in the case of individual dryland areas, continently, topographically, anti-cyclonic subsidence and oceanic cold currents. The nomenclature for aridity zones used are those commonly employed in scientific analysis. Although, at the global scale, each zone inevitably embraces a range of environmental conditions, the following general criteria is used to characterize them:

(a) Hyper-arid Environments (P/PET < 0.05): Covers 978.1 million hectare area or 7.52 percent of the global land surface and have very limited and highly variable rainfall amounts both inter-annually up to 100% and on a monthly basis such that there is no seasonal rainfall regime. In virtually all cases, there data are available, year-long periods without rainfall have recorded. These areas offer very limited opportunities for human activity.

(b) Arid Areas (0.05 < P/PET < 0.20): Covers 1569.2 million hectare or 12.06% of the global area. Mean annual precipitation values up to about 200 mm in winter rainfall areas and 300 mm in summer rainfall areas, but more importantly inter-annual variability in the 50-100% range. Pastoralism is possible but without mobility or the use of groundwater resources are highly susceptible to climate variability.

(c) Semi-arid Areas (0.20 < P/PET < 0.50): Spread over 2305.3 million hectares or 17.71% of total land area Highly seasonal rainfall regimes and mean annual values up to about 800 mm in summer rainfall areas and 500 mm in winter regimes. Inter-annual variability is nonetheless high (25-50%) so despite the apparent suitability for grazing of semi-arid grasslands, this and other sedentary agricultural activities are susceptible to seasonal and inter-annual moisture deficiency.

(d) Dry Sub-humid Areas (0.50 < P/PET < 0.65): Dry sub-humid area covers 1294.7 million hectares of land i.e.. 9.95% of the global land. Less than 25% inter-annual rainfall variability and rain-fed agriculture is widely practiced.

(e) Cold and humid lands are spread over in 1765 million hectares and 5100.4 million hectares of areas constituting 13.56 and 39.20 % of the total global land.

2. Problems of Delineating Dryland Boundaries

Dryland boundaries are neither static nor abrupt. This is not surprising given the high inter-annual variability in mean rainfall and the occurrence of drought, which may last for periods of several years at a time. Attempt to locate boundaries on the ground or define them in terms of features such as natural vegetation are likely to fail. Physical changes are likely to be gradual and they will be modified by human induced processes like grazing, deforestation and burning. Identification of climatic changes from shifts in boundaries, therefore, needs to proceed with caution given the dynamism that is inherent in dryland climate regimes and the fact that drought can cause spectacular but temporary changes in natural vegetation.

It is also worth mentioning that individual aridity zones do not represent homogeneous climates. Specific P/PET ratios need to be derived from a wide range of values for individual meteorological parameters. The zones, therefore, represent one way of characterizing climate, a way that has particular relevance to desertification studies.

(a) Climatic Variability in Drylands

The study of desertification is hampered by the normal variability of dryland areas. Accurate identification of the causes of desertification and thus suitable strategies for its treatment, can only be made by paying close attention both to the human use and possible mismanagement of resources, also to the way, in which dryland eco-systems and their resources respond to climatic variations.

The inherent variability of dryland environments is very largely governed by the variations in climatic parameters that characterize such regions. Chief among these climatic parameters are precipitation, the input moisture into the desert eco-system. While many dryland areas receive important inputs of moisture from dew and some other from fog. rainfall is the key source of moisture in most of the world's dryland regions. However, the "effectiveness" of rainfall that is the amount available for plant growth or other uses, is also dependent upon the main output from the ecosystem, evapo-transpiration, which is governed by parameters such as vegetation cover and type, wind speeds and most importantly temperature. Hence, variability of rainfall and temperature in the world's drylands as a contextual background provide significant variations observed.

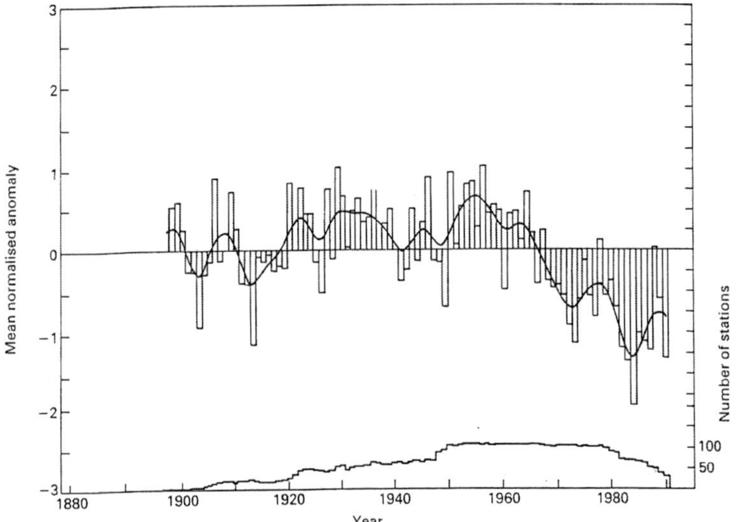

Figure 5.2 : *Time series of annual rainfall-The Sahel (1897-1990)*

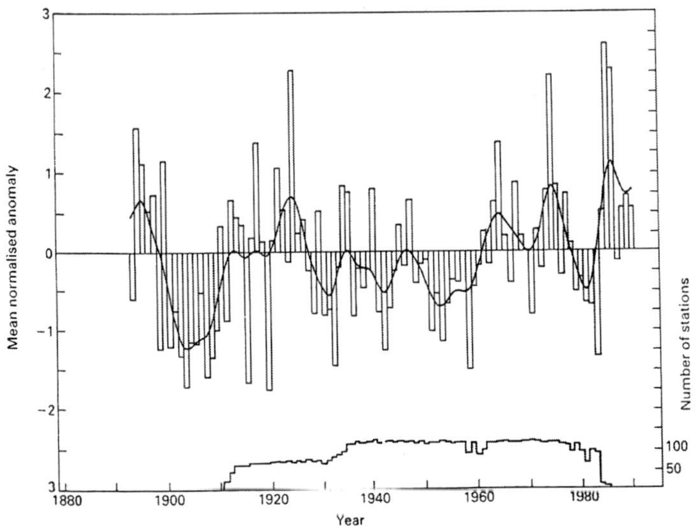

Figure 5.3 : *Time series of annual rainfall-Northeastern Brazil (1893-1990)*

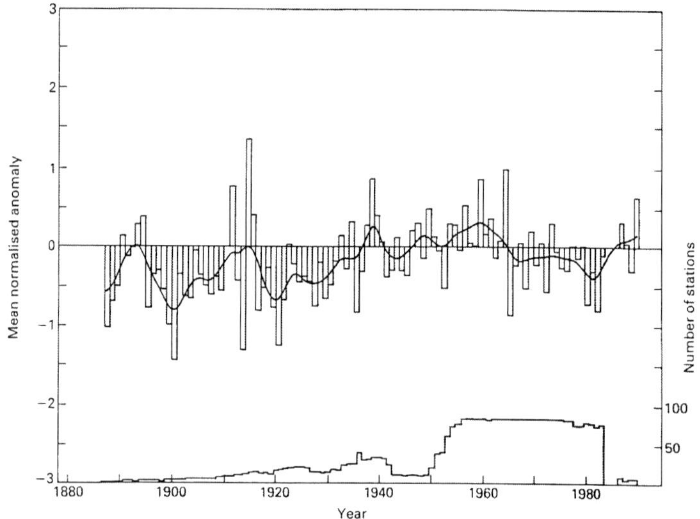

Figure 5.4 : *Time series of annual rainfall-North China (1887-1990)*

Figure 5.2, 5.3 and 5.4 show time series graphs of annual rainfall for three dryland regions: the Sahel from the Atlantic coast to 35° E; the north-eastern region of Brazil fro 44° W to the Atlantic coast and from the Equator to 10°S and North China from 100° E to the China Sea Coast and from borders with Mongolia and Russia to 35°N. The figures in the form of graphs for each area have been derived as follows:

(i) Time series of rainfall for each area is for the duration of 1897/1893/1887 to 1990 respectively,

(ii) The annual rainfall series for each station in the area is normalized by taking away the long-term mean from each value and the difference is then divided by long term standard deviation,

(iii) Number of station in each area for the corresponding years is also shown in the graphs for which the data is made available from the total area,

(iv) Long term period on which the mean and standard deviation are based in each area is 1951-1990, the period of the climate surfaces used for the Annual Aridity Index, which has given the dryland area boundaries.

(v) Normalizing gives a set of data series that are more readily comparable as each series has a mean close to zero and standard deviation close to one.

(vi) The spatial mean rainfall anomalies are derived by averaging the values for all stations in the area with data.

(vii) Although, the number of stations with data throughout each period of the three areas is not constant, the time series are the best available and since they represent a spatial average for each area, they reliably reflect the annual rainfall anomalies in the regions.

(viii) The smooth curves on each time series are ten-year Gaussian filters, which suppress variations of less than ten years. Therefore, they give a reliable indication of trends in rainfall anomalies.

(ix) All the three graphs clearly illustrate that dryland regions experience prolonged periods of above and below average rainfall, as well as year-to-year fluctuations.

(b) Changes in Climatic Surfaces 1930-1959 and 1960-1989

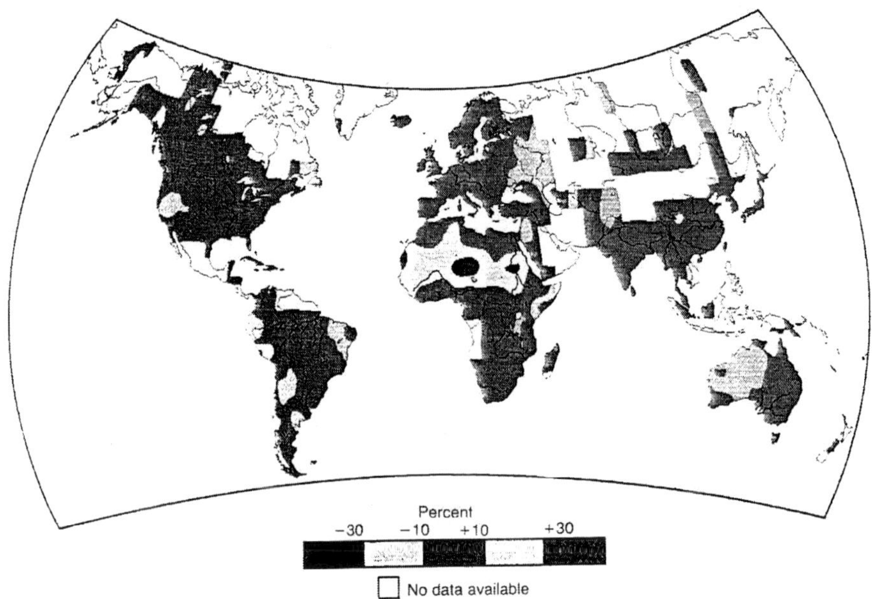

Approximate equatorial scale 1:235 million

Source: CRU/UEA

***Figure 5.5 :** Mean annual precipitation change from 1930-59 to 1960-89*

Figure 5.5, 5.6 and 5.7 give a spatial perspective on the climatic variability of drylands. These maps show change in global climatic surfaces between two periods of thirty years each- 1930 to 1959 and 1960 to 1989 for precipitation, temperature and the Aridity Index. The rainfall graphs, for the Sahel and north-eastern Brazil are confirmed in Figure 5.5, which shows that these regions have been drier and wetter respectively during the second, most recent periods. Most of the Sahel west of Ethiopia has been considerably drier, with annual precipitation declining by between 10% and 30%. In southern Somalia, by contrast, an area where climate is dictated by a largely separate set of influences, annual precipitation has been greater by more than 30% in concluding decades. A similar trend is illustrated for north-eastern Brazil and for the Middle Asian region of the former USSR east of Caspian Sea. Much of the central and western Australia and parts of the drylands of the American south-west have experienced precipitation between 10% and 30% greater than the first thirty year period. Other dryland areas of southern Africa and north China received roughly similar annual precipitation total during the two periods.

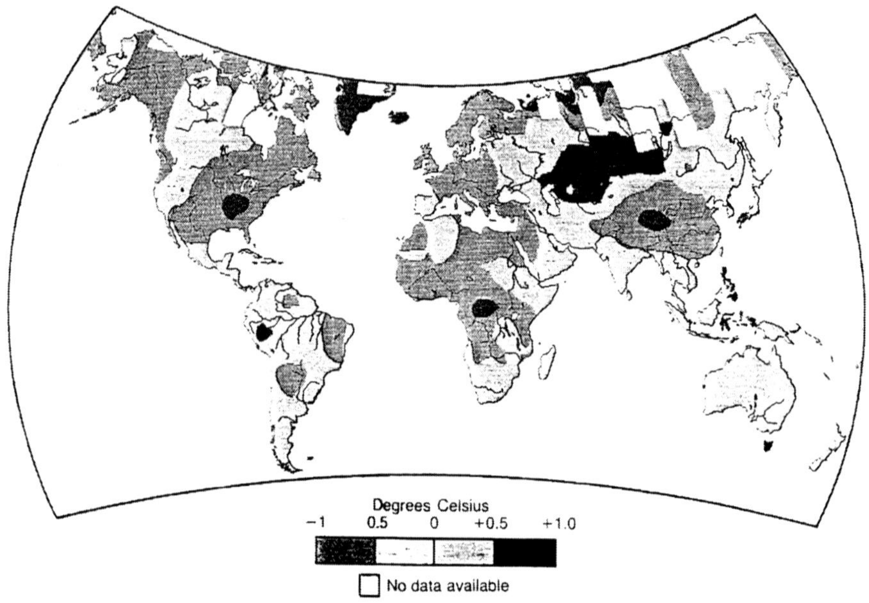

Approximate equatorial scale 1:235 million

Source: CRU/UEA

***Figure 5.6** : Mean annual temperature change from 1930-59 to 1960-89*

Figure 5.6 shows mean annual temperature changes between two periods of thirty years and indicate a number of dryland areas, where temperatures have been significantly different. The region of north-eastern Tibet and the eastern portion of the Kunlun Mountains were between 0.5 °C and 1.0 °C colder, while much of the Middle Asian Republics of the former USSR were between 0.5 °C and 1.0 °C warmer. The global mean temperature change between the two periods, derived from the data set used to produce Figure 5.6 is very small, 1960-1989 being 0.7 °C warmer than 1930-1959. Figure 5.8 implies a slightly greater temperature difference between the two periods, but this is due to larger size of data set used.

Figure 5.7 shows changes in Aridity Index values, witnesses the composite picture of differences in precipitation and temperate in terms of moisture availability. The most striking area of relative moisture deficiency is the Sahel from Ethiopia westwards and parts of Meghreb. Southern African drylands are also moisture deficient since precipitation was slightly changed, temperatures were little greater. Areas of northern Mexico, by contrast, where precipitation inputs were little changed, but temperatures were slightly colder, experienced greater moisture availability, as did north-eastern Brazil due to greater precipitation and slightly cooler temperature.

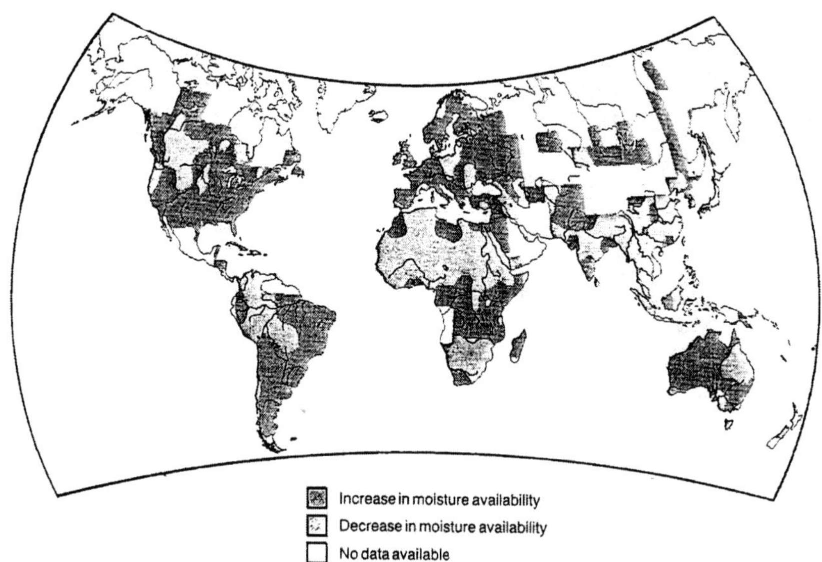

Approximate equatorial scale 1:235 million *Source:* CRU/UEA

Figure 5.7 : *Mean annual aridity index change from 1930-59 to 1960-89*

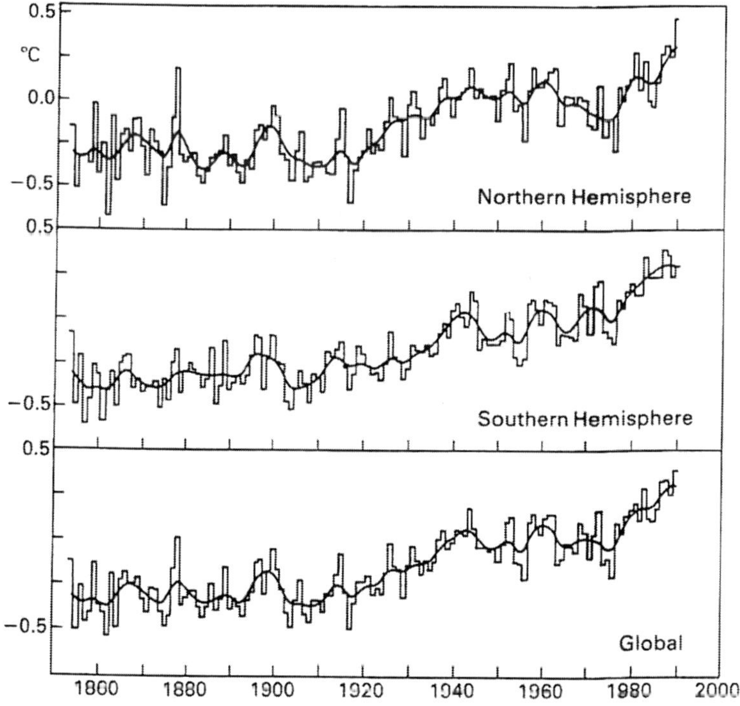

Figure 5.8 : *Mean annual aridity index change from 1930-59 to 1960-89*

(c) Global Warming

Our ability to forecast dryland climate changes is very limited in view of various factors responsible to govern the system. Meteorological forecasting in the tropics is only possible for two to three days in advance, compared to seven to ten days in temperat latitudes. However, attempts to forecast regional climates within a year, from season to season, are being worked on with some limited success. On the longer timescale of decades, however, it is not possible to produce objective predictions of future regional climates anywhere in the world. In addition to the problems of predicting natural climatic variability, there is the very real possibility that regional climates may be significantly changed as a result of increasing levels of carbon dioxide and other greenhouse gases in the atmosphere. Figure 5.8 indicates the phenomenon of global warming for the two hemispheres and the whole globe. While not all climatologists are convinced that the warming trends in these graphs are conclusively attributed to human activities, the implications for dryland climates should this trend continue are great. Computer simulations from Global Circulation Method for Sahelian climate in a "warmer World" disagree, some suggesting a drier climate, others a wetter one.

Whatever the affects of global warming on the regional climates of drylands, many scientists are arguing that the anthropogenic impact can no longer be ignored as a new input to the global and regional climate systems, the resulting changes of which also have to be incorporated into future efforts to ameliorate the problems of land degradation and desertification in the world's drylands.

5.2 SOIL DEGRADATION / LAND DEGRADATION

Desertification and land degradation are the end-results to poverty. The natural and human efforts coincided together play significant role and a vicious circle is formed, where it is difficult to differentiate the effects to desertification and land degradation. Virtually, all the issues are inter-related to each other. The rainfall and temperatures are directly or indirectly the results of human efforts in the form of deforestation and rapid growth of population are the primary causes of poverty and subsequent are improper use of water and land resources.

1. Severity and Distribution of Desertification and Land Degradation

The soil degradation severity embodies the degree of degradation and its spatial occurrence within individual units. It provides a comparative framework for the overall degradation problem throughout the world's drylands.

(a) Methodology

GLASOD (Global Assessment of Human-Induced Soil Degradation) defines soil degradation as "human-induced phenomena, which lower the current and/or future capacity of the soil to support human life." Two categories of human-induced soil degradation processes are recognized. The first deals with degradation by displacement of soil material, principally by water erosion and wind erosion. The second deals with internal soil deterioration by physical and chemical processes. This latter category does not include cyclic fluctuations of soil conditions of relatively stable agricultural systems, in which the soil is actively managed to maintain its productivity, nor it does include gradual changes in the chemical composition as a result of soil forming processes.

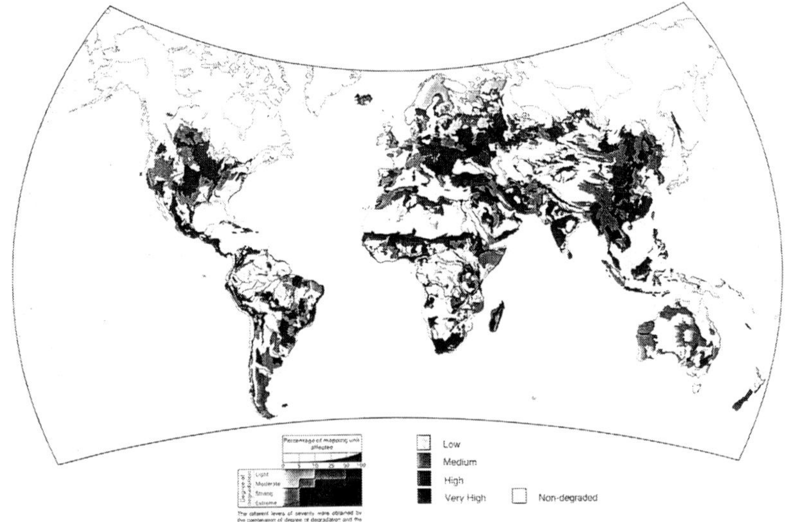

Approximate equatorial scale 1:115 million

Source: UNEP/ISRIC

***Figure 5.9** : Soil degradation severity*

Figure 5.9 shows the severity of human-induced soil degradation on the global scale. Soil degradation is assessed over the recent past averaged over the last five or ten years. Although GLASOD has specifically dealt with human induced soil degradation by processes instead of natural degradation. It is necessary to emphasize the difficulties involved in distinguishing between human-induced and naturally induced degradation processes. This problem is often acute in dryland areas, where the environment is naturally highly dynamic. During a drought period, in which vegetation cover is naturally less extensive in comparison to non-drought periods, the distinction between naturally enhanced soil erosion and that due to human activities is inevitably be subjective to some degree.

Areas that are non-degraded are either "stable" or are being degraded by natural processes, unaffected by human actions, Stable regions may be those, where human intervention us minimal due very low population densities or regions, where soil improvement or protection programmes have been successfully implemented. Some of the areas, with low population densities may be of little value for human use. Many of these zones are in hyper-arid areas, as the central Sahara and parts of Arabia.

Severity is calculated according to combination of the degree of soil degradation and the percentage of the area affected. The degree, for which soil is degraded, has been estimated in relation to changes in agricultural suitability, in relation to declined productivity and in some cases, is classified according to the following system:

(b) Degree of Degradation

There is no sign of present degradation from water or wind erosion, from chemical or physical deterioration: all original biotic functions are intact. Such land is considered stable:

(i) *Light:* The terrain is suitable for use in local fanning systems, but with somewhat reduced agricultural productivity. Restoration to full productivity is possible by modifications of the management system. The original biotic functions are still largely intact.

(ii) *Moderate:* The terrain is still suitable for use in local farming systems, but with greatly reduced agricultural productivity. Major improvements are required to restore productivity such as draining for waterlogged or salinized land or contour banks for eroded land. The original biotic functions are partially destroyed.

(iii) *Strong:* The terrain is not reclaimable at the farm level. Major engineering works are required for terrain restoration. The original biotic functions are largely destroyed.

(iv) *Extreme:* The terrain is not reclaimable and impossible to restore. The original biotic functions are completely destroyed. While a moderately degraded soil may require major structural alternations to restore productivity under one particular land use, for a less demanding land use, the alterations required may be less expensive.

(c) Relative Extent of Degradation: This classification is expressed using five categories as under

(i) Infrequent: Up to 5% of the unit is affected.

(ii) Common: Between 6 to 10% of the unit is affected.

(iii) Frequent: Between 11 to 25 % of the unit is affected.

(iv) Very Frequent: Between 26 to 50% of the unit is affected.

(v) Dominant: More than 50% of the unit is affected.

With four categories for the degree of degradation and five categories for the extent of degradation, twenty combinations are possible. These combinations have been grouped in to four severity classes given in Figure 5.9 with different shading.

Table 5.2 shows that about more than one billion hectares or approximately 20% of the global extents of susceptible dryland soils are currently being degraded by human activity. Figure 5.10, the pie chart indicates that in the light, moderate, strong and extreme categories more land area is susceptible dryland is being degraded than non-degraded regions categorized under "Others". All the figures are given in percentage terms of each category. In the total column figures in million hectares are given to witness actual area of respective category.

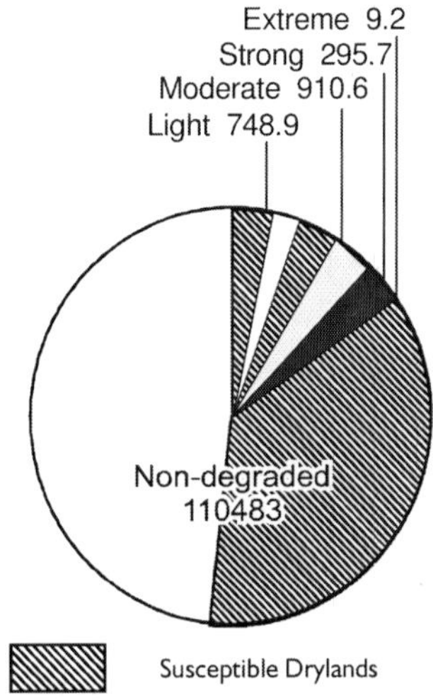

Figure 5.10 : *Globle land area by degree of soil degradation (millions of hectares)*

Table 5.2 : Soil Degradation by Continents

(% of respective category)

Region	**S O**	**Light**	**Mode-rate**	**Strong**	**Ex-treme**	**Total Degra-ded**	**Total Non-degraed**	**Total (Million hectare)**
Africa	S	15.75	13.97	23.92	38.04	16.26	8.75	1286.0
	O	7.44	7.09	17.86	18.48	8.90	13.62	1679.6
Asia	S	20.92	18.68	14.55	5.43	18.85	11.78	1671.8
	O	18.40	19.13	21.85	0.00	19.17	19.98	2584.2
Australasia	S	11.16	0.26	0.37	4.35	4.45	5.21	663.3
	O	1.74	0.18	0.27	0.00	0.78	1.84	218.9
Europe	S	1.84	8.86	0.61	33.70	5.06	1.81	299.7
	O	6.24	7.01	3.01	0.00	6.08	4.81	650.8
North America	S	1.79	6.46	2.47	0.00	4.05	5.91	732.4
	O	0.73	5.90	6.60	0.00	4.01	12.49	1458.5
South America	S	5.58	3.41	2.10	0.00	4.02	3.95	516.0
	O	8.41	9.05	6.39	0.00	8.37	9.84	1251.5
World	S	57.05	51.65	44.01	81.52	52.70	37.42	5169.2
	O	42.95	48.35	55.99	18.48	47.30	62.58	7843.5
Total (M.Ha)		**749.0**	**910.6**	**295.6**	**9.2**	**1964.4**	**11048.3**	**13012.7**

S= Susceptible O= Others. *Source:* GLASOD

The characteristics of degraded land in terms of total global land area and its continental position is as under:

(i) Total degraded land in the world is about 15.10% of the total global land out which 7.96 % is in susceptible areas and remaining 7.14% in other areas. Thus 84.90 percent land is non degraded land in global terms.

(ii) In Africa continent, total degraded land is 16.66 % of the total continental land of which 10.77 % is in susceptible areas and remaining 5.89% in other areas. Out of total 2965.6 million hectares of continental land, 83.34% land is non-degraded.

(iii) In Asia continent, 17.55 % land is degraded comprising of 8.70% in susceptible areas and 8.85 is other areas. Out of total 4256.0 million continental land. 82.45% is non-degraded land.

(iv) In Australia continent, out of total 882.2 million continental land, 88.34% is non-degraded land and 11.66% is degraded land of which 9.92 in susceptible areas and 1.74% in other areas.

(v) In Europe continent 76.99 % of 950.8 million hectare area is non degraded and 23.01% is degraded land with 10.45 % in susceptible and 12.56 % in other areas.

(vi) In North America continent. 7.25% is degraded out of total 2180.9 million hectare area comprising of 3.64% in susceptible area and 3.61% in other areas. 92.75% land is non-degraded.

(vii) 13.77% land of South America continent is degraded out of total 1767.5 million hectare area of which 4.48% is in susceptible area and 9.29% in non-susceptible areas. Therefore. 86.23% is non-degraded land in the continent.

(viii) Out of total global susceptible land of 5169.2 million hectare land, 20.03% is degraded land. Concentration of such lands is in Africa and Asia continent.

(ix) Land classified under non-susceptible area in the world is 7843.5 million hectares of which 11.85% is degraded land.

(x) Out of the total global land of 13012.7 million hectares, 1964.4 million hectares is degraded land, which is 15.10 % of the total global land.

(xi) Out of the total degraded land of 1964.4 million hectares throughout the world, 38.12% is light degraded land, 46.36% is moderate degraded land, 15.05% is strong degraded land and 0.47% is extremely degraded land. This possible is clearly visible in Figure 2.7.

(xii) Position of degraded land classification out of the total global land of 13012.7 million hectares, 15.10% is total degraded land of which 5.76% is light degraded land, 7.0% is moderate degraded land, 2.27% is strong degraded land and 0.07% is extreme degraded land.

2. Soil Degradation in Susceptible Drylands

Almost 20% or 1035.2 million hectares of the total susceptible area of 5169.2 million hectare of global area is recorded as experiencing soil degradation, which is concentrated in semi-arid and dry sub-humid land area. Hyper-arid areas are excluded from the main consideration; they generally have very strong desert characteristics and nutrient deficient soils, giving limited potential for degradation and they offer limited opportunities for degraded inducing human land uses. The Nile Valley is an exception to this as this area concentrates on areas where natural soil properties are susceptible to degradation and where land use activities can trigger this damage. Such areas are termed the susceptible drylands.

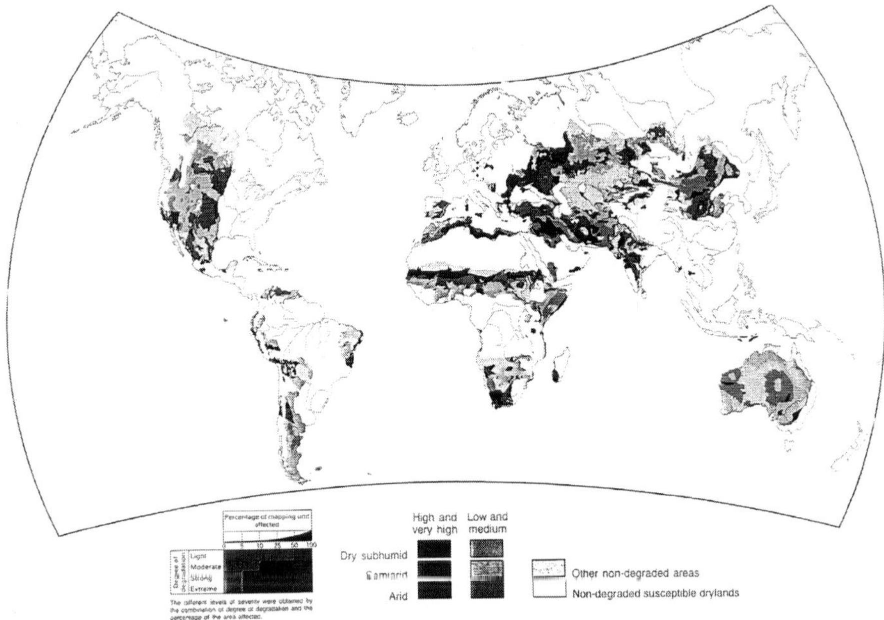

Approximate equatorial scale 1:115 million *Source:* UNEP/ISRIC, CRU/UEA

***Figure 5.11** : Soil degradation severity in susceptible drylands*

An important characteristic that makes dryland soils especially vulnerable to degradation is the slowness of their recovery from a disturbance. Since water is often not available or only in very limited quantities in a few places, new soil is formed at low speeds. Salts, once accumulated, tend to remain in situ. The build up of organic matter, which is usually consumed at high speed under warm conditions, takes a long time and moisture deficiency discourages recolonization by plants that have been removed or damaged. Hence, the ability of soils in dry areas to recover from negative changes in the soil/ land surface is lower than in humid areas. In other words, dryland soils have a low resilience. Figure 5.11 possesses all requisite details.

(a) Susceptibility to Degradation

In drylands, susceptibility to degradation by specific processes varies spatially and is affected by a range of environmental characteristics. Susceptibility to wind erosion, once a protective vegetation cover has been disturbed, is affected by the particle size distribution of soil. Soils with a high clay content naturally offers greater resistance to deflation that silty or sandy soils because of particle cohesion. With mixed grain sizes, selective winnowing of the deflatable component can dramatically reduce soil fertility and resilience. In southern Africa, the Sahel belt, central and eastern Asia and Australia, extensive areas of ancient wind-lain sediments in drylands are particularly vulnerable to deflation following damage to the protective vegetation cover.

Topography affects susceptibility to water erosion. High relative relief, in Ethiopian Highlands, encourages run-off even when moderate areas of the ground surface have lost the protection of a vegetation cover. Water erosion is rarely spatially continuous and usually manifests itself in scattered patches through rilling and gullying. The soils with most potential for loss of nutrients are those that naturally contain higher nutrient concentrations in the first place. These are the very soils, which offer some of the greatest potential for agricultural development. Other characteristics like Salinization occur locally, where water is available, either from groundwater sources or from perennial rivers with headwaters in more humid areas, as in the case of parts of the Colorado River basin of the USA and the Indus Valley in Pakistan. Overall susceptibility to degradation, in total or with regard to specific processes, is by no means, even throughout the drylands, reflecting a range of natural antecedent conditions.

(b) Spatial Patterns:

About 1035 million hectares of the world's susceptible drylands are affected by soil degradation, of which 87.08 % are in the light and moderate categories. The distribution by continent and by aridity zone is given in Table 5.3. High and very high severity degradation can be seen to be a phenomenon that is particularly significant in Africa, whether considered in relative or absolute terms. When considered in terms of total dryland area, soil degradation affects 319.4 million hectares in Africa out of a total of 1286 million hectares or about a quarter of the drylands. A similar proportion is affected in Asia, but in Europe, 99.4 million hectares out of total 299.7 million hectares dryland or a third, are degraded to some degree. 10.85% of North America, 15.33% of South America and 13.19% of Australia drylands are affected with degradation.

Information given in Table 5.3 is for susceptible dryland areas and percentage distribution of the total susceptible dryland global area to witness the proportion and magnitude of problem in each continent in light and moderate category and strong and extreme degradation.

Table 5.3 : Soil Degradation in Susceptible Dryland Areas
(Area in million hectares and percentages to total figure of susceptible dryland area of 1035.2 million hectares)

Continent	AZ	Light and Moderate		Strong and Extreme		Total	
		Area	%	Area	%	Area	%
Africa	D	25.2	2.43	12.1	1.17	37.3	3.60
	S	69.9	6.75	39.6	3.83	109.5	10.58
	A	150.2	14.51	22.3	2.15	172.5	16.66
Asia	D	70.6	6.82	7.7	0.74	78.3	7.56
	S	124.2	12.00	17.2	1.66	141.4	13.66
	A	131.9	12.74	18.8	1.82	150.7	14.56
Australasia	D	4.2	0.41	0.6	0.05	4.8	0.46
	S	32.9	3.18	1.0	0.09	33.9	3.27
	A	48.9	4.72	0.0	0.00	48.9	4.72
Europe	D	59.0	5.70	2.3	0.22	61.3	5.92
	S	30.8	2.98	2.6	0.25	33.4	3.23
	A	4.8	0.46	0.0	0.00	4.8	0.46
North America	D	15.0	1.45	3.2	0.31	18.2	1.76
	S	50.9	4.92	2.3	0.22	53.2	5.14
	A	6.3	0.61	1.6	0.15	7.9	0.76
South America	D	21.4	2.07	2.3	0.22	23.7	2.29
	S	43.9	4.24	4.0	0.39	47.9	4.63
	A	7.5	7.24	0.0	0.00	7.5	7.24
Total		**897.6**	**86.71**	**137.6**	**13.29**	**1035.2**	**100.0**

Source: GLASOD, D= Dry Sub-humid, S= Semi-arid, A= Arid areas.

Regardless of how much of the areas are affected, the map indicates that the distribution of the severest degradation in Africa, Asia and North America occurs on the wetter margins of the drylands. This is probably a consequence of two main factors. Human populations are generally higher in such areas than in the core of drylands due to greater resource base. Additionally, such areas are likely to experience the expansion of agricultural practices from neighbouring humid lands, the nature and intensity of which increases the vulnerability of degradation. The level of overall severity is consistently lower in Australia, a probable result of the dominance of low intensity agricultural practices in dryland areas.

(c) Relative Importance of Different Degradation Types

At the world scale, the GLASOD database indicates that dryland degradation is dominated by water to the extent of 45% and wind erosion by 42%. Chemical deterioration accounts for 10% and physical deterioration is about 3%. This situation, however, varies according to bio-climatic zones. Given likely variations in vegetation cover, it is not surprising that 60% of soil degradation in arid zones is by wind erosion, a figure that falls to 21% in dry sub-humid areas. The reverse trend is found for water erosion, to which the wetter dryland areas are inevitably more susceptible, accounting for 63% of dry sub-humid degradation, but 29% of that in arid areas. These and other trends are explored in greater detail in the analyses of different degradation types at the global scale.

5.3 WATER EROSION

The spatial and temporal limitations of dryland rainfall events do not discount the significance of erosion by water. In drylands, limited natural vegetation covers, relatively thin soils and high rainfall intensities all promote run-off during precipitation events and therefore, erosion occurs. By reducing natural plant densities, replacing a perennial plant cover with a seasonal crop, ploughing steep slopes and causing surface compaction by livestock trampling or the use of mechanized farming implements, human actions can readily lead to accelerated water erosion in drylands.

(a) Methodology

The definitions and criteria used by GLASOD for the designation and identification of water erosion are as under:

(i) *Light:* For soils with a rooting depth exceeding 50 cm, part of the topsoil has been removed. Shallow rills with a spacing of 20-50m spacing may be present. Thin soils have rills at least 50m apart. In areas under pastoralism, the perennial plant cover or the original or optimal plant cover extends over at least 70% of the surface.

(ii) *Moderate in Deep Soils:* All the topsoil are removed from deep soils. Rills may be present, less than 20m apart. Gully development occurs at a spacing of 20-50m. Thin soils lose part of the topsoil and are likely to have rills with a 20-50m spacing. Perennial/ original/ optimal vegetation cover in pastoral areas reduced to 30-70%.

(iii) *Strong:* All the topsoil and part of sub-soil removed from areas of deep soils, with moderately deep gullies less than 20m apart. All the topsoil is removed from areas of thin soils, exposing bedrock, weathered bedrock or a hard pan. In pastoral areas, the perennial/ original/ optimal vegetation cover is less than 30%.

(iv) *Extreme:* The general criteria that the land is unreclaimable and impossible to restore, applies.

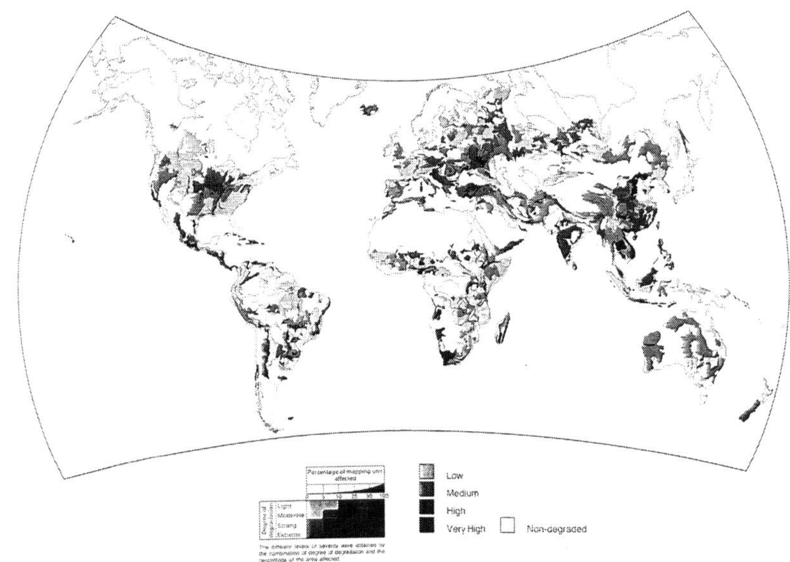

Approximate equatorial scale 1:115 million

Source: UNEP/ISRIC

Figure 5.12 *: Water erosion severity*

Figure 5.12 shows land units in which human induced water erosion occurs. It neither necessarily means that water erosion is the only form of degradation occurring in a particular unit, nor that water erosion is the dominant degradation process in a unit.

(b) Nature of Dryland Erosion

The soil commonly unable to absorb all the energy that mass provided by raindrops hitting the ground surface. A plant cover acts as an interceptor, reducing direct crop impact on the soil surface and reducing the energy available for erosion. Bare or partially exposed soils are vulnerable to the dislodgement of individual particles by drop impact in the form of splash erosion and the subsequent generation of overland follow, which transports the particles down slope. The ability of a soil to absorb raindrop energy is, therefore, dependent on antecedent environmental conditions and the nature of the rainfall event.

The ability of a soil to resist erosion is dependent upon the range of particle sizes, the chemistry and organic content as well as the nature and extent of the vegetation cover. Sandy soils have a relatively large pore space and, therefore, a greater infiltration capacity than silty soils. Clay rich soils have a tendency to form clods and are also susceptible to surface sealing, together with their low pore space, this makes them susceptible to rapid overland flow generation. The depth of a soil also influences its ability to absorb and store rainfall, while overland flow is more likely to be generated on steeper slopes. Dryland rainfall events are often noted for their intensity, which may exceed soil infiltration rates, further encouraging overland flow, which can attain higher velocities than in non-dryland areas.

(c) Processes of Water Erosion

Water erosion means increased loss of topsoil as well as water and nutrient storage capacity. It has downstream effects of flooding, siltation and eutrophication, but river erosion and sedimentation are not covered here. Apart from river erosion, water erosion takes various forms designated as follows:

(i) *Sheet-wash or Inter-rill Erosion:* It occurs as a continuous film of water, when the ground surface or as a myriad of small interconnected rivulets on rougher surfaces. Sheet-wash is effective in eroding particles loosened by drop impact and normally affects upper parts of hill-slopes. Where sheet-wash operates across an entire slope, it can lead to the reduction in fertility in upslope locations and a corresponding increase at the slope foot. Plant productivity can mirror this pattern. Sheet-wash can be difficult to monitor and therefore, may remain under estimated.

(ii) *Rilling:* It results from the concentration of over-land flow. The depth of water in rills is greater and more turbulent than in sheet-wash, giving the potential for larger particles to be entertained. Rills develop into networks that can, over time, extend laterally and up slope. They can, however, be removed by ploughing and need not be an obstacle to agriculture though they will reappear unless remedial action is taken to deal with their cause.

(iii) *Gullying:* It is caused by widening and deepening of rills or by change in surface conditions on a slope leading to a sudden increase in runoff. A Gully can be defined as having a steep head and sides, wider than 0.3m and deeper than 0.6 m. Development can be rapid and not only do gullies act as effective conduits for the removal of soil from fields, but they obstruct movement and inhibit the use of mechanized farming methods.

(iv) *Piping:* It is erosion through the development of sub-surface tunnels. This can occur naturally, particularly in certain types of susceptible soils, but it is readily enhanced by a reduction in surface vegetation and a loss of internal binding by roots.

(d) Dryland Water Erosion Assessment

GLASOD data indicate that human-induced water erosion affects about 140 million nectars in dryland sub-humid areas, 213 million hectares in semi-arid areas and 113 million hectares in arid areas. Translated to percentages of the total land area by aridity zones, this indicates that 11 % of the sub-humid zone, 9 % of semi-arid zone and 7% of arid zone are subject to water erosion degradation, with an additional 1% of hyper-arid areas affected. Water erosion affects 48% of the total degraded land area in the susceptible drylands. It is thus a substantial problem, though one that varies significantly both in terms of distribution and degree of severity.

Only European and African susceptible drylands contain areas that have experienced extreme water erosion, affecting more than 2 million hectares in each case. In Europe, this especially affects Murcia in southeast Spain, where water erosion severity is very high under both traditional vine and almond stands and more intensive land use.

When overall severity is considered, water erosion is a particular degradation problem in parts of Africa's Sahel belt, throughout the drylands of South Africa, in Spain and in China. In many cases, this can be attributed to the inappropriate use and intensification of agricultural practices on steep slopes, as in the Loess. Plateau area of China, the northern Ethiopian Highlands, the Adamawa Highlands of Cameroon. In other cases, severity can be related to specific changes in human activities. In Yemen, high water erosion severity is related to the recent breakdown of traditional soil management terraces and their associated cultivation systems, leading to irreversible erosion damage.

5.4 WIND EROSION

The movement of soil particles by wind occurs naturally in many environments. It is most pronounced and presents the most serious problems, however, in the world's drylands due to their normally sparse vegetative cover and low erratic rainfall.

(a) Methodology

Wind erosion is one of the two-soil degradation processes involving displacement of soil material covered by GLASOD have been sub-divided into three effects. These are: (1) loss of topsoil, defined as a loss by which a uniform part of the topsoil is removed; (2) terrain deformation, in which there is an irregular or uneven displacement of soil material resulting in major hollows, hummocks or dunes; and (3) the over blowing of sediment affecting physical structures such as roads and buildings. The guidelines followed in the compilation of the GLASOD data describe the following criteria for designing the degree of wind erosion:

(i) *Light in Deep Soils:* Topsoil is partly removed or few (10-40% of the area) shallow (0-5 cm) hollows: in shallow soils; very few (< 10%) shallow hollows; in pastoral country; groundcover of perennials of the original/ optimal vegetation (>70%).

(ii) *Moderate in Deep soils:* All topsoil or with common (40-70% of area) shallow (0-5 cm) hollows, or few (10-40%) moderately deep (5-15 cm) hollow; in shallow soils; topsoil partly removed or few (10-40%) shallow (0-5 cm) hollows; in pastoral country; ground- cover of perennials of the original/ optimal vegetation from 30-70%.

(iii) *Severe in Deep Soils:* All topsoil and part of sub-soil removed or with many (>70% of area) shallow (0-5 cm) or common (40-70%) moderately deep (5-15 cm) or few (10-40%) deep (15 cm) hollow/ blowouts; in shallow soils: all topsoil removed: bedrock or hardpan is exposed. In pastoral country: ground-cover of perennials of the original/ optimal vegetation is <30%.

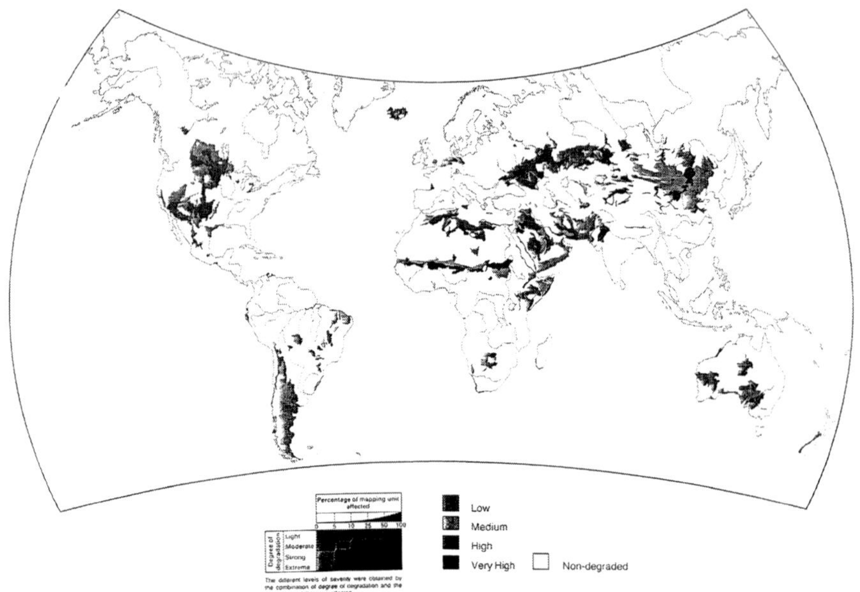

Approximate equatorial scale 1:115 million

Source: UNEP/ISRIC

Figure 5.13 : *Wind erosion severity*

Figure 5.13 shows land units in which human-induced wind erosion occurs. It neither mean that wind erosion is the only form of degradation occurring in a particular unit, nor that wind erosion is the dominant degradation process in a unit.

(b) Nature of Wind Erosion

Occurrence of wind erosion depends upon mutual interaction between the ability of the wind to entrain particles, its "erosivity" and the susceptibility of the soil surface to movement, its "erodibility."

A wind's erosivity is determined by factors as its speed, turbulence, frequency, duration and direction. The erodibility of a soil surface is determined by he properties of the soil itself, such as particle size, cohesive properties, moisture content and organic matter and the state of variables that protect the soil surface. Important aspects of the soil surface with regard to wind erosion include surface roughness, the presence of stones and boulders and most importantly the percentage cover, density, structure and orientation of vegetation.

(c) Process of Wind Erosion

The wind erosion of soils has many environmental impacts, most importantly for the farmer, but also for a range of other human activities. The problems caused by wind erosion occur both on and off-side and can be looked at conveniently according to the three fundamental phases of Aeolian activity: deflation, transport and deposition.

Soil deflation, the most important aeolian process in soil degradation, preferentially removes the finest particles. Since two of the most important influences on soil structure are the ratio of sand, silt and clay and decomposing organic matter, removal of silt, clay and organics has deleterious effects on structure. These particles also exert important influences upon the soil's ability to retain moisture; hence their removal reduces a soil's moisture-retention capacity. The fine silt and clay particles also have the maximum concentration of nutrients attached to them, removal reduces fertility.

Abrasion by moving soil particles is probably the most serious problem associated with the transport phase of wind erosion. Soil clods may be blasted by bouncing particles, impoverishing soil structure and rendering it more erodible and crop are abraded and in extreme cases cut from their roots. Structures as walls, telegraph poles and fences can also be affected. Large volumes of eroded material in dust clouds present visibility problems to road and air transport and can adversely affect radio and satellite communications. The inhalation of fine particles, some of which may be disease-carrying pathogens, can create hazards for human and animal health.

Material deposited by wind can bury and kill plants, fill ditches and block roads, runways and pipelines. Salts transferred by deflation eventscan increase the salinity of ground water and be highly destructive of buildings. There may, however, be positive aspects to the deposition of wind-eroded soils. Their high nutrient contents provide additional fertility to marine and terrestrial ecosystems, the latter both in terms of soils and as direct nutrient inputs through leaves to certain plant types like rice, wheat and grasses.

(d) Dryland Wind Erosion Assessment

Human-induced wind erosion is very largely confined to the world's drylands. Some regions immediately become apparent as suffering from a particularly acute wind erosion problem. These are the Sahel, the Maghreb, a broad belt of land stretching from the northern shores of the Black Sea to the north of the Caspian Sea, Mongolia, China, Mesopotamia and parts of the Levent and the Thar Desert in India. The importance of treating the is appreciated, when looking at the large zone of high wind erosion severity on the Arabian Peninsula. This large area lies in the hyper-arid heartland of Saudi Arabia, where human population densities are very low and the problem is confined to highly localized situations.

On the global scale, human-induced wind erosion affects 432.4 million hectares or 8% of the susceptible dryland area. In terms of the total degraded area of susceptible drylands, wind erosion is an important degradation process on 41.77% of the area. In absolute terms, wind erosion problems are most extensive in Africa and Asia, both showing affected areas as 159.9 and 153.2 million hectares respectively. Australia, Europe. North America and South America possess 16.0, 38.6, 37.8 and 26.9 million hectare degraded areas respectively. The 16 million hectares affected area in Australasia are located in Western, South and Northern Australia. In South America, 26.9 million hectares affected areas are largely in Argentinian regions of the Pampas and Patagonia.

The human activities most commonly responsible for inducing wind erosion are change or removal of vegetation cover and destabilization of natural soil surfaces. There is a variety of causes for such environmental damage. Vegetation cleared for agriculture and building and used for fuel and fodder. It is also modified by cropping practices. Stable soil surfaces may be disturbed by ploughing, by trampling of animals, by off-road vehicle use, construction or military maneuvers.

The ploughing up of grassland for grain cultivation has created massive wind erosion problems in the Great Plains of North America, notably in the 1930s and in the wheat belts of Southern Russia and Kazakhstan in the 1950s. These continued activities are largely responsible for the human-induced wind erosion in these areas today. Similar activities go a long way to explain wind erosion areas of Canadian Prairies. The situation in Africa is more complex. Overgrazing by livestock, excessive cultivation of dryland soils and cutting wood for fuel and extensive use of fire are commonly quoted as causes of land degradation in the Sahel.

5.5 CHEMICAL DETERIORATION

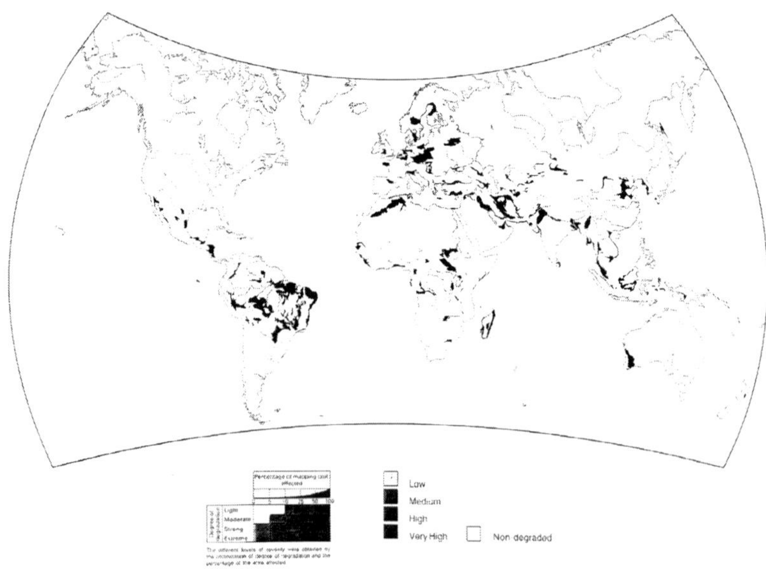

Approximate equatorial scale 1:115 million

Source: UNEP/ISRIC

Figure 5.14 : *Chemical deterioration severity*

The potential for chemical deterioration is likely to increase in situations where land use methods become more intensive and in some cases reliant on the addition of artificial fertilizers and particularly irrigated water. It is thus a problem that is unlikely to be of major significance throughout the drylands but rather one that can hit specific localities particularly hard. Such localities may also be in hyper-arid regions, such as in desert oases and in the Nile Valley.

GLASOD methodology recognizes several processes in the chemical deterioration category: Salinization, nutrient and organic matter loss, acidification and pollution. The explanation below concentrates on those mechanisms particularly important in drylands, although the FIgure 5.14 makes no distinction according to process.

(a) Nutrient Depletion

Soil nutrient depletion is a further component of chemical degradation relevant to some dryland locations. It can occur through over-cultivation and the inadequate application of replacement nutrients. This can also occur with natural vegetation clearance, burning in savanna areas of poor soils, where much of the nutrient store is held in natural vegetation rather than the soil. It also occurs, where replenishment of soil fertility by annual flooding is prevented by dam construction and the control of river flood regimes. Nutrient Depletion can also be related to water erosion, where increased run-off removes nutrients in solution or to wind erosion.

(b) Salinization:

High evaporation rates contribute to the surface accumulation of salts that leach out of the system in wetter environments. Salinization, therefore, refers to the surface or near surface accumulation of salts, mainly chlorides, sulphates and carbonates of sodium and magnesium. Salt accumulation reduces soil pore space and the ability to hold soil air and nutrients. High salt concentrations are toxic to many plants especially during the seedling stage. Plants suffer from salt burn due to excessive salt concentration, causing an inability to take up the moisture necessary for their growth. The Salinization problem also embraces alkalization, the excess accumulation of sodium, which rises in association with the dissolved salt load of irrigation water.

Though, there are extensive areas of natural salt-affected soils, Salinization is a problem, when human activities cause soil salinity to rise significantly. Drylands can be susceptible to this problem for several reasons: the predisposing climate; the need to irrigate soils to generate or increase crop yields and the occurrence of suitable terrain for irrigation in flat, low-lying situations, the localities with a natural predilection to salinity. The area of irrigated land worldwide has increased from about 8 million hectares in 1800 to cover 250 million today according to the FAO, much of it in drylands, creating an enormous potential for the problem.

In water-deficient drylands irrigation-induced Salinization occurs due to poor drainage. If water is applied in amounts exceeding those used by plants, the local water table may be raised. This in turn mobilizes stored soil salts, which then accumulate in the root zone or at the surface through capillary rise or "evaporative pumping." Even without water logging, evaporation concentrates the salts carried in irrigation water and left in the soil after water use by plants. In many places, this is enhanced, where groundwater is used for irrigation as it is often more saline than water from surface sources. Salinization can also occur without irrigation. If natural vegetation is replaced by a crop that uses less water, root-zone water logging and subsequent evaporative pumping can ensue. Here, the distinction between evapo-transpiration by plants and evaporation from the soil surface is important.

(c) Methodology

The degree of Salinization has been taken as the relative change in the status of soil salinity over the last fifty years. Salinity can be measured in terms of electrical conductivity. Four classes of salinity ranging from non-saline, slightly saline, moderately saline and severely saline have been defined at the international level. Human induced Salinization has been identified by changes in the soil salinity status.

Light Salinization is a change of one class e.g., from slightly to moderately saline. Moderately Salinization is a change of two classes e.g. from non-saline to moderately saline. Severe Salinization is a change of three classes, e.g., from non-saline to severely saline. 'Extreme Salinization is a change from non-saline to totally unproductive conditions.

(d) Dryland Chemical Deterioration Assessment

The GLASOD study indicates that in the susceptible drylands, 23 million hectares in the dry sub-humid zone are affected. 41 million in semi-arid areas and 37 million in arid areas, with an additional 11 million hectares in the hyper-arid zone. In percentage terms, 22% of semi-arid areas, 2% of dry sub-humid and arid areas and 1% in the hyper-arid areas are affected. Map No. 2.9 shows all land units in which human-induced chemical deterioration occurs. It neither mean that chemical deterioration is the only form of degradation occurring in a particular unit, nor that chemical deterioration is the dominant degradation process of an unit. In total, some 10% of the soil degradation experienced in the susceptible drylands can be accounted for by chemical processes, of which undoubtedly, the most dominant are nutrient depletion and Salinization.

Very high severity of chemical deterioration in the susceptible drylands occurs as pockets in Senegal and Sudan in Africa, in Iraq and Syria in the Middle East and in the Republics of Soviet Middle Asia. The Aral Sea Basin and the Euphrates Valley are particularly affected. In Syria, it has been estimated that about half the irrigated lands are so severely affected that crop production is substantially reduced. Downstream in Iraq, over 60% of the land irrigated under the 1953 Mussayeb Project was affected by Salinization by 1970 and subsequent reclamation attempts have only produced mixed results. Salinization has been a persistent problem in Uzbekistan, since irrigation commenced in 1902. The problem is particularly severe on the Golodnaya Steppe, near Tashkent, where over 80% of the irrigated area is now saline.

Chemical deterioration reaches high severity status in many areas. UNEP-DC/ PAC 1990 statistics indicate that irrigation supports nearly 75% of Pakistan's cropland, but the Indus Basin is now badly affected by Salinization. The construction of the Aswan High Dam, which has no provision for the passing sediment, has deprived locations downstream of nutrient replenishment from river silt.

Chemical deterioration is not restricted to developing countries. The lower Colorado River, South-western USA, supports several irrigation schemes. The Wellton-Mohawk project in Arizona, completed in 1952, has suffered particularly from water-logging and high salinity and though a drainage scheme has been implemented, the return of saline waste water to the Colorado has generated its own set of problems. The severity of soil degradation through chemical deterioration also reaches a moderate degree in part of western Australia. The Murray River catchment, near Perth, has undergone clearance of natural perennial vegetation and replacement with cereal crops and grazing land, which do not use available moisture leading to surface water-logging and salt accumulation.

5.6 PHYSICAL DETERIORATION

Six forms of in situ soil degradation by physical processes are recognized in the GLASOD methodology. These are the sealing and crusting of topsoil, the compaction of topsoil, deterioration of soil structure due to dispersion of soil material by salts in the sub-soil sodication, waterlogging, aridification and subsidence of organic soils.

(a) Compaction, Sealing and Crusting

The compaction, sealing and crusting of soil surfaces each occurs under virtually all climatic and physical conditions. The most common causes of compaction are the use of heavy machinery and trampling by livestock on soils with a low structural stability. Livestock trampling can also lead to crusting, although the most common cause of crusting and sealing is due to clogging of soil pores by fine-grained silt and clay particles dispersed by raindrop impact. This effect commonly occurs in areas, where vegetation does not adequately protect the soil surface from the impact of raindrops. The sparse vegetation cover may be naturally occurring, but equally it could result from human clearance, in which case it is included in the GLASOD classification. Soils with a low humus content, poorly sorted sand fractions and appreciable wilt content are particularly vulnerable. Crusting may also be a product of Salinization.

The effects of sealing, compaction and crusting are numerous. They hinder the tillage of arable soils and impede or delay the emergence of seedlings and the penetration of roots. Oxygen and carbon dioxide exchange between the root zone and the atmosphere is retarded. Soil's water infiltration capacity is diminished, affecting soil moisture properties, causing increased surface run-off and often higher erosion. It is interesting to note, however, that this last-mentioned effect of compaction, increased run-off, is sometimes actually encouraged by local occupants under certain circumstances. Such action is not uncommon in drylands, where water is of course a scarce resource. The clearing of surface gravel deposits to expose finer soil beneath to raindrop impact and consequent crusting and sealing is an ancient dryland farming technique designed to enhance run-off, which can be channeled to cropland. This practice has been carried out in the Near East for many centuries, whereas today a greater range of compaction techniques using heavy machinery and water-repelled materials is available for the same agricultural ends, as well as livestock, industrial and urban reservoirs. This human modification of land surface is often referred to as "water harvesting".

(b) Sodication

Sodication is a physical consequence of Salinization, largely dealt with in GLASOD as a chemical degradation process. Sodication occurs when saline water in a soil, often from irrigation in drylands, is concentrated by evapo-transpiration, leaving sodium ions dominant in the soil solution, because calcium and magnesium components tend to precipitate as carbonates. The sodium ions tend to be absorbed by aggregates of very fine clay particles or "colloids", which consequently become broken down or "deflocculated." The result is a structureless soil, which is almost impermeable to water and unfavourable to root development.

(c) Water-Logging

Water logging includes flooding by river water and submergence by rainwater, when caused by human intervention in natural drainage systems. It usually results from over-irrigation and poorly managed irrigation systems and leads to severe loss of soil air content and accumulation of toxic substances with consequent effects on plant growth. Water-logging also causes problems of sodication and Salinization.

(d) Aridification

Aridification of soils is the human-induced change of the soil moisture regime towards a more water-deficient soil system. It may be caused by lowering the local groundwater levels, other than by deep groundwater extraction. Such depletion may occur in areas where river or lake water is used for domestic purposes. Aridification may also occur in areas where natural vegetation is replaced by a crop that needs greater moisture for successful growth. The replacement of dryland grasses with wheat is the causation impact of such situation.

(e) Subsidence of organic Soils

The subsidence of organic soils, due to excessive drainage and / or oxidation, is only included in areas where agricultural potential of the land is adversely affected. This occurs when peaty materials become susceptible to oxidation after drainage has lowered the water table, leaving the peat susceptible to oxidation and deflation, hence lowering the land. Peat is often highly productive in the early stages of post-drainage, so that its loss effects fertility. Conversely, in many cases, drainage may lead to subsidence, but also to an increase to the agricultural potential of the land, in which case it is not shown on Figure 5.15.

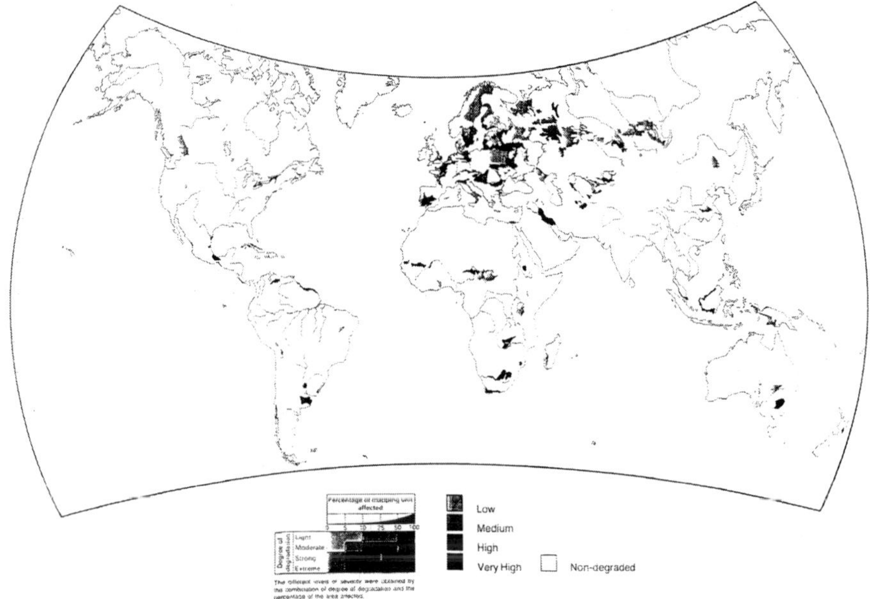

Approximate equatorial scale 1:115 million *Source:* UNEP/ISRIC

***Figure 5.15** : Physical deterioration severity*

(f) Dryland Physical Deterioration Assessment

In terms of global susceptible drylands affected, physical degradation is the least important of the four degradation types identified in the GLASOD methodology. Physical degradation has affected 34.7 million hectares or less than 1% of of the global susceptible dryland area and is mostly confined to croplands. Most of the physical deterioration in the dry sub-humid - 13 million hectares or 1% of the area and semi-arid-15 million hectares or 0.7% zones, with arid areas experiencing relatively little degradation by physical process- 7 million hectares or 0.4%.

In Africa, most of the 13.9 million hectares affected are degraded to a moderate or strong degree. These areas are in the Cape region of South Africa and in the central parts of the Sahel around Lake Chad and in the western parts of the Sahel, largely in western Mali and on the Senegal /Mauritania border.

Asian drylands affected by physical degradation are almost exclusively degraded to a light or moderate degree. An area of high severity physical degradation due to compaction and crusting occurs in Lower Mesopotamia, while medium and low severities are found in Afghanistan, parts of middle Asia, areas of Siberia, west of Lake Baikal and on the floodplain of the lower Volga River.

Susceptible drylands of 9 million hectares affected in Europe are all degraded to a light or moderate degree primarily by compaction and crusting. These areas are located in southern Spain and to the west of the Black Sea. Although, the area of susceptible dryland affected in North America is small to the extent of one million hectares, these regions are all degraded to a very high severity by water logging. They are located in the dryland regions of Mexico, in Sinaloa State on the Pacific coast and in Tamaulipas State on the Gulf of Mexico. Medium severity physical degradation characterizes the affected areas of Argentina in South America and low severity occurs on the Darling River in New South Wales, Australia.

5.7 CAUSES OF SOIL DEGRADATION AND LAND DEGRADATION

The phrase "human-induced soil degradation" implies by definition a social problem. Many types of soil degradation occur with or without human interference, but their inclusion is dependent upon circumstances, which natural processes have been initiated or accelerated by human mismanagement in a particular area. Hence, it is crucial to identify the nature of the land use that has led to soil degradation so that appropriate ameliorative management responses can be planned. Such responses must aim to make land use sustainable. This may either mean a complete change of land use or a modernization of the existing one.

1. Nature and Effects of Soil Degradation

GLASOD database has identified one or two of the five degradation causes for each land unit:

- (i) Deforestation and removal of the natural vegetation
- (ii) Overgrazing
- (iii) Agricultural activities
- (iv) Over-exploitation of vegetation for domestic use, and
- (v) Bio-industrial activities.

(a) Deforestation and Removal of Natural Vegetation

Deforestation and removal of the natural vegetation cover is defined as the total removal of natural vegetation is defined as the total removal of natural vegetation from stretches of land. In most cases, this natural vegetation is forest or woodland. Vegetation is cleared to make way for a new land use, which depend upon a vegetation cover in the case of agricultural uses such as cropping or cattle raising or new commercial forestry plantations. Conversely, some clearance is undertaken to remove vegetation permanently, like road construction or urban and industrial development.

There is a wide range of consequences stemming from such action. In areas, where the vegetation removal is permanent, erosion can be very high during the period between vegetation clearance and construction, but after construction, the new land use effectively terminates the resource potential of the soil. Nonetheless, the new land use may affect surrounding soils. The impermeable land surfaces of urban areas, create greater run-off than from natural soils and this run-off, channeled through storm drains, can be highly erosive, when it reaches uncovered soils or natural water courses.

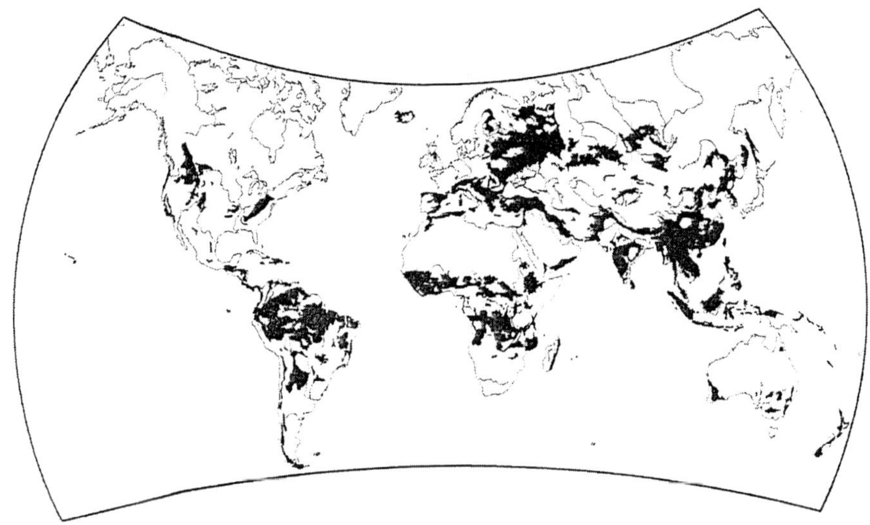

Approximate equatorial scale 1:115 million *Source:* UNEP/ISRIC, NOAA

***Figure 5.16** : Area affected by deforestion*

High rate of erosion and other degradation processes, such as compaction, can also occur in the interim period between clearance and the planting of a new vegetation cover. Once a new cover is established, different forms of degradation may occur as the soil establishes a new equilibrium with its vegetation overlay. Differences in vegetation quality and coverage may mean increased erosion. Different plants require different types and amounts of nutrients and consequently their decomposing leaf litter similarly returns different types and quantities of materials to the soil. These areas are depicted on Figure 5.16.

(b) Over-grazing

Overgrazing includes both the actual overgrazing of vegetation and other effects of livestock trampling and consequent compaction. Overgrazing and other forms of degradation due to livestock occur in areas either, where too many head are being grazed or where the wrong sort of animals are being used. The most common consequence of overgrazing is decrease in the vegetation cover, leading to increased erosion by water or wind. These effects may be extensive or concentrated in certain areas around waterholes. A widespread effect of overgrazing is the encroachment of unpalatable or noxious shrubs into grazing lands, but such encroachment certainly influences grazing potential,)t is not identified in GLASOD as degradation since soil itself is not affected. These areas are depicted on Figure 5.17.

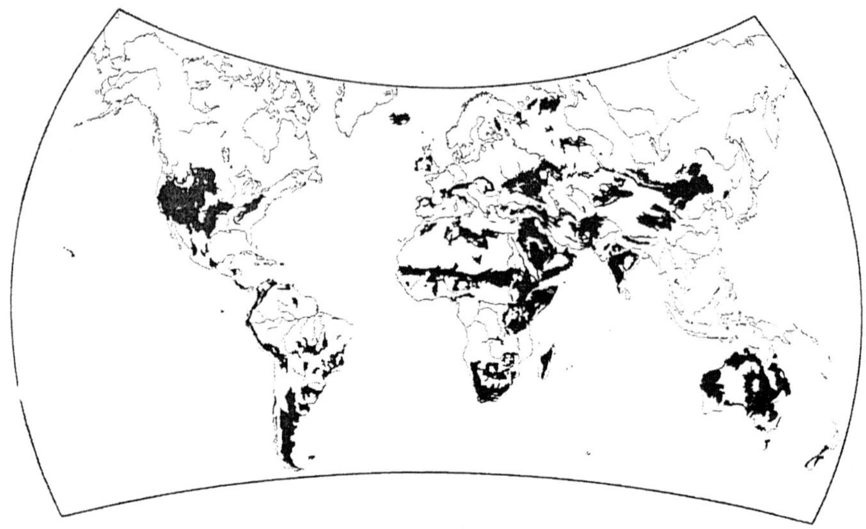

Approximate equatorial scale 1:235 million Source: UNEP/ISRIC

Figure 5.17 : *Areas affected by overgrazing*

(c) Agricultural Activities

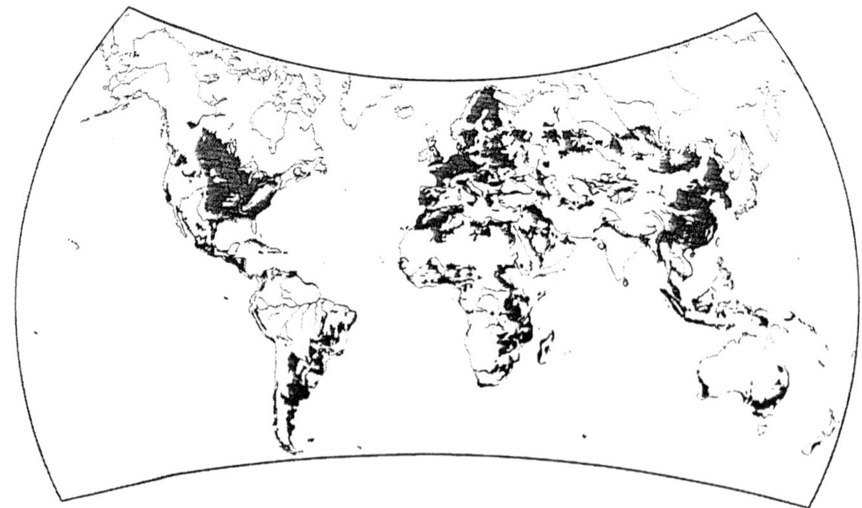

Approximate equatorial scale 1:235 million Source: UNEP/ISRIC

Figure 5.18 : *Areas affected by agriculture activities*

Agricultural activities lead to degradation, encompass all aspects of improper agricultural land management as depicted in Map No. These areas are depicted on Figure 5.18. They include a wide variety of mismanagement types, which result in various forms of degradation. Among these, are the improper use of irrigation water and poor drainage leading to problems such as Salinization, the absence of anti-erosion measures, the shortcoming of fallow periods in shifting cultivation resulting in soil exhaustion, the insufficient or excessive use of fertilizers and improperly timed use of heavy machinery.

(d) Over-exploitation of Vegetation for Domestic Sue

This encompasses the use of vegetation for such purposes as fuel-wood fencing and construction as shown in Figure 5.19. In contrast to the deforestation cause, over-exploitation for domestic purposes does not usually lead to complete removal of all vegetation. The use of vegetation is, however, of a degree that is beyond the natural capability of vegetation to renew itself and thus results in a degraded vegetation cover. This exposes soils to increased erosion and in the longer term rob the soil of inputs of nutrients and organic matter from decomposing leaf litter.

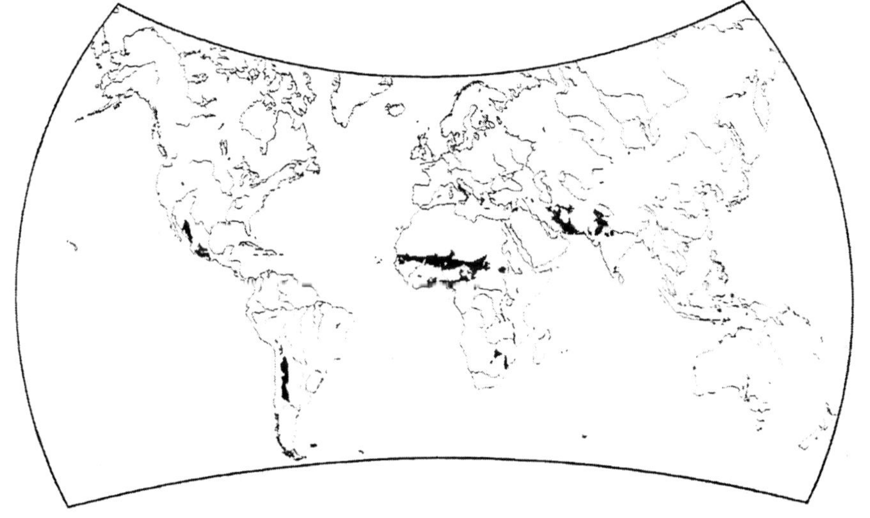

Approximate equatorial scale 1:235 million *Source:* UNEP/ISRIC

Figure 5.19 : *Areas affected by overexploitation of vegetation for domestic use*

(e) Bio-Industrial Activities

These activities lead to all forms of pollution outlined in chemical deterioration section. These include, the accumulation of urban and industrial wastes, the excessive use of pesticides, oil spills and acidification by airborne pollutants. Such causes are associated with intensive agricultural, industrial and urban land use and in practice this cause shows very little impact on the world's drylands.

2. Causes of soil Degradation in Drylands

Overgrazing is the most important cause of degradation in the susceptible dryland areas of Australia, Africa, Europe and Asia, where it affects 90.23%, 57.80%, 41.55 and 32.08% of the total degraded susceptible dryland areas respectively. It is a problem in all susceptible dryland areas of Australia and Africa. In Europe, it is largely confined to the Ukraine and in Asia, the Arabian Peninsula parts of Soviet Central Asia and Pakistan and Afghanistan are particularly affected.

In European and Asian susceptible drylands, deforestation closely follows overgrazing as the second most important cause of degradation, affecting 39.13% and 30.11% of all degraded susceptible dryland areas respectively. European areas affected include southern Spain, Sicily and southern Greece. In Asia, the main deforestation problem areas are located in a broad belt from Turkey along the Zagros Mountains and into Pakistan. Deforestation is the primary cause of degradation in south America 40.71% of the 79, 1 million hectares of drylands damaged. These are largely concentrated in northeast Brazil and on the coasts of Venezuela and Colombia.

Agricultural activities represent the most important cause of susceptible dryland degradation in North America, where they account for 52.07% of the degraded dryland area. These degraded zones are located in northern Mexico, the Great Plains of the USA and the Canadian Prairies. Agricultural activities are also important cause of degradation in Asia, where 26.11% of the degraded dryland areas exist, besides Africa 19.47%, Europe 18.41% and South America 14.66% areas are affected. Among the Asian areas, worst affected are Mesopotamia, parts of the Arabian peninsula, some areas of Siberia and Middle Asia, Mongolia and northern China. In Africa, areas of north of the Sahara are worst affected.

Over-exploitation of vegetation for domestic use only accounts for more than a tenth of susceptible degraded drylands in Africa- 16.91%, South America - 11.50% and Asia - 11.42%. In Africa, this problem is prominently located in the Sahel, in South America is largely confined to northern Argentina and southern Bolivia. Bio-industrial activities account for very small area of dryland degradation in Asia- northern Iran and the European parts of Russia. Region-wise details of soil degradation in susceptible and other areas are given in Table 5.4.

Table 5.4 : Main causes of soil degradation by regions

(in Susceptible Drylands and Other Areas)
(in Million hectares).

Regions	Aridity Zone	Deforesta-tion	Overgrazing	Agricultural	Over-Exploiation	Bio-Industrial	Total Degraded	Non-Degradad	Total
Africa	S	18.6	184.6	62.2	54.0	0.0	319.4	966.6	1286.0
	O	48.2	58.5	59.2	8.7	0.2	174.8	1504.9	1679.7
Asia	S	111.5	118.8	96.7	42.3	1.0	370.3	1301.5	1671.8
	O	186.3	78.5	107.6	3.8	0.4	376.6	2207.5	2584.1
Australasia	S	4.2	78.5	4.8	0.0	0.0	87.5	575.8	663.3
	O	8.1	4.0	3.2	0.0	0.1	15.4	203.5	218.9
Europe	S	38.9	41.3	18.3	0.0	0.9	99.4	200.2	299.6
	O	44.9	8.7	45.6	0.5	19.7	119.4	531.4	650.8
North America	S	4.3	27.7	41.4	6.1	0.0	79.5	652.9	732.4
	O	13.6	10.2	49.1	5.4	0.4	78.7	1379.8	1458.5
South America	S	32.2	26.2	11.6	9.1	0.0	79.1	436.9	516.0
	O	67.8	41.7	51.9	2.9	0.0	164.3	1087.3	1251.6
Total	**S+O**	**578.6**	**678.7**	**551.6**	**132.8**	**22.7**	**1964.4**	**11048.3**	**13012.7**

Source : GLASOD

5.8 SOIL DEGRADATION SEVERITY AND VEGETATION

Links between desertification, land degradation and vegetation are complex. Many explanations for desertification and land degradation have included vegetation degradation as a major direct component of the problem, but for a number of practical and environmental reasons it should be treated on separate basis from soil degradation.

Vegetation and soils display differing resilience to disturbance. Although, vegetation communities can be readily disturbed and degraded, recovery rates are relatively fast, though slower in drylands than in other environments, when compared to soil degradation. An effectively permanent undesirable change to vegetation can occur only when the soil resources itself have been severely degraded. In other situations, where rangeland degradation has occurred, it often appears to be the land use system that maintains the vegetation in a degraded state, rather than a change of soil characteristics that makes vegetation degradation irreversible. Both soil and vegetation degradation are of great importance, however, they are not necessarily coincident as indicators of land degradation.

Dryland vegetation generally displays sensitivity to natural climatic variability. Reducing production and ability to lie dormant are two strategies displayed by arid-land plants for survival at times of moisture deficiency. In fact, many dryland vegetation communities respond to rainfall, when it occurs but otherwise display very limited or no productivity. In may thus be difficult to distinguish, if a particular vegetation state at a particular time is a consequence of natural short-time climatic variability or a longer-term trend induced by soil degradation.

These factors contribute to the absence of an adequate global database of vegetation degradation, but tow useful relationships between soil degradation and vegetation can be determined. First, vegetation cover may help indicate vulnerability to soil degradation. A limited plant cover or low biomass, whether natural or resulting from human disturbance, can often indicate a greater susceptibility to degradation by soil displacement. Second, vegetation cover or community type may indicate that soil degradation has occurred or is taking place. The latter is difficult to determine meaningfully, except through detailed ground survey over a timescale that exceeds natural climatic variability, but useful information relating to the former relationship can be controlled at the global scale.

1. Characterizing Vegetation

Vegetation can be characterized in many ways, say for example by plant community composition. Production is another useful way of characterizing vegetation in a form that allows large scale inter-regional comparisons. Primary vegetation production can be defined in two ways:

(a) Gross Primary Production: The energy fixed plants during photosynthesis.

(b) Net Primary production: This equals the energy fixed by plants in photosynthesis minus the energy lost by transpiration. Net primary production can be measured on the ground by biomass, the volume of plant matter per unit areas. Although, this is a static measure and is less environmentally significant in degradation studies that the maximum potential production for a given environment, it can be estimated from remotely sensed data allowing vegetation patterns to be determined.

2. Global Vegetation Assessment:

The nationalized Difference Vegetation Index (NDVI) is a qualitative index of photo-synthetically active vegetation. The Global Vegetation Index (GVI), supplied by NOAA, is a weekly NDVI value, available with global coverage, derived form surface reflectance data generated from NOAA Advanced Very High Resolution Radiometer (AVHRR) satellite wage-band information. GVI values have a 16 km x 16 km spatial resolution and result from a screening process in which the cloud-free parts of all the images for a week have been combined to produce a cloud-free weekly scene. GVI values relate to the capacity of plants to photo-synthesize and relationship of plant canopy to evapo-transpiration rates. The index can, therefore, be used as an indicator of vegetation biomass, though several studies suggest that the correspondence of the index with actual values in weal under some environmental conditions particularly outside dryland areas.

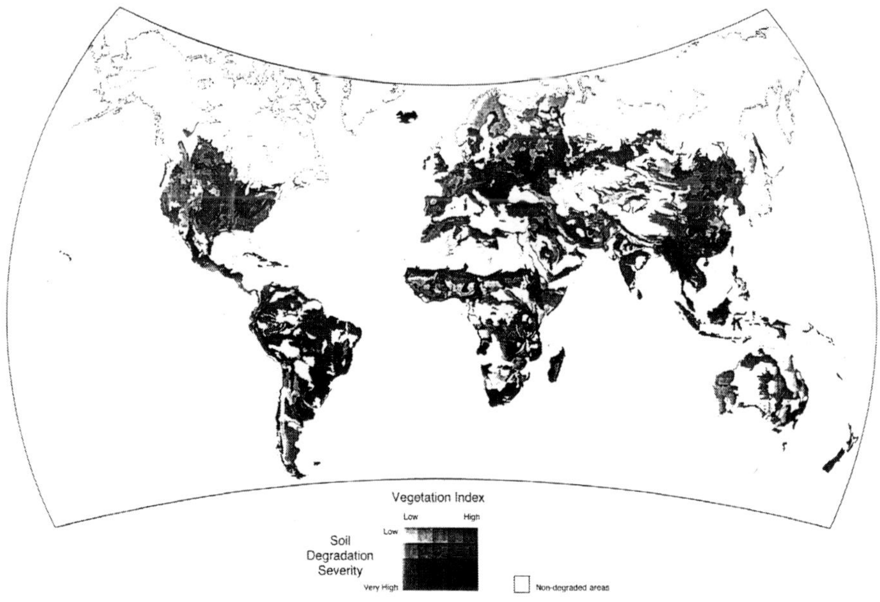

Approximate equatorial scale 1:235 million

Source: UNEP/ISRIC

Figure 5.20 : *Soil degradation severity and vegetation*

In the context of this study, a global GVI surface has been produced taking the mean of individual monthly maxima for the time-band 1983 to 1990. The continuous spectrum of GVI values is represented on Figure 5.20 as a scale from low to high, indicating biomass differences. It is further mentioned that, as the map of aridity zones, GVI values are not static, as they respond not just to human disturbance, but to natural meteorological variability, particularly seasonal variability in drylands.

3. Interpreting Soil Degradation and Vegetation Map

Figure 5.20 provides an integrated assessment of overall soil degradation from the GLASOD survey and vegetation production from the GVI. The four degrees of degradation severity have been combined with five grades of he vegetation index to produce twenty-colour grid showing general relationships between the degree of degradation in susceptible areas and vegetation production.

The meaning of relationship between the two variables requires careful interpretation. Neither casual links can be inferred from this map between vegetation production and soil degradation, nor one between the severity of degradation and types of vegetation. The map neither gives indication of vegetation quality, nor vegetation influenced by human action. It might generally be assumed that a high vegetation, but in semi-arid rangeland areas, it could represent the replacement of palatable grassland with scrub vegetation through the process of bush encroachment, which may in fact offer protection to the ground surface against further soil erosion. High values can, therefore, also include disturbed vegetation communities.

Use of the map can be made to highlight areas within which susceptibility to soil or vegetation degradation may exist or have been realized. The most productive and least degraded environments are represented by colour from the top right of the key. With high GVI Values, such areas also possessed the least susceptibility to soil erosion.

Conversely parts of the map represented by colours from the bottom left of the key most likely contain eco-systems displaying the greatest overall response to damage by human actions. Soil degradation is high, creating situations unfavourable to biotic processes, with plant communities appearing to have realized this negative potential by having low biomass. Caution needs to be exercised, when interpreting the spatial extent of such conditions though, range of scenarios that any degradation severity category can represent. Mapping units possessing colour from the top left part of the grid key may well contain areas susceptible to wind and water erosion. This is because, degradation severity during the GLASOD survey period was low, but with a low vegetation index also indicating groundcover was probably restricted, creating soils vulnerable to wind and water erosion.

4. Spatial Patterns of Drylands

The map allows several general interpretations to be made cautiously. High severity-low biomass areas are found throughout the susceptible drylands of North Africa, parts of southern Africa, Arabia, Iran, Pakistan, the southern steppes of Middle Asia and to a more limited extent in Spain. These are susceptible drylands where the greatest overall environmental degradation has occurred. Although, many mapping polygons in Australia, North America, with exception of the lower Colorado River basin and the Alberta badlands, and the Argentinain Pampas contain areas that have experienced limited soil degradation, low to medium biomass values suggest that these are locations, where potential for degradation by erosion exists, especially if land usage were to intensify. Finally, the susceptible drylands in the rain-shadow areas of southeastern Peru possess the unusual situation of lands that have been highly degraded but which also appear to have high GVI values. The ability of the map to highlight such an unusual situation points to a use in indicating areas, where the status and origins of degradation warrant more detailed investigation.

Chapter **6**

Combating Processes of Desertification and Land Degradation

The direct cause of desertification is poor land use. It is imperative to improve the productivity and sustainability of each major type of land use in the best suited manner to control the desertification. The efforts must be concrete and sincere with application of preventive and curative measures. The measures can be divided into physical and biological processes incorporating the social and economic dimensions of desertification and land degradation. The recommendations of UNCOD for improving land use and combating desertification are as under:

(1) Increasing the productivity of rain-fed cropping on good land for prevention of soil erosion and expansion of cropping on the marginal land; improving the management of irrigation cropping and rehabilitation of failed irrigation schemes to reduce Salinization and water-logging.

(2) Improving range management and development of new livestock breed to increase productivity and reduce pressure on land; intensifying rain-fed cropping on better lands and restoring traditional fallows for pasture and grazing of livestock in the dry season.

(3) Restoring tree and wood-land cover to stabilize cropping and pastoralism by reducing soil erosion and providing more supplementary fodder, halt widespread deforestation for fuel wood and reclaim desertified land for making it more productive.

(4) Introducing soil conservation practices and stabilizing sand dunes.

On the other hand, climatic fluctuation, underdevelopment. social and economic factors and misguided government policies play important roles in causing desertification and need addressed by formulating strategy to bring it under control. Climatic fluctuations and the uncertainty is inevitable in future and only effective response to make land use more productive and sustainable ascertaining it being resilient to these changes. All the physical techniques necessary to control desertification are available and their application us needed with concerted efforts.

Underdevelopment cannot be solved overnight but there is no reason to ignore the social and economic factors, as many agricultural and forestry projects are focussed on purely physical techniques. There is a necessity to ascertain the admissibility of all agricultural and forestry project to qualify on physical and biological factors as well as its social and economic dimensions. Therefore, involvement of local people in design and implementation of the project is necessary with a view to improve the standard of living which is sometimes at too much risk and stand a much better chance of success. The well-designed projects can improve the land use and overcome some of the constraints of underdevelopment.

Although, well designed projects keeping in view all the environmental and ecological aspects can be helpful for long-term success and widespread adoption of the techniques, which can be severely limit the misguided government policies like emphasizing of irrigated cropping to exclusion of rain-fed cropping are not modified.

The plan of action must advocate the improvement of land use throughout the drylands, not just in severely desertified areas or on the fringes of deserts. It is important to emphasize that one of the great misconceptions about desertification is that it occurs in the driest areas on the fringes of deserts, implying that preventive action is not needed anywhere else. This understates the role of cropping and gives the impression that the predominant cause of desertification is overgrazing, because livestock raising is the main land use in the driest areas. The reality is that desertification is caused by imbalanced land use over a wide range of climatic zone, with rain-fed and irrigated cropping in more humid areas having a major role in the process by displacing other land uses on to marginal lands that are highly vulnerable to desertification. Improving the sustainability and productivity of cr and needs to be accompanied by better land-use planning and minimization op growing, which forms the focus is vital, if desertification is to be controlledof displacement.

In general all the factors causing desertification and land degradation need to be practiced to make the degraded areas revitalized. The factors causing land degradation are many in view of multi-disciplinary problem. It is therefore, desirable to classify these issues into following two broad categories:

(a) Combating the Physical and Biological Processes of Desertification and Land Degradation.

(b) Combating the Social and Economic Dimensions of Desertification and Land Degradation.

The issues to be incorporated in the action plan for combating desertification and land degradation are as follows:

A. Biological and Physical Components

(i) Introducing methods of planning land use in ecologically sound ways.

(ii) Improve livestock raising by means of new breeds of livestock and better rangeland management.

(iii) Improve rain-fed cropping by introducing more sustainable techniques.

(iv) Rehabilitate irrigated cropping schemes that have failed owing to water logging, Salinization and alkalinization.

(v) Manage water resources in environmentally sound ways.

(vi) Project existing trees, woodlands and other vegetative cover and restore tree cover and vegetation to denuded lands.

(vii) Establish woodlots as sustainable sources of fuel-wood and encourage the development of alternative energy sources.

(viii) conserve flora and fauna.

(ix) Ensure the fullest possible public participation in measures to combat desertification.

B. Socio-Economic Measures:

(i) Investigate the social, economic and political factors connected with desertification.

(ii) Introduce measures to control population growth, as appropriate.

(iii) Improve health services.

(iv) Improve scientific capabilities.

(v) Expand local awareness of desertification and skills with which to combat it by training and education, both by means of mass media and courses at various educational institutions.

(vi) Assess the impact of settlements and industries on desertification and keep desertification in mind, while planning or expanding new settlements and industries.

6.1 IMPROVING RAINFED CROPPING

Rain-fed cropping is still the dominant form of crop production in most dryland areas. In the Sahel, it accounts for 95% of cereal production and 96% of cultivated area. It is unlikely that this situation will change very much in the medium term. Therefore, urgent action is needed to improve the productivity and sustainability of rain-fed cropping, as both meet food production targets and to bring desertification under control. The Club-du-Sahel states, " Cereals will still be grown throughout the Sahel mainly using rain-fed farming methods. It is probable that to attain food self-sufficiency the share of development assistance for rain-fed farming cereals and especially for intensified farming will have to be increased substantially." (Club-du-Sahel 1983).

The Sahelian nations, like other arid areas in the world, have increased crop production in the last few decades by expanding area under cultivation, rather than by increasing yields. Most of the newly cultivated marginal lands, however, are not suited for cropping, as fertility and yields usually fall off sharply after a few years of cultivation. As a result, grain production has not kept pace with the growth of population, food self-sufficiency has fallen and imports have risen. If present trends continue, rain-fed crop production in the Sahel will increase by only 30% in twenty years and population will have increased by 75% and each farmer will have to feed an average 3.6 persons instead of 2.8 persons today. The poor performance of rain-fed cropping is not necessarily the fault of the techniques or those practice it. Much of the blame lies with government agricultural policies to encourage rain-fed cropping. It is essential to modify these policies to encourage its productivity, otherwise cereal imports will quadruple and more land will be degraded. There are five priority areas, where action could be taken to improve the productivity of rain-fed cropping:

(a) Developing improved drought resistant crop varieties.

(b) More extensive use of fertilizers.

(c) Improving cropping systems.

(d) Better use of water.

(e) Improve land use planning.

1. Developing Improved Drought- Resistant Varieties

Breeding improved varieties of staple crops like sorghum and millets will allow farmers to gain higher yields than they can gain with traditional varieties under average rainfall conditions, although it is important to retail drought resistance of the latter varieties to prevent crop failure in dry years. Development of staple crops in the Sahel has got off to a slow start and while, a number of promising improved sorghum varieties are currently under trail in Burkina Faso. millet improvement programmes have so far not been successful. Many Indian millet varieties were tested but found to be unsuited to the harsher Sahelian environment and none performed better than the leading local variety.

One reason for the lack of progress is that local crop improvement research only started relatively recently. The International Crop Research Institute for the Semi-Arid Tropics (ICRISAT), the scientific body with responsibility for developing staple crops for drylands were only based in India, even though most of its member countries were African. The first ICRISAT research institute in West Africa, the Sahelian Centre near Niamey, Niger, did not open until 1981 and scientists there found it difficult to apply the techniques that had been learnt in Asia. The first new varieties developed by ICRISAT were next to useless for local farmers, since they were bred at research stations, which had good soils, plenty of water and mechanical ploughs, quite unlike the conditions on most of the farms in the region. Crops were also well fertilized and weeded in ways beyond the means of the ordinary farmer. When these new varieties were tried out on real farms, the yield obtained were 40-60% below those of traditional varieties. Fortunately, the lesson has been learnt that there is now greater awareness of the need to breed crops, which could grow under actual farm conditions, giving higher yields than traditional varieties on poor soils with erratic rainfall and much higher yields, if grown on irrigated lands treated with fertilizers.

Even, if researchers are successful in developing improved varieties, poor soils and erratic climate make it rather doubtful that African farmers will be able to repeat the "Green Revolution" seen in Asia, where the soils are good and the climate more reliable. The use of selected seeds, fertilizers and pesticides has increased yields of cotton in the Sahel fivefold in last forty years, but this is the only rain-fed crop the region, which has benefited from more intensive cropping. The picture is very different for subsistence food crops. The sowing of selected sorghum and millet is exceptional and the yields of these crops have either remained unchanged or declined. Farmers are reported to be suspicious of new varieties, particularly regarding to their yields and resistance to diseases and pests. They were also discouraged by the high costs of seeds. In Niger, selected seeds were used for only 3% of the area under millet and 0.3% of the area under sorghum. In Burkina Faso, selected seeds were used on less than 1% of all cereal cropland. Some progress has been made in persuading farmers to use selected seeds of groundnuts and maize, but average groundnut yields in the Sahel were low and hardly-rising even though, most are grown with selected seeds (Club du Sahel 1983).

Sorghum is the most important cereal crop in Sudan, accounting for 75% of its cereal production and one-third of its cropland. Most sorghum is grown in rain-fed conditions and only 10% of the 3.2 million hectare under sorghum is irrigated. Sudan's first commercial hybrid sorghum variety was released by Sudan Agricultural Research Corporation and ICRISAT in January 1983 after six years of research. "Hybrid Sorghum No. 1" proved high yielding, early maturing and suitable for both irrigated and rain-fed cropping. Four seasons of tests on Sudan's Central Plains gave yields 58% higher than local varieties under rain-fed cropping and 52% higher in rain-fed conditions.

Development of hybrid began in 1977 with collaborative programme involving ICRISAY. Texas AandM University and Purdue University. An Ethiopian. Dr. Gebisa Ejeta, was the principal scientist (Ejeta and Woods Thomas 1984). The first large scale trials took place on about 2000 hectare of land in 1984. during a severe drought, but yields were still impressive. Compared with the national average yield per hectare of less than 1.25 tonnes per hectare in normal years, the hybrid gave 5 tonnes per hectare under irrigation and 2.9 tonnes per hectare under rain-fed cropping.

The use of hybrid seeds means a change in routine for farmers, since new seed has to be purchased each year rather use the saved seed from the previous crops under traditional system. Thus, if Hybrid Sorghum No. 1 is to become widely grown in Sudan, it would require a substantial local seed production facility. The US Agency for International Development (USAID) has supported a pilot project, which bean in 1983, to multiply and distribute sufficient seed of the new hybrid for initial needs and lay the foundations for a local commercial seed industry. With the establishment of local seed preparation industry, sorghum production has been increased substantially.

There has been encouraging progress with another staple crop, cowpea. This leguminous crawling plant, which can grow in areas with as little as 200 mm annual rainfall, is a prolific nitrogen fixer and its stalks and leaves are also used for fodder. Improvement work began at the International institute for Tropical Agriculture (IITA) in Ibadan. Nigeria in 1970. The work was made difficult by the low yields of traditional varieties and their susceptibility to a wide range of pests and diseases. Crossing various strains to produce high yielding varieties resistant to pests and diseases was. therefore, quite a complex task. IITA had developed varieties resistant to twelve different pests and diseases and yielding about 2 tonnes of beans per hectare, compared with the tradition production of 0.3 to 0.6 tonnes per hectare. One variety, TVx 3236, has been taken up with great enthusiasm by farmers in Kano state in northern Nigeria and is giving yields three times more than the traditional varieties. Other cowpea varieties have been adapted to local conditions and introduced to farmers in Botswana. Burkina Faso. Cameroon. Togo and Zambia (Harrison 1986).

Crop development work in Tunisia has focussed on wheat. Five bread wheat varieties and five durum wheat varieties were developed by Tunisian Wheat Development Programme, with funding and technical assistance fro the US government, the Rockfeller. Ford Foundation and the International Maize and Wheat Improvement Centre (Anon 1983). Farmers accepted the improved varieties and even though the programme officially ended in 1977, while other varieties continued to develop. Annual per capita production of cereals rose from 104 kg between 1970 and 1980. For a cost of only $ 3.5 million, the programme saved $ 125.9 million in foreign exchange due to increase in production.

Another measure to improve the productivity of rain-fed cropping is introduction of new drought resistant crops. There are usually crops that have been cultivated in fairly limited areas of the tropics for many years but until now have not been the object of scientific improvement efforts. The US National Academy of Sciences had recommended few previously under-exploited crops deserve to be cultivated on a wider scale. These include grain crops like grain amarinth (Amarynthus Hypochondriacus). Pima-Papago 60 day maize (Zeamays). vegetables like Hopi mottled Lima bean (Phaseolus Lunatus) and Papago Cushaw squash, Cucurbita Mixla (NAS 1975). Persuading farmers to adopt new crops are likely to be a slow process but one. which can pay dividends in terms of higher income and better nutrition.

Increasing crop yields by the use of more productive varieties have insignificantly benefited in view of lack of grain storage facilities to protect grain from pests and spoilage and to build up a reserve for drought years due to crop failure. Mauritania has constructed a national network of grain silos in a $ 10.5 million programme co-financed by African Development Bank and the Netherlands government through the UN Sudano - Sahelian Office (UNSO).

2. Increase in Use of Fertilizers

Application of fertilizers is helpful in increasing the crop yield even of conventional varieties of millet and maize and prevent fertility of poor soil from being depleted by intensive cropping. Phosphate fertilizers are especially valuable in Sahel. where soils are generally deficient in phosphorus. Adding phosphorus stimulates root growth and thereby increases the effective drought resistance of crop, since root can reach deeper moisture levels in dry periods. Use of 15 kg. per hectare of triple super phosphate doubled millet yields in Niger from 0.55 to 1.1 tonnes per hectare and only 10 kg. per hectare tripled cowpea yields (Harrison 1986).

Despite these potential gains, fertilizer use for rain-fed crops tends to lag behind the more profitable cash crops due to high cost involved. In early 1970s, fertilizers in the Sahel were essentially restricted to cash crops e.g., cotton, groundnuts and irrigated rice, although there are now encouraging signs of a growing use of fertilizers for rain-fed cereals too. The fertilization of cereals more than doubled in Senegal between 1972 and 1979; in Burkino Faso. where fertilizer use began much more recently, it increased by a factor of 24 over the same period, although still only 3% of all cereal cropland in the country is treated with fertilizer. In the Sahel. pesticides and fungicides continue to be exclusively used for cotton.

There are a number of obstacles to the wider use of fertilizers in the Sahel. Despite the abundance of natural phosphates in the region, there is only one fertilizer factory at Dakar in Senegal and so the majority of fertilizer is to be imported. While there is a good system for distribution of subsidized fertilizers for cotton cultivators, low cereal price make it uneconomic for farmers to buy fertilizers for food production and for government subsidies for those (Club du Sahel 1983). The precarious nature of dryland farming means that yields and income are uncertain even under the best conditions, because the use of fertilizers can lead to only marginal increase in crop yields, where annual rainfall is low and erratic, especially under less than 400 mm rainfall. It costs a lot of money to buy fertilizers and seeds of improved varieties causing recurring debts to farmers. Generally more than 25% of the cultivators are defaulters of crop loans in view of this situation.

Views differ on the effectiveness of green revolution. Cropping through improved seeds and fertilizers converts many cultivators in to tenant farmer or casual agricultural labour due to burden of loans. Thus farmers are generally unwilling to commit themselves to invest for intensive cropping without any firm commitment form government for attractive crop price and reasonable credit terms. The situation is different in Asian drylands, where marginal farmers use fertilizer intensively than large farmers.

Soil fertility and crop productivity could be increased much more by application of organic manures or deep tillage. Restoring the traditional fallow system would mean that livestock would be feeded with crop residues in return for depositing their dung in the fields. An alternative to plant fast growing trees like Gliricidia sepium to produce plentiful foliage to feed animals (Falvey 1982). But farmers are reluctant to apply this system when fodder is scarce, even though they are aware with its benefits.

3. Improving Cropping System

The type of cropping system employed would make a vital contribution to increase yields over the longer term. Traditional dryland cropping systems of four year rotation sequence of fallow, millet, cowpeas and groundnuts traditional in the Sahel were sustained for being designed as resilient to tremendous variability in rainfall common in these areas. However, such systems were broken down as fallows are reduced to grow more food for raising population and maintain overall production as yields decline due to poor rainfall and depleted fertility.

Inter-cropping, the growing of two or more crops together in the same field, has long been practiced in the Sahel and yields are superior to those of monocultures. Growing millet and cowpea together, gives a total yield 50% higher than the crops are grown separately. Inter-cropping can be more efficient than monoculture if the crops are carefully chosen to have different heights and root depths to have most efficient use of available sunlight and water. The cowpea is nitrogen-fixing plant, it acts as fertilizer to other crop, leading to higher yields. The denser vegetation cover of multiple crops reduces soil temperature and erosion and increases water filtration. Weeds have less opportunity to get established, saving labour cost for seeding.

Inter-cropping also provides farmers with a form of cross insurance against crop failure. The growing period of one crop is usually different from the other and in case of late rains for one crop, reduces its growth, rains may arrive in good time for the other crop. According to a study by ICRISAT, the probability of a disastrous crop failure is one year in five, when pigeon pea is grow alone, one year in eight, when sorghum is grown alone, one year in thirteen, when both the crops are grown together in separate farms, but only one year in thirty-six, when they are inter-cropped (Harison 1986). African farmers are highly knowledgeable about inter-cropping, but further scientific research could improve the efficiency of this technique.

Many farmers in the Sahel still cultivate the soil by hand. The use of animal traction could make a substantial improvement in farm productivity, but its introduction has been slow and confined to relatively few areas. The Club du Sahel (1983) estimated that it is only used on about 5% of all cereal lands in Burkina Faso and Niger. Animal traction has become widespread in cotton growing areas like southern Mali but even there, its use for cereal cultivation is sporadic.

4. Efficient Use of Water

Many dryland farmers do not make full use of even the small amount of rainfall, which occurs on their lands, since most of it runs off before infiltrating in the soil. The productivity of rain-fed cropping could be considerably increased and soil erosion decreased by the introduction on to farms of water harvesting systems, which retain rainfall for a longer time on the soil surface so that more of it percolates through to crop roots. A system the "stone lines" project in the Yatenga area of Burkina Faso was introduced, where years of poor rainfall had lowered the water table by up to 5 meters and the sandy soil has become so crusted by overuse that grass could not grow even during rainy season. Local farmers traditionally used lines of stones placed along the contours to hold back rainwater before it san down the slope. With the help of UK based voluntary agency OXFAM and project director Peter Wright, made the technique even more effective. In the new technique, the lines of stones divide fields in to a number of catchment bands and water to held back for a distance of up to 15 meters behind each stone line. The stones were placed in a shallow trench, 5-10 meter deep and piled to form a barrier 15-25 cm high and 20-30 cm wide. Earth from the trench was placed in front of the stone line to give additional protection against water running on to it. Small gaps were left in the stone lines to pass water through the next band.

The greatest challenge was finding a cheap way to mark out the contours, for the very slight slopes were difficult to assess by eye. Peter Wright devised a simple water level by using a 10-20 meter length of transparent hosepipe filled with water. The hose was moved until the water level at each end coincide with calibration marks. Farmers were shown to master the technique in a two day course to see scale models of the new water-harvesting system, learnt from the experience of other farmers already using it and actually see it operating in the field. OXFAM trained 600 farmers of thirty villages the system of water use.

The results of this simple technique were amazing. Crop yield increased by 50% and soil erosion was substantially reduced. In the severe drought year of 1984, farmers in the village of Gourga were able to grow enough food to last them for three months, while farmers without this system application had nothing to eat. With the aid of technique the people of Gourga were able to turn some of the most degraded land into common field. The technique is now spreading informally from farmer to farmer and from village to village (Harrison 1986).

5. Integrated Farm Improvement Scheme

The World Bank was hesitant for a long time about funding the dryland agricultural development schemes, but in an attempt to identify the most promising techniques for improving rain-fed cropping it decided to fund a six year experimental project in 1979 in Baringo, in a highly eroded area on the slope of the Kenyan Rift Valley, where virtually no vegetation remain at the end of dry season and severe dust storms blow for up to a month before rains. The result was a project which supported the development of small farms and livestock owners using simple, low cost techniques, which integrated a number of the approaches advocated above, including the cultivation of improved varieties and better use of available water.

A prominent component of the project was the construction of soil conservation works, mainly different types of earth walls or bunds to prevent soil erosion in an area that had already lost much fragile soil through overgrazing and deforestation and check dams to prevent the further growth and development of gullies, which were widespread in the area. The bunds were spaced every 1.5-3 meters and enabled farmers to make the maximum use of the available rain, on an average only 600 mm per year, fro growing crops. These were planted only on bottom part of the terrace behind each bund, where most of the water collected and plant growth was better than further up the slope. The bunds, together with improved structures for tapping water from rivers, led to a major improvement in the availability of water for crops. The cost of $ 20 per hectare was paid back within one season by sorghum yields that were two to three times those obtained before, even though only about 60% of the total area of land use was planted. Once the bunds had been constructed the system required less labour than in the traditional hand cultivation of soil. Although, the techniques used here are applicable in other areas, they really require soil similar to the alluvial soils at Baringo. These are not hard at the start of the wet season and so rainwater can infiltrate easily instead of running off along the surface.

Farm productivity also benefited from the introduction of improved varieties of sorghum, cowpeas and tepary beans (Phaseolus acutifolia). Care was taken to ensure its local adaptability and farmers actually intended to grow. Two nurseries were also established to develop 2,50,000 tree seedlings per year. Although both indigenous and exotic species were sold, Leucaena, Prosopis and Sassia were more popular with farmers, who apparently found it easier to come to terms with planting exotic trees rather than conventional.

The Baringo project was unusual and without well-defined task. The project aimed to know the application of technologies being carried in other parts of the world, but there was no expectation of any financial return from the assistance. The project involved cooperation of eight different ministries of the Kenyan government but most of the decision making process was decentralized to district and project levels. Management of the project was in the hands of the Kenyan from the beginning with minimal number of expatriate experts, who spoke Swahili and in view of language problem for interaction with the local people the project took inordinately longer time.

6. Improvement in Land use Planning

Desertification is a sign that land uses have got out of balance with the land's ability to support the people. For getting additional crop production, the easiest way adopted by cultivators to expand the area by removing the vegetative cover, but such lands could not sustain after few years and became degraded. The Club du Sahel had made the surprising but categorical assessment that 13-14 million hectare land under cultivation had quarter of area of potential cultivation. The distribution of unutilized land did not match the distribution of people, who needed for subsistence cropping, hence the unwise cultivation on marginal lands in some regions was proved futile effort at the cost of environmental degradation.

Club Du Sahel had suggested that pressure on over-cultivated lands could be alleviated by encouraging people to migrate to develop new lands for cropping in under-populated areas in Mali, Chad, Senegal and Burkina Faso. but still intensification of cropping system on existing farm land was the topmost priority. If the government agencies could have strengthened the capacity of the people and regulated land use. by intensifying rain-fed cropping which was comparatively suitable strategy rather than disturbing lands fit for extensive and occasional grazing.

7. Government Policy Reforms

Increasing the productivity and sustainability of rain-fed cropping on the more fertile lands can discourage the need to cultivate on marginal lands, which often leads to desertification. Increase of 75 to 100% on existing fertile drylands is possible, according to Club du Sahel (1983). by use of improved techniques. Most of the schemes taken up were for improving the productivity rather than to control the desertification. The general lack of action of widespread neglect of rain-fed cropping by the governments, though the farmers had started practicing in drylands. Desertification is, therefore, a part of reflection of major flaw in agricultural policy and needed to be rectify quickly in the interest of farmers and increasing food production as for controlling desertification.

Improving rain-fed cropping greatly depend on the determination of governments to modify food and agricultural policies. Government controlled low crop prices to provide political support of urban population for survival of the government, but increasing of cost of fertilizers, pesticides and improved varieties of seeds certainly increase the cost of crop production and government need to enhance the prices up to the minimum level the farmers are able to sustain cultivation. Once the production cycle is maintained at a reasonable cost, it would help in production of crop and remunerative return to cultivators to continue production through rain-fed cropping. Without positive action regarding the policy reforms, the present stagnation in rain-fed cropping is bound to continue.

Fortunately, there are signs that governments in Sahel are at last giving attention to food crop production to cater the growing population. Mali had introduced programmes to increase millet and rice production; Senegal is also trying to increase millet and sorghum yields in the groundnut producing area of Sine-Saloum and to eliminate the country's rice deficit. Niger, Burkina Faso and Chad have national programmes to expand cereals research and governments encouraged to be far more open to new and more appropriate agricultural policies.

6.2 IMPROVING IRRIGATED CROPPING

Irrigated cropping have been given priority over rain-fed cropping in the agricultural development of most dryland countries in recent past and its impact are visible in the crop production. Irrigation practices are carried out through rivers, reservoirs and underground aquifers to compensate for the lack of water n dry seasons or droughts. The greater availability of water especially in critical parts of the growing season leads to two or three times more yields in comparison to rain-fed cropping. Irrigation has helped people in Ajmer district of Rajasthan in India, where groundwater was exploited by digging new wells, deepening of existing wells and installation of diesel and electric pumpsets. The efforts enabled in growing more economic crops and increasing yield. Small and marginal farmers were benefited by coverage of larger area under irrigation (Patel 1983).

Experts favour more irrigation to meet food production for growing population to overcome droughts. In dryland areas, irrigation facilities are very rare and possible through canal transport from water sources or wells in local areas. These efforts have successfully increased food production by 40% in 20% croplands. use of water saving devices has been successfully introduced with subsidized devices. This system helped in spread of rain-fed cropping on marginal lands, which results in desertification.

However, irrigation is a cause of desertification in view of inadequate design and management leading to water-logging, saliization. alkalization and sedimentation resulting in decline of crop yields due to severe soil degradation. Irrigated cropped areas have their own problems. Land degradation is a common phenomena in such areas after few years of cultivation. Irrigated cropping requires considerable investment in construction and high running costs. For keeping such croplands economic viable to grow profitable cash crops for export rather than food crops for domestic consumption. Here, the role of government is important to regulate policy reforms for water management and growing specific crops to meet the country requirements of food-grain and cash crops in rationale manner. The policy reforms needed to allow the cropping in marginal lands in the better interest of the country and growing population but not at the cost of desertification. There are three main methods to improve irrigated cropping and control desertification:

(a) Manage existing and future irrigation projects better to prevent further water-logging. Salinization and increase food production while maintaining the potential utilization at the maximum possible extent on sustainable basis.

(b) Rehabilitate existing irrigation project, where the production and productivity have severely reduced by water-logging, Salinization or mechanical problems.

(c) Switch from large scale irrigation schemes with their management problems to small scale irrigation schemes, which are easily managed and possess better potentiality to increase production of staple food crops.

(d) Cash crops and luxury crops are also essential for increasing the domestic product and meeting the country requirement. Therefore, an equilibrium need to be maintained in food grain, cash crops and luxury crops for enhancement of industrial production and export promotion if possible.

1. Irrigated Cropping in Asia and Africa

Asia is the world's most irrigated region with China. India and Pakistan accounting for almost half of all irrigated cropland, but the effectiveness of irrigation has been undermined by extensive water-logging and salinity problems. India and Pakistan account for up to 40% of all desertified irrigated cropland. Due to desertification and poor irrigation management, efficiency of irrigation of irrigation is very poor, being less than one third of Israel and Jordan (FAO 1985-a, Bansil 1983).

Africa, on the other hand, is the least irrigated part of the developing world. Only 5% of the cropland is irrigated, compared with 8% for Latin America and 29% for Asia. Main reason for this situation is lack of underground and surface water supplies. Most of the underground aquifers in West Africa are either small or deep and expensive to tap. In Mali. FAO has put the cost of groundwater at $ 15400 per hectare using diesel pump. 80% higher than for a large public surface irrigation scheme in Niger at $ 8400 per hectare and more than five times the cost of a small scale irrigation scheme in Mauritania with $ 3000 per hectare (FAO 1985-a).

Irrigated cropping of surface water for many African irrigation schemes is still vulnerable to drought. Only one-third of all the irrigated cropland in Sahel has fully guaranteed water supplies in dry season because river flows fall sharply at that time (Club du Sahel 1979). River flows in the Asian and South American drylands are more continuous because they originate from watersheds covered with forests on which rain falls throughout the year and the "sponge effect" of the forests ensures fairly even flows. Most rivers in Africa flow instead from sub-humid uplands, like the Ethiopian highlands, which experience a seasonal drought and are highly populated, poorly forested, over-cultivated and overgrazed. River flows from these watersheds remains variable even in ideal circumstances, where watersheds have a good vegetative cover, but devegetation and chronic population pressure mean that most of the rain floods down to the lowlands in the rainy season and little is stored in the watersheds to be released in the dry season. The soil erosion from hillsides by floods end up as sediment in reservoirs and irrigation canals, reducing their capacity and overall efficiency of irrigation schemes (Freeman 1986).

The giant 8,80,000 hectare Gezira irrigation scheme in Sudan depends on the level of the Blue Nile and rainfall in the Ethiopian highlands. In 1984, the rains were the lowest since 1901 prevented 126,000 hectare of wheat panting and production loss of 150,000 tonnes. Of the 9 million hectare irrigated cropland in Africa, half is in five countries of North Africa and two-fifth in Madagascar. Sudan and Nigeria. The Sahel has only a small area of irrigated cropland, just 392,000 hectare in six countries in the region (FAO 1985-a). In the first half of 1970s, Africa's irrigated area expanded by 4% per annum, but the rate of increase fell to 1.7 percent in the second half of decade (Harrison 1986). Fewer than 5000 hectare additional irrigated cropland are established every year in Sahel and 4000 hectare are for rice cultivation (Harrison 1986). This increase is just to equal to the area going out of production due to water-logging and Salinization.

2. Management of Irrigation Schemes

Significant money is spent for expanding irrigation facilities in developing countries. Investment in irrigation was running at $ 10-15 billion per annum and World Bank devoted 28% of all agricultural lending (Carruthers 1985). In the Sahel. irrigation accounts 10% of all foreign aid, in Pakistan for 10% of total public investment and in Mexico for 80% of all public investment in agriculture since 1940 (Repettp 1986).

Large scale irrigation projects cost quite high money for construction of water storage structure and canal network. Management of these schemes is generally placed in the hands of large and extremely inefficient irrigation authorities. High-level responsibilities are usually shared between various government ministries with insignificant coordination. Poor management results improper delivery of water and other inputs in time and tail end farmers generally deprived with irrigation benefits. In due course of time water-logging and salinity problems play their role which generally end in reduction of benefits.

In the $400 million Rahad irrigation project, covering an area of 4000 sq. km. in eastern Sudan, cotton yields dropped by more than 50% over the first two years of operation. Responsibility for managing the highly mechanized project, which began in 1972 was vested in a bureaucratic corporation, which according to Paul Harrison (1986): "The corporation took all decisions on crops to be grown, cotton, ground nut etc. without fallow, the type and quantity of inputs and the nature and timing of operations. It controlled directly no less than 90% of all the tenants" costs. Tenants did not comply with the corporation's directives faced fines or eviction. The rationale was that experts must know better than illiterate farmers. The experience proved that the opposite is true. Corporation services like tractors, ploughing, seeding and weeding were often late, due to breakdowns, shortages of spare parts and fuel, lack of trained drivers and mechanics."

Large scale projects are often too inflexible to respond market forces as quickly as individual farmers. In Sudan, the Corporation persisted in growing cotton, even though falling world prices made it increasingly uneconomic and farmers wanted to grow more profitable sorghum instead. Even judged on its own merits, overlooking the harmful environmental impacts, irrigated dropping does not look very attractive from economic perspective, perspective and hardly justifies the investment allocations for it. Initial crop yields never reach project levels and often decline over time due to poor management reduces the efficiency of irrigation. On the other hand salinity and water logging rise with time and the project becomes a liability.

Central Soil Salinity Research Institute at Karnal compared yields immediately before and after the commencement of six Indian irrigation projects in 1960s and 1970s. In only two prohects, rice yield exceeded 1 tonne per hectare after start of irrigation and in only one the yield exceeded 2 tonnes per hectare. Yields initially increased by 30-40% in two projects and 10-20% in another two projects, but in one project the yield actually fell down by 4% after canal irrigation. In three out of six projects, yields declined in comparison witnessed in the beginning, in two yield remained constant and only in one project, yield actually increased (Joshi and Agnihotri 1984).

The construction of large scale irrigation projects frequently runs over time and budget. Nine projects completed financed by Asian Development Bank took 72% longer to complete and cost 66% more than actually planned. Governments find it extremely difficult to collect all water charges from farmers getting benefits. In Pakistan gross receipts represented only 13% of the fiscal cost of public irrigation services (Repetto 1986). This adversely impact over the maintenance of project.

3. Improvement in Maintenance of Equipment and Canal

There are various reasons for non-fulfillment of expectations by existing irrigation projects, but lack of maintenance of machinery has been state by Club du Sahel to be "first problem encountered in all Sahelian states without exception. This is caused in part by a shortage of trained personnel and frequent breakdown of pumping and control machinery. Most of the items are imported and in case of breakdown, it is often difficult to make repairs even if trained personnel are available. Shortage of foreign currency delays the purchase of essential spare parts, sometimes for years. Fuel shortage occurs for the same reason. The pumping stations often have very short live than originally planned.

There are no simple solutions for maintenance problems, the economic difficulties the dryland countries are facing are likely to continue in the future. One solution is to promote domestic manufacture of pumps to save foreign money, but for few years the functional performance of local machinery is not so reliable. Donor countries provide assistance in the form of loan to sell their own machinery and equipment and they do not agree to get the purchases done from third country or even discourage locally manufactured machinery and equipment. Training of local technicians for repair and maintenance is necessary for functioning the machinery and equipment.

Irrigation canals also suffer from poor maintenance and a large proportion of the water never reaches to cultivated fields due to seepage through unlined canals, holes in canal banks and sedimentation. Seepage caused loss of more than 20% irrigation water in Indian state of Punjab before proper lining of canals (Gupta 1982). Canals become infested by weeds like water hyacinth, cat-tails, water lettuce and water fern. The weeds obstruct flow of water, provide favourable environment for hosts of water borne diseases like dysentery, cholera, malaria, filariasis, schistosomiasis and bilharzias and result in substantial loss of water by evapo- transpiration before it ever reaches the fields. In one irrigation scheme. Gandak command area, it was estimated that the amount of water transpired by weeds would be sufficient to irrigate 117,000 hectare of wheat and 467.000 hectare of rice (Joshi and Agnihotri 1984). Determining the responsibility for maintenance and operating canal may cause endless disputes between irrigation authorities and farmers. Since the farmers were not involved in design and management of large irrigation project, they did not feel any sense of responsibility like smaller projects.

4. Suitable Project Design

The design of irrigation project needs considerable improvement. The omission of adequate drainage from project specifications leads inevitably after a few years of operation to water-logging and Salinization resulting in decline in yields. Projects constructed without adequate preparatory assessment fail to cater the needs of farmers. Greater attention needs to be paid to social aspects of project design. Large scale irrigation projects usually imposed on cultivators with no consultation with the beneficiaries in the background of their local knowledge. Ensuring the widest possible distribution of benefits from the project is also important, especially the small and marginal farmers.

5. Greater Farmer's Involvement

Involvement of local farmers is crucial to the success of irrigation projects and can pay great dividends as their knowledge, may prevent costly mistakes in both the technical and social design along with the involvement of cultivators never thought of (Barnett 1984). In large scale irrigation projects, farmers are usually not allowed to be tenants and so become justifiably alienated when their land is compulsorily purchased with little or no compensation (Famoriyo 1984). This is the reason shy farmers are disinclined to maintain properly the parts of the system for which they are responsible. In Rahad project, village production council was supposed to give farmers the opportunity to take part in the decision making process, but they declined with a mention "they are agents for the Corporation's orders and not channels for feedback" (Tiffen 1984). In Tanzania, most of the small-farm irrigation schemes initiated and constructed by governments have been unsuccessful, largely due to the lack of active participation by farmers (Mbwala 1980).

Paying attention to farmers" needs is essential. Peasant farmers must be assured that benefits will accrue to them, for without a specific political commitment by the government, it is unlikely that they will have the same access as larger farmers to the benefits of irrigation projects (Metral 1984;

Edwards and Vincent 1983). If farmers needs are ignored, then projects may not be fully implemented. In some areas, only one crop is grown ever year, instead of two needed to make the scheme economic, but farmers decline due to high input cost and other incentives. In Mexico, farmers continued to grow traditional low yielding maize crop varieties on irrigated fields and their yields were only half than those expected. On the other hand, the Ghab Regional Pilot Project in Syria, implemented by government agencies between 1976 and 1979 was mainly profitable for local peasants: the acquisition of land and water was guaranteed, financial and technical support was provided and government intervention did not stifle competition between farmers.

6. Rehabilitation of Existing Irrigation Projects

It is in the interest of governments to try to maximize the returns on their investments by making existing irrigation projects work as efficiently as possible. It is stressed that priority must be on the rehabilitation of existing irrigation projects that have fallen into disrepair and are not working at full capacity. Construction of new ones should be taken new areas while ascertaining all the drawbacks experienced in earlier systems. In India and Pakistan, whose agricultural sectors are heavily dependent on irrigation, there is simply no alternative to large scale rehabilitation if yields are to be improved and worrying rate of desertification is to be halted.

In Africa, rehabilitation may require the repair of pumps, canals and other system infrastructure and better overall management of system. In other areas, extensive reclamation of saline and waterlogged lands are necessary. The techniques for reclamation of saline lands are well known and include the installation of pumps and drainage works to drain the woil and reduce waterlogging. Leaching salts from soil with fresh water and use of gypsum . improving the soil cover with vegetation or mulch to suppress capillary rise and surface evaporation: deep ploughing cultivation or loosening to improve the depth of rooting zone and preliminary planting of highly salt-tolerant trees and shrubs to restore nutrient cycling and drain the soil (NAS 1974). Alkaline soils are more difficult to reclaim rather than saline soils.

Draining the soil is the first priority as inadequate drainage is usually the original cause of waterlogging and Salinization. Open drains are the simplest to construct but typically occupy 15% of the total area of the cropland, which they serve and so present a major cost to farmers in terms of total income foregoing from cropping. If poorly maintained, they serve as breeding grounds for mosquitoes and the snails, which are vectors for bilharzias. Tile drains are less costly than open drains, especially if plastic piping is used. Vertical tube-wells are theoretically much cheaper than horizontal drains, where the groundwater level is appropriate, but in practice, they have proved difficult to operate (Carruthers 1985).

Pakistan began a series of Salinity Control and Reclamation Projects (SCARPs) in 1961. improving drainage channels and sinking tube-wells. Forty-two of these projects were completed at a cost of over $ 2 billion, but did not prove very successful. Drainage initially reduces water-logging and crop yield increases, but when more water is applied to increase yield further, the operation and maintenance of scheme declines and water-logging returns (Repetto 1986). Water-logging was at critical levels in 36% of all SCARP areas covering a total area of 4.3 million hectare (World Bank 1984). Salty groundwater had severely reduced the average life of tube-wells from initial estimate of forty to fifty years. Average crop yields were only about 80% of those projected at the start of project. SCARP programme had difficulty in collecting the full water charges from farmers; the government had effectively subsidized the programme by about $22 million a year. The economices of the programme were severely affected by big increase of energy cost (Johnson 1982).

The World Bank and the government of Canada attempted to reverse the cycle of failure by funding yet another SCARP project at Mardan in North Pakistan. The area was intensively cultivated but yields were significantly low as 60% of the irrigated croplands were water-logged or salinized. The project aimed to rehabilitate 29,000 hectare of affected lands by building new drains and widening and deepening 320 km of canals. The emphasis was on using gravity rather than electric power, for drainage, since Pakistan existing 20,000 existing tube-wells were estimated to account for one-quarter of the country's total electricity consumption. The project was a new departure as it used horizontal drains rather than vertical tube-wells for drainage (Johnson 1982). The problem with vertical drains was that any salts washed out of the soil stay in the system. The only way to ensure salts did not return to the soil was to flush them out to rivers or oceans by horizontal drains and canals. Initial reports were positive but water table started falling. However, it remained to be seen, whether the new SCARP project could avoid the mistake of previous project (CIDA 1986).

India possesses considerable experience in the reclamation of saline lands. Salt tolerant trees were planted in a number of states to reduce salinity and water-logging and this technique proved cost effective in large areas of barren lands for treating. Gypsum was added in the soil before planting trees, which increased transpiration of water and tree growth as well helped in to lower the water table where it was too high. The shade of trees reduced surface evaporation, thereby retarding the upward movement of groundwater and deposition of salts near the soil surface. Fallen leaves increased organic matter and improved soil structure, which further helped in penetration of roots open up to soil, improving permeability and assisting the leaching the salts. A system developed in Haryana state using Prosopis Juliflora. Acacia Nilotica and Eucalyptus spp. to reclaim highly saline barren land. These multipurpose trees are attractive because they can also be cropped for fuelwood, fodder and other products.

Governments have to bear considerable proportion of the cost of drainage ad reclamation of irrigated lands, but determining cost sharing cost by cultivators is bound to cause political difficulties. Even without government assistance, some private farmers in Egypt are reported to have been digging deep open drains on their farms, an activity made possible by low labour costs at certain times of the year and subsidized energy process (Carruthers 1985). Cost-benefit analysis both in India and Sri Lanka have shown that the value of potential increases in crop yields made reclamation of saline and alkaline lands not only socially desirable but also economically feasible (Joshi 1983; Herath 1985).

7. Small Scale Alternatives

The sheer complexity of large public irrigation projects makes them a challenge even for the most skilled manager (Ali 1980). The only solution to the problem of poor management is to switch the emphasis from mammoth projects to small scale projects, especially those, which build on local people's knowledge and expertise. In sub-Saharan Africa, an estimated 2.5 million small farms irrigated using traditional labour-intensive techniques based on shallow wells in river basins. Improving the productivity of these farms proved major impact on food production. Small scale schemes can be administered by local representative committees, without the need for large bureaucracies; the produce also marketed locally, trained farmers in irrigation skills helped other farmers to learn by experience (Makadho 1984). The decentralization approach to irrigation lead to more efficient farming and less wasteful of water where management of tube-wells and pumps remained in hands of very few people. Organization problems were also minimized and consensus developed amongst farmers to use water on necessity. In Indian state of Uttar Pradesh, private tube-well irrigation had significantly increased agricultural production, cropping intensity and higher income in comparison to public tube-well and canal irrigation (Thakur and Kumar 1984).

There was 2.2 million hectare land with potential for irrigation in the major river basins of Sahel, more than half of this area was in Chad. The Club du Sahel had increased the area under irrigation to 1.2 million hectares cultivating wheat, rice and sugarcane. Intensifying rain-fed cropping is also vital to meet the demand of growing population. Rehabilitation of existing irrigation projects is extremely necessary for enhancement of food production in South Asia.

A successful small scale project with the assistance of FAO/ UNDP for village irrigation programme on the Senegal River near Matam on the border between Senegal and Mauritania. Farmers had traditionally planted crops on land close to riven which was naturally irrigated by flood waters. The project aimed to help Halpulaar people to use water pumpbed from river to make productive land, which otherwise was not possible as annual flood water did not reach there. The programme began in 1974 on an experimental basis with just two schemes of 7 and 8 hectare. Soon other farmers also associated and participated in maintenance and operation of the scheme. The cultivators were able to produce two crops with significant higher yield.

8. Role of Cropping in Desertification Control

Making rain-fed and irrigated cropping more productive and sustainable on the most fertile lands play a crucial role in bringing desertification under control. This also increases food grain production in drylands, a policy goal high on the list of economic priorities of many African and Asian countries. This is a strategy political feasible as well as technically desirable. Nevertheless, it demands a number of shifts in agricultural policies.

1. High priority need to be given to improvement of rain-fed cropping by allocating more funds for the development of high-yielding varieties and more productive cropping system and ensuring that crop prices are sufficiently attractive to give farmers incentive to crop more intensively.
2. New large-scale irrigation projects should only be introduced after the most careful preliminary social and technical evaluation has determined their feasibility.
3. Existing irrigation projects fallen in to disrepair either due to water-logging and salinity or mechanized problems should be rehabilitated carefully.
4. The emphasis should be on small scale irrigation schemes which have short gestation period and cost can provide greater benefit to farmers, more easily managed and less susceptible to water-logging and Salinization.
5. Land-use planning techniques and skills in dryland areas should be strengthened so that the lands best suited to cropping can be identified and cropping is intensified as far as possible only on those lands, it is sustainable.
6. Farmers should be given incentives to practice less intensive types of farming, not likely to cause desertification.

Controlling desertification through these methods is likely to be economically beneficial to most of the cultivators involved in dryland agriculture and this demonstration will be helpful for sustainable development and environmental conservation.

6.3 CONTROLLING DESERTIFICATION BY IMPROVED LIVESTOCK RAISING

Nomads and herdsmen received much blame for desertification worldwide and specially in Sahel in the early 1970s. It was felt that they had let their herds grow too large during good rainfall years and had led to overgrazing. However, it was recognized later that overgrazing was as much the result of external factors as dynamics of pastotalism. The area of pasture available for grazing had been reduced by the expansion of rain fed cropping on the marginal rangelands normally used for grazing. By increased cultivation of cash crops like ground nuts on fallow lands, traditionally grazed by nomad herds in the dry season. Simply trying to regulate grazing in isolation from other agricultural sectors is therefore not sufficient to prevent overgrazing and control desertification.

Curbing overgrazing is high priority and another priority issue is to restore self sufficiency in animal products in drought affected areas and supply the food needs of growing population. Since the start of drought in 1969, the Sahel's meat exports started falling and domestic consumption of meat is now only three-quarters of the pre-drought period. Milk imports have increased six fold and the region is now net importer of milk and milk products. Imports of all livestock products to sub-Saharan Africa had reached to $ 2000 million in 1980, twice the value of all foreign aid to livestock sector over past twenty years (Anteneh 1984).

Rising demand for livestock products caused pressure for overgrazing which was far worst than the imports of milk and milk products. If the herd size increased sharply and the area of pasture available for grazing remained restricted by continued cropping of marginal and fallow lands, there was possibility of severe desertification. The situation was quite complex as the government had to suffer in both the situations. There were following five possible ways and experts agreed on the value and feasibility of each of them:

1. Improving the quality' of animals by disease control and selective breeding.
2. Increasing the off-take i.e., the number of animals sold each year for slaughter, so that the stock level falls and become more in line with sustainable earning capacity.
3. Improving rangelands by measures including reseeding, allowing time for regeneration, planting new fodder/ forage crops and other similar measures.
4. Improving infrastructure by digging wells, making other improvements along routes to market and establishing feedlots and abattoirs.
5. Changing the organization of pastoralism by taking measures to restrict or regulate grazing in certain areas, encouraging the sedentarization of nomads and their herds, establishing ranches and developing regional livestock raising schemes by stratification or other means.

1. Improving Animal Quality

Raising the quality of livestock by disease control and selective breeding increases the average yield per animal. During 1960s and 1970s, the production of meat, silk and other products in the Sahel grew rapidly than in previous decades and most of the livestock increase was due to higher livestock numbers rather than to increased productivity per animal (Anteneh 1984). Disease control is the oldest component of livestock programme in many countries and is rapidly accepted by herdsmen. Vaccination and castration of low bread helped to reduce and led to livestock overpopulation, overgrazing but droughts helped in controlling the animal population. The World Bank supported livestock disease control programme in 1960s among the Karamojong people of Uganda. The programme was also intended to improve the marketing of animals thereby counteract the increase in overall number of livestock, which was inevitable due to health improvement. However, the marketing scheme failed and herds increased in size and overgrazed the land. The Karamojong ended up worse off than before the programme started.

Other animal health programmes attempted to increase cow fertility, milk production and calf survival rates, but the methods used had depended largely upon vaccinations, mineral and vitamin supplements and concentrates. Those were quite expensive and often beyond the means of owners once external assistance was stopped. A much more basic approach was adopted in Ethiopia where herders from the Borana tribe were trained as "barefoot veterinarians" to give basic injections and cure simple health problems. Experts from the International Livestock Centre for Africa based in Addis Ababa were also introduced fodder legumes to help improve calf nutrition, lowering the mortality rate and raising weaning weight. The legumes, grown as part of the crop rotation on farms to coup up the drought were fed to cattle mixed with maize stalks. The lower mortality rate of male calves meant more animals could be sold for meat while leaving the level of milk production unaffected. This programme was succeeded as diet of the pastoralists depends far more on milk than on meat (Harrison 1986).

Selective breeding has received less attention than animal health. Considerable research has been directed towards the breeding of a tsetse-tolerant cow that could open up to livestock raising large areas of the more humid regions of West Africa that are currently unsuitable because of tsetse infestation. The hardy zebu cattle common in the Sahel, are highly susceptible to the disease trypanosomiasis carried by the tsetse fly, when they are moved south to the higher rainfall zones.

Some exotic sheep breeds, which have been introduced into the Sahel have failed because the animal could not adapt to the environment, while other apparently more suitable breeds have not been introduced. These include the long legged Kabashi desert sheep from Sudan , which is an excellent walker, produces good quality meat and needs watering only every three or four days and the small Moroccan ewes bred at the Royal Farm in the Souss region, which have a high twin pregnancy rate but have not been introduced very widely.

Goats have been received little attention from improvement schemes so far, despite their prominent role in livestock husbandry. There were more than 143 million goats in Africa in 1978, mostly in Chad, Ethiopia, Kenya, Mali, Mauritania. Niger.

Nigeria, Senegal, Somalia and Sudan (Wilson 1982). Goats are usually found in areas with 350-550 mm annual rainfall at stocking level of 20-25 head per square kilometer. They account for 31% of all meat production but only 16%

of the total live-weight biomass. One of the main reasons for their neglect is the goat meat consumed domestically and is not as significant as beef in international trade.

Another neglected animal is the camel. Daniel Stiles (1983) of UNEP had suggested that helping nomads to herd more camels and fewer cattle might be in the interest of both the pastoralists and the environment. As demand for camel grows, so will the need for better quality herds. Camels are far better adapted than cattle to dry degraded lands. They eat grass but prefer browsing shrubs and trees for leaves, twigs and berries and so do not remove the soil's protective grass covering, as would a herd of cattle. Camels spreads out when they browse and their soft hooves do not cause as much damage to soil as do hard hooves of cattle. Camel can go for more than a week without water and even in dry periods continue to produce 5-10 of milk a day compared to less than half a litre for a cow. Milk is the main food for many nomadic peoples and so camel provides them with an insurance against drought. Camels are much more efficient at converting vegetation into milk A family of six need a herd of twenty-eight camels to provide its subsistence instead of sixty four cattle (Stiles 1983).

Camels cost more than cattle, so a project funded by the West German government for helping the Samburu people of northern Kenya to buy camels. The Samburu exchanged their cattle for camels, which the project bought in Somalia and money from the sale of cattle sued to buy more camels. Camels suffer from a number of diseases and like cattle, they are susceptible to trypanosomiasis carried by the tsetse fly. Some new initiatives were in hand to tickle the problems. The Food and Agriculture Research Mission, a non-governmental organization had begun a "flying vet" service for camels in Kenya while in 1986 UNEP held a camel workshop for Samburu people to learn like the Rendille, who had herded camels for hundreds of years and developed techniques for preventing and curing diseases (Hilsum 1987).

2. Reducing Stock Levels

It could be argued that if pastoralists decreased the size of their herds, there would be more food and water for remaining animals, productivity would increase and overgrazing and desertification would decrease. The average proportion of all animals sold each year was estimated to be around 9% for Burkina Faso, Niger and Mali in the late 1960s, a few percentage points higher than in the early 1950s, but very low by western standards (Jerve 1981). Raising fewer animals can smooth out the dramatic swings in livestock numbers that resulted from large herd sizes during good rainfall years, which were then reduced by high mortality rates in times of drought.

Estimating the sustainable carrying capacity of a region is difficult. One suggestion was that livestock numbers should never significantly exceed those, which could be maintained during the worst years and that was best to stock at 80% of average capacity in order to keep at least 20% forage reserve for drought years (USAID 1980). On that scale, the 1982 Sahel herd size of 23 million was

5% above the safe sustainable level. Such exercises remained largely academic and attempts to reduce stock levels by increasing the number of animals sold each year always seemed to lead to problems. "Stock reduction scheme did not work as the herdsmen operated on fiat, created resentment and antagonism among, a traditionally independent people," according to Walter Goldschmidt (1981) of the University of California at Los Angeles. Nomads had fiercely and violently defended their right to retain their mobility and to determined herd sizes, the two cornerstones of nomadic pastoralism.

The plan to increase off-take remained theoretical as it did not fit in with the herdsmen point of view, as the herdsmen were means of storing wealth, rather than so much potential beef had to be sold and were more in the form of capital than a source of income (Jerve 1981). The need to conserve such capital, especially as an insurance against bad years, was paramount. The herdsmen objected the strategy to reduce the number of stock as it could reduce the chance of a viable herd remaining after the drought had ended.

Herdsmen used to barter their livestock, animals and milk in exchange for grain from farmers. Cash need not and often did not change hands. In such system, the exchange rate for livestock and grain was disadvantageous to herdsmen and more animals were required to obtain same amount of grain. Conditions were changed, as farmers had to sell some of the grain in national and international markets, leaving less available for barter. From herdsmen point of view, the herd represented the basis of an alternative form of economics. It was also central to all social negotiations, recording kinships and creating patterns of alliance, dependency and support. "A herd is the monument and obituary of its owners and a close reading of the animals in the kraal is a record of the major social interactions of a herder's life," writes Walter Goldschmidt. describing the Sebei people of Uganda. The same is true for pastoral societies. Cattle were commonly given as bridal payments and the Sebei calculated lineages of cattle to express social rather than biological continuity. "A Sebei wants to retain at least one representative of important sources of such cattle lineages, e.g.. one from each aunt's, sister's and daughter's bride payment" (Goldschmidt 1981).

Such attitudes are quite logical to the herdsmen but difficult to represent in mathematical equations of livestock supply and demand; hence the problems experienced in introducing a more rational off-take system based on estimated carrying capacity. More mundane reasons can also be behind limitations on off-take. Herdsmen, who look after livestock belonging to farmers receive their payments largely in the form of milk from those animals. This leads to over-milking, high calf mortality and low rates of reproduction and few animals could be sold for slaughter.

3. Improving Rangelands

If there were fewer livestock, the valuable perennial grasses and browse shrubs would have a chance to regenerate. Reducing livestock numbers in the Badiya El

Sham area in Syria, which receives only 150 mm rainfall a year, resulted in the re-emergence of desirable perennial grasses and shrubs. In Gabes, in southern Tunisia, which receives 195 mm annual rainfall, controlling stock number and rotating areas of rangeland used for grazing led to improved vegetative cover. This reduced the area, which a ewe needed to graze every year from 7 to 2 hectare. Similar results have been achieved in Jodran by-protecting rangelands and allowing only occasional light grazing (FAO 1985-b).

In the absence of natural regeneration, the improvement of rangelands requires seeding with desirable fodder plants. The introduction of two new forage grasses was one of the major accomplishments of the huge $ 103 million seven year. Drought Prone Areas Project to stabilize and increase agricultural production in six drought prone areas in west and central India. The project, which ended in 1981, was funded by the World Bank. The two grasses, Stylosanthes hamata and Stylosanthes scabra. came from Australia and have the potential to increase fodder production substantially, providing a low cost way to improve rangelands rapidly. Stylosanthes could be established on the better wastelands just by broadcasting seed, while on badly eroded sites, simple furrows are required. The plant sets seed quickly, even under low rainfall and its early growth is unpalatable, premature grazing is disturbed (World Bank, personal communication). Tunisia has created 138,000 hectare pastoral reserves in the governorates of Gabes and Medenine by planting forage species and protecting them from over-exploitation (UNEP 1985). Reseeding rangelands had been successful in areas of USSR, Australia and Pakistan receiving less than 300-400 mm rainfall, but other countries had experienced difficulties (FAO 1985-b).

Trees are another valuable source that has been sadly neglected in livestock development projects. Lessons learnt in Indian project was the reintroduction of traditional fodder trees and shrubs like Prosopis and Ziziphus in rangelands was just as important as establishing perennial forage grasses. The potential of many of these multipurpose trees, cropped for fodder, fodder and a variety of other products, had really recognized and their use is being encouraged by prominent organizations like FAO. International Livestock Centre for Africa and US national Academy of Sciences. Research to develop this potential is one of the most rapidly growing fields in tropical forestry.

Forty-four species of the genus Prosopis are typical multipurpose trees. They are highly drought-resistant, saline-tolerant and members of Leguminous family bear pods containing up to 14% protein and 44-55% carbohydrate. Most species are nitrogen fixers. Prosopis can grow in areas receiving less than 100 mm of rainfall per annum and even in highly drought-resistant, where Acacia Tortilis could not survive. Quick growing roots reach as deep as 80 meters beneath the surface to tap underground water, but the tree can also grow in hyper-arid areas, where hardpans prevent roots from reaching groundwater. There are some evidences that prosopis can take in water from leaves. Its saline tolerance can used to make salinized lands productive again.

Foliage and pods of Prosopis trees can be used as fodder. In the arid and semi-arid lands in India, Prosopis Juliflora and Prosopis cineraria trees are lopped for fodder every winter without harming future growth. The leaves and pods are also browsed by sheep and cattle and Prosopis is preferably used as mixed diet, in Chile the pods have formed the sole diet of sheep without ill-effects and supported more than ten sheep per hectare (Pedersen 1980). Plantations of Prosopis nigra and Prosopis alba have been established in Chile and Brazil, giving an annual fodder production of 1-2 tonnes from 50-105 trees per hectares. Animals are excluded from the plantations for the first five years and afterwards their browsing has to be carefully managed. Prosopis species are also planted in India and Pakistan (FAO 1985-b).

Acacia Albida is another valuable fodder tree. It is highly drought-resistant and in some dryland areas the only greenery remaining in dry season. It is also a soil improving, nitrogen-fixing tree, with additional advantage of a relatively open canopy during the crop growing season that allows it to be grown together with field crops like millets. Far from having any detrimental effects, crop yields under and around Acacia albida have been found to be equivalent to those in fertilized fields. Another outstanding fodder tree is the carob Ceratonia siliqua, which has a long history of use in countries around the Mediterranean and in other semi-arid and sub-humid areas, where it grows on thin soils of degraded lands and rocky hillsides. An evergreed tree growing up to 10 meters in height, the carob's pods are poor in protein but rich in sugars, besides being harvested for local livestock, they are also used for commercial feeds (Winer 1980). Two equally versatile multi-purpose trees widely used as sources of fodder in sub-humid areas are Leucaena and Gliricidia (Falvey 1982).

Fodder trees and shrubs need to be protected from uncontrolled browsing to avail long term benefits. The concept of "fodder reserves" adjacent to boreholes has been advocated, but without proper protection the trees can be over-browsed or cut for fuel-wood. The feasibility of such reserves is increasing and the World Bank and USAID fund for pastoral associations, which shoulder the responsibilities to manage specific boreholes and protection of planted areas.

Another possibility is to restore village fallow land, lost to cash-crop cultivation. More and more farmers are switching from cash crops to staple crops following stimulated by appropriate incentives. The land is to be replenished with fodder grasses, trees, shrubs and traditional symbiosis between pastoralists and farmers restored. Another possible way to improve animal feed value and the productivity of the herds by using oilcakes produced by farmers as supplementary' feed (Khogali 1983).

Long term management of fodder trees and shrubs is easier in India, where animal keeping in villages is widespread and there are community grazing lands, but protecting these land from unauthorized browsing is still major problem. In the State of Gujarat, large scale social forestry scheme has planted multi-purpose fodder trees and shrubs on marginal village lands. Fodder is often a more attractive product than fuel-wood. In India and Nepal, people are more easily motivated to plant trees to harvest fodder from them. Villagers plant fodder trees and protect from herds, but the situation is different in Sahel. where pastoralists are nomadic, who did not stay at one place long enough to manage fodder reserves. Pastoral associations can overcome this problem, while the herds continue to move around the region, the responsibilities of herders are well defined.

Many pastoralists, like the Borana in Ethiopia, have been forced by dire circumstances to grow crops, while others have been pressurized by strong hand of government to become completely sedentarized. In both the cases, possibility exists to intercrop valuable fodder plants with regular food crops. Sedentarization has led to severe overgrazing around some villages due to high concentration of livestock those were previously scattered over large areas of rangelands. In such cases high-yielding fodder trees and other sources of forage can be the only solution to support intensive livestock production without causing further environmental degradation. In Africa, many villages are still skeptical about animal keeping. Efforts to widen the use of animal based ploughs instead hand-wielded hoes have made little headway, despite economical disadvantages.

From a broader perspective, the pressure on rangelands could be alleviated by-intensifying field cropping on existing arable lands (Khogali 1983). A more radical measure of controlling the encroachment of cropping on to traditional rangelands, had been tried in Syria, where the government had restricted rain-fed cropping in semi-desert grazing areas and encouraged cooperative approach to the control of grazing. In Niger, cultivation has been prohibited by law north of latitude 15*N, and farmers south of that line could sue herders for compensation if their crop was damaged by livestock.

4. Digging of More Wells

Digging more wells for improving livestock is another measure. In dryland areas problem of underground water already exist and increasing more number of wells certainly accelerate desertification process. The phenomenal growth in the number of watering points, especially along routes to markets, has concentrated herds in limited areas along routes and around wells, causing degradation and desertification through trampling, browsing and grazing. Although, herdsmen clamour for the construction of new wells, they are often aware of such dangers and prefer privately owned wells to which access is limited. The Bororo herdsmen from the Bernmou region of Niger had urged the government not to sink new wells in their traditional grazing areas, since that could attract outsiders and overload rangelands. They had suggested instead that well be sunk in Tuareg country to the north, as Bororo could use those in dry season, but would not have to reciprocate by sharing their own wells. On the other hand, the Tuareg also appreciate the threats which wells pose to their rangelands.

Despite the great profusion and clear warnings at UNCOD of the dangers of too many water points, Sahel government assisted by aid agencies, continue to sink new wells. Since the mid 1970s, almost $ 7.5 million were channeled through UNSO to well sinking projects in Gambia. Mauritania and Burkina Faso. Ground water studies are necessary to be undertaken before starting any project of construction of new wells.

In India, ground water studies have been undertaken for ascertaining level of underground water and as findings of such studies, all the areas were classified in to white, gray and dark zone. In areas demarcated as dark zone, construction of wells and electrification of existing wells had been banned to regulate the underground water resources and combat desertification.

Once the water points are established, regulating its use can prove very difficult and in some cases, can lead to violence. According to Peter Hopcraft of the World Bank. "In areas, where bore holes have been installed, attempts have sometimes been made to turn off the pumps before grazing destruction sets in. The results of these attempts have been generally been such anger that pumping equipment is deliberately destroyed. This has led to proposals that pumps and equipment be removed to the next block of grazing areas, again forcing the livestock to move at water points. In general, as anyone who has been thirsty and without water in hot and arid environment knows, tempering with what is regarded as a possible water supply in an explosive activity'."

There are some encouraging signs of a change in approach to wells. UNSO financed a project to rehabilitate degraded rangelands surrounding boreholes in Gambia. Villages were given the responsibility for taking care of boreholes and rangelands in their vicinity and in return they received the produce from orchards of cashew and other trees, which were planted close to boreholes. Cattle from co-operating villages were circulated around a number of boreholes to lighten the pressure on rangelands (UNSO. personal communication).

5. Regulating Nomads

Nomadic grazing is a sound technique for using scarce pasture resources in dryland areas, where rainfall fluctuates widely from place to place and mobile herds have freedom to search for areas where rain has fallen and grass has grown. Maine-Sora in the south-east Niger has an average rainfall of 432 mm per annum. In 1949, it received 230 mm of rain, but only 67 mm fell at Diffa, 50 km further east. Herds would have suffered if government grazing regulations had prevented them from moving out of Diffa that year.

Mobility also means better nutrition, if herds could arrive in a place. When the grass is still green, digestibility is of the order of 70% compared with 43% for dry grass. Without mobility, herds would gain ever-poorer feed value from the pasture as its digestibility decreased throughout the season. Cattle which annually trek more than 1000 km between Malian Delta and the Mauritanian Sahel return to better physical shape than the milking cows and calves, which remain in the village throughout the year.

Despite the advantages of nomadic grazing, there have been continuing attempts by governments and aid agencies to regulate it. mainly by reducing herds mobility. Some attempts arise from a desire for a more scientific approach to livestock raising, which it is hoped may prevent overgrazing. Others result from a determination by governments to bring these wandering people more under control, whether for altruistic reasons such as providing education and health facilities or merely to tax them easily. There are four types of interventions:

(a) Grazing Controls,

(b) Sedentarization,

(c) Establishment of Ranches,

(d) Stratification.

(a) Grazing Controls:

Despite strenuous efforts over the last many years, most attempts to control grazing have been failed. This is largely due to the ignorance of experts or the neglect of traditional grazing controls exerted by pastoralists. It may be imagined that grazing lands belong to community and they are mostly unregulated, but this is not usually the case. Michael Horowitz, a social anthropologist at the State University of New York at Binghamton writes.: there is an emerging awareness among scientists that strict regulation of access to scarce resources of water and grass and limits on herd size may well be the rule rather than the exception among herding societies in semi arid areas" (Horowitz 1979).

The imposition of grazing controls by governments has. therefore, aroused resentment and resistance. One type of control is the grazing fee, which did not limit herd's freedom to move, has the motive of forcing the herdsmen to dispose of non-productive animals. In view of Peter Hopcraft, "Enthusiasm for the notion of a grazing fee should not be anticipated, especially from the owners of relatively large herds of cattle. Paying for a resource that has such a long and culturally embedded history of being freely available or at least free, even if not always available, is an innovation that is likely to have its legitimacy seriously questioned. Grazing fees were an integral part of a number of grazing schemes in the colonial period and they undoubtedly generated considerable antipathy.

Another type of control is the division of a region into blocks of rangelands. These are then used successfully as a part of rotational scheme in which each block is left for a fallow period during which pasture regenerates. One way to manage such a scheme is to turn off or physically remove the pumps from water holes in a particular block, where forage is approaching exhaustion. This could provoke violent reactions on the part of herdsmen. In Potkot and Samburu blocks of Kenya, the system failed dismally. Enforced sedementarization limits the freedom of movement, which is key component of nomadic life.

(b) Sedentarization

Sedentarization or settlement of nomadic herdsmen is one option for improving livestock raising in arid and semi-arid areas, but in practice, it is difficult to consider it in isolation from the strong political pressures against nomads, which in some countries had led to forcible sedentarization. Many nomads of the Fulani in Mali settled spontaneously in the aftermath of Sahel Drought. They treated it as a temporary measure till their herds build up again, but many had returned to their former way of life.

Sedentarization has a number of disadvantages in practice. The increased concentration of livestock around a village often degrades nearby pastures, which could be exacerbated by an increase in herd size as the settled nomads tend to become more dependent on grain, require less milk and meat. The resultant impact could be selling of animals for slaughter. Resettlement schemes in Sierra Leone, Burkina Faso and Nigeria had not very successful as their animals had suffered from health problems in new locations and conflicts arose between nomads and local peasants. Nomads traveled longer distances had experienced particular difficulties in adjusting in settled life.

Fulani in Central Nigeria had successfully settled nomads by combining livestock raising with cropping. Each household cropped an average 0.9 hectare with sorghum and maize accounting for 70% of the total area. Forage crops were grown with cattle manured fields. Nomads may suffer a drop in income in settled life, as they had to pay for fodder, which was free during movements. The economy of nomads could be dwindled.

(c) Ranching

Introduction of intensive livestock ranching in developing countries has rarely succeeded. After second world war, development usually meant transferring western technology without adaptation to developing countries. Many aid agencies eventually tried to take Texas to Sahara. The President of Niger and other Sahelian heads of state talked of exporting beef all over the world, but reality was less rosy. Marketing of production was hampered as ranches were too far from large commercial centers.

(d) Stratification

Stratification is a far more elaborate form of reorganization of livestock management. The basic idea was to rear the calves in rangelands of north Africa and then move to more humid pastures of South Africa for fattening. This was the basis of the SOLAR project (Stratification of Livestock in Arid Regions) and practiced as one of six trans-national project. With a few very local exceptions, stratification has been a failure. The life-style and culture of nomadic pastoralists severely limit the rate of off take from the herd and herdsmen were not willing to sell large numbers of young stock for fattening. The planned reduction in herd size could not be achieved and there was no improvement in living conditions of nomadic people as envisaged.

Some success was achieved with attempts to stratify Kenyan beef industry by breeding herds and producing store cattle on dryland ranges and then finishing animals for slaughter on intensively managed pastures on better lands in special feedlots (Creek 1984). The feedlots increased income and employment but they were owned by large farmers or state enterprises and the benefits were not widespread.

6. Prospects for Livestock Development

Many projects for livestock were tried but could not yield much success. There were various reasons but important were the project preparation done by outsiders totally unaware about the geophysical conditions of Africa and the cultural background of herdsmen. The implementation was faults' due to non-involvement of local people. This is a truth of many external aided projects as their experts prepare plan in their own setup without considering grass-root situations. The recipient government is also inclined to receive external aid without understanding the consequences of the project over the people and area. This is also a crucial factors that regulations are made to satisfy donor conditions, but systems cannot be regulated by force or against the wishes of local beneficiaries for acceptance of the project.

The lack of success is not surprising. Nature does not run like clockwork in the Sahel or other arid areas. It is difficult, if not impossible, to devise a block grazing scheme which can be flexible enough to give levels of productivity and insurance against poor rainfall similar to those of the traditional system. The schemes are devised on the basis of inappropriate and inadequate policies and very often coordinate with the actions of pastoralists. Poor project administration results in to failure (Haldeman 1985).

Therefore, in the interest of the people and appropriate development of livestock, it is essential to understand the situations first in which the people survive in natural degraded lands. In an area, where monsoon failure is a common and natural phenomena, dependence on agriculture is quite uncertain. The livestock rearing is the principle livelihood means for survival of the people in such areas and the local governments as well the donor agencies should consider the reality while preparing any plan for livestock development keeping in view the social, economic, biological and physical situations of people and area. The environmental and ecological aspects are the primary' parameters for success of such projects for sustainability.

In principle, it is possible to improve the livestock production in way which do not result in overgrazing rangelands and which are compatible with the traditions and wishes of the pastoralists. However, major problems await solution in terms of achieving the right balance between conservation of environment and the constraints of land tenure; balancing the individual and community interests of pastoralists; overcoming the cultural and economic conflicts between pastoralism and commercial beef production and decision making for the best path for livestock development.

7. Livestock Dilemma

Technically, it seems quite simple to control desertification caused by overgrazing, wither by reducing the number of animals or by improving the quality" of rangelands. All the attempts with meagre resources and inadequate management illustrate the importance of social controls over resources. Reducing the extent of overgrazing requires greater recognition by governments of the importance of pastoralism in dry areas and corresponding action to prevent the incursion of rain-fed cropping on the marginal rangelands. which reduces the area available to herds and is a major indirect cause of overgrazing.

6.4 CONTROLLING DESERTIFICATION BY PLANTING TREES

Tree plantation plays a vital role in establishing more productive and sustainable land uses in dryland areas. Trees protect land from soil erosion, reclaim already degraded land and helps to prevent further deforestation by providing sustainable supplies of fuel-wood, fodder and other products. The rate of deforestation is 4 million hectares per year in the world's drylands of which two-third in Africa. Trees are being planted in a wide variety of ways. In early projects, forest departments established large-scale fuel-wood plantations, usually on government forest reserve lands to counter the growing scarcity of fuel-wood. But the planting rates could not keep pace with rising demand for fuel-wood, the major source of energy in drylands. The focus was switched on social forestry schemes, where trees were planted on village forest or community lands.

In view of success of the schemes, forest departments started inspiring local communities to plant trees and spread of improved wood and charcoal stoves to make the most efficient use of wood from remaining forests and new plantations. Alternative means of fuel-wood energy were attempted world wide to ease the crisis of inordinate cutting of trees and it was deemed proper to invent local technology with cost effective to cope up the rural masses to switch over the alternative sources. The success areas were biogas based on cattle dung; solar energy based cookers, kerosene oil and liquefied gas. The biogas and solar energy cookers could not yield much success even after government incentives. The kerosene and liquefied gas are in short supplies to provide all the local people. In response to growing shortages of fuel-wood stress was focussed on fast growing tree plantations.

1. Large-Scale Fuel-wood Plantations:

The initial response of forest department in Sahel to rising fuel-wood shortages was to +-establish large-scale plantations of fast growing tree species using conventional forestry techniques on government owned land. These project were not very successful, mainly due to slow growth of trees in areas with poor soils and low rainfall. This led to general pessimism about the feasibility of large-scale fuel-wood plantations in Sahel.

In a project funded by the World Bank in Niger. 779 hectare of fuel-wood plantations were established between 1979 and 1981 within the boundaries of forest reserves. This was 10% higher than the target area and was achieved in two years, one year earlier that the originally intended. Neem (*Azadirachta indica*) accounted for 71% of the area and Lucalyptus camaldulensis for the remainder. However, the rates were lower than anticipated due to poor soils and no preliminary soil survey was carried out (*World Bank*).

Mali forestry department established fuel-wood plantations in the La Faya ad Monts Mandingues forest reserves between 1980 and 1985, in an attempt to keep up the fuel-wood demand in the nearby capital city of Bamako, which was expected to rise from 0.3 million cubic meters in 1982 to 1.5 million cubic meters by 2000. Gmelina arborea plantations accounted for over 70% of the total area of 2273 hectare area and the remainder consisted of neem (*Azadirachta indica*) and Eucalyptus spp. in both pure and mixed stands. However, this World Bank funded project also railed to meet its target as plantations could be established in only two-third of the intended area.

USAID funded project in Senegal was supposed to establish 3000 hectare of fast growing Eucalyptus plantations in Bandia fuel-wood area between 1979 and 1983 to provide much-needed fuel-wood to the cities of Thies and Dakar. Here too, two-third of the target area was planted. Mortality rate was higher in view of low rainfall and annual growth rates were only 1.5 cubic meters per hectare against the targeted 10 cubic meters per hectare expected, mainly because the species used were not appropriate to the site (USAID 1983-a).

2. Social Forestry

Social forestry is the only viable long term strategy to meet afforestation targets with involvement of local people for tree plantation outside regular forest areas. This is also relevant due to rapid rising demand of fuel-wood and the fact that national forestry departments are already stretched by their modest afforestation commitments. Even if government forestry departments were able to plant the required number of trees, without wide popular support the young trees might be lopped or felled for fodder and fuel-wood by farmers and nomadic herdsmen. The forest officers who were earlier little more policemen, had to change overnight into patient inspirational tree gurus is quite a challenge. Experiences of various countries for developing the social forestry scheme are narrated below:

(a) China and South Korea

In the developing world, the most spectacular examples of afforestation carried out largely by ordinary people in China and South Korea. Between 1950 and 1983, China's forests were increased by 60% to 122 million hectare. This required a phenomenal afforestation rate of 1.5 million hectare per year and included massive projects like the "Green Wall" against the Gobi Desert, a huge shelterbelt of 1500 km long and 12 meters wide which 700,000 farmers established in just two seasons. Although China was able to mobilize large number of people to plant trees, the programme could not yield into total success. When the vast areas of plantations are established over a short period, no sufficient care is possible in earlier years resulting in high failure rates. In 1987 large areas of forest including part of green belt were destroyed by fire. Preliminary estimates spoke of almost 700,000 hectare being lost in Heilongjiang province alone, but no data were available for other affected areas in Inner Mongolia (Thompson 1987). However, China intended to increase its forest cover from 12% to 20% by 2000, but ultimately achieved 30%.

South Korea was faced with a shortage of fuel-wood, more than 20,000 villages formed cooperatives to plant over 1 million hectare of fruit, fuel and timber trees in just five years, from 1973 to 1977 (Arnold 1983; Gregersen 1982). South Korea is a highly capitalist country but like China, has a tradition of organized community activities. There already existed a national movement to improved rural living and elected Village Forestry Associations were established in each participating village. The government enacted legislation to make land available for the new wood-lots together with existence of a grass-root organization, which the forest service could advice and provide with seedlings and other materials, had a great deal to do with the success of the programme.

In other countries, the social conditions for afforestation are not as favourable as in China and South Korea, so it is unlikely to achieve similar results of afforestation. However, social forestry schemes may well prove to be the practical options, if forest cover is to be restored. There are two main components in most social forestry scheme, tree plantation on village lands for forest department personnel with hired labour, preferably from local village and true community forestry in which it is villagers themselves, who plant and tend trees. Often these schemes also promote farm forestry on private farms.

(b) Social Forestry in India

Social forestry was pioneered in India in the western state of Gujarat. In 1969, the state forest department set up a separate Extension Forestry Wing to plant trees outside regular forest boundaries and its first activity was to establish multi-row wind-breaks along-side roads and canals, with shade trees nearest the road and faster growing trees for fuel-wood and timber behind them. The forest department owned the land and planted trees, but nearby villages had grass cutting rights, protected the samplings and in return took a share in profits. These mini plantations were not planned or planted with much public participation but "they marked a critical psychological turning point. The new plantations also gave many people their first positive impressions of the forest department. They saw that their activities could benefit them directly." By 1982 tree plantation along road and canal side in 14750 km. equivalent to 29.500 hectare of forest. The trees were harvested on a rotation of 7-10 years and benefits shared equally between forest department and local people (Karamchandani 1982; Foley and Barnard 1984).

Next phase was started in 1974 with the "Supervised Woodlots" scheme to establish small plantations of about 4 hectare on community grazing lands in 28.000 hectares in 4000 villages. The demonstration impact was quite encouraging and village councils started contacting forest department to benefit their villages with similar activities. In 1980 forest department started another scheme "Villager Self-help Scheme" followed by another scheme of tree plantation on cultivators land in 150,000 hectares, where each landless family was assigned 37.5 hectare @ 2.5 hectare per year in fifteen years viewing the performance of earlier years. This got overwhelming participation of local people as each family was given Rs. 250 per month for protection and growing trees.

The projects were part of National Forestry Programme implemented in total country and Gujarat state remained pioneer throughout the process. The country-wide response of the local people remained quite encouraging as people took overwhelming interest for participation in the schemes. There was one more component, which helped people to encourage were individual benefits where no investment was required. Poor people who had no land or other capital assets for livelihood took it as a relief to get their livelihood through their labour support.

(c) Social Forestry in Africa

Social forestry projects in Sahel were taken up in 1980 after witnessing the unsatisfactory results of large plantations. Some progress was made in Burkina Faso and Niger, where people paid for planting trees, but community based projects, with involvement of villagers to work together to plant trees on village lands for joint benefit could not succeed significantly. Church groups and other non-governmental organizations in Sahel were often active than governments promoted afforestation and financed a number of small-scale projects.

The most popular social forestry measure in the Sahel was "greenbelt" a shelterbelt of hardy, drought resistant trees, which protected town or village from harsh winds, swirling sand storms and encroaching sand dunes, where people were benefited to harvest fuel-wood, poles and fodder. Greenbelt around Ouagadougou, capital city of Burkina Faso in 500 hectare and 300 hectare belt around Niamey, capital city of Niger had successful amongst others. The town of Bouza in Niger is famous for its cool tree lined streets.

Tanzania began a community forestry programme in 1967 but little progress was made during ten years. In one African village woodlot programme was successfully implemented in 260 hectares funded by UK government's Overseas Development Administration, the Anglo-American corporation and the government of Lesotho. The majority of trees in woodlots were species of Eucalyptus and Acacia. Trees were planted at altitudes of up to 2000 meters and coniferous trees grew faster than Eucalyptus although they did not coppice and were susceptible to fire (UNEP 1985).

The Eastern Refugee Reforestation Project in Sudan was implemented with paid and voluntary planting components, provided an example of difficulties faced in African social forestry. The project area had a 250 and 500 mm annual rainfall and located on the clay plains of Kassala province in eastern Sudan, which was completely deforested by-rapid expansion of mechanized sorghum cultivation and led to extensive soil erosion and gully formation. In addition,

400,000 refugees were streamed into the province from Ethiopia and made it their home. The major objective of multi-facet USAID and CARE funded project was to restore tree cover to improve the environment and quality of life in the area, provide renewable sources of fuel-wood to local people, increase employment opportunities and encourage the adoption of sustainable agro-forestry practices.

By 1985, two years into the project, progress was mixed. More than 1200 hectare of fuel-wood plantations were successfully established, despite the ravages of drought of 1984. which left many seedlings dead. The replanting measures were ruined by heavy rainfall. Out of 93 km of windbreaks planted, only 28% were successfully established. owing to incorrect seeding, competition from weeds and inundation by heavy rains. The voluntary planting component was less successful. The extension programme certainly led to plantation of many trees in house compound. According to an evaluation by USAID team, the focus shifted from agro-forestry to house forestry due to inexperienced staff and unwilling experts discussing their difficulties openly. The village woodlot scheme was a failure as trees suffered from shortage of water and farm grazing by local goats. Wire fences were insufficient and villagers were unwilling to supply volunteer guards for the woodlot. There was a general lack of interest of villagers to manage woodlot for common benefit.

Some of the social aspects of the project were successful. Refugees accounted for 90% of the casual labour force and benefited enormously from being employed to plant trees in wind breaks and plantations. However, the tree-planting programme confused them, since there was no clear distinction between relief and development aspects. Many refugees accepted trees for planting in their compounds just they accepted relief food, rather than out of any strong commitment to afforestation and were subsequent surprised at how much cost them to buy water to irrigate the seedlings.

One of the problems with social forestry' in African countries in contrast to India was that the projects funded by donor agencies largely focussed on growing trees for fuel-wood in the belief that those would correspond to a real local need. Unfortunately the situation was quite different as Tom Catterson, Fran Gulick and Tim Resch pointed out, "the fuel-wood model accounts imperfectly for disappearance of trees in the desertification equation" and community forestry would be unlikely to succeed unless it was supportive to local people's main priority of growing food. Therefore, the people required planting those trees, which provided food, fodder, shelter and fuel-wood. Village woodlots were exacerbate to distinguish between agriculture and forestry, but the people wanted much greater integration between the two practices.

3. Community Forestry

Community forestry schemes were formulated with the general assumption that community involvement was essential if countries in Sahel and other dry regions could plant enough trees to control desertification and satisfy need for fuel-wood. The ultimate goal of most social forestry programmes was therefore to achieve true community forestry by mobilizing a larger number of people to plant trees than would be available from the forestry' department alone, to ensure trees once planted would be protected from damage and pre-mature harvesting and improving the livelihoods of local people by increasing the productivity of land and widely distributing the benefits obtained thereby. The reasons for lack of success were more social than technical.

(a) General Awareness Creation amongst Women

The community forestry projects face numerous obstacles. It may take years to make aware local community to understand need for tree plantation to reach a point at which they ask to join a scheme. This requires exposure to general publicity through posters, media and patience of social forestry officers in discussing matters with villagers. Visits to other villages already joined the scheme to see tree planting in practice. This preliminary phase, although protracted, can well determine the likeliness of success of the scheme before beginning. The women folk face the problem in collection of fuel-wood and could better understand the project rather than men. Women are the very categories who have most to gain from tree growing for subsistence purposes and arguably for cash too. Shepherd made ain interesting point that since women already tend to co-operate with each other for subsistence cropping purposes, involving them more closely in social forestry project could greatly increase their success rate.

(b) Ensuring Wide Distribution of Benefits

Ensuring that as many sections of the community participate in and benefit from community forestry projects is not only desirable but absolutely essential. Project contains a component to protect young trees by fences, guard or other means, from browsing by animals or felling by humans. Technically such a component is understandable and necessary, but socially it is not sufficient, as access to resources is closely associated with equity and project benefits are shared between all sections equally. Without careful consultation, some people are likely to feel aggrieved or dispossessed by exposure of land for forestry and they could try to continue to use the land for grazing or other purposes. All sections of the community generally have established rights and privileges upon village common lands, but often the people most affected are from poorer sections, who depend heavily upon the community lands for grazing their animals or gathering fuel-wood or fodder for their own use and for selling at local market.

Equity has a major role in determining the acceptance of social forestry projects. Social projects in India have been criticized by development economists and non-governmental organizations, who claim that they benefit rich and disadvantage the poor. This is a common objection to any government policy or programme in developed as well as developing countries. It arises from a combination of idealism and false expectations. The term social forestry and community forestry may give the impression that they would help to redress the economic imbalance between rich and poor inside developing countries, but it is doubtful, whether such interpretations were ever contended by foresters in the early development of social forestry. They thought that the terms simply described a new type of forestry which involved ordinary people rather than professionals and probably the most they expected was that local people would derive some benefits in terms of improved earnings, supplies of wood, minor forest produce and better environment. The idea that social forestry would also be a form of social engineering was far from their minds, there is no reason the forestry sector be better at redistributing wealth than the agriculture sector.

(c) Social Structures: Ideas *vs.* Reality

Some of the critics of social forestry expect more altruistic behaviour from people in developing countries than is evident in their own. The success of projects can be threatened by political or economic schisms inside the village, which are totally unconnected with tree planting. According to Erik Eckholm, Gerald Foley, Geoffrey Barnard and Lloyd Timberlake, "In few, if any village does all the residents have identical interests and priorities. Community programmes must be based on compromises on matters of deep concern to all. Rigid social stratification, such as India's caste system, makes any community' action extremely difficult. Community' programmes may presuppose the existence of competent and representative local organizations that can plan and implement projects and distribute benefits fairly. But often, villages have no organizations with such capabilities. The main lesson from the past is not that community programmes should be abandoned, rather their design and implcmcntation should be improved" (Eckholm *et. al.* 1984).

Another complication arises in India, where village forests are often Panchayat forests. A Panchayat is the administrative body in charge of usually more than one village. There are endless possibilities for one village to feel aggrieved about the project if it feels disadvantaged relative to others. In one Panchayat, in Madhya Pradesh, the grass from a young plantation area was distributed among all people in the Panchayat at a nominal price but money raised was used to electrify just one of the five villages in the Panchayat. People from another village felt aggrieved and released their cattle into the plantation.

(d) Economic Incentives

It is unrealistic in most cases to expect villagers to co-operate merely for the sake of a community ideal. As Ton Catterson (1984) remarked, "the attractive solution of shared work for shared benefits in pursuit of a solution to a common problem, the raison d'etre of village forestry, remains an elusive, Utopian goal." This helps to explain the types of social forestry have most successful so far, have either been those with a low degree of community involvement or project on private farms, where the prime motive is profit rather than some hazy ideal of improving the life of the community. The commercial incentive is often the strongest stimulant to tree growing. Indian farm forestry has been criticized on the grounds that the richer farmers benefit most and that such schemes lead to fall in employment, but while richer farmers may well be the first to establish the woodlots. they are often able to take risks with a locally unproven concept. The smaller farmers are able to follow their experience that large farmers are making money. Certainly, the latest studies of farm forestry in Gujarat show a fairly wide spectrum of participating farmers.

(e) Political Support is Vital

Political support by government can differentiate between success and failure of social forestry projects. Tree planting as reached the level of mass movement in Kenya and this is due to advocacy of government officials from the president downwards. Frequent coverage of the issue in newspapers, radio and television have created awareness among the people to understand the necessity of tree plantation. The number of nurseries and seedlings raised are unable to keep up pace with the growing demand for tree plantation on farms and house compound. Every district commissioner, chief secondary school and branch of Kenya's political party is required by presidential order to have nursery. In addition voluntary groups are also engaged in development of nurseries.

(f) Conflict over Land

Lack of available land is often a constraint on community forestry projects. While a village may have a sizeable area of degraded grazing lands, individual members of community have different views over its use. This is classic problem of the commons, which recurs throughout tropical land-use matters. Generally protests are made by people who continue to avail benefits from that lands. Such conflicts are very difficult to resolve in communities, which do not have homogeneity of ethnic, economic or social interest or which lack social cohesion.

(g) Matter of Trust

Community forestry projects sometimes fail to live up to expectations due to limitations of those responsible for implementing. Often, foresters deputed to organize schemes are trained in technical aspects but possess no ability to deal the people in favour of the project. The most crucial and determining factor the success of community forestry programme is the degree of trust between villagers and forest department representatives. In one village of Gujarat, it took five years to establish trust of the community. In Pakistan there were more than 50,000 cases of forestry offences pending in the local courts, when the government decided to launch the project. Once such suspicions have been finally solved it is not vital to initiate the project in an area. When villagers in a number of West African countries did not receive the benefits to which they were entitled after woodlots had been harvested, it became unlikely that they would participate in any future projects.

(h) Social Forestry as Fuel-wood requirement

Village woodlot projects were failed for a much more fundamental reason that local people require other products far more urgently than fuel-wood. Fuel is seldom the only product people got from trees and often not the most important. This has been widely recognized the people intending to participate in the project want food, fodder, minor forest produce and fuel-wood from trees. The project able to suffice these entire requirements can sustainably implement. In case of African countries the experiences of externally funded projects were quite sour and they need to be convinced about fulfillment of their entire requirement before initiating such projects.

(i) Time Delays Before Villagers Benefit

Even if a social forestry project can yield in a desirable mix of products, the delay before harvesting them might be too long to satisfy the needs of local people. The project should therefore, be designed to ensure the short-term costs of enclosing village land, while trees become established are kept as low as possible. The short term benefits like wages for planting and protecting trees need to be maximized. Grass may be collected from inside the enclosure for fodder instead of grazing animals as before. After a few years, livestock can often be allowed to return to plantation at certain times of the year, when trees are not at risk. The provision of paid, rather than voluntary, employment with either wages or food for work may be significant contributing factor to the success of directed village forestry schemes. It could be relatively lack of success of truly community based projects where major part of the cost is borne by the village in return of long term benefits.

(k) Challenging Future

Community forestry projects are likely to become gradually more successful in the future, but only when social foresters become more skilled in their jobs. Projects are designed to take more account of the needs of local people and the communal activities of villages grow as part of the general social and economic development process. In the meantime, the continuation of other types of social forestry like farm forestry may increase general awareness of the value of tree planting. The main lesson from the past is not to abandon the community projects but the problems can overcome by improved training and detailed cost sensitive research and consultation prior to initiating and managing a forestry project.

Chapter 7

PART–A

Desertification and Land Degration in India

Arid lands make up about 60 percent of the Indian territory and represent very diversified natural conditions, simultaneously, various levels of socio-economic development. Indian experience is extensive in irrigation development, the basis of sustainable farming in arid regions. Similar to many other developing countries, dry regions of India are least developed in socio-economic terms. Since planning was adhered to in India, regional economic policies pursued the goal of developing these arid regions. Research on water resources assessment and rational dry land use, desertification control and scientific validation of regional planning was undertaken on large scale.

In 1952, the Central Arid Zone Research Institute (CAZRI) was established in Jodhpur, the centre of the Thar Desert. In 1972, the International Crop Research Institute for the Semi-Arid Tropics (ICRSAT) was established in Hyderabad.

7.1 GEOGRAPHICAL BACKGROUND

Arid lands in India constitute a significant portion of the geographical area of the country. It extends over wide latitude covering both peninsular and extra-peninsular regimes as depicted in Figure 7.1. The northern portion of the arid zone is the continuation of the major Afro-Asian hot desert region. Biogeographically, the biota of this tract holds 'western' affinities to a great extent. Taking into account the manifestation of aridity in the natural setting and ecological features, the core region of the arid tract, including a small extremely arid area, are broadly restricted to Rajasthan and its neighbouring states of Gujarat, Punjab and Haryana. The focal centers for manifestation of desertification are most pronounced in Rajasthan, especially in view of intrinsically 'desertic' nature of this zone.

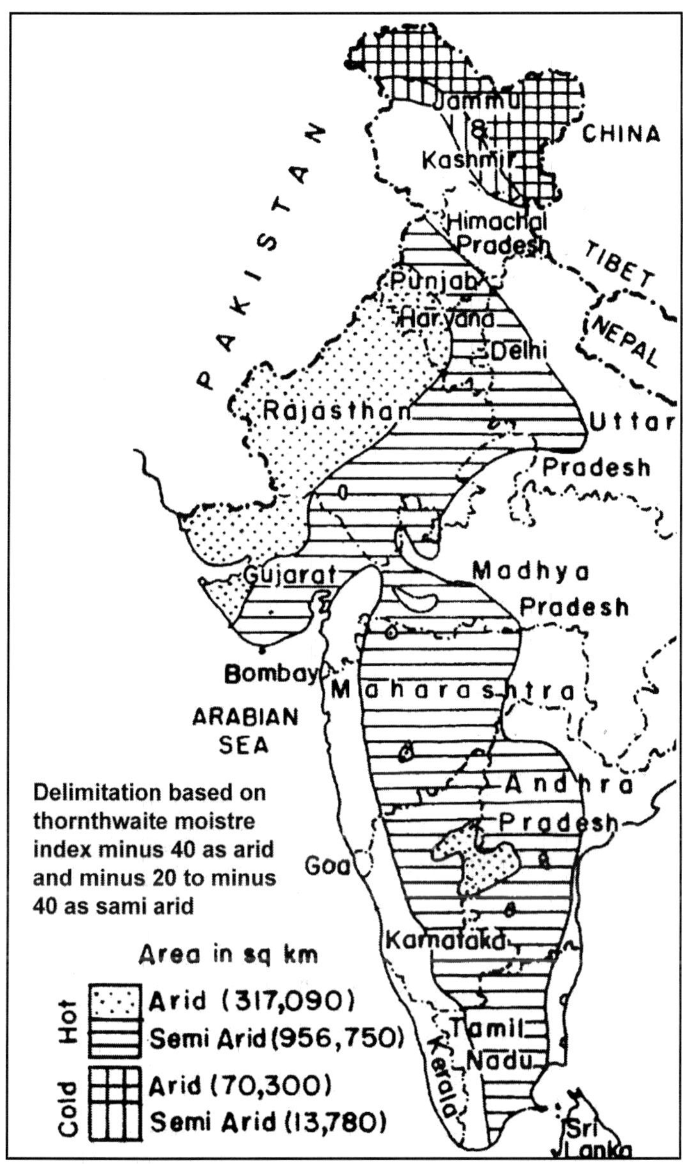

Note : The external boundary and the coastlines of India appearing on the map are neither verified nor authenticated by the Survey of India.

Source: Central Arid Zone Reasearch Institute, Jodhpur

***Figure 7.1** : Schematic map of India showing Arid and Semi-arid Zones.*

1. Geology

Detailed geological mapping with evaluation of economic mineral deposits and rock formations was carried out by several generations of workers. Table 7.1 gives the stratigraphic succession and the spectrum of geological formations in the tracts.

Table 7.1 : Geological formations and stratigraphic succession in the Arid Zone of Rajasthan

Era	Period	Series/ Group	Geological Formation
Quaternary	Recent		Alluvium and Blown Sand
Tertiary	Eocene	Mandai Series Akli-Kapurdih Series. Jogira-Mar-Banda	Sandstone Sandstone and bentonitic clay Fuller's earth
		Series Khuliala-Palana Series	Limestone
	Cretaceous	Abur- Fatehgarh Series	(Deccan Trap) Sandstone, limestone, clay and lignite.
Mesozoic	Jurassic	Parihar Series Bedesir Series Baisakhi Series Jaisalmer Series Lathi-Mayakor Series	Feldspathic sandstone Ferruginous sandstone Calcareous sandstone Fossiliferous limestone and sandstone. Sandstone and shale
	Perma-Carboni-ferous	Badhaura Series	Sandstone and boulders
Palaeozoic	Vindhyan Super Group	Bhander Group Rewa Group Kaimur Group Samri Group	Sandstone, shale and limestone Sandstone and shale Sandstone, shale and conglomerate Limestone, shale, sandstone, grit and basic flows
Proterozoic	Delhi Super Group	Ajabgarh Group Alwar Group Raiale Group	(Acid, basic and ultra-basic intrusives and extrusives) Quartzite, mica-schist, calc-schist, gneisses, marble and basic flows. Quartzite, conglomerate and mica. Marble, dolomite and conglomerate
	Aravalli Super Group	Jalore Group Udaipur Group	(Granite, basic and ultra-basic intrusives) Quartzite, mica-schist and phyllite. Quartzite, conglo-merate, metevoleanics, phyllite and greywacke marble.
Archacan	Pre-Aravalli		(Granite and basic intrusive) Phyllite, slate, schist, gneisses, granite and magmatites

Source: Geological Survey of India.

2. Physiography

The terrain conditions vary widely, ranging from ergs or typical dune fields with inter-dunal plains to rocky and gravel pediments and restricted fluvial tracts with flood plains, terraces ad abandoned river channel The effects of Aeolian action and aridity are manifest in the dunes, sand sheets, tafoni, bornhardts, pediplains, pediments, wadis and playas or saline basins. In addition, there are considerable tracts in Nagaur, Jodhpur, Pali and Jalore districts revealing interplay of fluvial and Aeolian regimes with extensive flat alluvial plains, particularly in the vicinity of the rivers and streams of the Luni system. Luni is the only river system with an integrated drainage, rising in the Ajmer hills and releasing its load into the Great Rann of Kutch. The rest of the drainage features are limited to ephemeral wadis and streams losing their identity in the desert sands. Of these, prominent are the Kantli, Mendha and few others in he zone of internal drainage.

The geological investigations reveal the recurrence of aridity in parts of the terrain in the geological eras manifest in the gypsum deposits of Bikaner and Nagaur in the Trans-Aravalli Vindhyan pre-Cambrian formations, occurrence of evaporite sequences, halite and potash, at depth in Churu district and buff coloured sandstones of the Vindhyan age. In the perspective of the Thar Desert, the palaeoclimatic vicissitudes of the Quaternary epoch, when man emerged and started interacting with nature, are most significant. The quaternary era commenced in Rajasthan under strictly non-marine conditions. The palaeoclimatic phases and geomorphic features of this era are summarized in Table 7.2 and described as follows:

(a) *Humid-Cold Phase:* Studies indicate the presence of a widespread fluviatile environment in the tract, marked by a widespread boulder conglomerate and sandstone of the earliest Quaternary phase. The Luni, Ghaggar and their tributaries existed in this period. Typical alluvial sequences of terrace and flood plain deposits have been located in these basins from the earliest Holocene.

(b) *Arid Phase (First Generation):* The early and middle Holocene 8000-3000 BC was marked by evidence of aridity and Aeolian activity of a wider extent than the present phase. Stable dunes with considerable palaeosoil development are found even in Alwar, Jaipur, Bharatpur and other districts of eastern Rajasthan. These evidences are also clearly noticed in sections of the Luni, Mendha and other streams and in the lake deposits of Sambhar. This phase of aridity is marked by the presence of typical desert landforms, viz., dunes, pediplains, pediments and bornhardts and choking of river courses and cut-offs.

(c) *Warm Sub-humid Phase:* The sub-recent to recent phase 3000-1000 BC in these tracts was marked by accelerated fluvial activity with an amelioration of the climate towards warm humidity, indicated by the cutting of the older Aeolian sequences and more widespread fluvial deposits, including sands of coarser variety, conglomeratic lenses and clays. The humid phase, with enhanced fluvial activities, led to present geomorphological configuration of the Luni basin and its surrounding areas, in so far as the riverine situation is concerned with braids, flood plains and back swamps. The Rangmahal culture thrived in north-west Rajasthan during this period.

(d) *Recent Arid and Semi-arid Phase:* The last tow millennia witnessed another onset of aridity in the region. This phase has left the Luni and other river systems as relics of the earlier humid phase as evidenced by drifting sands, abandoned channels, active dunes and sand sheets, choking of stream courses, development of wadis and playas and other desert features. This phase is, however, restricted to the west of the Aravallis.

Table 7.2 : Palaelclimatic phase and geomorphic features of the quarternary era

Period	Geology	Geomorphic Features	Climatic Phases
THIRD PENEPLANATION IN PROGRESS (Under Aeolian Environment)			
Recent	Younger aeolian sediments	15. Sand heaps and sand ripples 14. Playas 13. Wadis 12. Active dunes of second generation	Recent arid and Semi-arid
Sub-Recent to Recent (3000-1000 BC)	Younger alluvium	11. Present river channels and braids 10. Present flood plains 9. Back swamps	Warm sub-humid
Early to Middle Holocene (8000-3000 BC)	Older Aeolian sediments	8. Stable dunes of first generation 7. Pediplains 6. Pediments 5. Bornhardts comprising basement rocks	Arid
SECOND PENEPLANATION (Under Aeolian environment)			
Upper Pleistocene to Early Holocene (10000 years BC or more)	Older alluvium	4. Left out channels 3. Older flood plains 2. Terraces	Humid cold
Pleistocene	Sandstone and conglo-merate	Indicate the ancient river bed, but do not form any landforms	
FIRST PENEPLANATION (Under fluviatile environment)			
Pre-Cambrian	Delhi super Group and Post Delhi	Hill ranges, discontinuous hills and outcrops.	

Source: Geological survey of India.

7.2 PROBLEMS OF DESERTIFICATION AND LAND DEGRADATION

Using the degree of aridity, Indian territory can be divided into four categories according to the UNEP classification. These areas have been depicted on Figure 7.2 for visual presentation.

1. Arid lands covering the north-western margins of the Thar Desert bordering Pakistan contains about 5 percent of the overall dry areas of the country. This region has the maximum moisture deficit.

2. Semi arid lands constitute about 25 percent of total dryland area. This region has significant moisture consisting of Central Rajasthan, the Deccan part of Maharashtra, part of the Karnataka and Andhra plateaus.
3. Sub-humid lands of 30 percent spread over the regions of moderate moisture deficit - Central India, areas adjacent to Deccan.
4. Transitional lands with adequate moisture supply is 40 percent, the regions transitional to moisture sufficient- Chhota Nagpur, Chattisgarh and Baghelkhand plateaus.

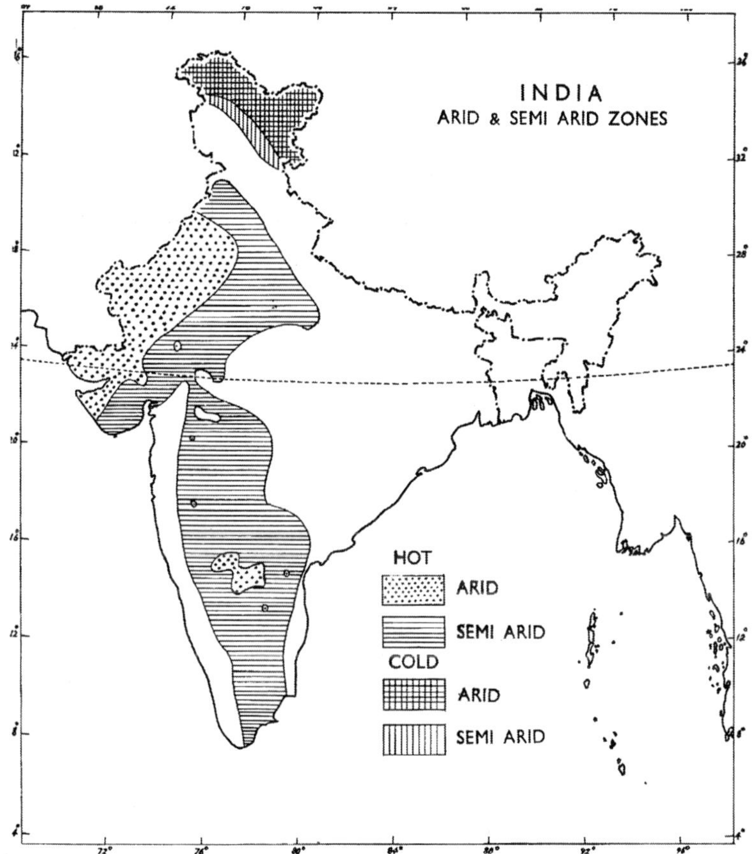

Note : The external boundary and the coastlines of India appearing on the map are neither verified nor authenticated by the Survey of India.

Figure 7.2 *: Arid and semi-arid zones*

Source: CAZRI

Aridity tends to decrease in the north-western direction. However, the zone of unstable moisture stretches across a rather wide belt in the peninsular part of India, the Deccan. Western Ghats in the states of Maharashtra, Madhya Pradesh, Andhra Pradesh, Karnataka and Tamilnadu. The driest territories of India are the least developed in socio-economic terms. Rajasthan, Kutch and Kathiawar, located in the most dry zone, were a conglomerate of multiple small feudal princedoms in the colonial period. They were isolated from the mainland and were very backward. Nowadays, this is the zone of low agricultural productivity, mostly rain fed farming and trans-human pastoralism.

High population growth rates, under conditions of a poorly diversified economy in the arid regions and the limited capacities for rural population migration into towns, result in the increased pressure on the unstable ecosystems of these regions. This process is accelerating. Thus, beginning from 1901, the population growth rates in Rajasthan comprising of the greater part of the Thar Desert, are surpassing the national average. This tendency became prominent particularly in 1950s, reaching its highest level in 1971-81 of 32.97 percent against the national increase of 25 percent. The population of Rajasthan had increased from 10.3 million in 1901 to 56.5 million in 2001 registering an increase of 549 percent against national increase of 431 percent. The average population density per square kilometer in the state has increased from 30 persons in 1901 to 165 persons in 2001. In general, India holds the first place in the world for population density in arid regions.

Such socio-demographic processes coincide with active cultivation of marginal lands and the reduction of grazing lands coupled with the simultaneous growth of cattle herds. Several increases of petroleum products in the 1970s led to greater cutting of trees and shrubs for firewood for domestic purposes. Greater use of dry manure deprived fields of a major source of organic fertilizer. This is the anthropogenic factors of low level of socio-economic development for widely adopting primitive methods interfering normal cycles of rehabilitation of recovery of natural resources, thus stimulating the desertification processes. Estimates show that on over 70 percent of arid regions in India, landscape degradation is progressing as result of various factors.

India has a vast span of hot and cold deserts prone to desertification and land degradation. The cold desert occurs in the Ladakh region of Jammu and Kashmir and spread over an area of about 70,000 sq. kms. About 317090 sq. km of area is arid, restricted to north-west, semi arid lands are distributed towards south as given in Table 7.3.

Table 7.3 : Area of Arid Zone in India by States

State	Area in Sq. kms	% to total arid area
1. Rajasthan	1,96,150	61.86
2. Gujarat	62180	19.61
3. Punjab	14510	4.58
4. Haryana	12840	4.05
5. Maharashtra	1290	0.41
6. Karnataka	8570	2.70
7. Andhra Pradesh	21550	6.79
Total	**317090**	**100.00**

Source: ICAR 1977.

Losses which are sustained in the country due to land degradation have acquired dramatic scales and are threatening the whole of our economic progress in general. According to the Ministry of Agriculture (1980), nearly half (150 million hectares) of the overall assessed dryland area of 317 million hectares has severe erosion problems. The assessments of the early 1970s indicated that about 6 billion tones of topsoil were annually washed off. Nearly 60 percent or 175 million hectares are rapidly being degraded comprising of agricultural lands. Deforestation is also increasing as out of 75 million hectares of forest lands, less than half is actually by forests, 20% forest land is affected by erosion. Practically only 12 percent of the country's land is covered under forests.

Grazing lands are in the worst condition as a result of the excessive pressure of numerous cattle herds. Notwithstanding the official classification of "13 million hectares as permanent rangelands" These are practically devoid of any vegetation due to overgrazing or other uses. Rangelands in India are sources of soil and vegetation degradation, affecting adjacent areas as well.

Gully erosion on the Chhota Nagpur Plateau has turned about one-sixth of the farmlands into unproductive barren lands in only three-fourth of the century. Vast expansions of lands on this Plateau and in Central India as well are becoming covered with a compact laterite crust, which even a plow cannot break.

When the vegetation is destroyed, the soil wash off and the general decrease of biological productivity go on particularly rapidly in mountain regions. In the Western and Eastern Ghats, "the mountain-forest landscape are replaced by rocky grass-bushy mountain-savannah landscapes, undoubtedly of a secondary origin. Being suitable only for grazing, mainly for sheep and goats, they are referred to as low-agricultural lands. The destruction of primary forests in mountains entails soil wash-off and weathering. As a result, the landscape degradation became irreversible. The restoration of their former potential on the basis of wide-scale Afforestation measures on denuded slopes presents great difficulties (Nikolayev, Ryadchikov, 1967).

Desertification processes" in mountain highlands have implications. Even in the colonial period, the high authorities had to pay attention to mass soil degradation in the Himalayas piedmonts such as the Siwalik ridge. For instance, in the Hoshiarpur District (Punjab), afforestation and cultivation of flat regions in the 19th century led to active gully erosion and as a result, sand flow "cho" rushing to the plains from Siwalik, changing river channels, covered fields. More frequently the diagnosis is put; "In Nepal, the Himalayas are dying, in India they are seriously ill." The degradation of this highest mountain system, which has a water regulating effect on the vast south Asian sub-continent, may entail the increased risk of natural disasters, such as large-scale floods and droughts.

The estimates have indicated that flood intensity in India has doubled in the recent decade; "While in the early 1970s recurring flooding affected on an average 20 million hectares, then in early 1980s this figure became 40 million."

(Indian Sixth Plan 1981). In 2004 this figure further increased to danger level by surpassing to more than 80 million hectares of area. Floods occurred more often in arid regions, in 1978,a devastating flood took place in the basin of basin of the seasonal river Luni in the Thai-desert. In 1981 Jaipur the capital of dry state of Rajasthan, suffered from a destructive flood. Nearly in the same period, 1979-1980, unprecedented drought struck the usually wet regions in the north and east of India, damaging nearly 38 million hectare of arable land. The survival of 130 million head of cattle and over 200 million people was at shake.

The idea of Special National Institute for Natural Calamities Control at the Planning Commission was considered at that time. Awareness has grown that "time is ripe for formulating a long term strategy for forecasting, preventing and alleviating the natural calamities" as well as for developing novel idea and programmes. Obviously one of the major thrust in research in this field is the mapping of regions affected by natural calamities and the mapping of the distribution of various types and stages of degradation with emphasis on areas nearing in irreversibility transition into barren lands.

India, similar to other countries of the tropical and sub-tropical zones, possesses infinite thermal resources for a year-round farming almost over the whole country, except for highlands. The major natural constraints is water deficit in dry seasons. Much of the country's regions suffer from recurring droughts that bring much damage to the economy.

India is one of the world's leaders in irrigation. Irrigated land potential, during the years of independence has increased from 22.6 million hectares in 1950-51 to 56.6 million hectares in 1979-80. The irrigation area was further increased through micro level efforts like watershed development approach and water saving devices helped in enhancement of the irrigated area. Irrigation is furnished on nearly one-third of the overall harvested area and a second sowing is done over one third of the area. The share of large irrigation projects in the total irrigation potential amount to 43 percent, while half of the remaining share are small irrigation facilities utilizing surface and ground water potentials. However, India is self sufficient in food crops and able to meet the needs of alarmingly growing population.

The drawbacks of irrigation system have occurred during the past and resulted severe loss to the economy of the country. These drawbacks are narrated hereunder:

1. Low productivity in irrigated lands the average yield of cereals is 1.7 tonnes per hectare while on test fields it is 4 to 5 tonnes per hectare.
2. Delay in completion of many large irrigation projects like Indira Gandhi Irrigation Project of Rajasthan, Nagarjuna Sagar of Andhra Pradesh, Malaprabha of Karnataka, Tava in Madhya Pradesh etc.
3. Improper irrigation rates and poor drainage and impact on cultivated lands in the form of water-logging, Salinization and alkalization.

4. Due to enhanced erosion, the silting of water bodies has accelerated which is particularly dangerous for large headworks for reservoir clearing.
5. Large attention paid for construction of major, medium and minor irrigation projects but less attention paid for sound land management and proper use of soils causing land degradation.

The new strategy evolved for maximum utilization of the potential created and intensification of agricultural production by putting to better use of land and water resources in regions of irrigated farming. In the areas of 76 irrigation projects, special management bodies were set up to devise and bring about integrated development plants, calling for land consolidation, improvement of irrigation distribution network and improvement of irrigation system performance. In addition provision of efficient drainage, rational cropping rotation and development of road network were proved enhancing the performance of the irrigation system.

The Central Water Commission of India had undertaken an assessment of the seasonal regimes of the water budget in individual river basins with special emphasis on drought affected areas, as a possible alternative, surface flow transfers from water abundant to water short areas was studied for transporting water. Three groups of arid and semi-arid regions may be distinguished by water supply (Levintanus, 1981), generating various possibilities for further development. These areas are depicted on the Figure 7.3 and narrated below:

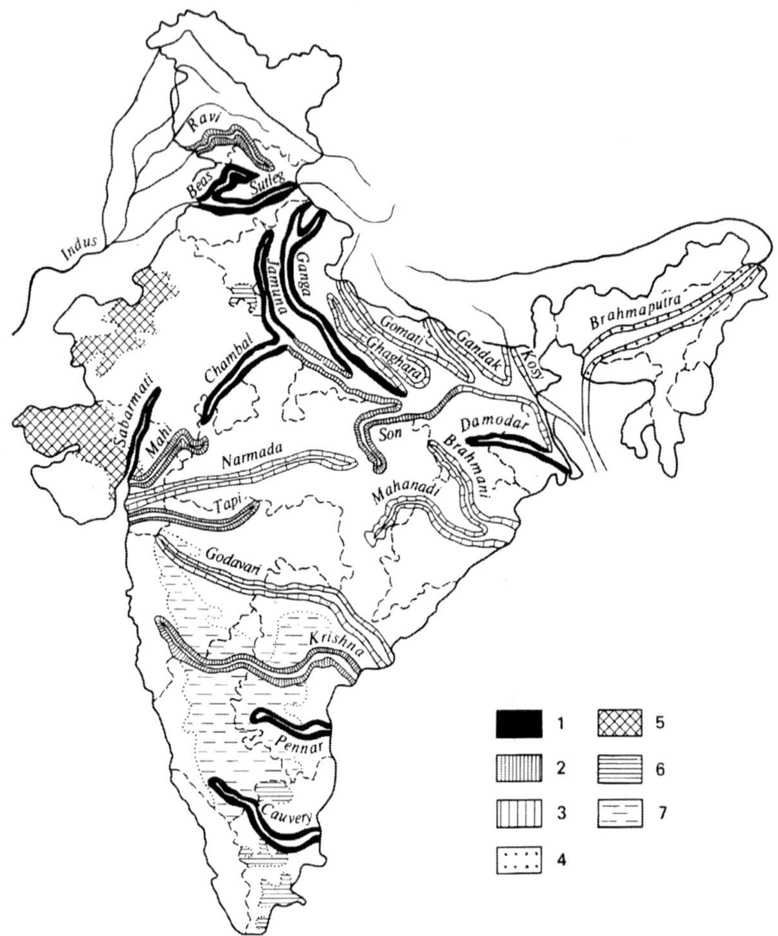

Note : The external boundary and the coastlines of India appearing on the map are neither verified nor authenticated by the Survey of India.

Figure 7.3 : *Development of water resources and water availability of Arid regions of India (after report of the Irrigation Commission, 1972)*

* 1- complete utilization of water resources will be reached by 1981-1985; 2- by 1990; 3- by 2000; 4- after 2000; 5- arid regions of absolute water deficit; 6- of relative water deficit; 7- with available water-supply but poorly utilized

1. Regions of Absolute Water Deficit

Western Rajasthan, Kutch Peninsula and Kathiawar are categorized as absolute water deficit regions. High aridity is manifested by insufficient precipitation and low level of river flow. On the completion of Indira Gandhi Canal Project, the water potential of the Indus (20 percent of flow to which India has a right in pursuance of the International Agreement) is fully exhausted. The main objective was to more efficient water use on the limited water resources. Possibilities for the partial transfer of Narmada and Mahi was also materialized since most regions are arid and semi arid. But these systems could yield very scanty results and remaining areas are still facing acute water shortage.

2. Regions of Relative Water Deficit

South Haryana, West Tamil Nadu and South Gujarat (Saurashtra), where accelerated irrigation construction has led to almost total exhaustion of available water resources. For these regions, the most burning issue is the problem of optimal management of existing water facilities on the one hand and reducing the water requirement of agriculture and industry.

3. Regions provided with Water Resources but use less optimally

The West Madhya Pradesh, inner areas of Andhra Pradesh, Maharashtra and Karnataka are under this category. The general low level of economic development and the long-standing disputes between states on water sharing of common river basins are responsible for slow development of water potential.

The most urgent problems are optimal utilization of water resources by all available means as 40 percent water in irrigation networks is lost to evaporation and by more complex means e.g., by way of integrated development of river basins. These problems still await their solutions. India like other countries, faces the need for scientific validation of grandiose projects of inter-basin flow transfers, their ecological-economic evaluations and assessments e.g., identification of priorities, in terms of available local and regional resources.

Even if the irrigation is maximized, half of the Indian regions and many of the dry regions can not be provided irrigation. The target set in the country to carry out "Green Revolution" in the zones of rain-fed farming has yielded good results. However, its resolution presents difficulty both in ecological and social terms. The diversification of production on the basis of industrialization emerges most instrumental in curbing the growing pressure on ecosystems and the desertification process. This would help to divert part of the population from agriculture to create conditions for further agriculture development on an industrial basis and to attain a general acceleration of socio-economic progress.

Industry, existing in dry areas, is to date inadequate in scale and fragmented. In rural areas, handicraft industry prevails, which gives sustenance to relatively closed rural communities. In Rajasthan, there are some centers of mining industry in Zavar, Khetri-singhana, Kho Dariba and recently located oil deposits and processing industry using local resources. Yet in all industrial output, the share of the state is only 2 percent. The largest cities like Jaipur, Jodhpur, Udaipur, Bikaner, Ajmer and Kota as former capital of princedoms, have no close linkages with surrounding rural areas.

Hyderabad may be taken as a classical example of the production and territorial isolation of metropolitan centre and its less developed agrarian surroundings. It is the capital of Andhra Pradesh state, the "City-Millionaire", located in the centre of Telengana, a poor dry region. In the years of independence, the new large port of Kandla was set up in the south-east of the Kutch Peninsula as a "successor" to Karanchi, in Pakistan. However, the activities of a well-equipped port are constrained by the economic backwardness of its hinterland.

In general, the increase in industrial production and the development of the inter-sectoral integration on a regional basis, between industry branches and among industry, agriculture and other sectors, are the target of developing areas. They contribute to the consolidation of the production base of urban areas, expanding their capacity to absorb rural migration and intensifying links with agrarian regions, the latter would grow more rapidly, if the life of medium and small towns is revived.

Insufficient power-generating resources and a large energy deficit are a hindrance to industrialization and general socio-economic progress in arid regions. Efforts are taken by India in consolidate the energy generation base by various means like atomic energy, solar and wind energy have done commendable work, but the problem of energy supply has not been resolved. The need to develop the energy for industrial, agricultural and domestic requirements as well as to develop non-conventional energy sources are the requirement for development of arid regions.

The efforts made by Central Arid Zone Research institute and International Crop Research Institute for the arid and semi arid regions have done significant work in the field of dry farm practices, which have yielded successfully. India has amassed extensive knowledge on the strategy and methods of desertification control for optimal land and water resources use, improved land use, enhancement of crop productivity, which merits special study and dissemination. Experience, however, has shown that the mass poverty of small land holders and landless peasants is the major constraint for introducing innovations, which are often capital extensive. The way out of the vicious circle of socio-economic backwardness, poverty and the degradation by nature may be found not only in the improvement of desertification control strategies and in irrigation development.

A number of area development programmes being implemented in India with the aim of bringing new life to economic backward dryland regions. Drought Prone Area Programme, Desert Development Programme, Command Area Development Programme and individual beneficiary programmes like Swarn Jayanti Rural Self Employment Programme are meant to benefit the poor people of arid and semi arid regions.

4. Population Demography

Human population of the country as per 2001 Census is 1027 million, of which 27.78% is urban population and 72.22% people live in rural areas. Decennial growth rate during 1991-2001 was 21.35%, witnessing a declining trend from previous two decades during 1971-1981 and 1981-1991 of 25.00% and 23.51%

respectively. Density of population has witnessed an increasing trend in view of population growth. During 1981, 1991 and 2001 the density of population was 221, 267 and 324 persons per sq. km respectively. A matter closely related to the problems of environmental security in India is the issue of scarcity of natural resources *vis-á-vis* human carrying capacity. Arid zone standards, drylands of India have a high population density. The Thar Desert in the Indian Arid Zone is considered one of the densest deserts of the world. The Great Gangetic Plains has an average density of 456 persons per sq. km. The rate of population during the decade 1981-1991 in dryland regions has been 29% of the country (MOEF 1996). Population of India has increased from 16.2% to 16.7% of the world population in 2001. Salient features of population in India are given in Table 7.4.

Table 7.4 : Demography Trends During 1951-2001

(Population in Million)

Census year	Total Population	Rural	Urban	Density	Sex ratio
1951	361	298	62	117	946
1961	439	360	79	142	941
1971	548	439	109	177	930
1981	685	524	161	221	934
1991	846	628	218	267	927
2001	1027	742	285	324	933

Source: Census of India 2001

Out of total 452 districts in the country as of 1991, 224 showed population growth rate higher than the national average of 23.85% as of 1991. Seven districts of Rajasthan showed very high rate of growth of 30 to 35 percent. The state of Uttar Pradesh is the most populated in the country with 166 million population and more than doubled from 54.59 million in 1951.

5. Livestock

Man and livestock have long had a close association since the early history of mankind. The livestock population in India is one of the highest in the World. Information given in Table No. 7.5 indicates information of livestock and cattle population as per 5-year cattle census and indicating percentage growth over past census. Total livestock includes cattle, buffaloes, sheep, goats, horses, pigs, donkeys and mules, camels, yaks, mithuns etc.

Table 7.5 : Livestock Population in India

Year	Total Livestock		Cattle	
	In Thousand	% increase	In Thousand	% increase
1977	369,654	—	180,140	—
1982	419,742	13.55	192,453	6.83
1987	445,286	6.09	199,695	3.76
1992	470,860	5.74	204,584	2.45

Source: Agricultural Statistics at a Glance, 2001, Directorate of Economics and Statistics, Ministry of Agriculture.

Over the four decades, during 1951-1992, the years for which livestock censuses are available, the number of bovine increase substantially viz., 45 percent sheep and goat population were 39 million and 47 million respectively in 1951 and the same grew to 51 and 115 million respectively in 1992, thereby increasing by 39% to 144% respectively over the period. The annual rate of growth during these two periods for bovine, sheep and goats were 0.9, 0.64 and 2.20 % respectively over the period.

Livestock population and density in the arid region is very high as detailed in Table 7.6. The region also has a number of wild life such as Chinkara, Black-buck, Peacocks and the Great Indian Bustard, which is an endangered bird species.

Table 7.6 : Poplation of Livestock in the hot Arid Region

Livestock	Population	%
1. Camels	1,001.000	0.23
2. Sheep and Goats	155.910,000	35.89
3. Cattle	199,695,000	45.97
4. Horses, Donkeys and Mules	1,922,000	0.44
5. Buffaloes	75,677,000	17.42
6. Others	175,000	0.05
Total	**434,380,000**	**100.00**

Source: Indian Council of Forestry Research and Education, 1999

Livestock is an important source of supplemental income of rural households. National Council of Applied Economic Research survey indicated that income from livestock averages 22.5 percent of total household income (Ministry of Agriculture 1996). Livestock in the Indian context is thus a major instrument for improving rural employment, particularly rural self-employment. In 1991, 65 percent of the total Indian working force of 185 million people was dependent of agriculture, 80 percent of these involved in livestock production, either as producer or as workers. The family women carry out 60% of work related to care and management of livestock in households. The total value of livestock asset in 1993-94 was Rs. 669 billion. This has increased to Rs. 1302 billion in 1999-2000 as per the latest estimates of Central Statistical organization.

Livestock production in India is entirely dependent on crop residues and crop by-products. The total supply of feed and fodder in 1993 was straw 398 million tones, green fodder 573.50 million tonnes and concentrates 41.98 million tonnes (MOEF 1993). Estimates of feed and fodder past, present and future portray huge deficits in demand and supply, the livestock population increased across all species between 40-300 percent and output of the sector tripled or quadrupled, all taking place within the supportedly diminishing feed and fodder supply base. The fodder deficit is met by grazing on common property resources, overgrazing by herds has progressively rendered them marginal and wastelands.

The annual erosion of top soil at 6000 million tonnes, reduction of water points 55 to 90 percent and changing plant association, making them unsuitable for bovines and fit only for sheep and goats.

The analysis of land use classification of the later half century is given in Table 7.7. This information gives changes in land use pattern and pressure on land in view of growing population. Population growth during past fifty has increased from 361 million in 1951 to 1027 million in 2001 registering 284.49 percent. The change in land use pattern witnesses the resultant impacts to cater the food requirement of population.

Table 7.7 : Land Utilization trend in India

(In Million Hectares)

Classification	1950-51	1960-61	1970-71	1980-81	1985-86	1990-91	1997-98
1. Geographical Area	328.73	328.73	328.73	328.73	328.73	328.73	328.73
2. Reported Area	284.32	298.46	303.76	304.15	304.66	304.87	304.92
(i) Forests	40.48	54.05	63.91	67.47	67.04	68.00	68.86
(ii) Not available for cultivation (a+b)	47.52	50.75	44.64	39.62	40.71	40.90	41.56
(a) Non-agricultural uses	9.36	14.84	16.48	19.66	20.54	21.20	22.53
(b) Barren and uncultivable	38.16	35.91	28.16	19.96	20.17	19.70	19.03
(iii) Other cultivated land (excluding fallow land) (a+b+c)	49.45	37.64	35.06	32.31	31.11	30.50	28.36
(a) Permanent pastures and other graz-ing land	6.68	13.97	13.26	11.97	11.97	11.80	10.91
(b) Misc. tree crops and groves.	19.83	4.46	4.30	3.60	3.45	3.70	3.57
(c) Culturable wasteland	22.94	19.21	17.50	16.74	15.69	15.00	13.88
(iv) Fallow land (a+b)	28.12	22.82	19.88	24.75	24.88	23.40	24.12
(a) Fallow other than current	17.44	11.18	8.76	9.92	10.02	9.60	9.76
(b) Current Fallow	10.68	11.64	11.12	14.83	14.86	13.80	14.36
(v) Net Area sown	118.75	133.20	140.27	140.00	140.92	142.20	142.02
(vi) Gross cropped area	131.89	152.77	165.79	172.63	178.83	185.50	190.76
(vii) Area sown more than once	13.14	19.57	25.52	32.63	37.91	41.62	48.74
(viii) Net irrigated area	20.85	24.66	31.10	38.72	42.08	45.14	54.57
(ix) Gross irrigated area	22.56	27.98	38.19	49.78	54.65	59.64	72.78

Source: Agricultural Statistics at a Glance 2001, Ministry of Agriculture.

Above table details out the changes in land use pattern during past 47 years and attained a landmark in food production to cater one billion souls starting from a food deficit country. The efforts made in the agriculture are remarkable but at the cost of various adverse impact of desertification, Salinization and alkalization due to human efforts. The analysis of land use pattern during 1950-51 to 1997-98 gives significant features of the changes:

(a) Forests area has increased from 40.48 million hectares to 68.86 million hectares registering an increase of 70.01 percent. Area under forest has been reached to 22.58% of the total geographical area of the country.

(b) Area not available for cultivation has decreased from 47.52 million hectares to 41.56 million hectares i.e., 12.55% This is due mixed impacts of urbanization and other factors indicated in (c) and (d) below indicating positive and negative efforts.

(c) Out of (b) the area put to non agricultural uses has increased from 9.36 million hectares to 22.53 million hectares witnesses 140 percent increase which is mainly due to urbanization, railway and roadways and other buildings. Land acquired for these purposes was mostly good cultivated land but infrastructure facilities are bound to be created at the cost of environment and ecology.

(d) Barren and uncultivated lands have decreased from 38.16 million hectares to 19.03 million hectares which is evident that mostly marginal and degraded lands have been put under agricultural uses. These areas are mostly included under cultivation in view of pressure on land and the people were compelled to use such lands though, not able to provide significant yield.

(e) Area classified under other cultivated lands generally not used for cultivation purposes has decreased from 49.45 million hectares to 28.36 million hectares is the result of mixed actions narrated in subsequent points but the net decrease is the result of change in use mostly under cultivation.

(f) Permanent pastures and other grazing lands have increased from 6.68 million hectares to 10.91 million hectares. Though the classification category has demarcated the lands but significant area is under encroachment of the local people and generally lacks in development.

(g) Area under miscellaneous tree crops and groves has significantly decreased from 19.83 million hectares to only 3.57 million hectares is evident that the land has been put under cultivation in view of scarcity of land.

(h) Culturable wasteland has also decreased from 22.94 million hectares to 13.88 million hectares and the shifting of such lands to cultivation purposes is the only area, where people are compelled to use it after treatment.

(i) Area demarcated under fallow land is the combination of permanent fallow land and current fallow witness mixed changes, but overall decrease of 4 million hectares seems to be shifted to cultivation use. However, this is an area, which can be treated and put to cultivation use with significant treatment.

(j) Area classified under permanent fallow land has reduced from 17.44 million hectares to 9.76 million hectares is the impact of population pressure, where people are compelled to use such marginal lands for their livelihood.

(k) Current fallow land has increased from 10.68 million hectares to 14.36 million hectares can be the impact of treatment of permanent fallow lands and keeping current fallow in view various natural factors.

(l) Net area sown has increased from 118.75 million hectares to 142.02 million hectares and is the impact from changed classification from other categories and compelling pressure on people to attain their livelihood. The net area sown is 46.6 percent of the total geographical area of the country used to feed over one billion population of the country.

(m) There has been a significant increase in gross cultivated area from 131.89 million hectares to 190.76 million hectares. This is evident that the area sown more than once was 13.14 million hectare in 1950-51 has been increased to 48.74 million hectares in 1997-98. This is a mixed reaction of changes partly due to increase in irrigation facilities and others as dryland farm practices.

(n) Net irrigated area has increased from 20.85 million hectares to 54.57 million hectares is the clear indication of increase in irrigation facilities through large, medium and minor irrigation schemes. The contribution of wells is quite significant in irrigated area and in view of over-drawl of excess water through underground resources causing increased under dark zone at the alarming rate.

(o) The gross irrigated area has doubled with irrigation facilities. The increase from 22.56 million hectares to 72.78 million hectare is the impact of double or multi cropping facilities made possible from land under cultivation and varieties of crops sown.

The change in the land utilization has been significant to cope up the increasing demand of the people to bring maximum possible land under cultivation and in view of increasing demand of land for various uses. On the one side the changed land practices are for meeting food and other requirement of the people and to provide source of livelihood. But the cost at which all these activities are being carried out have severely affected the ecology and environment. The process of desertification has been increasing year after year.

7.3 PROCESSES AND CAUSES OF DESERTIFICATION IN INDIA

The major causes of desertification and land degradation in the country are deforestation, unsustainable agricultural and water management practices, land use changes for development and industrialization. The major process of land degradation is soil erosion due to water and wind erosion, contributing to over 71% of the land degradation in India. Soil erosion due to water along contributes to about 61.7% and that by wind erosion 10.24%. The other processes include problems of water logging, salinity-alkalinity. The process of desertification is impacting every aspect- loss of agricultural productivity, loss of natural resources- forest and vegetative cover, bio-diversity and soil changes, socio-economic conditions such as economic losses, problems of sustenance, decline in quality of life etc. Causes of desertification and land degradation are listed hereunder:

1. Unsustainable Agricultural Practices

Unsustainable agricultural practices include excessive use of fertilizers, pesticides, frequent cropping patterns, inappropriate technologies, choice of crops / plants etc. Non-point sources of pollution is a problem in areas with wide application of fertilizers. A cycle has been developed where the cultivator is required to use more fertilizers and pesticides to maintain the fertility of the soil.

2. Unsustainable Water Management Practices

Poor and inefficient irrigation practices, over abstraction of ground water, particularly in the coastal regions resulting in saline intrusion into aquifers, etc. are some of major unsustainable water management practices, which has led to problems of desertification in such regions. Over-abstraction of ground water without compensatory recharge has led to depletion of groundwater table.

3. Land Use Change

Diversion of land from forestry and agriculture to other land uses has been the principal causes of land degradation. Diversion of forest lands for non-forestry purposes was curtained with the enactment of Forest (Conservation) Act, 1980 with the objective of arresting diversion of forest land for non-forestry purposes. Wherever diversion of forest land is unavoidable, for developmental projects like energy, infrastructure, transportation etc, compensatory afforestation on non-forest land is mandatory. However, loss of prime forests could have an impact in the long term stability of the forests. The other land use change is due to encroachments, through violation of forest boundaries, illegal farming in forests. Due to their illegal status, they are unable to receive extension services and improve their farming systems, further accelerating land degradation. The encroachment of forest land and the socio-economic pressure to regularize them, continue to be the most pernicious problem of forest protection.

4. Deforestation

It is difficult to separate the causes from the effects of deforestation and forest degradation. Some direct causes of deforestation are land clearances for agriculture including shifting cultivation, other land use changes include unplanned urbanization, land transfers of encroachments, over-grazing, uncontrolled and wasteful logging, illegal felling and excessive fuel wood collection.

5. Industrial and Mining Activities

Industrial effluents and mining are also gradually emerging as important agents of desertification. In most of the cases the root of the problem is the mismanagement by land users and poor implementation of pollution control regulations. Industrial effluents and their discharge into inland waters and irrigation with poor quality water in many parts of the country are rendering stretches of land in some of the states as degraded. Industrial effluents from textile, printing and dyeing industry and their discharge into streams and rivers, which are non-perennial with no flow during the lean season severely contaminates them. Use of such waters for irrigation has affected agricultural land as well. Besides productivity decline or complete loss, progressive degeneration of bio-diversity is yet another major consequence of land degradation. In many areas, the groundwater has been polluted. Some of the most affected areas are found in Pali, Jodhpur, Balotra in Rajasthan due to dyeing industry, Bicchari also in Rajasthan due to discharge of highly toxic effluents, large scale tracts of land have been rendered unfit in industrial estates such as Vapi, Ankleshwar in Gujarat, Pattancheru, Bollaram in Andhra Pradesh, which house a large number of chemical manufacturing units. Vaniyambadi in Tamil Nadu due to effluents from leather processing industrial units.

Mining is a factor of desertification in the country, especially with unplanned open cast mining and dumping of mine refuse in the vicinity of agricultural lands. Despite guidelines and regulations for undertaking adequate environmental measures mining operations, open cast mining of sandstone, limestone, marble, gypsum and clay is largely practiced by small scale entrepreneurs, who do not take post mining mitigative operations. Consequently, such areas are gradually turned into wastelands. China clay, fullers earth, calcite and gypsum generate fine particles, which are washed down the slopes with run-off and get deposited in the adjoining cultivated fields. This eventually leads to problems of water logging and salinity.

Disposal of solid and toxic wastes onto lands is severe in Vapi, Ankleshwar in Gujarat, Pattanchery and Bollaram in Andhra Pradesh, Pali and Balotra in Rajastha where large tract of land have been rendered useless due to disposal of toxic industrial wastes. The costs for reclamation of such lands is enormous.

6. Demographic Pressures

The general problem of arid areas with large population is essentially one of human ecology. The inherently limited resources within arid and semi-arid regions set the ultimate limit of production are finally dependent. Furthermore, erratic rainfall results in widely fluctuating production leading to scarcity, which imposes stress on these populations. In general, the population density of both human and livestock in arid regions is much higher than the national average. The decennial growth rate of population during the decade 1981-91 in the desert region was 29% as against 23% for the country. The livestock population also increased from 9.4 million in 1951 to 15.4 in 1971. The density of livestock on grazing lands has consequently increased. The increase in cattle, buffaloes and camels has been very high in this region.

As population increases, the demand on natural resources is further magnified. This has led to further intensive use of land and other natural resources in drier regions. The consequence is an imbalance between human and animal population on the one hand and plants, water and land resources on the other. As the demand by the first persists and increases, the resources tend to become depleted and as depletion proceeds, the stress upon them becomes even greater. Thus, a process of progressive degradation of resources is set into operation, which intensifies with drought. If not checked timely and effectively, it leads to loss of vegetation, leading to loss of biodiversity. The barrenness of the land effects the hydrological cycle, which can affect the rainfall pattern of the region. In semi-arid and sub-humid regions of the country also, there are some areas where population density is one of the highest in the world.

7.4 PROCESSES OF DESERTIFICATION

The different processes involved in land degradation include wind erosion, water erosion, salinity and alkalinity, water logging etc. The soil erosion by water and wind account for 87% of the area affected by soil degradation. It has been estimated that between 1977 and 1997, the area affected by erosion has almost doubled.

1. Wind Erosion

Wind erosion is the major process of land degradation in the hot arid regions of the country affecting 10.46 million hectare area. These include the States of Rajasthan, Haryana, Gujarat and Punjab covering an area of 28,600 sq. km of which, 68% is covered by sand dunes and sandy plains. Wind erosion is also prevalent in he coastal area, where sandy plains dominate and the cold desert regions of Leh in Jammu and Kashmir (Prasad and Biswas 1999).

2. Water Erosion

Water induced soil erosion is the most serious process of land degradation and desertification in the country affecting 107.01 million hectares of country's geographical area. In the Indian context, it results in loss of topsoil and terrain deformation, ravines gullies etc (Ministry of Agriculture 1985). The broad types of water erosion are given in Table 7.8. Soil erosion through accelerated sheet-wash and rill/ gully development occurs in mainly in the Saurashtra and Kutch uplands and along the eastern margin of the Thar in Rajasthan, where the average rainfall varies from 350 mm to 500 mm. The major reasons are increased cultivation of marginal land with high slopes and shallow soils, destruction of natural vegetation for fuel and fodder, over-grazing and other environmentally destructive uses. In the Aravalli Hill ranges along the eastern margin of the Thar, the hill slopes are being regularly denuded of natural vegetation cover for fuel-wood and fodder. Consequently the soils are being washed out by sheet rill and gully erosion, so much so that in many areas there is hardly any soil left to start a replantation programme. In Kutch region, the problem is partly related to a slow natural upliftment of the terrain over the centuries, leading to a change of base level and increased erosion.

Table 7.8 : Types of Soil Induced Erosion

Types of Erosion	Predominant in
Sheet Erosion	Indo-gangetic plains, Northern Himalayas, Nilgiris.
Gully and Ravine Erosion	Major areas of Central India in the Indo-gangetic plain along the Yamuna and Chambal. Also some areas in Western, Central and Eastern India and in the foothills of Himalayas.
Torrent and Stream Channel Erosion	Foothills of the northern and north-eastern Himalayas, Assam valley.
Erosion due to Shifting Cultivation	North East Hill Region

Source: Ministry of Environment and Forests, Government of India, 2001.

Run-off-induced erosion is quite prevalent in most states of India located in arid, semi-arid and dry-sub-humid regions. The annual water erosion rate ranges from less than 51 tonnes per hectare per year for dense forests to more than 80 tonnes per hectare per year in the Shivalik hills (Prasad and Biswas 1999).Sheet erosion affects red soils comprising Alfisoils, Ultisoils and Oxisoils 4-10 tonnes per hectare per year and black soils constituting Vertisoils and Vertic soils 11-43 tonnes per hectare per year. Gully erosion severely affects hilly areas more than 33 tonnes per hectare per year, while hill slope erosion affects 13 million hectare area with more than 80 tonnes her hectare per year. Run-off erosion has resulted in extended and deep gullies in vast areas of Uttar Pradesh Agro Ecological Sub Regions 9.2, Madhya Pradesh AESR 5.2, Rajasthan AESR 2.3 and 4.2, Bihar AESR 13.1, Gujarat AESR 5.1, Maharashtra AESR 6.1 and 6.2, Andhra Pradesh AESR 7.1 and 7.2, Tamil Nadu AESR 8.3 and Karnataka AESR 3.0, 6.4 and 8.2.

Among the soil groups undergoing serious erosion by water are red, black and the mountain soil regions. The vertisoils in peninsular India are highly erosive because of a high run-off rate, 40-50% of the total rainfall, high erosive slope and poor permeability. The laterite soils suffer from rill erosion losing almost 40 tonnes per hectare per year in the absence of adequate soil conservation measures (Prasad and Biswas 1999).

Ravines are a kind of network of gullies formed as a result of water erosion and they are found generally in deep alluvium and entering a nearby river. It is estimated that about 3.97 million hectare area is affected by ravines in States of Uttar Pradesh, Madhya Pradesh, Gujarat, Rajasthan, Punjab, Maharashtra and Orissa, in the region along with the banks of river Yamuna, Chambal, Mahi, Tapi and Krishna. Ravines have been classified into shallow -3 to 6 m, medium - 6 to 9 m and deep >9 m. Further they have also been classified based on heir suitability for reclamation for operational convenience (ICAR 1999).

3. Soil Salinity and Alkalinity

Vast areas in the otherwise productive Indo-Gangetic plain cutting across the states of Haryana, Punjab, Uttar Pradesh and some coastal regions of Gujarat have lost their productivity due to soil salinity-alkalinity. These soils are characterized by excess soluble salts with sodium carbonate in substantial quantity. Consequently, the soils accumulate sodium on the exchange complex thus resulting in poor physical properties including low infiltration rates. In many areas, a layer of calcium carbonate concretion (kankar pan), which is normally found at a depth of 1 m, acts as a barrier for root penetration into the soil. The soil pH is very high, adversely affecting germination, plant growth and nutrient availability to plants. The process of Salinization sets in due to:

(a) Irrigation with ground water containing excess of carbonate and bicarbonate ions- secondary Salinization.

(b) Run-off from adjoining undrained basins, and

(c) Rise in ground water table as a consequence of mismanagement of irrigation command. This is man-made problem. In addition, there is natural salinity in depressions in landscaping of lower elevations.

4. Water Logging

Water logging is estimated to affect about 8.52 million hectares of the land surface. The problem is severe in the Indira Gandhi Canal Command Area in Rajasthan, where excess irrigation in the soils having gypsum-rich barriers at shallow depth and wrong drainage planning are the major cause for degradation, leading to saline sodic water and salt rich hard pan, Some areas of Uttar Pradesh, Haryana and Punjab under agriculture also have this problem. According to a World Bank study, India loses 1.2 to 2.0 million tonnes of food grain production every year due to water logging (ICAR 1999).

7.5 DESERTIFICATION PROBLEMS IN THAR DESERT

Besides the past geomorphological history, encompassing geo-tectonic and organic movements in the Thar of the Great Indian Desert, there are three major desertification factors, followed by a number of processes that consequently bring about the deterioration of desert eco-systems as described in Table 7.9.

Table 7.9 : Processes and Consequences of Desertification

Factors	Processes	Consequences
1. Climatic Changes	Establishment of aridity and droughts	Dust storms, soil depletion of crop yields, degradation of vegetation, livestock migration, nomadism
2. Increase in human population	Cultivation of marginal lands Increase in irrigation cropping Exploitation of woody biomass	Decline in crop productivity, soil erosion and loss of soil fertility. Water logging, salinity problems, over-exploitation of ground water, introduction of new weeds and pests. Degradation of forests, suppression of natural regeneration, diminution of wildlife.
3. Increase in livestock population	Overgrazing	Degradation of vegetation resources, dimini-shing livestock production, livestock migra-tion and inducement of nomadism

Source: ICAR (1977).

The analysis of various climatic parameters of the Thar Desert has revealed that aridity index line (of 80) shifted erstwhile between the first and second decades of the 20th century, especially in Ganganagar, Bikaner, Churu and Jodhpur districts. The deterioration was not notices in Barmer and Jalore districts during that period. Thereafter, there were some small to and fro shifts of this line during different decades from 1920 to 1970. Such strips were more marked in the north of Churu district and in the south of Jalore district. In Churu district, the maximum dryness was recorded in the decade 1951-1960, when the aridity index line (of 80) attained its easternmost position. In case of Jalore, the eastern shift was maximum in 1961-1970, thereby showing this region to have become much drier recently (Krishnan 1977).

The human population, starting with a base of roughly 3.56 million in 1901, registered a linear increase to 10.23 million by 1971, almost three-fold increase in the arid districts of Rajasthan. The growth rate of 158 percent in the hot desert was more than that of the country as a whole i.e., 132 percent. The density of population varied from 157 to 4 persons per sq. km. As a consequence, in western Rajasthan, the cultivation of marginal lands increased by 44.6 percent during 1961-1971. Pastures and other types of lands declined by 16.8 and 6.96 percent during the two decades. The increase in rain fed farming on marginal lands had not only resulted in a decline of crop productivity per unit area but had resulted in an over-exploitation of ground water.

The increasing human population is a serious stress, particularly on the vegetation resources of the desert. The trees, shrubs and even their roots were indiscriminately cut by the rural population for fuel, top feed , thorn fencing and the construction of thatched huts. It was estimated that the requirement of the people in the desert in respect to woody biomass had increased from 1.85 million tones in 1951 to 3.33 million tones in 1971. All the available air-dried seeds and pods of the trees were used as delicacies. The seeds of Acacia Senegal (Kumat), the fruits of Capparis Decidua (Ker) and the pods of prosopis Cineraria (Sangri) were harvested for human consumption. Almost all the fruits of Ziziphus Numularia (*beg jhardberi*) growing in accessible parts of the desert were harvested.

The seeds of grasses, e.g., Panicum Turgidum, P. Antidotale, Cebshrus Biflorus and Echinochlea Colonum were mixed with millets for chapatti (unleavened cakes), especially during the drought period. The grass seeds are supposed to add to the nutritive value of the food. The intensity with which seeds collection was made for direct human consumption throughout the desert regions, seriously affected the natural process of regeneration of the desirable plant species in the inhospitable terrain.

Inspite of the low productivity of the arid lands, Rajasthan sustained a fairly high population of livestock. Paradoxically, along with the reduction in the grazing area during last decades, the livestock population registered an alarming increase, from 9.4 million in 1951 to 15.5 in 1972. The livestock density per 100 hectares rose not only in the desert districts from 72 in 1951 to 175 in 1971 but adjoining districts as well, but their enhancement was no so spectacular i.e., 25 percent as against 293 percent in desert districts.

The goat and sheep population ranged from 57.1 to 69.3 percent during 1951-1971. Owing to continued droughts during 1967-1971 with consequent migration and mortality, the number of hardy animals, e.g., the goat increased by 34%. The data thus not only reveal goat and sheep population in the arid Rajasthan but also point their increase during the years of drought. The pressure of livestock on the grazing lands resulted in the depletion of vegetation resources and at certain habitats in the desert; the natural succession trends have been reversed. Due to severe depletion of food species, livestock productivity has declined and has induced migration and nomadism in the human population.

A general approach to improve productivity of arid lands was to bring in irrigation water, chiefly for raising crops. The non-rational bringing of water into desert also adversely affected eco-systems. It is evident that after enactment of canal system in western Rajasthan particularly in Ganganagar district, the water table rose between 1 to 9 meters per year. In addition, the perched water table rise along with the sub-surface gypsum bed created severe water logging problems.

For collecting rainwater for subsequent irrigation, village ponds and dams across prior channels were constructed in many parts of Thar Desert, have also resulted in deterioration of the desert ecosystem. They have pushed up the water table and have spread Salinization. Likewise, it is revealed that about 9290 sq. kms or 4.35 percent of western Rajasthan has already been affected by the processes of desertification and 162,900 sq. kms or 76.15 percent of area has been categorized as highly and moderately vulnerable whereas 41,692 sq. kms or 19.5 percent of the area is considered to slightly vulnerable to various processes of desertification.

7.6 STUDIES AND EFFORTS MADE BY GOVERNMENT OF INDIA

Environmental conservation has been an integral part of the Indian ethos. Like international efforts, the Government of India has made concerted efforts on the basis of studies and findings of various reports and international agencies measures. The Ministry of Water Resources of the Government of India had prepared National Water Policy in 1987 for optimum and rationale use of water resources pertaining to surface, underground and rain water management and various schemes were initiated as follow up of the national policy. The National Land Use Policy 1988. The policy was intended to have a dynamic conservation, sustainable development and equitable access to benefits of intervention as its trust. Land being a state subject, much legislation on land has been enacted by various states as per their requirement and situations. The objectives of the National Land use Policy outlined are as under:

(1) To meet the consumption needs of growing population by increasing productivity of the integrated land resources of the country.

(2) To prevent any further deterioration of land resources by appropriate preventive measures.

(3) To restore the productivity of degraded land by appropriate package of practices.

(4) To provide the necessary technology and extension support to all concerned and to the farming community in particular for obtaining maximum production through increased productivity.

(5) To allocate land for different uses based upon land capability, land productivity and National production goals.

(6) To install an efficient and effective administrative structure for prescribing and regulating land by all concerned, including the government department and to revitalize the land use boards in this respect.

(7) To involve the community for adoption of appropriate land use for increased productivity, by ensuring that the land use policy provides adequately for the consumption and energy needs and generally improves their income levels and provides them a better quality of life.

(8) To create greater awareness of advantages of national land use policy at all levels through appropriate education, extension and training programmes.

(9) To restructure the livestock production programmes in such a manner that the livestock population is gradually limited to economically productive stocks and to prevent degradation of grassland by promoting measures of increasing stall-feeding of livestock in rural areas.

(10) To provide for optimum use of land under agriculture by promoting mixed farming systems in which the production programme will include the production of fodder and tree crops also on marginal land and sub-marginal land.

(11) To motivate farmers by organizing input supplies and marketing support for encouraging them to cultivate he appropriate crop or fodder or trees in conformity with the land use policy.

(12) To take up on priority basis the completion of land and soil surveys and to complete the inventory of land resources on the basis of the prescribed land use classification so that allocation is based on a reliable database.

(13) To examine the legal support available for enforcement of land use policy in the form of existing state and central laws and to consider the need for comprehensive legislation to provide some teeth to any machinery entrusted with implementation of the national land use policy.

(14) To coordinate the formulation and implementation of water resources management policies, forest management policies and urban planning within the overall resource allocation needs dictated by a comprehensive national land use policy.

(15) To prepare a plan of action at all levels covering a time frame relevant to all the objectives specified above and to continuously monitor action taken thereon in all effective fashion.

National Forest Policy, 1988 envisages for the massive need based and time bound programme of Afforestation and tree planting with particular emphasis on fuel-wood and fodder development, on all degraded and denuded lands in the country (NFAP, MOEF GOI 1999). The policy also makes it necessary to encourage planting of trees alongside of roads, railway lines, rivers, streams and canals and other unutilized land and state/ corporate/ village and community lands and institutional or private ownership. It also provides that green belts and woodlots should be raised in urban/ industrial areas as well as in arid tracts to check erosion and desertification as well as to improve the micro-climate. Main objectives of the National Forest Policy 1988 are as follows:

Increasing substantially the forest / tree cover in the country through massive Afforestation.

- Increasing substantially the forest / tree cover in the country through massive Afforestation.
- Increasing productivity of forests.
- Meeting the requirements of fuel-wood, fodder, minor forest produce and small timber and bamboo of the rural and tribal population.
- Efficient utilization of forest resources.
- Conservation of the natural resources, which is a heritage.
- Involving people through massive peoples' movement.
- Checking soil erosion.
- Checking the extension of sand dunes.

The national Forest Policy, 1988 recognized environment stability to be the main objective of forest policy and the derivation of direct economic benefits subject to this main objective. It look cognizance of basic needs of fuel-wood, fodder, minor forest produce and construction timber of tribal and poor and accepted these as the first charge on forest. The policy further states that one of the essentials of forest management is increasing forest cover on semi-arid, arid and desert tracts. It called for involvement of people in protection, regeneration and development of forests. In these respects the policy had marked a paradigm shift in the objective of forest policy and forest management.

The National Livestock Policy Perspective, 1996 presents a strategic framework with six identified thrust areas as given below:

(a) Use national and global market pull creativity to provide the energy and impetus for sectoral growth.

(b) Enhance quality and economic efficiency at the sector level through promotion of appropriate institutions and mechanisms.

(c) Use positive discrimination to support livelihood-intensive institutions, technologies and programmes so that future growth of livestock sector does not occur at the cost of rural livelihood.

(d) Redesign and revitalize the research establishment, extension and input supply mechanism to serve as the protective armour around small producers and women.

(e) Promote research, experimentation, and propagation of intermediate production units that blend advanced techniques with traditional modes of production.

(f) Monitor environmental implications of the livestock sector growth and involve participatory local mechanisms for common property resources management.

The goals of policy perspective were (i) to transform the national and global market pull into an engine for generating sectoral and national competitive advantage, (ii) to enable small producers all over the country to participate in the process, growth and globalization and (iii) to ensure the ecological sustainability of the growth of livestock sector. The strategy includes providing credit support to landless to own bullocks under the anti-poverty programmes. It also promotes programmes like joint forest management, community fodder farms and grazing land protection committees as a method of sustainable use of village common property resources. It recommends the conservation of livestock germplasm in order to preserve genetic diversity. The policy perspective also mentions the issue of feed and fodder requirements of the large and growing cattle population of the country.

National agricultural policy, 2000 seeks to promote technically sound, economically viable, environmentally non-degrading and socially acceptable use of country's natural resources viz. land, water and genetic endowment, to promote sustainable development of agriculture. The policy particularly emphasizes integrated and holistic development of rain-fed areas by conservation of rain water by vegetative measures on watershed basis and augmentation of biomass production for maintenance of ecological balance and for higher income generation through agro and farm forestry with the involvement of the watershed community and by evolving technology, extension and credit support packages. It indicates the broad policy framework for management of grazing land for increasing availability of animal feed and fodder, promoting use of unutilized wastelands for agriculture and Afforestation, reclamation of degraded and fallow lands as well as problem soils for optimization of their productive use; generation and transfer of technology; improving input use efficiency; pooling and evaluating traditional practices, knowledge and wisdom; promoting investments in agriculture; strengthening institutional infrastructure, ensuring better risk management and introducing management reforms, to achieve the objectives (Ministry of Agriculture, 2000).

National Population Policy, 2000, while continuing with two child norm, takes a comprehensive look at issues determining population growth. Health care of mother and child, health and sanitation as well as compulsory education figure high on its agenda. It aims to achieve reduced population growth through decentralized decision making. The policy aims at achieving a stable population by 2045. The new policy is significant that it has been endorsed at the highest level, which should generate the political will to carry forward a programme that is already in place. The policy has taken shape through a series of consultations with all stakeholders, the state health and family welfare ministries, state secretaries, professionals medical associations and medical experts, NGOs, media persons, corporate sector and the civil society towards advocacy and dissemination of the policy. The policy also focuses attention on the states of Uttar Pradesh, Bihar, Madhya Pradesh, Rajasthan and Orissa involving voluntary associations and the Panchayati Raj institutions. As a follow up, State Population Commissions have been set up in ten states.

National Convention Strategy and Policy Statement on Environment and Development, 1992. Environmental management has been accepted as a major guiding factor for national development. For this purpose, instruments like suitable legislation, fiscal incentives, agreement and treaties, educational programmes and information dissemination through publicity are used. The strategy includes use of modern sciences, new technology, traditional technology with active involvement of the people.

The Ministry of Environment and Forests had adopted the National Conservation Strategy and Policy statement on environment and development in June 1992. it lays down the guidelines that help to weave environmental considerations into the fabric of the national planning and development processes. Policy prepared with a view to integrating environmental consideration into development process.

Environment strategy is to serve as management guide for integrating environmental concerns with development imperatives. The primary purpose of the strategy and policy statement is to reinforce the traditional ethos and to build up conservation consciousness in society, which may live in harmony with nature and make efficient use of resources. The strategy and policy statement lays down comprehensive action points in respect of sectors as agriculture, irrigation, animal husbandry, forestry, energy generation, use and conservation, industrial development, mining and quarrying, tourism, transportation and human settlements to ensure that conservation and enhancement of the environment is taken due care of sustainable development. The strategy was presented as the framework of sustainable development for the country and in Earth summit held in Rio in 1992.

7.7 INSTITUTIONAL FRAMEWORK

As part of process of formulation and implementation, the Government of India has established a number of boards and corporations for proper conservation and management of natural resources. These are National Land Use and Wastelands Development Council (NLWC), the National Wastelands Development Board (NWDB) and the National Afforestation and Eco-development Board were constituted for addressing land degradation issues at national and state levels. Functions of above bodies are as under:

FUNCTIONS OF NLCB, NWDB AND NAEB

1. National Land Use and Conservation Board is chaired by Union Minister of Agriculture	Formulation and implementation of a national land use policy. Concerned with conservation and optimal utilization of the land resources. Guidelines and financial support to state Land use Boards (SLUBs)
2. National Wasteland Development Board chaired by Union Minister for Rural Development	Set up in 1985 in the MOEF and presently functioning under the Department of Land Resources under the Ministry of Rural Development. Development of wastelands in the country to sustainable use and increasing biomass available, especially fuel-wood and fodder
3. National Afforestation and Eco-Development Board chaired by Union Minister of Environment and Forests.	Set up in 1992 in MOEF. Promotion of afforestation, trees planting, ecological restoration and eco-development activities in the country. Regeneration of degraded forest areas and land adjoining the forest areas, national parks, sanctuaries and other protected areas and ecologically fragile areas.

Rural development and poverty alleviation programmes are implemented on a decentralized basis throughout the country. There are a number of project implementing agencies at the local levels. These include the District Watershed Development Committees, the District Rural Development Agency, the District Forest officers and the Panchayati Raj institutions. The governing body of the DRDA include Members of Parliament, Members of Legislative Assembly, District level officers of the development departments and representatives of the weaker sections of the society. At block level, the Block Development Officer and at the village level Gram Sevak or village level workers are responsible for implementation of the programme. The District Collector and district administration helps in the overall administration of the area.

Watershed projects under the Desert Development Programme, Drought Prone Areas Programme and integrated Watershed Development Projects Programme are sanctioned in favour of the District Rural Development Agency or Zila Parishad, the district level body of elected Panchayat representatives. Various line departments of the state governments, non-governmental organizations as Project Implementing Agencies (PIA), implement the projects. Institutional framework at various levels of functioning is given below:

Levels	Institutional Framework
Central	Central government- Ministries, Departments- Environment and Forests, Rural Development, Agriculture, Human Resources Development, Women and Child Development, Social Justice and Empowerment, Health and Family Welfare, Water Resources, etc. RandD organizations and autonomous organizations of the central government- Central Ground Water Authority (CGWA), Central Ground Water Board (CGWB), Central Water Commission (CWC), Command Area Development Authority (CADA), Indian Council of Agriculture Research (ICAR), Indian Council for Forestry Education and Research, Central Pollution Control Board, NAEB, NLCB, NWDB, NRCD etc.

State	Department of Forest and Environment, Rural Development, Agriculture, Water Resources/ Irrigation, Animal Husbandry, Social Welfare. State level organizations such as State Land Use Boards, State Renewable Energy Development Agencies, State Pollution Control Boards and State/ Regional centers of RandD and other institutions of the Government.
District	District Collector/ District Magistrate, District Forest Officer, District Rural Development Agency, District Watershed Committee, District Social Welfare officer, Krishi Vigyan Kendra etc. Community Health Centres, Regional and Zonal centers of the central and state departments and institutions.
Block	Panchayati Raj Institutions, Community Health Centres, Block Development Officers.
Village Panchayat	Panchayati Raj institutions, Gram panchayat, janpadh, Zila panchayat, Gram Sabha (Village council), Village Development Committee and implementing agencies.

Combating desertification is a multifaceted issue and requires involvement of large number of institutions for addressing various issues according to their respective field of specialization. The following institutions and government departments are engaged in research, surveys, investigations and pilot experiments on various aspects of dryland problems in India:

1. The Central Arid Zone Research Institute (CAZRI)

CAZRI established at Jodhpur is working under the Indian Council of Agriculture Research (ICAR), New Delhi. The Institute has its various functional divisions: (i) Basic Resource Survey Division covering geology, geomorphology, hydrology, ecology, cartography and photo interpretation, (ii) Plant Studies Division covering range management, plant protection, horticulture, silvi-culture, toxicology and agrostology, (iii) Division of animal studies covering animal ecology, rodent control, animal nutrition and physiology, (iv) Division of Soil-water-plant relationship covering agronomy, plant physiology, water management, soil chemistry and soil physics. (v)Division of Economics and Extension covering sociology, agricultural economics, (vi) Division of Extension and Training covering extension, training, information dissemination and audio-visual aids, (vii) Division of Wind Power and Solar Energy covering climatology, wind power and solar energy utilization. In addition, the institute in collaboration with other agencies played a leading role in coordination and operational research projects in the field of dryland agriculture, improvement of millets, water management, soil salinity, drip and sprinkler irrigation and arid-land management.

2. The Central Sheep and Wool Research Institute

An associated institution of ICAR, carried out pilot studies with demonstration units for extension work on various aspects of improvement and promotion of livestock breeding, production of wool and management of grasslands. The Indian Grassland and Fodder Research Institute, Indian Veterinary Research Institute, national Dairy Research Institute and Central Soil Salinity Research Institute are some of the other institutions under the ICAR.

3. Geological Survey of India

The GSI through its Quaternary Geology, Geomorphology and Environmental Geology Division (Desert Geology) has taken up six specific arid zone projects on (i) Sand advance, (ii) Desert Evaluation, (iii) lake salinity, (iv) Luni Basin, (v) Ghaggar basin and (vi) Rann of Kutch. Through a programme of detailed geomorphological mapping of the arid tracts by means of intensive fieldwork, this division is providing basic information on geology, geomorphological and geo-chemical aspects of the arid zone relevant to the movement of sand, Salinization, relationship between land use and geologic and geomorphic characteristics and palaeoclimatic evaluation. This is in addition to the preparation of detailed geological maps.

4. The Botanical Survey of India and Zoological Survey of India have taken up systematic exploration in the arid and semi-arid tracts in order to build up a detailed inventory on the flora and fauna in the desert region.

5. The Central Salt and marine Chemical Research Institute at Bhavnagar is primarily concerned with problems of desalinization of water, utilization of solar energy and development of economic products from naturally occurring brines in the Rann of Kutch and the off-shore areas of the arid zone.

6. The Central Ground Water Board is engaged in complete ground water surveys of the dryland areas and other parts of the country. Through geo-hydrological surveys and explorations in the arid zone, various areas on sub-districts have been classified into white zone, gray zone and dark zone detailing out water potential in respective areas. The areas demarcated as dark zone have been banned for further digging of wells, tube-wells for irrigation purposes.

7. The Indian Meteorological Department Poona provides a network of basic meteorological data pertaining to rainfall, temperature, wind parameters and evaporation for the arid and semi-arid tracts of the country. This department is carrying out research on various aspects of desert meteorology with special reference to Thar Desert. A drought research unit conducts studies on all climatic aspects of chronically drought prone areas of the country. It also studies agro-climatology of drought, drought predictions, fluctuations in aerial extent of drought and synoptic features associated with drought, dust storms and rainfall trends.

8. The Central Water Commission, Ministry of Water Resources has undertaken detailed studies on appraisal and promotion of irrigation facilities in drought prone areas. It has also been associated with the activities of the Technical Advisory Committee on agro- climatic study of droughts. Presently one prestigious assignment of study of availability of flood waters for water deficient areas, where the excess water of rivers of the country is planned to be transported to other parts of the country facing water problems.

7.8 ASSESSMENT OF DESERTIFICATION AND LAND DEGRADATION

Assessment of desertification and land degradation of the entire country had been carried out by a number of organizations. These have helped understand the extent, nature and severity of desertification and land degradation in the country. There are large variations in these estimates due to the methodologies adopted. Estimates of desertification and land degradation made by these agencies vary considerably e.g. 53.28 million hectares to 184.9 million hectares. These wide variations in estimates of degraded lands are due to different approaches in defining degraded lands or guess estimates arrived through discussions with limited fieldwork using manual methods and/ or adopting various criteria for their delineation.

One of the assessments, which is in use is that of the Department of Agriculture and Cooperation, ministry of Agriculture, which indicates that nearly 173 million hectare of the total country's geographical area of 328 million hectare is under desertification and land degradation of some form or other. Drought is a regular occurrence in some regions of the country. A number of organizations are involved in drought monitoring and early warning systems of natural calamities. Soil survey and desertification monitoring and assessments are being carried out by central and state level organizations as per source of Department of Agriculture and Co-operation, Ministry of Agriculture (1999) are as under:

A. Central Level Organizations

1. All India Soil and Land use Survey by Department of Agriculture and Co-operation, Ministry of Agriculture.
2. National Bureau of Soil Survey and Land use Planning by Indian Council of Agriculture Research.
3. Central Arid Zone Research Institute (Arid Zone Mapping), Indian Council of Agriculture Research.
4. National Remote Sensing Agency, Department of Space.
5. RRSSC, Department of Space.
6. Space Application Centre, Department of Space.
7. National Atlas and Thematic Organization.

B. State level organizations

1. State Departments of Agriculture.
2. Soil Survey Units.
3. Agriculture Universities.
4. Remote Sensing Application Centres.

Soil maps are prepared on different scales from very small of 1:1 million to very large 1:4000 to meet the requirements of analysis at various levels. Small-scale maps are needed for macro level analysis at the national level for planning developmental programmes at the national scale. Mapping at a scale of 1:250,000 scale are done at the state and regional levels. Maps of 1:50,000 scale are useful for district level. Soil maps of 1:8000 or larger scales are required for micro level analysis of specific purposes such as soil conservation, reclamation, soil fertility studies, revenue taxation, crop suitability etc. The scale and utility of mapping are summarized in Table 7.10.

Table 7.10 : Soil Resource Maps used for Various Purposes

S. No.	Information Required	Scale
1.	National Scale	1: 1,000,000
2.	State level	1:250,000
3.	District level	1:50,000
4.	Command Areas and Pre-irrigation survey	1:4,000-1:8,000
5.	Farm level/ Micro-watershed	1:4,000-1: 8,000
6.	Soil Conservation Planning/ Implementation	1:4,000-1: 8,000
7.	Reclamation of Salt-affected soils	1:4,000-1:8,000
8.	Optimum Land use planning at:	
	(i) District Level	1:50,000
	(ii) Village Level	1:4,000-1:8,000

Source: Department of Agriculture and Co-operation, Ministry of Agriculture.

A number of surveys and assessments have been carried out by different agencies over the past several decades on the extent and type of desertification and land degradation occurring in the country. The major survey / estimates are summarized in Table 7.11.

Table 7.11 : Major Surveys / Estimates of Degraded Lands

S.No.	Agency / Organization	Year	Extent (in M.ha.)	Criteria for delineation
1.	National Commission on Agriculture (NCA)	1976	175.00	Based on secondary data only.
2.	Ministry of Agriculture, GOI	1985	173.64	Based on land degradation statistics for the states.
3.	National Bureay of Soil Survey and Land Use Planning	1994	187.70	Mapping in 1:4.4 million scale based on Global Assessment of Soil Degradation (GLASOD) guidelines.
4.	Ministry of Agriculture, Department of Agriculture and Co-operation	1994	107.43	Based on the land degradation statistics for the states.

Source: Compiled from Velayutham, 1999 and DAC 1999.

The estimates of degraded lands made by these agencies vary considerably. These wide variations in estimates of degraded lands are due to different approaches in defining degraded lands or guess estimates arrived through discussions with limited field work using manual methods and / or adopting various criteria for their delineation. The causes for the discrepancies in the area under degraded lands as reported by the various agencies are:

a. Some organizations based their statistics only on estimates and not on mapping.
b. The objectives and scales of mapping by various organizations were different.
c. Definition, criteria and mapping legend followed by different organizations are different, which were designed to make their specific requirements.
d. The data base used for mapping and the data collection are not consistent for the various cases, where the statistics have been generated by various agencies.

1. Use of Remote Sensing in Desertification and Land Degradation Assessment

For taking suitable remedial measures like reclamation of degraded land and proper utilization of wastelands for productive purposes, it is necessary to obtain timely and accurate information regarding the location and extent of wastelands. In 1985, the National Wastelands Development Board (NWDB) commissioned the National Remote Sensing Agency (NRSA) to prepare a wasteland map of the country, as well as for each state. The National Wastelands Identification Project was taken up in 1986 in technical collaboration with the Department of Space and Survey of India. A total of 146 districts showing more than 15% of their area as wastelands, was selected. These maps and census maps were used to transfer the village level on a scale of 1:50,000 where 1 cm on the map correspond 500 metres on the ground. Subsequently, in 1991, it was decided to take up additional districts, having wastelands in the range of 5-15% for mapping. Thus a total of 241 districts were covered for mapping of wastelands. In 2000, an update of the extent of the wastelands in the country was done through NSRA. The comparative estimates of wastelands in 1986 and 2000 reveal that 53.28 million hectares and 63.90 million hectare area was identified as wasteland.

Assessment of forest cover carried out by Forest Survey of India (FSI) revealed that 637,293 sq. km was under forest cover during 1999, which was 19.39 percent of the total geographical area of the country. Out of total 637,293 sq. km area, 377,358 sq. km was under dense forest, 255,064 sq. km under open forest and 4871 sq. km under mangrove class. The percentage of these classes of forests was 59.21%, 40.02% and 0.77% respectively. The FSI has prepared thematic map on a scale of 1: 50,000 by interpretation of aerial photographs. These maps depict forest types, species composition, crown density of forest cover and other land use. Every year, about 5000 aerial photographs corresponding to 260 topographical sheets on a scale of 1: 50,000 of the Survey of India have been interpreted. One work station has been provided to each zonal office to use satellite data of high resolution (IRS-1C/ ID -PAN data to prepare updated thematic maps with intensive ground truthing and support State Departments in preparation of Working Plans (MOEF, Annual Report 1999-2000).

The assessment made by the Department of Agriculture and Cooperation in 1985 revealed that out of the total geographical area of about 328.73 million hectares, approximately 173.64 million hectare comprising of 52.82 % are subject to moderate to severe form of degradation. Types and extent of land degradation is given in terms of affected area and corresponding to total geographical area of the country in Table 7.12.

Table 7.12 : Estimation of Land Degradation Status in India (In Million Hectares)

S. No.	Type of Land Degradation	Affected Area	% to Total Geographical Area of country.
1.	Water Erosion	107.12	32.59
2.	Wind Erosion	17.79	5.41
3.	Ravines	3.97	1.21
4.	Salt Affected	7.61	2.32
5.	Water-logging	8.52	2.59
6.	Mines and Quarry Wastes	0.04	Neg.
7.	Degraded Land due to Shifting Cultivation	4.91	1.49
8.	Degraded Forest Lands	19.49	5.93
9.	Special Problems	2.73	0.83
10.	Coastal Sandy Areas	1.46	0.45
	Total	**173.64**	**52.82**

Source: Department of Agri. and Cooperation, Ministry of Agriculture, 1985.

According to the information of the Ministry of Agriculture, total cultivated area of 142.02 million hectare in 1997-98, various changes affecting the addition and delineations, the increase of cultivated area of 1050-51 of 118.75 million hectare the calculation of net addition of 23.27 million hectare is not correct. The process of changes incurred in cultivated area during past 50 years is evident that some land being cultivated was shifted to other uses like construction of railways tract and roads, dams, reservoirs and other non-cultivation activities and total increase in area put under agriculture remained much more that 23.27 million hectares. This information is only possible by field assessment during past 50 years indicating the changes of each type of land in the base year and shifting to another use can give the true picture.

Therefore, the reported area under crops of 142.02 million hectares is combination of land under other uses, transferred for cultivation in past 50 years. Similarly the degraded area out of the present cultivated area is much more than 23.27 million hectare. Another indicator is irrigated area of 54.57 million hectare, which is combination of surface and underground water use. Undoubtedly, the cultivated and irrigated area has been increased during past 50 years, but the level of land degradation in cultivated land in total area and in dryland areas is our focus of the study.

Another factor is the growth of population of the country which has increased from 361 million to 1027 million from 1951 to 2001 and the country has successfully matched its food requirement with the increasing population. It is clearly evident that the rate of growth of production of agricultural crops is able to be matched with the growing population, the resultant factor is that there was no evidence of decrease in production or productivity in the country, which is one of the factors of desertification and land degradation processes. This is a general impact of the country as a whole and not specific to dryland areas in specific terms. Though the desertification and land degradation has been increasing silently unnoticed by politicians, administrators and policy planners of the country, efforts made for combating desertification and land degradation measures can only be accounted for such a situation.

Degree, Extent and Treatment of Desertification Hazards in India

Land degradation is a dynamic process. It is the result of both- natural and biotic forces operating on the earth. Natural calamities, viz, earthquakes, floods and droughts etc, over-exploitation of land resources, unwise land use and the consequences of high inputs agriculture on soil and environment are of great concern that sensitized the international bodies to combat land degradation and desertification in both developed and developing countries. Development of degraded lands in India is one of the options available to enhance food production and to restore the fragile eco-system. The scientific information and spatial distribution of various kinds of degraded lands is thus essential for formulation of strategic plan to arrest the menace of land degradation. The statistics on degraded lands published by different and non- governmental agencies not only vary widely but the nomenclature of degraded lands also differs considerably. It thus, necessitates rapid inventorying of degraded lands to generate realistic database on degraded lands using remote sensing techniques in the country as a first step of strategic planning. Subsequently, detailed information on soil and land characteristics are required to be generated, which is vital for development of location specific plan for reclamation of degraded lands.

All India Soil and Land Use Survey (AISLUS) has undertaken the task of land degradation mapping using remotely sensed data and developed a methodology accordingly. The mapping has been conceptualized as a four tier approach comprising kind of degradation, severity and degradation, degradation under major land form and land-use. Visual mode of interpretation technique based on image characteristics has been employed for mapping of degraded lands, image interpretation key has been formulated based on the spectral signatures of various causative factors of different kinds of degraded lands. The mapping legend has been made very systematic and commutative. The extent and spatial distribution of different kinds of degraded lands with degree of severity under major landform and major land use in a district could be derived easily.

Generation of realistic information on degraded lands of the country is utmost necessary. Visual mode of interpretation technique based on image characteristics has been employed for mapping of the degraded lands. Image interpretation key has been formulated based on the mapping legend. The degree and extent of desertification hazard of different kinds with degree of severity under major landform could be derived.

A significant work has been done in the field of desertification by eminent world personalities. Worth mentioning are Biswas, where causes and process of desertification – abiotic and biotic factors and measures to combat desertification are analyzed. Sehgal and Abrol has described that about 50.5% of the area under desertification through various factors. Jain has described the adverse geo-hydrological and geo-morphological conditions and pedogenic limitations of desertification. Shankarnarayan has described major problems of desertification in Indian sub-continent. Rao, et al have dealt with the role of space technology in assessment, monitoring and combating desertification. Kaul – has suggested measures for long-term sustainability. Chouhan has focused in his study on introduction of desertification situations, its indicators, process and causes, Satellite remote sensing and Measures for combating desertification. Above literature lack on focusing the pinpointed issue of extent and degree of desertification hazard, whereas the present attempt addresses the area and its location on the map as well as through narration to reveal the present extent and degree of desertification helpful for the planners and decision makers to take suitable measures.

7.9 Methodology

The present study has been substantiated by extensive fieldwork. The essential data have been collected from a wide range of sources. The remote sensing data available in black and white and false colour composites of IRS-LISS-III of different seasons on 1: 1 million and 1: 250,000 scale have been used. Survey of India topo-sheets (old and new) for the entire region both 1:4 miles or a: 250,000 scale, 1: 1 mile or 1: 50,000 scale and a bioclimatic map, have been used as base maps.

India has a geographical area of 329 million hectare, but land use statistics are available for 304 million hectare. Out of the total reporting area, 30 million hectare is classified as other uncultivated area and 41 million hectare is under non-agricultural use, making a total of 71 million hectare, which is not available for cultivation. The area under non-agricultural purposes is reportedly increasing at the rate of 0.3 million hectare per annum. The area classified as forests is reported to be 67 million hectare. Out of the total forest area, actual area with more than 10 percent canopy is about 64 million hectare and out of this, only 38 million hectare has a canopy cover of 40 percent or more. Thus about half of the forest area is in advanced stage of deterioration.

The area under arid, semi arid and sub-humid zones has been calculated as no such information was available to stress the areas of different bio-climatic zones to communicate the intensity and magnitude of problem through various methods by super-imposing over one map. The methodology has been transcribed in flow chart and Methodological Approach (Figure 7.4 and 7.5).

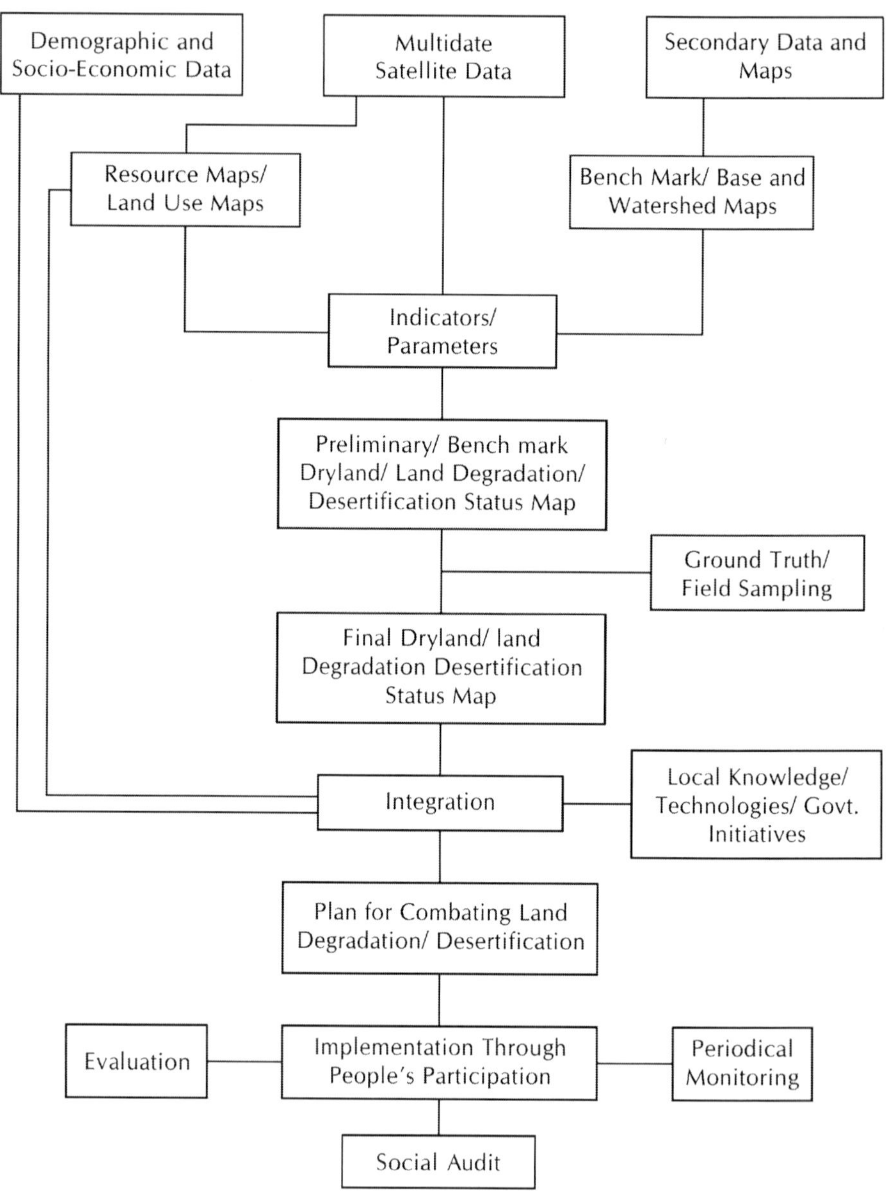

Figure 7.4 : *Research Design for Combating Desertification/ Land Degration in Drylands using Satellite Data & GIS*

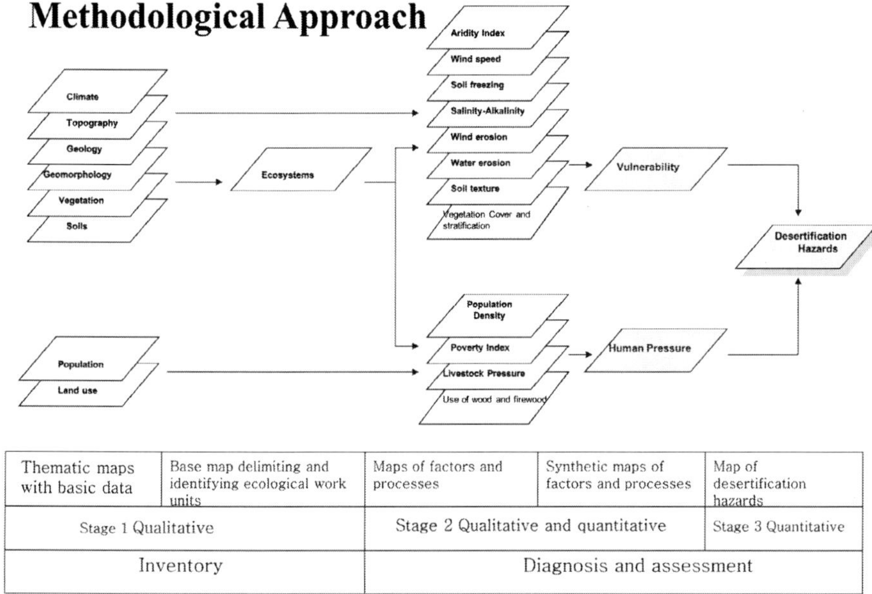

Figure 7.5 : Methodological approach

7.10 Degree and Extent of Desertification Hazard

In the absence of scientific measurements and monitoring of soil degradation/ land degradation for most of the world's land surface, the data that did exist for each unit were supplemented by expert opinions from the participants, who were selected for their intimate knowledge of local conditions. Using a standardized set of guidelines to ensure uniformity in reporting, soil degradation was assessed over the recent past, ideally averaged over the previous 5 or 10 years, but in some cases the time period extended over recent decades.

The involvement of a large number of scientists with personal experience of regional conditions is an important strength of GLASOD. The complex interplay of soil degradation/ land degradation processes and their spatial variability needs to be assessed according to local conditions and forms of land use. This point is well illustrated in the case of off-site effects of soil erosion. Soil is often lost from steep highland slopes and deposited in lower, move level sites. If the lower site contains a reservoir, the accumulation of sediment is a problem; it causes a loss of storage capacity requiring expensive remedial efforts.

This spatial variability in the operation of degradation processes also illustrates an important point concerning the scale of analysis. Scoones noted that despite indications of increased degradation in his study areas, livestock production had not declined, although this could have been because negative effects on production lagged the initial degradation. The scale of the GLASOD mapping polygons of large areas on the ground are coloured according to an average severity of degradation.

Areas identified as non-degraded in the GLASOD survey are designed wither 'stable' or 'non-used' by human populations. Stable areas are either those regions, where human intervention is minimal due to very low population densities, or regions where soil improvement or protection programmes have been successfully implemented. Non-used areas are those not used for agricultural purposes, permanent pasture or covered by forest or woodland. This category includes land with little value for agricultural use, but land in this category may also be potentially productive, but has been put to other land use e.g. build on areas and roads.

Severity is calculated according to a combination of the degree of soil degradation and the percentage of the area affected. The degree to which soil is degraded has been estimated in relation to changes in agricultural suitability, in relation to reduced productivity and in some cases in relation to its biotic functions. The degree of degradation is classified according to the following scheme:

(a) *None:* There is no sign of present degradation from water or wind erosion, from chemical or physical deterioration; all original biotic functions are intact. Such land is considered stable.

(b) *Light:* The terrain is suitable for use in local farming systems, but with somewhat reduced agricultural productivity. Restoration to full productivity is possible by modifications of the management system. The original biotic functions are still largely intact.

(c) *Moderate:* The terrain is still suitable for use in local farming systems, but with greatly reduced agricultural productivity. Major improvements are required to restore productivity, such as draining for water-logged or salinized land or counter banks for eroded land. The original biotic functions are partially destroyed.

(d) *High:* The terrain is not reclaimable at the farm level. Major engineering works are required for terrain restoration. The original biotic functions are largely destroyed.

(e) *Very High:* The terrain is not reclaimable and impossible to restore. The original biotic functions are completely destroyed.

The importance of the land use in question should be noted here. While a moderately degraded soil for example may require major structural alterations to restore productivity under one particular land use, for a less demanding land use the alterations required may be less intensive. The relative extent of degradation was expressed using five categories:

(i) **Infrequent** : Up to 5 % of the unit is affected.

(ii) **Common** : 6-10 % of the unit is affected

(iii) **Frequent** : 11-25 % of the unit is affected.

(iv) **Very Frequent** : 26-50 % of the unit is affected.

(v) **Dominant** : More than 50% of the unit is affected.

The degree and extent of desertification hazard has been depicted on the Figure 7.6. The map contains multi-facet aspects of desertification from various angles as per existence of problem and extent of the existence of the problem. It can be witnessed from the map that out of 29 states, 12 states have been affected with the desertification problem covering of an area of 531870.48 sq. kms which is about 16.18 percent of the total geographical area spread over in 12 states., viz, Andhra Pradesh, Gujarat, Haryana, Jammu and Kashmir, Karnataka, Madhya Pradesh, Maharashtra, Orissa, Punjab, Rajasthan, Tamil Nadu and Uttar Pradesh. Areas classified as arid, semi-arid and dry sub-humid with state-wise position of affected area are given in Table 7.13. It would be seen from the table that highest area under desertification exist in Rajasthan. Arid area of Rajasthan State is almost half of total similar category, whereas semi-arid area is maximum in Maharashatra. .Madhya Pradesh possesses maximum sub-humid area and that is the only category of desertification area in the state.

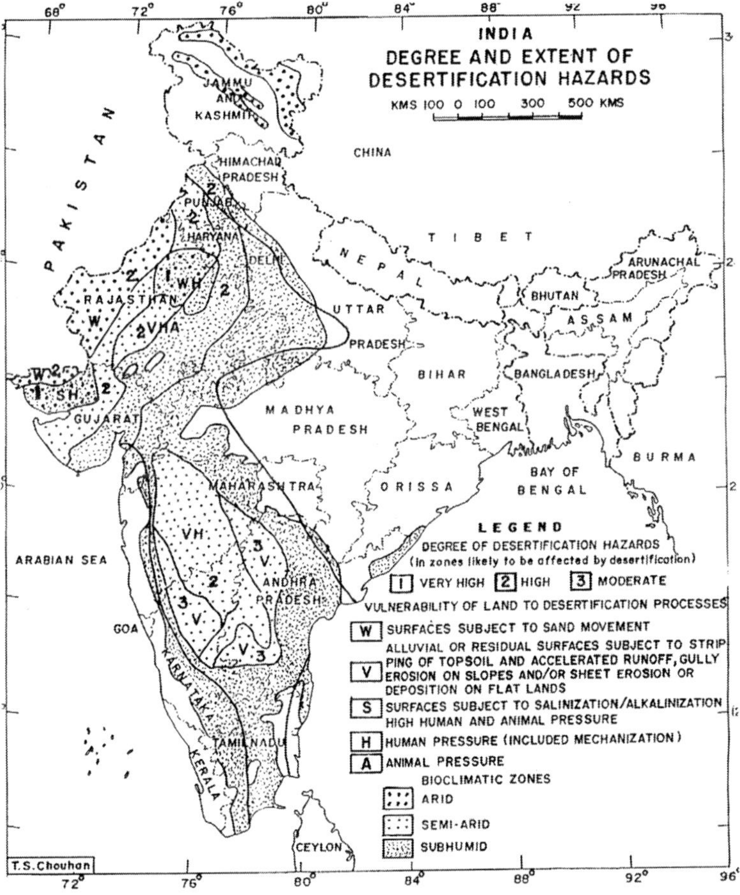

Note : The external boundary and the coastlines of India appearing on the map are neither verified nor authenticated by the Survey of India.

Figure 7.6 : *Dgree and extent of desertification hazars*

Table 7.13 : Degree and Extent of Desertification Hazard

(Area in Sq. Kms)

State	Total Geog. Area	Arid	Semi-arid	Dry Sub-humid	Total Dryland
Andhra Pradesh	275045.00	—	35754.12	48199.15	83953.27
Gujarat	196024.00	16561.24	18800.34	—	35361.58
Haryana	44212.00	—	4415.55	8811.99	13227.54
Jammu and Kashmir	222236.00	30862.17	—	9118.29	39980.46
Karnataka	191791.00	—	29115.08	31002.58	60117.66
Madhya Pradesh	443446.00	—	—	50032.76	50032.76
Maharashtra	307713.00	—	47468.09	23409.89	70877.98
Orissa	155707.00	—	—	801.38	801.38
Punjab	50362.00	—	6273.79	—	6273.79
Rajasthan	342239.00	47986.78	23556.08	34583.84	106126.70
Uttar Pradesh	294411.00	—	—	34803.94	34803.94
Tamil Nadu	130058.00	—	—	30313.42	30313.42
Total	2653244.00	95410.19	165383.05	271077.24	531870.48
India (Total Area)	**3287263.00**	**95410.19**	**165383.05**	**271077.24**	**531870.48**

Source: Area calculated by the author on the basis of Figure 7.6.

The information given in above table reveals that the states facing the problem of desertification cover 80.71% of the total geographical area of the country. However, the present level of desertification in these state is 20.15 percent of the total geographical area of the affected states. Table 7.13 gives the percentage distribution of position of desertification in each category in comparison to total geographical area of each state to understand severity of the problem. The percentage distribution of 12 states has been indicated out of the total geographical areas of these states. Similarly, the percentage distribution in the country has been shown to total geographical area of the country to stress the desertification problem.

The information given in Table 7.14 reveals that under arid areas the desertification situation prevail in Gujarat, Jammu and Kashmir and Rajasthan with maximum area in Rajasthan. The semi-arid area prone to desertification is in 7 states of the country with maximum percentage coverage in Maharashtra. Dry sub-humid area covers 10 states out of total 12, where Tamil Nadu has highest percentage area under the only this category. Though Madhya Pradesh has maximum area under dry sub-humid category, but it is lesser in percentage terms to its total geographical area.

Table 7.14 : Percentage distribution of desertification hazard to total geographical area of respective state.

State	Total Geog. Area	Percent of Total Dryland to state	Percent arid area to State area	Percent semi-arid area to state area	Percent dry sub-humid area to total state area
Andhra Pradesh	275045.00	30.52	—	13.00	17.52
Gujarat	196024.00	18.04	8.45	9.59	—
Haryana	44212.00	29.92	—	9.99	19.93
Jammu and Kashmir	222236.00	17.99	13.89	—	4.10
Karnataka	191791.00	31.35	—	15.18	16.17
Madhya Pradesh	443446.00	11.28	—	—	11.28
Maharashtra	307713.00	23.03	—	15.43	7.60
Orissa	155707.00	0.51	—	—	0.51
Punjab	50362.00	12.46	—	12.46	—
Rajasthan	342239.00	31.01	14.02	6.88	10.11
Uttar Pradesh	294411.00	11.82	—	—	11.82
Tamil Nadu	130058.00	23.31	—	—	23.31
Total 12 States	2653244.00	20.05	3.60	6.23	10.22
India (Total Area)	**3287263.00**	**16.18**	**2.90**	**5.03**	**8.25**

Source: Area calculated by the author on the basis of Fig. 7.6.

(A) Degree of Desertification Hazard

The degree of desertification has been classified under three categories as Moderate desertification, High desertification and Very High desertification. These areas have been demarcated as 3, 2 and 1 respectively in the map. The analysis of desertification in terms of geographical position reveals that the western part of the country from north to south is prone to desertification. The peninsular area is completely covered with exception to western coastal area and except some part f eastern peninsular of Tamil Nadu. Total eastern part of the country from north to south is completely free from desertification hazard Areas prone to very high desertification belong to semi arid areas of Rajasthan and Gujarat.

Information given in the figure is evident that the areas of various desertification categories have created significantly larger zones, where, rainfall deficiency has intensified the problem further. According to this classification, the position of each affected state has been described in Table 7.15.

Table 7.15 : Degree of Desertification in Affected States

State	Degree of Desertification
Andhra Pradesh	Areas under the categories of semi arid and dry sub-humid zones are presently under the moderate category of desertification.
Gujarat	Northern part adjoining to Pakistan under arid zone and middle part under semi arid zone is high desertification prone, whereas the north-western part situated in between is very high desertification category.

State	Degree of Desertification
Haryana	The state is high prone to desertification
Jammu and Kashmir	The northern and middle part of the state is cold arid area with moderate desertification category.
Karnataka	Barring the western coastal area of the state the total state is prone to desertification but presently in moderate desertification process.
Madhya Pradesh	Western part adjoining to Rajasthan and Gujarat is prone to desertification of moderate category.
Maharashtra	Most part of the state is under semi-arid and sub-humid zone but presently under the moderate process of desertification.
Orissa	Very small pocket of coastal region adjoining to Andhra Pradesh is sub-humid zone category with moderate process of desertification.
Punjab	Total state is under semi arid zone prone to high process of desertification.
Rajasthan	The western part of the adjoining to Pakistan is under arid zone and some eastern part of sub-humid zone are prone to high desertification, whereas the middle part of semi arid zone is very high prone to desertification.
Uttar Pradesh	Western part of the state adjoining to Rajasthan falling under the category of sub-humid zone is moderate prone to desertification.
Tamil Nadu	Barring east coastal and some parts adjoining to Kerala are under sub-humid zone and moderate category of desertification process.

(B) Vulnerability of Land to Desertification Processes

The desertification is the outcome of various factors relating to natural and human efforts. The desertification is constant impact of all these factors but the prime factor responsible for land degradation and desertification conditions have been classified into five following categories:

1. Surfaces subject to sand movement marked as "**W**" in the figure. This is significantly hazard disrupting the total process of development as sand movement disturb the transportation of all categories, communication network, problematic in view of sudden movement with the wind pressure.
2. Alluvial or residual surfaces subject to stripping of top-soil and accelerated run-off, gully erosion on slopes and / or sheet erosion or deposition of flat lands. Such problematic conditions have been demarcated as **"V"** on the figure.
3. Surfaces subject to salinization / alkalinization due to high human and animal pressure. Areas prone to such problems have been marked as "**S**" on the figure.
4. Human pressure including mechanization impacts responsible for desertification have been shown as "**H**" on the figure.
5. Animal pressure responsible for desertification conditions have been indicated through the sign **"A"** on the figure.

The above category of vulnerability of land subject to desertification are basically the prime and most acute factors causing desertification hazard, which indicates that other factors also play their role to deteriorate the productivity of land on regular basis. The occurrence of drought due to failure of monsoon further intensify the deterioration. Poverty, drought, land degradation and desertification are inter-related to each other specially in dryland areas of the country. When more than one factor play key role in degradation, the area becomes severely deteriorated and the main sufferers are the local residents, who are helpless, but to remain in the degraded areas for their livelihood. The social and economic factors of such degradation are far reaching in terms of resultant impact of their problems. Details of the factors vulnerable to land degradation and desertification are narrated in Table 7.16 along with the locations in the problematic states and causes playing important role for the process of desertification conditions. It would be revealed from the analysis that the causes of desertification are diverse in terms of intensity, impact and incidence.

Table 7.16 : Vulnerability of land to desertification process

States	**Factors Causing Land Degradation Process.**
Andhra Pradesh	The primary factor responsible for land degradation is alluvial or residual surfaces subject to stripping or topsoil and accelerated run-off, gully erosion on slopes and /or sheet erosion or deposition on flat lands. Northern and southern parts of the state are most vulnerable to the problem.
Gujarat	Northern part adjoining to Pakistan under arid zone is most vulnerable to surface sand movement. Whereas the middle part under semi arid zone marked as "SH" is prone to salinity and alkalinity through high human and animal pressure and human pressure which include the mechanization impact.
Haryana	The state is high prone to desertification and canal cultivated area is mostly affected by water-logging. Other well irrigated is prone to salinity and alkalinity.
Jammu and Kashmir	The cold desert zone is generally disturbed area due to international boundary with Pakistan. Soil erosion on slopes and deposition on flat lands are responsible for severity of the problem.
Karnataka	Various factors of category "V" exist in the state, which is very severe in the central part of the state.
Madhya Pradesh	Salinity and alkalinity is the severe problem due to uninterrupted drawl of ground water.
Maharashtra	Over-exploitation of cultivated land is the severe problem in rain-fed areas. Deficiency of water has further intensified the problem. Monsoon failure is common phenomena in dryland areas.
Orissa	Coastal area of more than 800 sq.km. is prone to desertification. With mixed impact of various factors.
Punjab	Water logging and salinity are severe problem factors since most of the state is canal irrigated and fertility loss due to these factors have severe affected the state.
Rajasthan	The western part of the adjoining to Pakistan is prone sand movement marked as "W". The adjoining semi arid land is mostly affected by sand movement and human pressure marked as "WH".

States	Factors Causing Land Degradation Process.
	The south eastern part of the state is prone to various factors of "VHA". The multiple impact of these factors followed by water deficiency and salinity/ alkalinity are the most problematic conditions of the area.
Uttar Pradesh	Western part of the state adjoining to Rajasthan is prone to desertification due to salinity and water logging in canal cultivated areas and salinity/ alkalinity in ground water irrigated areas. The ultimate impact is deficiency in productivity.
Tamil Nadu	The sub-humid areas prone to desertification face water deficiency in view of erratic monsoon and excessive use of ground water.

(C) Bio-climatic Zones of the Problematic States:

Bio-climatic zones of the country in the process of desertification have been classified into arid, semi- arid sub-humid according to biological and climatic conditions.

The north-west arid region prone to desertification is spread over in 95410.19 sq. kms of area of Jammu and Kashmir, Gujarat and Rajasthan. With 13.89%, 8.45% and 14.02 % of total geographical area of respective state. This is 2.90% of the total country area. There are intra-regional variations but major problem are wind and water erosion. In some pockets, recently brought under irrigation, problems of salinization and water logging are causing concern. Frequent droughts occur in this region- three times in five years on average. Serious problems include decreasing water availability and receding water table, brackish ground water, depleting forest cover, deteriorating grazing grounds and pastures, decreasing availability of fodder and drinking water, declining land productivity and yield of important crops, livestock suffering from scarcity of drinking water. In western parts of Rajasthan, shifting sand dunes is major problem.

Semi arid regions under the process of desertification cover geographical area of 13% of Andhra Pradesh, 9.59% of Gujarat, 9.99% of Haryana, 15.18% of Karnataka, 15.43% of Maharashtra, 12.46% of Punjab, 6.88 % of Rajasthan which is 5.43 % of the total geographical area of the country. The region is characterized by undulating topography, low to medium rainfall, high-density rainstorms, high volume and speed of surface flow, severe water erosion and very low level of rainfall utilization. Occurrence of intense drought and crop losses are frequent. Farmers are living in poverty. There have been reports of helpless farmers committing suicides due to crop failures. Low and fluctuating crop yields, over cultivation and shrinking pasturelands due to biotic pressures, receding ground water and depleting vegetative cover are assuming alarming proportions.

Sub-humid zones prone to desertification are spread over in 271077.24 sq. kms are in the states of Andhra Pradesh, 17.52% Haryana, 19.93 % , Jammu and Kashmir 4.10% , Karnataka 16.17%, Madhya Pradesh 11.28%, Maharashtra 7.60%, Orissa 0.51%, Rajasthan 10.11%, Tamil Nadu 23.31% and Uttar Pradesh

11.82% of their respective geographical area. The north-western and central sub-humid regions are prone to saline and sodic and water logged areas. The underground water is brackish with soluble salts. The north-western and north eastern sub humid regions suffer due to poor irrigation practices in the command areas. Ground water is sweet and sodification has taken place damaging physical conditions and chemical properties of the soil. Considerable areas have been rendered uncultivated.

Areas under sub-humid zones are spread over from Jammu and Kashmir to Tamil Nadu and of varying nature where the climatic conditions are different and similarly the topography and soil conditions differ. There cannot be one yard stick for measuring the situation prone to desertification in terms of degree and extent. Efforts have been made to analyze the situation of all the twelve states of the country prone to desertification at different stages of degree from very high to moderate desertification. Therefore, the area specific situations have been narrated to give the content and nature of the degradation situation.

Location of sub-humid given in the figure have also been described in the table 7.17, with the location of area in the state, situation of desertification in kms and the nature of the problem.

Table 7.17 : Sub-humid zones in the process of desertification

State	Bio-Climatic Zones
Andhra Pradesh	North, eastern coastal region and western parts of the state of 48199.15 sq. km area is in the process of desertification
Gujarat	Eastern part of the state from north to south is prone to sub-humid desertification.
Haryana	Eastern part of the state spread over in 8811.99 sq.kms is in the hub-humid desertification category.
Jammu and Kashmir	The sub-humid area of the stte spread over in 9118.29 sq. kms is in the process of desertification.
Karnataka	Barring the western coastal area of the state, the adjoining area of west and southern parts spread over in 31002.58 sq. km is sub-humid desertification prone.
Madhya Pradesh	Western part adjoining to Rajasthan and Gujarat spread over in 50032.76 sq. kms is sub-humid area prone to desertification.
Maharashtra	North eastern sub-humid region of the state spread over23409.89 sq. kms is prone to desertification.
Orissa	Very small pocket 801.38 sq. kms of coastal region adjoining to Andhra Pradesh is sub-humid zone is in the process of desertification.
Rajasthan	The sub-humid area of 34583.84 sq. kms situated in eastern part of the state from north to south is in the process of desertification.
Uttar Pradesh	Western part of the state adjoining to Rajasthan falling under the category of sub-humid zone spread over in 34803.94 is prone to desertification.
Tamil Nadu	Barring east coastal and some parts adjoining to Kerala are under sub-humid zone spread over in 30313. 42 sq. kms is in the process of desertification.

(D) Magnitude of the Problem and its Treatment

The above analysis of various factors of desertification have been analyzed to stress the magnitude of the problem existing in the country. It is evident that total area affected with the problem of desertification is 531870.48 sq. kms comprising of 95410.19 sq. kms arid zone, 165383.05 sq. kms of semi arid zones and 271077.24 sq. kms of sub-humid zone. Therefore the problem is not so simple, but necessary to be treated to make the total area free from the problem. There are different opinions amongst the specialists of various discipline, who have their authoritative approach. But the Ministry of Environment, Government of India has worked out estimates for treatment of areas at different levels of desertification process. While the norms calculated for construction of watersheds of 5000 hectare area each and the cost norms, the watershed approach is not the only feasible mode of treatment of the area. While learning the lessons from the past decade while construction of various watersheds in different climatic zones, it is difficult to select even a dozen watershed, which have succeeded in treatment of the area. However, our objective to adhere to the norms is to work out the cost of treatment of total area in the process of desertification According to the norms suggested, the cost of treatment has been worked out in Table 7.18.

Table 7.18 : Cost estimates for treatment of desertification

(Cost in Rs. per hectare)

Cost Component	In arid areas	In semi-arid areas	In sub-humid areas
1. Creation of awareness and capacity building of PRIs including entry point activities	1000	1000	2000*
2. Conservation measures – 70% for water resource development and 30% for soil conservation	5000	5000	5000
3. Production system	3000	3000	4000 **
4. Alternative livelihood systems	2000	2000	1000 ***
Total cost per hectare	**11000**	**11000**	**12000**

* Includes EPA

** Includes cultivation of medicinal plants

*** Includes processing and marketing of minor forest produce.

On the basis of above norms the treatment cost of arid, semi-arid and sub-humid areas as given in Table 7.13 is as follows:

(a) Treatment cost of arid zones spread over in 95410.19 sq. kms or say 9541019 hectares @ Rs. 11,000 is Rs. 10495.12 crores.

(b) Treatment of semi-arid zones of 165383.05 sq. kms or 16538305 hectares @ Rs. 11000 is Rs. 18192.14 crores.

(c) Treatment of sub-humid zones of 271077.24 sq. kms or 27107724 hectares @ Rs. 12000 is Rs. 29818.50 crores.

(d) Total cost of treatment of 53187048 hectares land i.e., Total of a+b+c = Rs. 58505.76 crores or say Rs. 585 billion

(E) Conclusion

The approach suggested and amount required is quite significant, which need to be managed in a time bound phase to cover all the degraded areas within a span of 10-15 years. Treatment of area in the process of desertification is necessary to meet the demand of growing population, employment generation, livestock development and environment development. However, the total programme require peoples participation in the decision-making process including empowerment in mode of execution and their cost sharing. Presently, treatment of individual land is generally avoided in view of personal benefit, but this ideology need to be changed as every land belong to the country and individual cultivator is unable to share the burden of total cost of the programme. However an indicative approach is necessary to identify the areas like those suggested for sub-humid regions.

The intervention approach suggested for programme execution starts from learning lessons from past efforts with rectification to unsuccessful efforts of watershed, irrigation practices and withdrawal of excess water from underground sources. The approach should associate the cultivators, understanding their problems and conventional approach for treatment of area with latest technological devices, adhering following measures:

1. Approach for Arid Regions

(i) Entry Point Activities and Capacity Building of PRIs

In most villages, farmers demand water sources for human and livestock drinking. Social capital building need to be promoted. In addition, sheds for schools, village link roads and culverts, community health centers, etc, need to be provided/ strengthened. Entry point activities need to be part of the programme and decided in consultation with local communities. The capacity building of the Panchayati Raj Institutions need to be developed to handle technical matters and financial management including group dynamics. Involvement of PRIs is essential for working out programme of common interest as the community land development, activities on common property resources, social forestry programme etc. require common understanding.

(ii) Conservation Measures

Water conservation measures based on indigenous systems like Nadi (Ponds), Tankas (covered dug-well), Khadins and nala ponding in small rivers or drainage systems with tunnel recharge wells need to be promoted. Farmers confidence increases with increasing availability of water. In addition a rationale pattern of water use need to be decided by the village community to benefit the maximum number of cultivators. The cultivators deprived from the benefit need to be acknowledged with the situation befitted to their interest. Appropriate low cost soil conservation measures based on local systems of trench and furrows, vegetative hedges of local grasses and shrubs. Kanabandi, a unique indigenous system of fixing brush wood barriers across the prevailing directions of wind to stabilize the sand dunes and save the crop from sand blows. Shelterbelt, wind breaks and strip plantation as per local requirement and feasibility.

(iii) Production System- Crop + Livestock Rearing

Mixed farming to minimize risk of erratic rainfall and diversified production system, mixed cropping producing millets, pulses and oilseeds, Agro-forestry and dryland horticulture, Strip and block plantation on desert front areas, grazing grounds, silvi-pasture systems.

(iv) Alternative Livelihood Systems of Rural Poor

Land based and non-land based activities like manufacturing of ethnic dresses, weaving of carpets etc. Livelihood for typically encompasses water, food, fodder and feed security with income generation through cash flows.

2. Semi- Arid Regions

(i) Entry Level Activities and Capacity Building of PRIs

Panchayati Raj Institutions and their community based organizations have to play a central role at field level. Their capacity building in to be built in technical, financial and institutional matters. NGOs and Government field functionaries to create awareness about the dangers of land degradation, help in organizing self help groups and other village institutions. The potential entry point activities can be creation of life saving irrigation during long dry spells in the cropping season.

(ii) Conservation Measures

Water conservation through life saving irrigation is the greatest need of the area. Collection of surplus surface flow in a network of small percolation ponds to recharge dug-wells, continuous contour trenches for enhancing the recharge should also be taken up. The utilization of harvested water during rainy season cropping and during the period of early withdrawal of monsoon and if possible, provision of pre-sowing irrigation for the next crops to be given priority. Thus water conservation would start the process of regeneration of eco-system.

(iii) Soil Conservation

Simple soil conservation measures in tune with traditional systems like small mud pebble bunds, boulder bunds, vegetative hedges and dead furrows in inter bunded areas need to be promoted.

(iv) Production System

Mixed farming with reliance on organic farming, LEISA (Low External Input Sustainable Agriculture) including Integrated Pest Management (IPM) and Integrated Nutrient Management. Agro-forestry and dryland horticulture.

(v) Alternative Livelihood system of Poor Households

Activities based on local resources, agriculture produce based activities and small-scale industries will little technical input for using improved design and quality to make the produce marketable. Livestock rearing for supplemental source of livelihood.

3. Sub-Humid Regions

(i) Capacity Building of PRIs and CBOs and Entry Level Activities

In view of low literacy level in rural areas, the capacity build need to be addressed with patience as tribals residing in these areas have tradition of community based behaviour. A well-structured awareness programme is needed to organize to bring them into self help groups in village institutions. The local population has good Indigenous Technical Know-hows (ITKs). The project need to build upon its systems and practices where voice of the local beneficiaries need to addressed properly. The capacity of PRIs and Community Based Organizations have to be nurtured and build in organizational, financial, record, book-keeping etc. slowly in tune with their absorption capacity.

(ii) Conservation Measures

Rain-water harvesting in village ponds, desilting of existing ones and construction of new ponds at hydrologically strategic points to enhance water availability for human beings and livestock be given highest priority. For life saving irrigation, the upgradation of traditional low cost Chuas (seepage wells) should be promoted and upgraded. Such wells save the crops during drought, provide pre-sowing irrigation for second crop and allow vegetable cultivation. Thus availability of water is helpful to build the confidence of local population and inspire them to move towards sustainable development.

(iii) Soil Conservation Measures

Soil erosion is serious problem in view of topographic reasons and as such continuous contour trenches, vegetative hedges across the slope, loose boulders, gully checks reinforced with vegetation and loose bolder checks in drainage lines to control soil erosion and loss of top soil.

(iv) Conservation of Bio-diversity

The tribal areas of this region are very rich in bio-diversity- land races and indigenous genes in cultivated and medicinal plants. Efforts need to be made to preserve bio-diversity to enable tribals to utilize genetic material for their benefit.

(v) Production System

Mixed farming- crop + livestock particularly small ruminants- goat, sheep rearing, poultry etc. for which the tribals have long traditions and local skills. Mixed cropping, coarse cereals, pulses, oilseeds adopted to ecological conditions. Low External Input Sustainable Agriculture with higher reliance on organic manures, IPM, INM etc. In addition promotion for cultivation of medicinal and herbal plants can be helpful in raising their income level substantially.

(vi) Alternative Livelihood System

Traditionally, the tribals have been supplementing their food, nutrition and income by collection of medicinal plants and minor forest produce, which need to be promoted. The processing and marketing is to be organized to save tribals from exploitation by middlemen. Other cottage industries including basket making and similar bamboo based decorative products need to be patronized. Sericulture, bee-keeping and lac cultivation can help tribals for traditional rearing useful insects through skill upgradation.

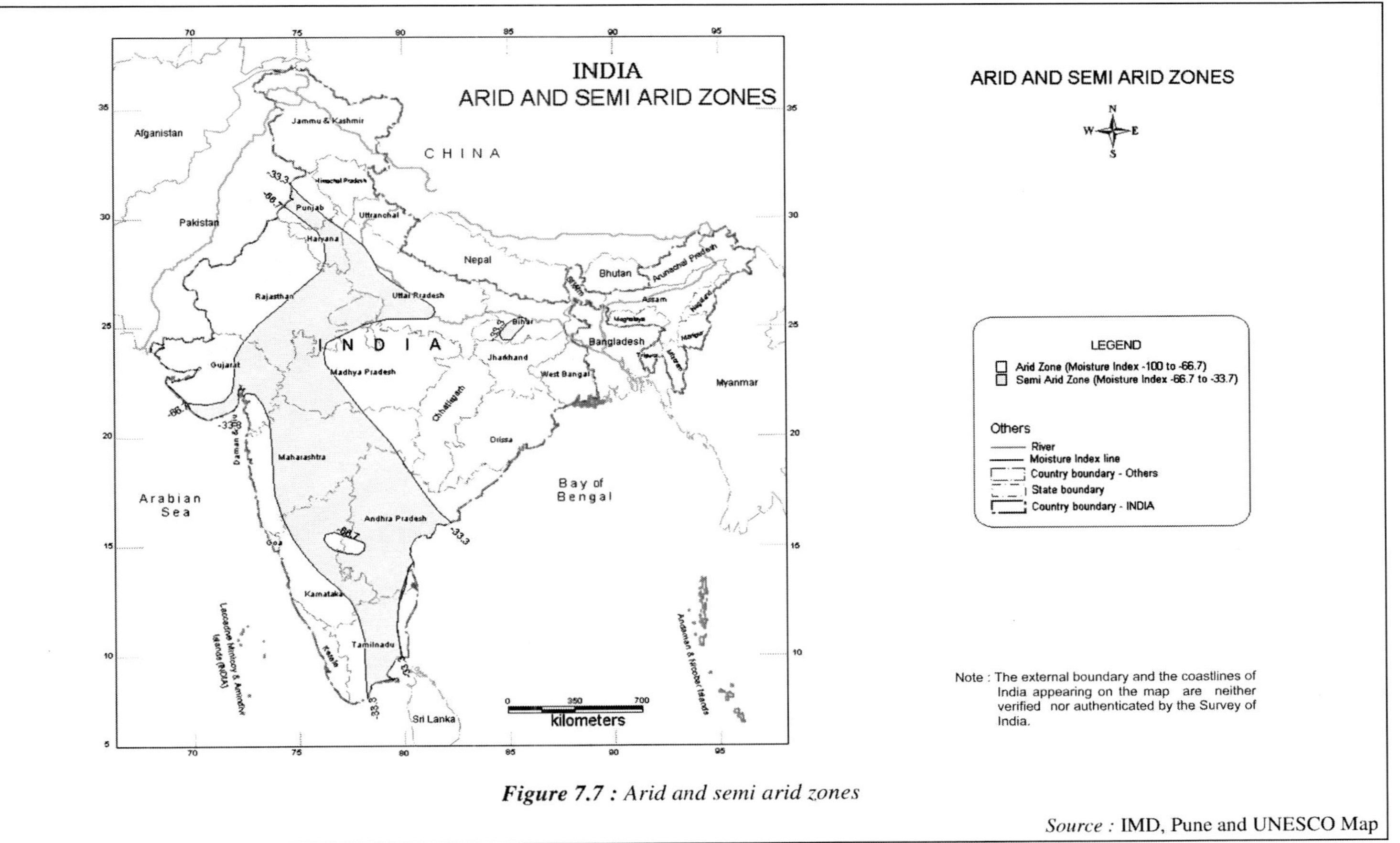

Figure 7.7 : Arid and semi arid zones

Source : IMD, Pune and UNESCO Map

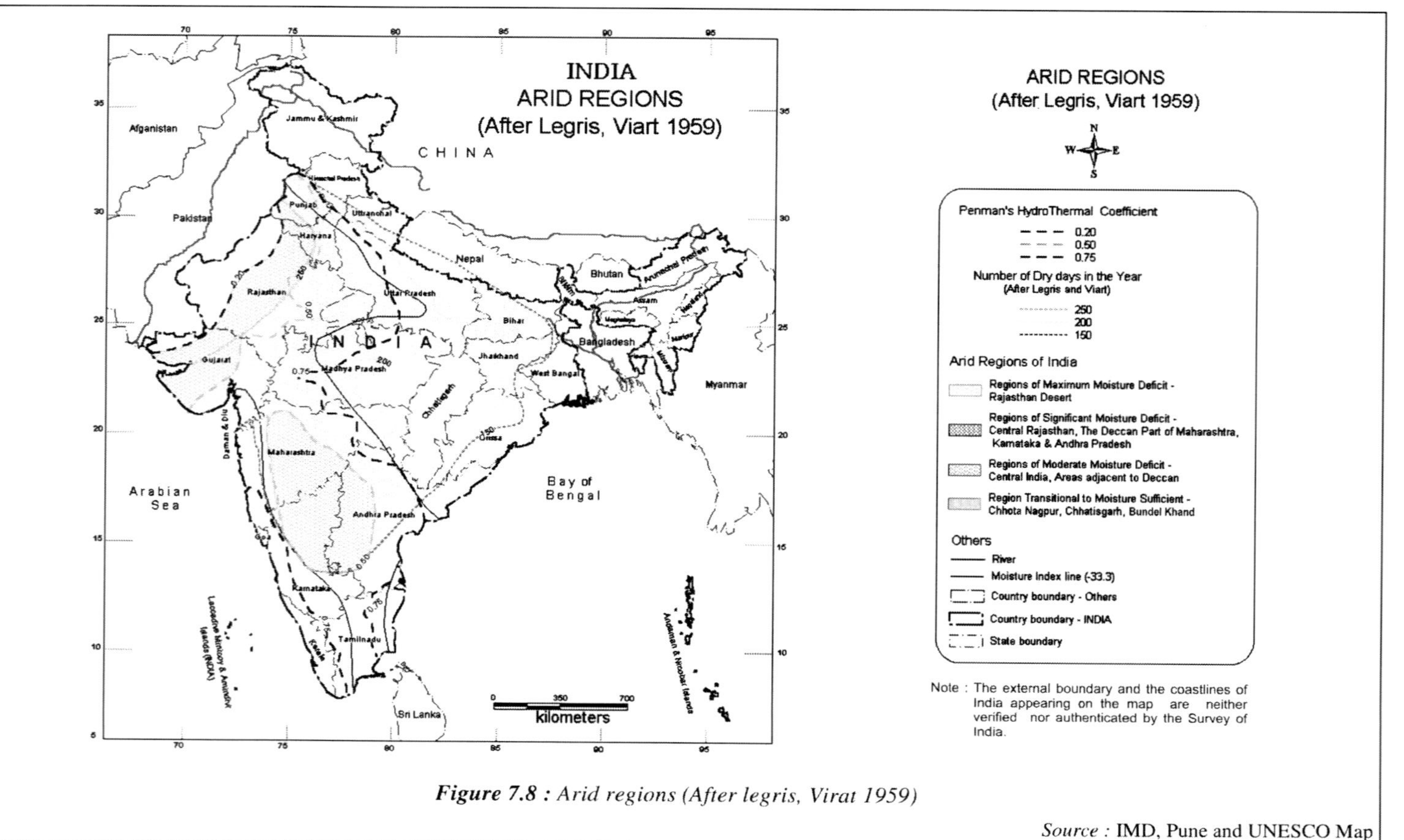

Figure 7.8 *: Arid regions (After legris, Virat 1959)*

Source : IMD, Pune and UNESCO Map

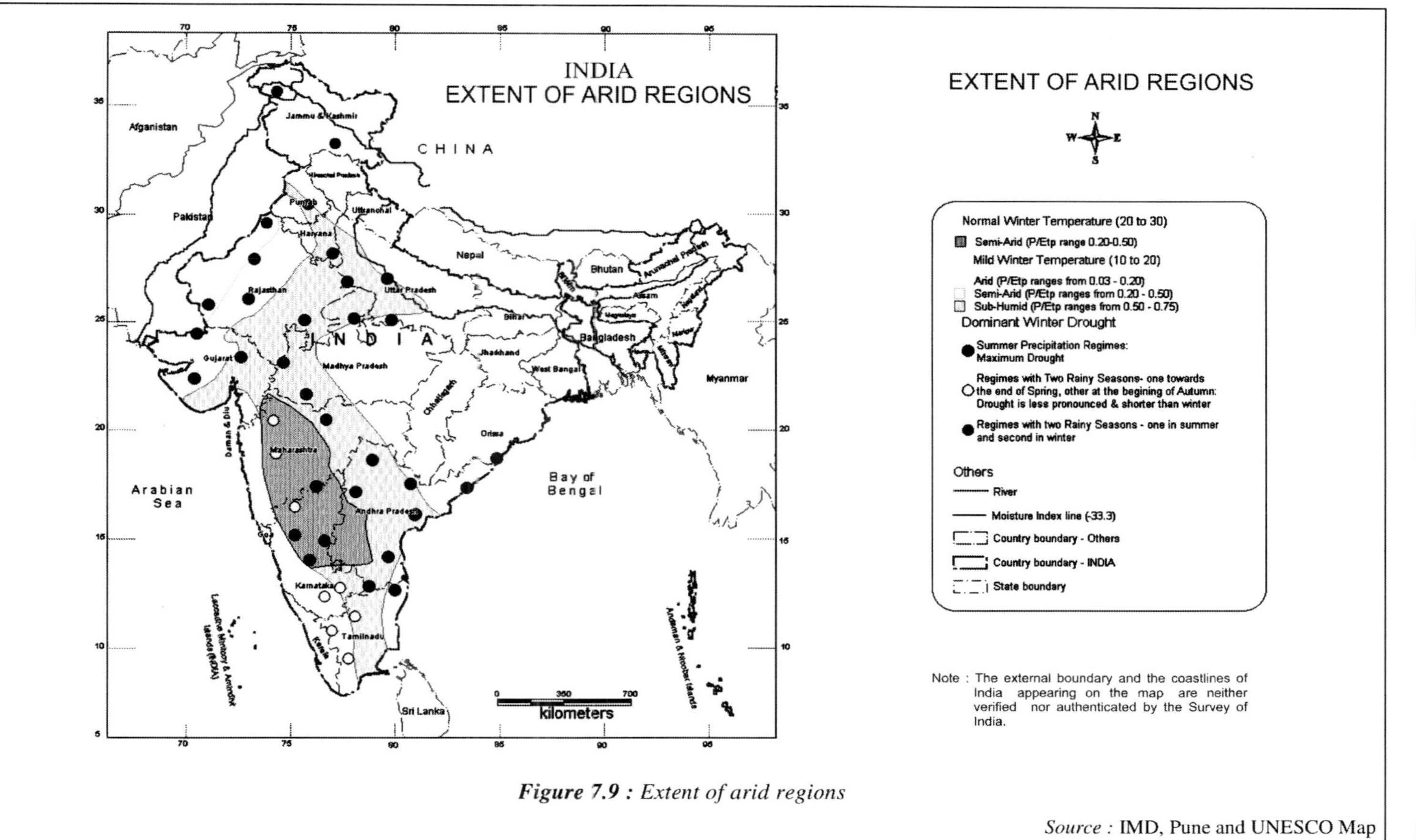

Figure 7.9 : Extent of arid regions

Source : IMD, Pune and UNESCO Map

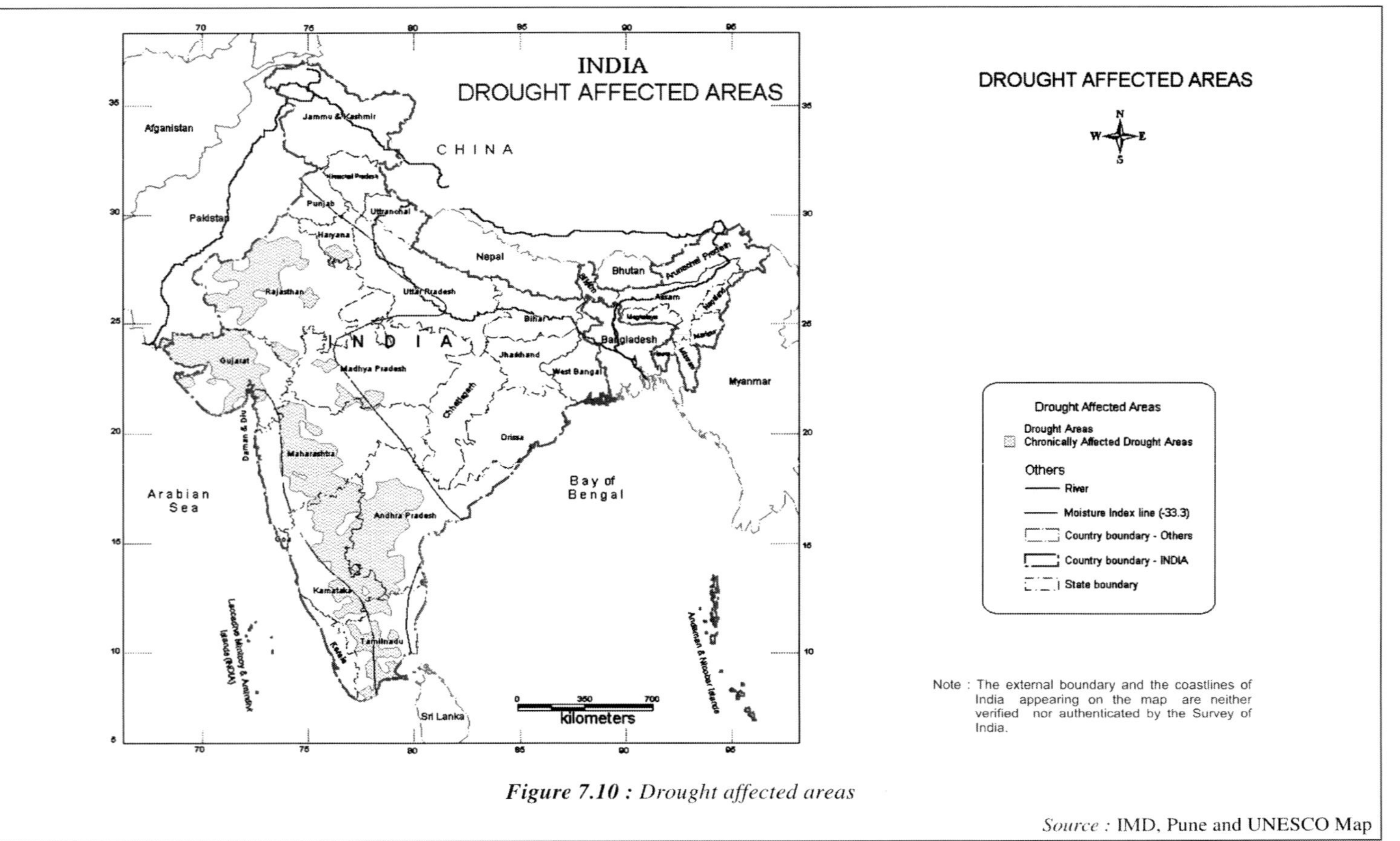

Figure 7.10 : Drought affected areas

Source : IMD, Pune and UNESCO Map

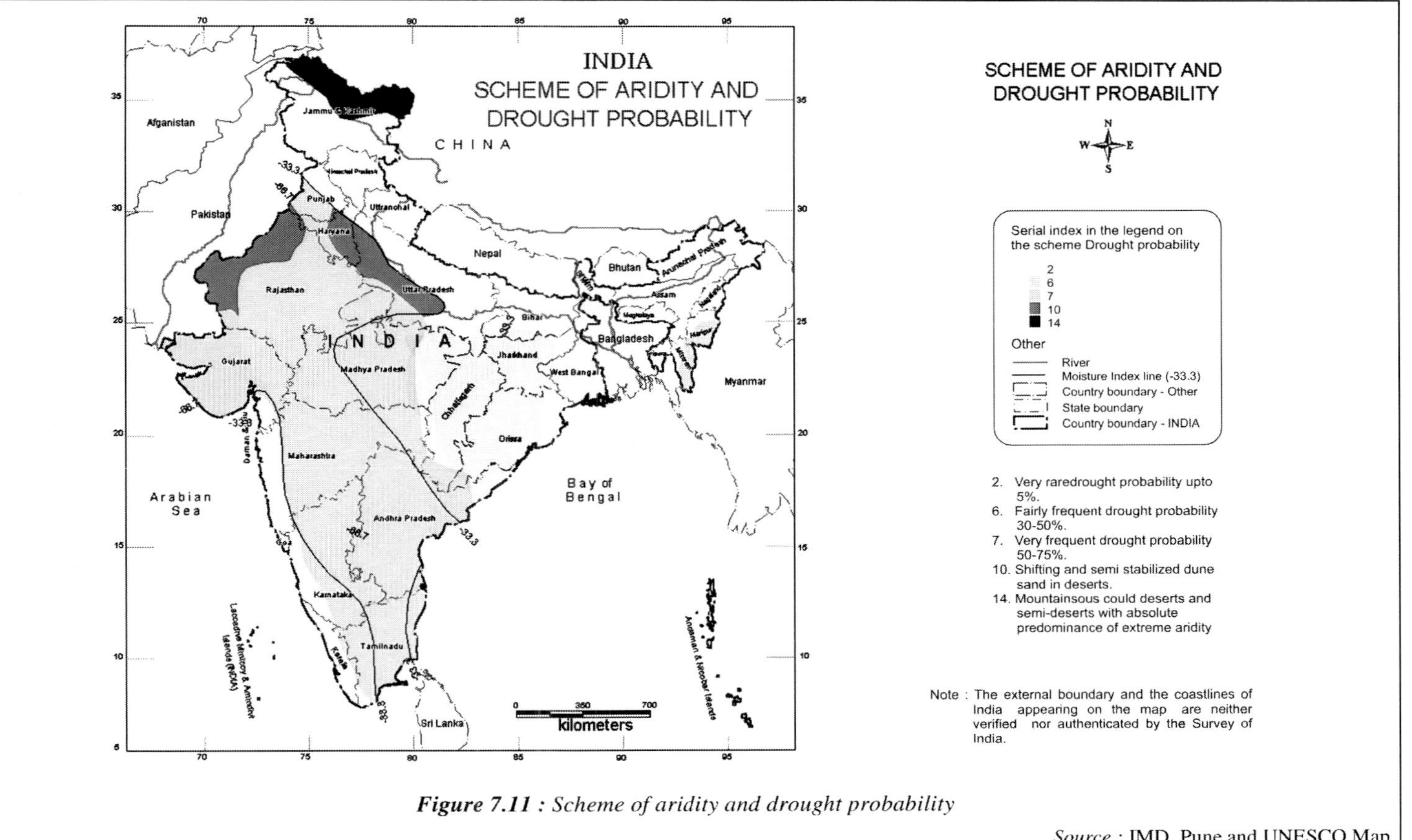

***Figure 7.11** : Scheme of aridity and drought probability*

Source : IMD, Pune and UNESCO Map

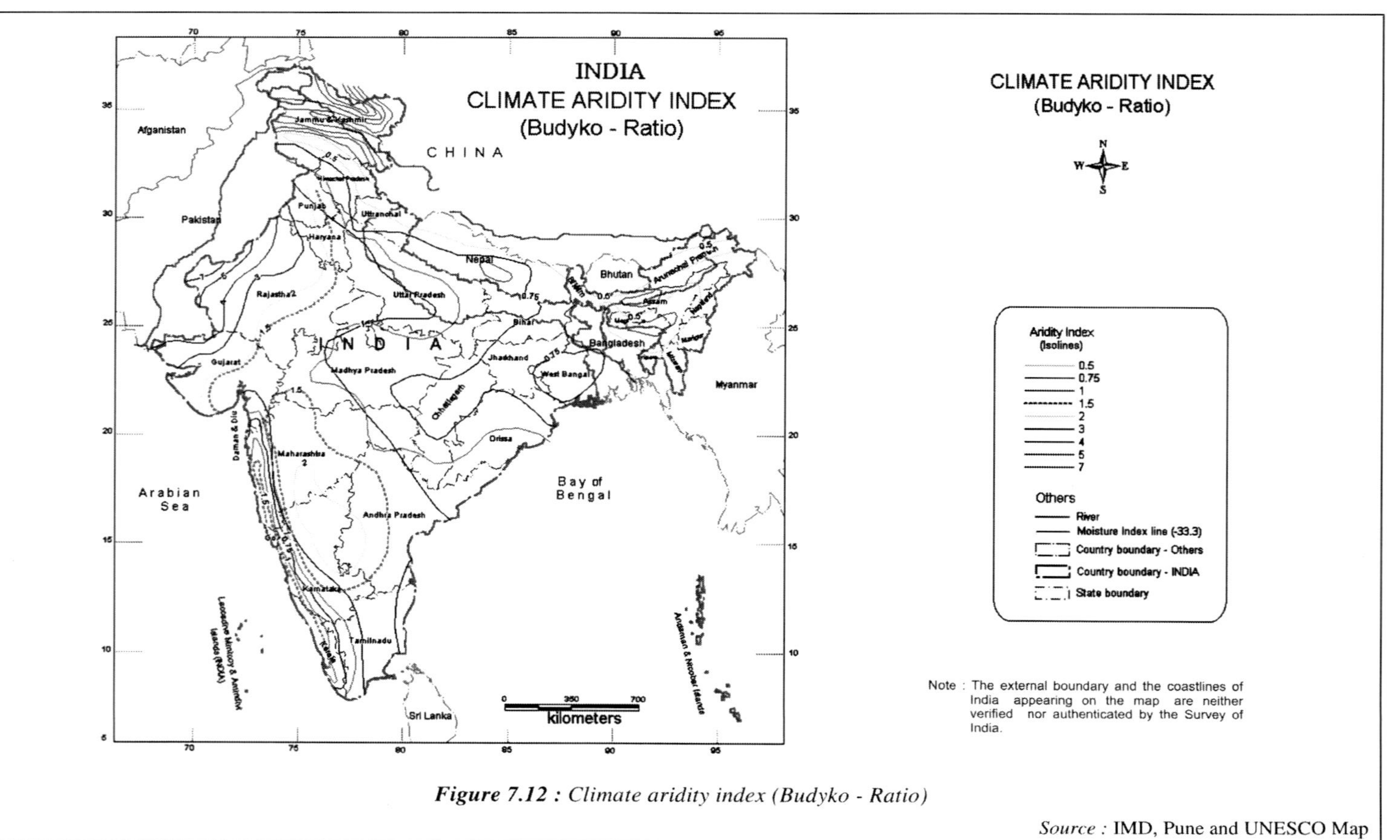

***Figure 7.12 :** Climate aridity index (Budyko - Ratio)*

Source : IMD, Pune and UNESCO Map

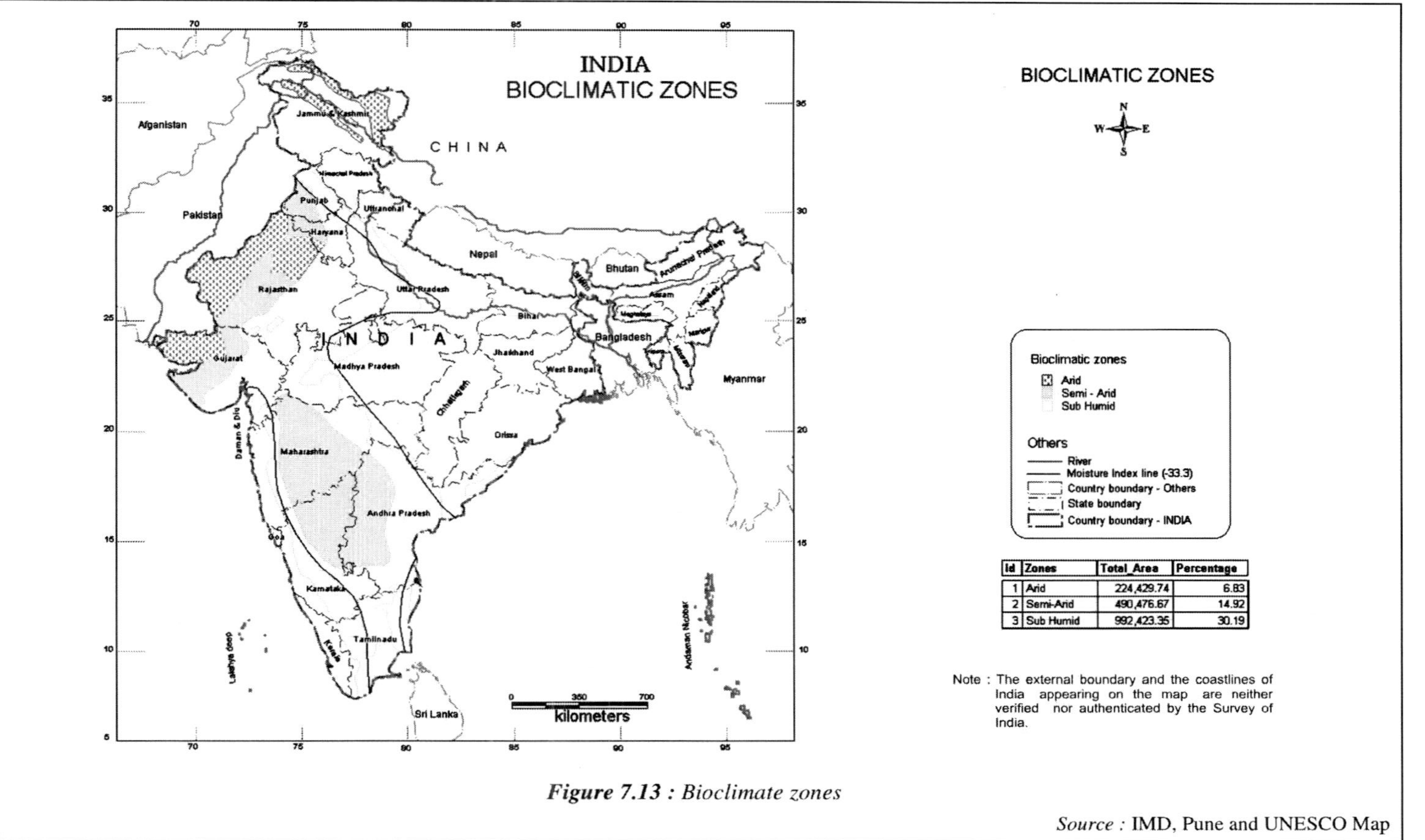

Id	Zones	Total_Area	Percentage
1	Arid	224,429.74	6.83
2	Semi-Arid	490,476.67	14.92
3	Sub Humid	992,423.35	30.19

Note : The external boundary and the coastlines of India appearing on the map are neither verified nor authenticated by the Survey of India.

Figure 7.13 : *Bioclimate zones*

Source : IMD, Pune and UNESCO Map

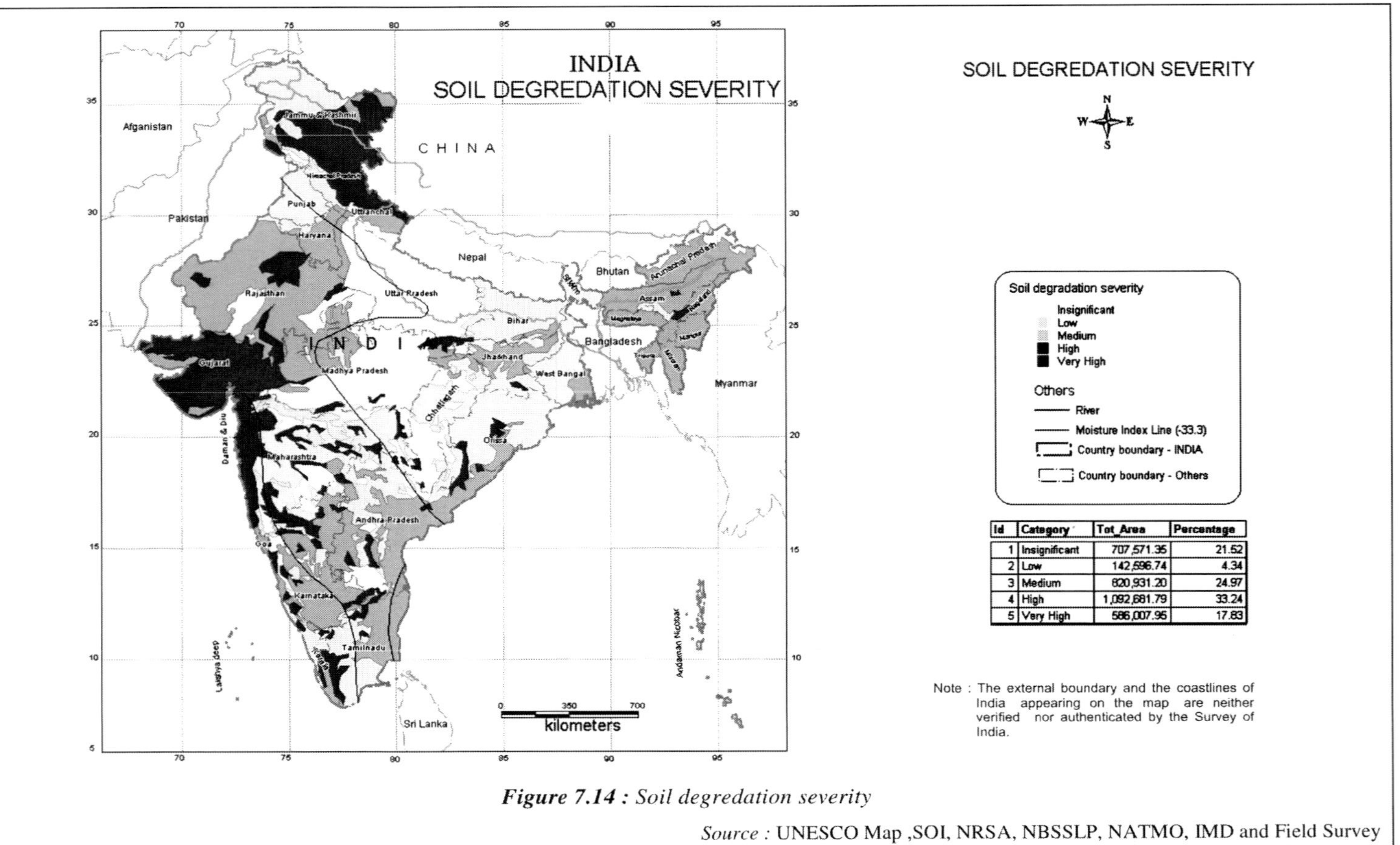

Id	Category	Tot_Area	Percentage
1	Insignificant	707,571.35	21.52
2	Low	142,596.74	4.34
3	Medium	820,931.20	24.97
4	High	1,092,681.79	33.24
5	Very High	586,007.95	17.83

Figure 7.14 : *Soil degredation severity*

Source : UNESCO Map ,SOI, NRSA, NBSSLP, NATMO, IMD and Field Survey

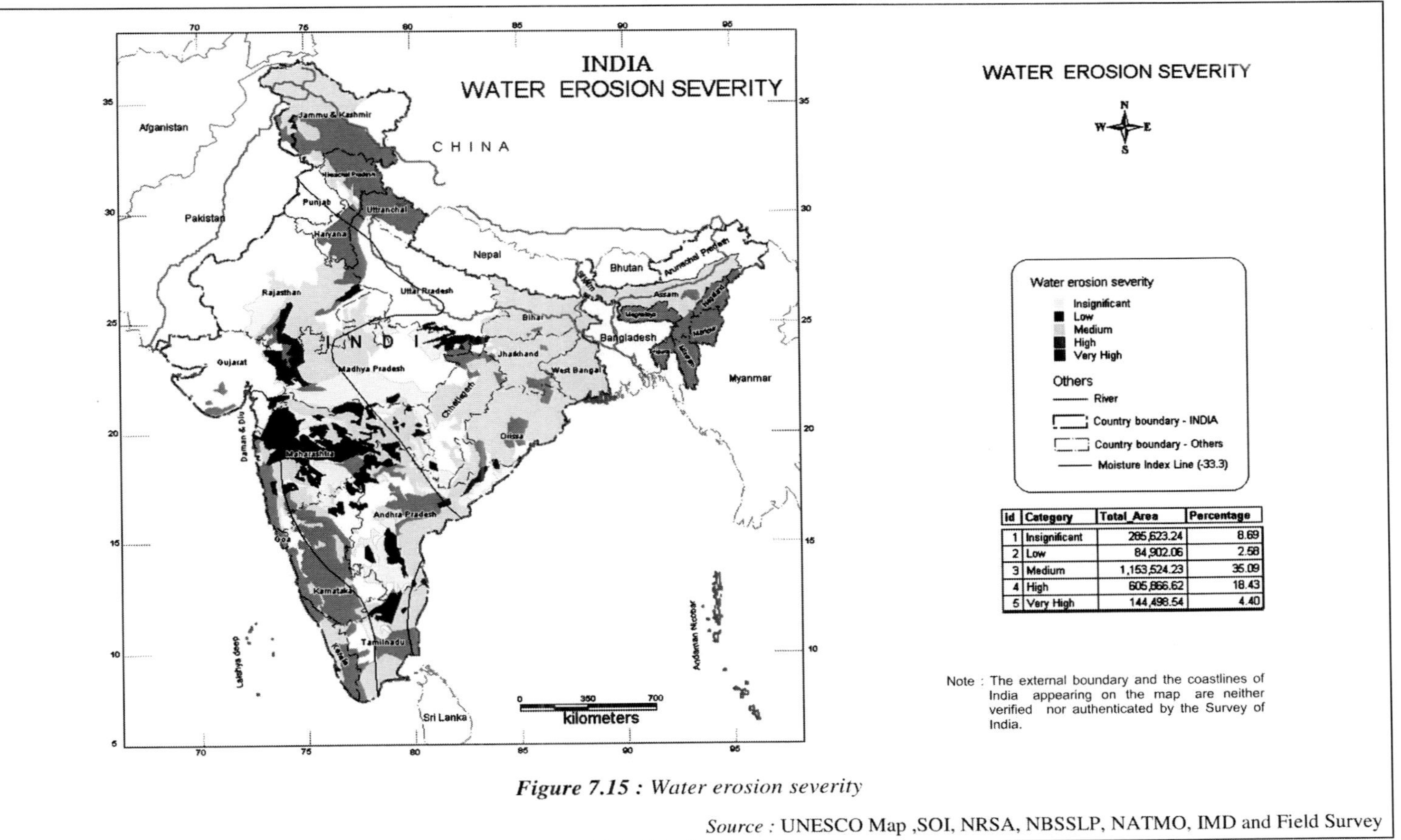

id	Category	Total_Area	Percentage
1	Insignificant	265,623.24	8.69
2	Low	84,902.06	2.58
3	Medium	1,153,524.23	35.09
4	High	605,866.62	18.43
5	Very High	144,498.54	4.40

Figure 7.15 : Water erosion severity

Source : UNESCO Map ,SOI, NRSA, NBSSLP, NATMO, IMD and Field Survey

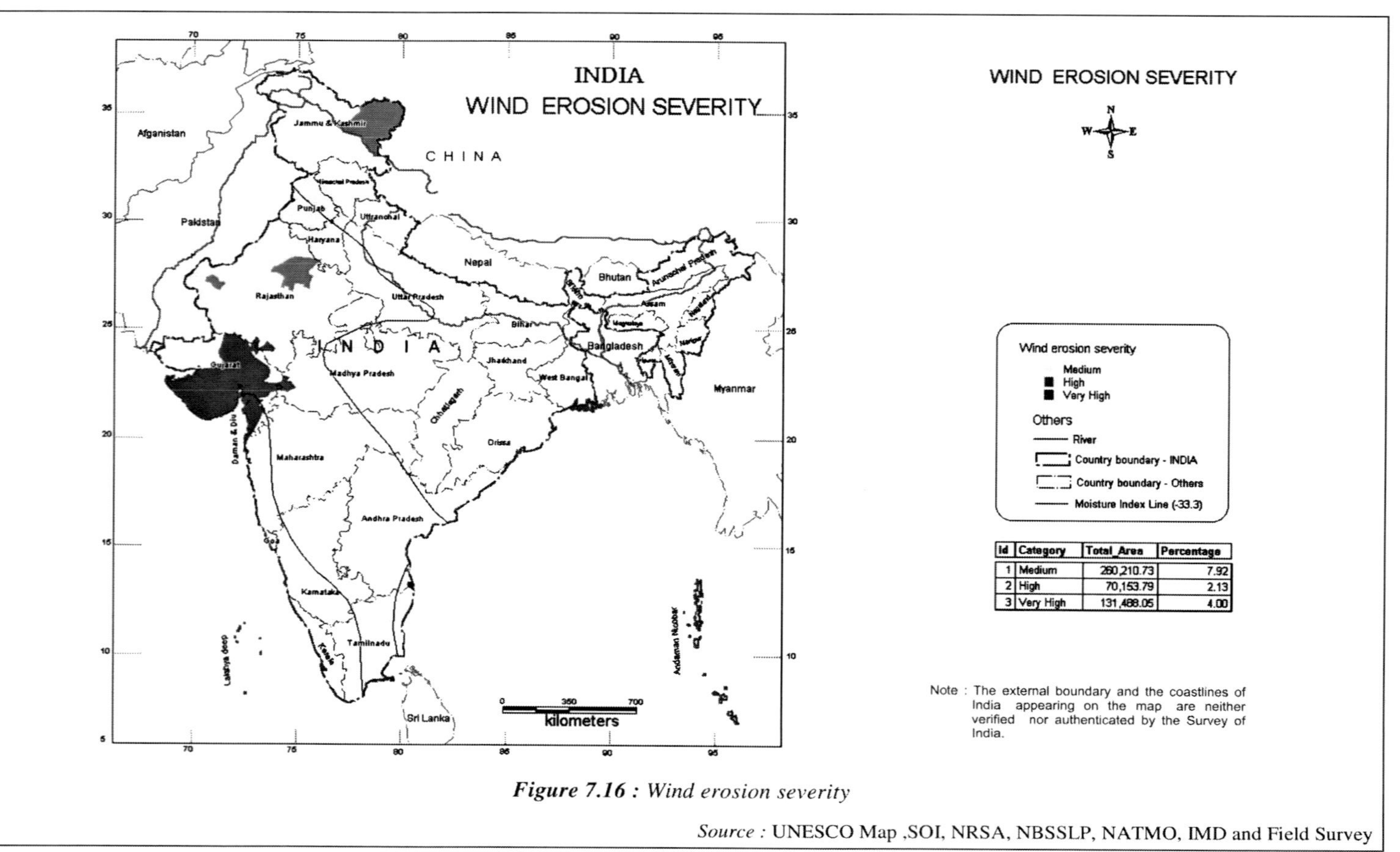

Id	Category	Total_Area	Percentage
1	Medium	260,210.73	7.92
2	High	70,153.79	2.13
3	Very High	131,488.05	4.00

Figure 7.16 : *Wind erosion severity*

Source : UNESCO Map ,SOI, NRSA, NBSSLP, NATMO, IMD and Field Survey

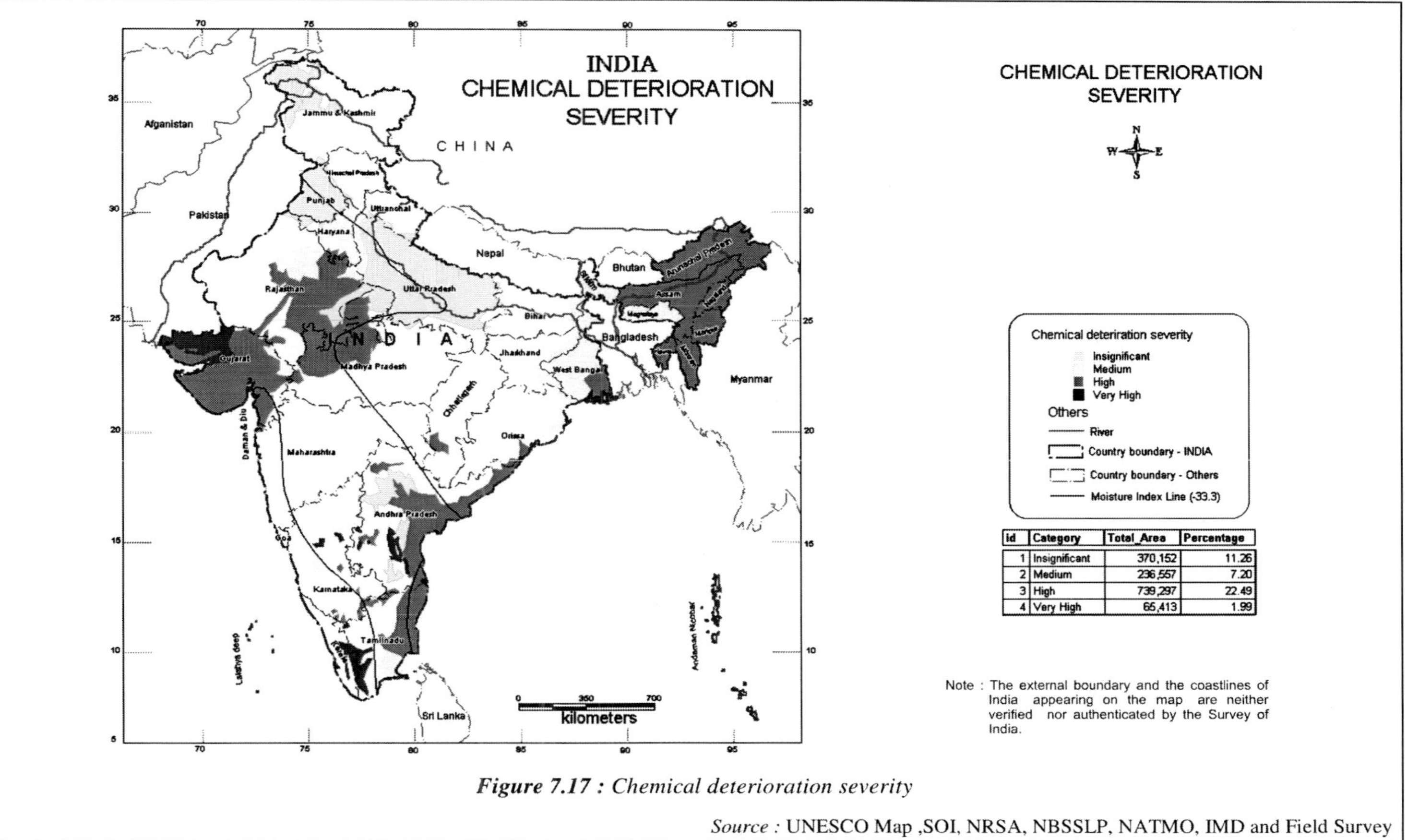

id	Category	Total_Area	Percentage
1	Insignificant	370,152	11.26
2	Medium	236,557	7.20
3	High	739,297	22.49
4	Very High	65,413	1.99

Figure 7.17 : *Chemical deterioration severity*

Source : UNESCO Map ,SOI, NRSA, NBSSLP, NATMO, IMD and Field Survey

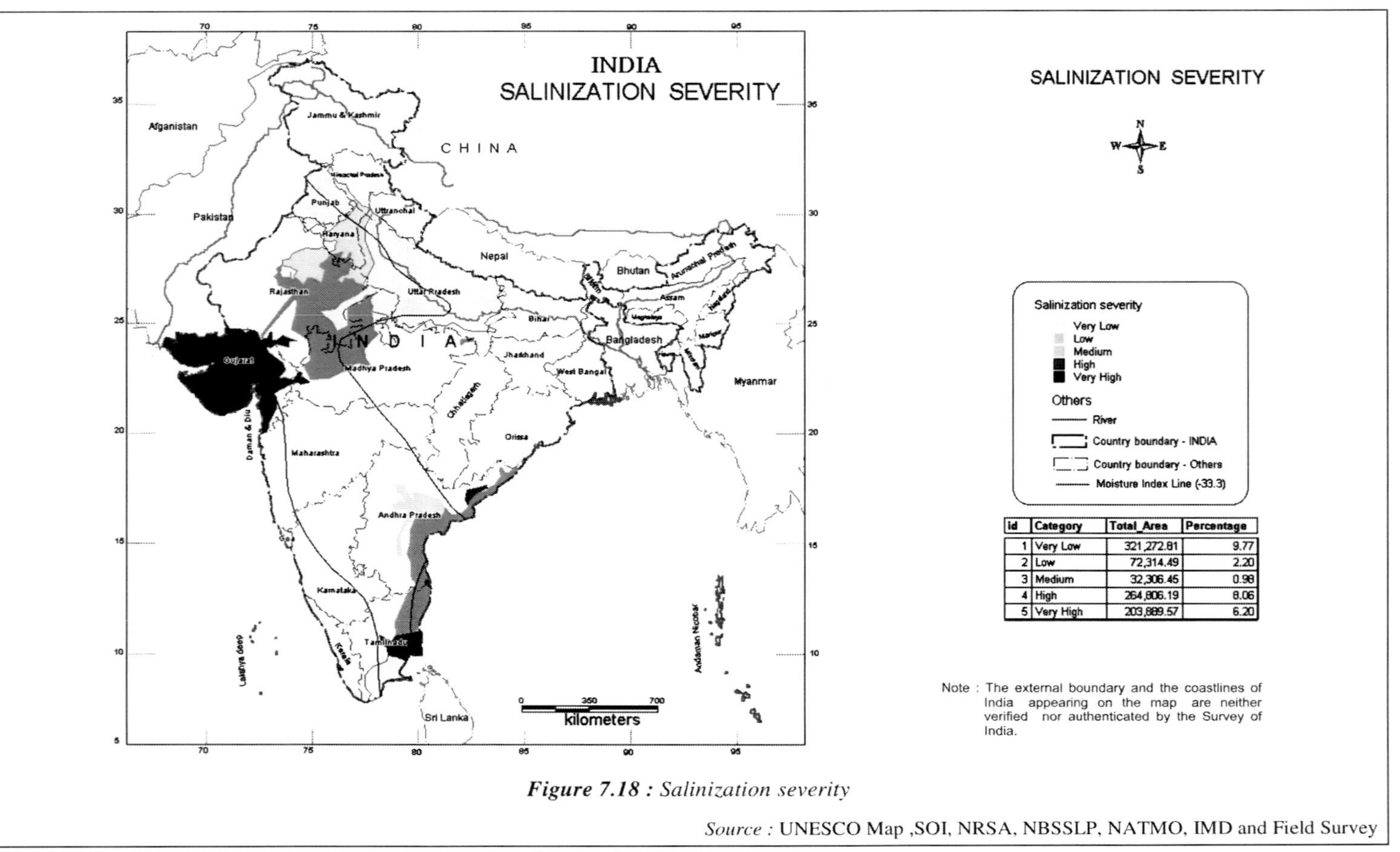

Id	Category	Total_Area	Percentage
1	Very Low	321,272.81	9.77
2	Low	72,314.49	2.20
3	Medium	32,306.45	0.98
4	High	264,806.19	8.06
5	Very High	203,889.57	6.20

Figure 7.18 : *Salinization severity*

Source : UNESCO Map ,SOI, NRSA, NBSSLP, NATMO, IMD and Field Survey

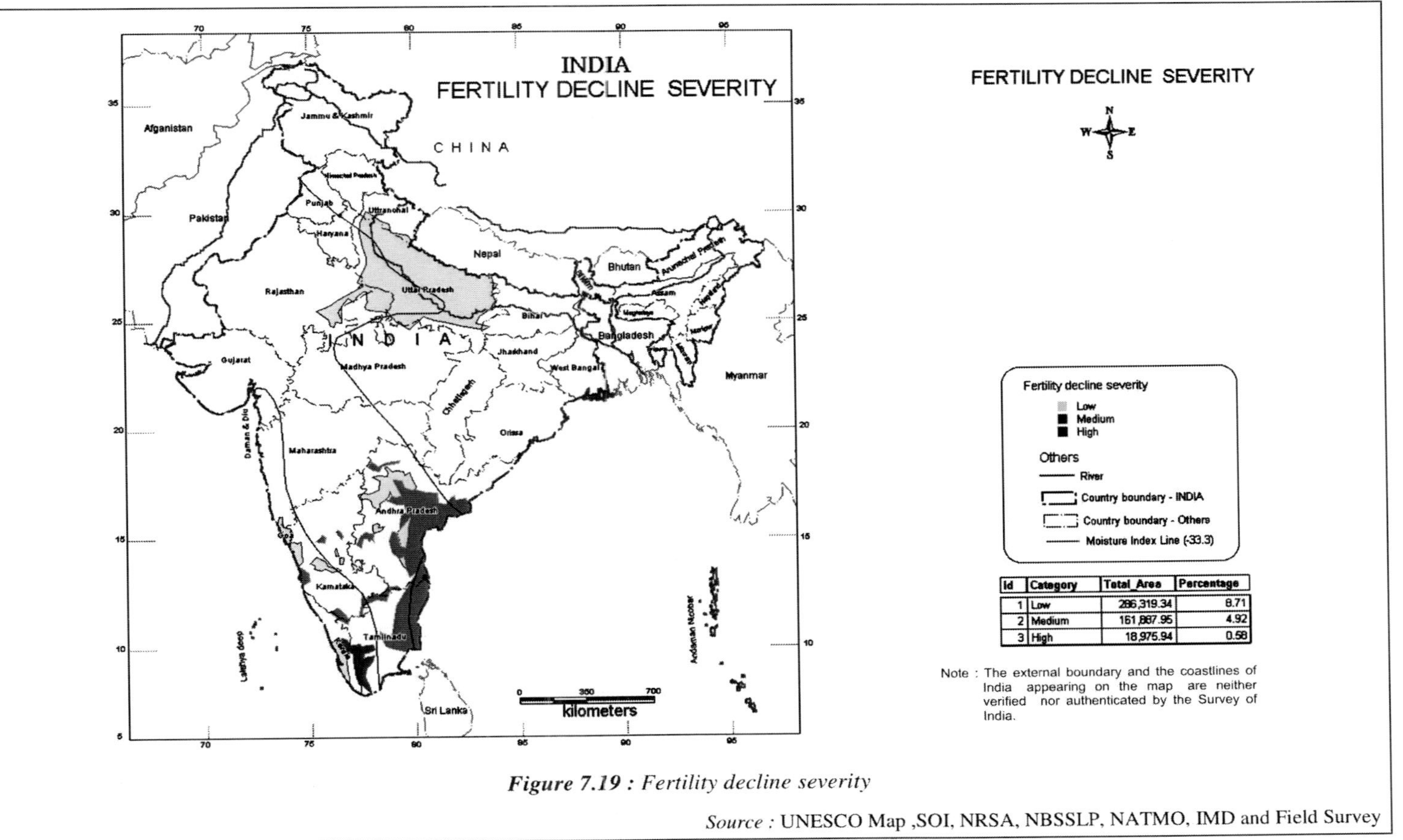

Id	Category	Total_Area	Percentage
1	Low	286,319.34	8.71
2	Medium	161,887.95	4.92
3	High	18,975.94	0.58

Note : The external boundary and the coastlines of India appearing on the map are neither verified nor authenticated by the Survey of India.

***Figure 7.19 :** Fertility decline severity*

Source : UNESCO Map ,SOI, NRSA, NBSSLP, NATMO, IMD and Field Survey

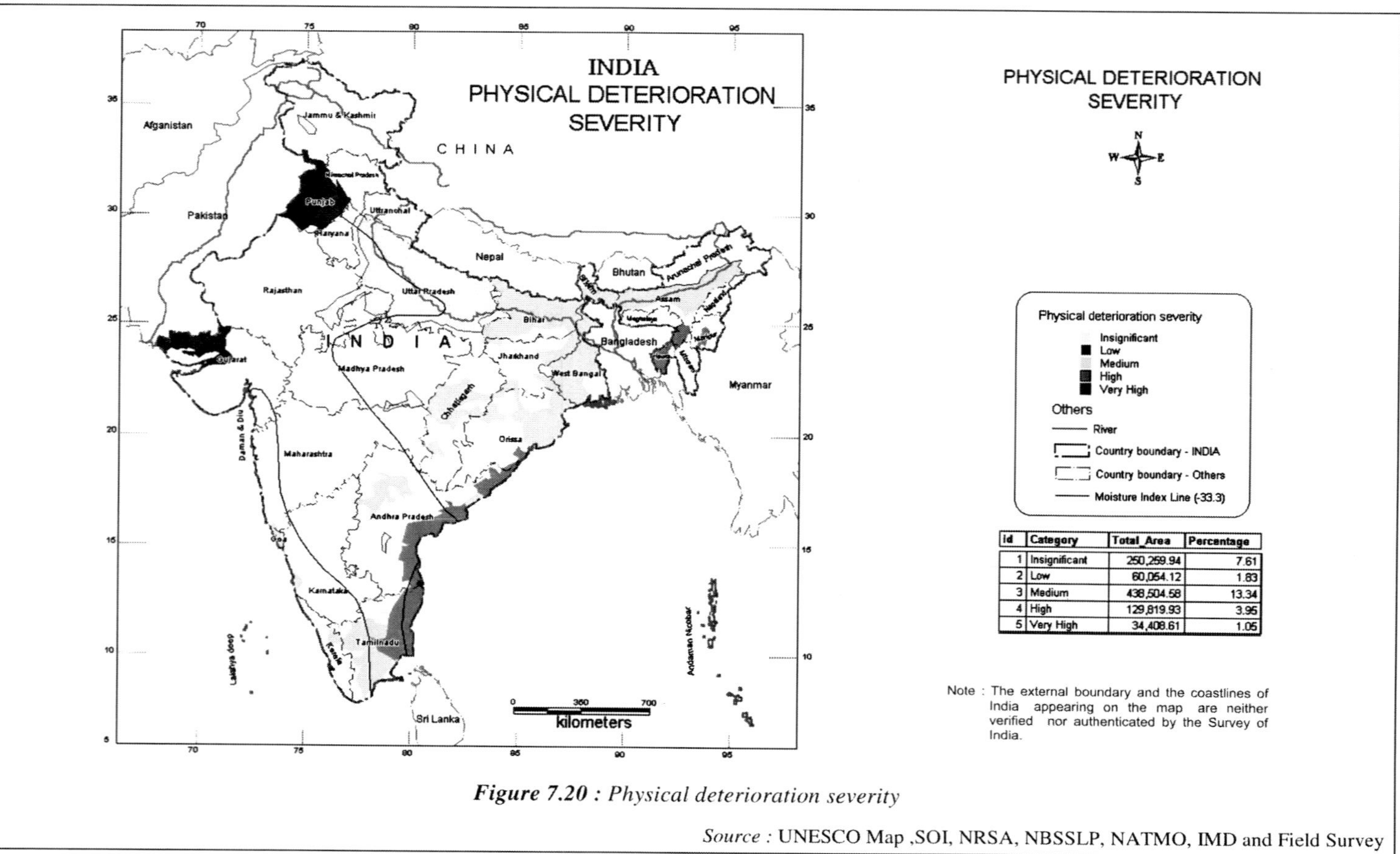

Figure 7.20 : Physical deterioration severity

Source : UNESCO Map ,SOI, NRSA, NBSSLP, NATMO, IMD and Field Survey

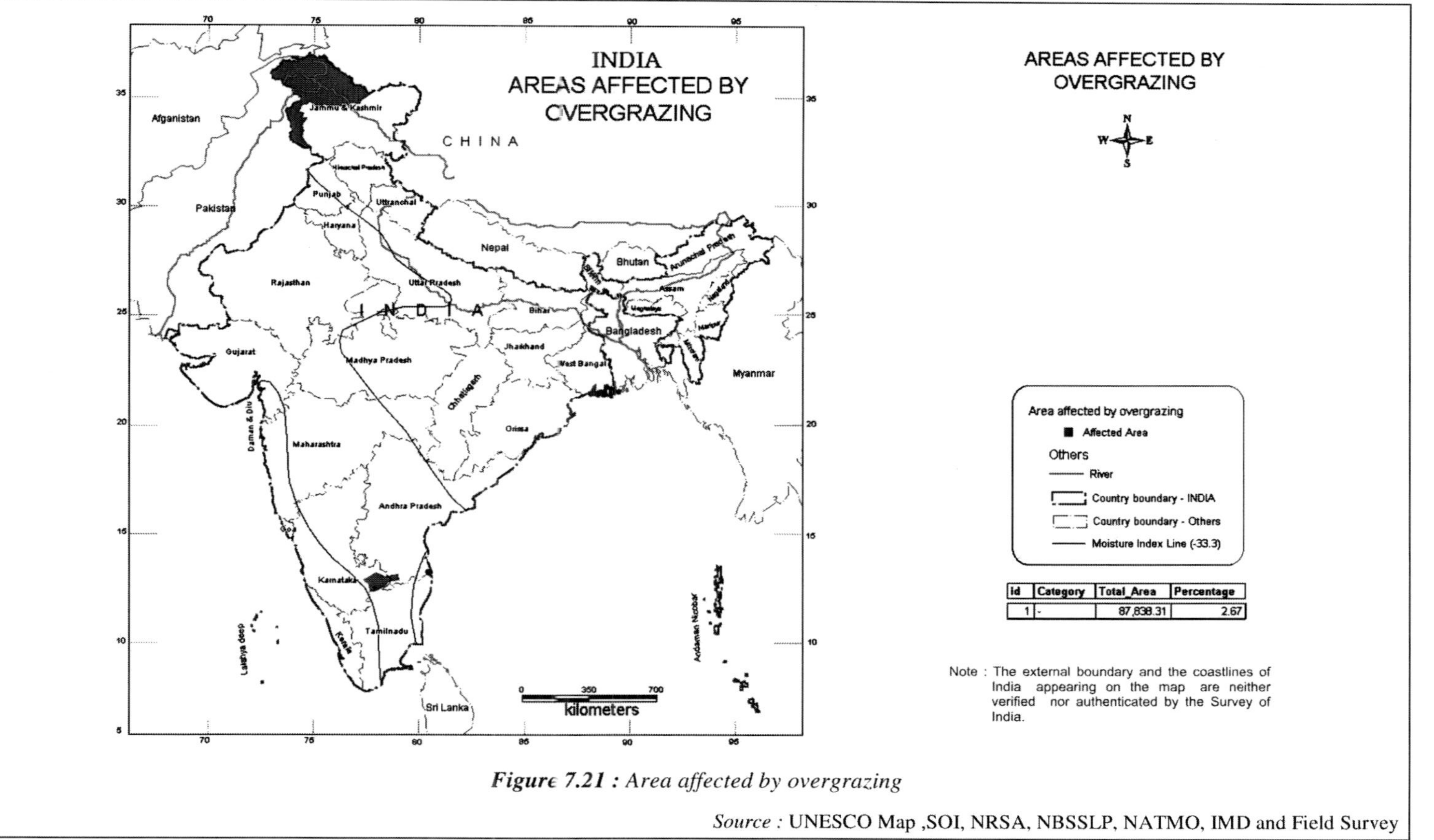

id	Category	Total_Area	Percentage
1	-	87,838.31	2.67

Figure 7.21 : Area affected by overgrazing

Source : UNESCO Map ,SOI, NRSA, NBSSLP, NATMO, IMD and Field Survey

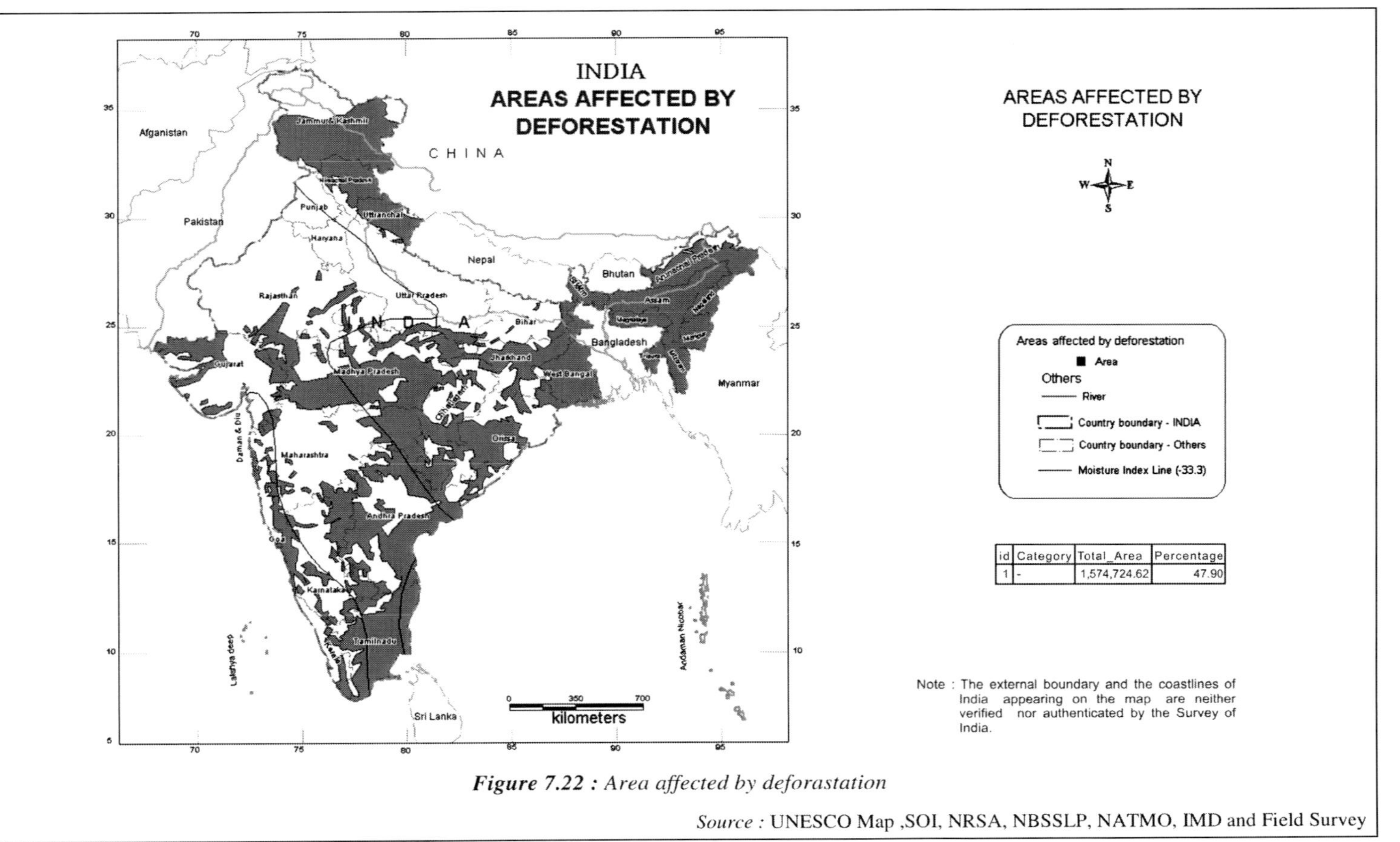

id	Category	Total_Area	Percentage
1	-	1,574,724.62	47.90

Figure 7.22 : *Area affected by deforastation*

Source : UNESCO Map ,SOI, NRSA, NBSSLP, NATMO, IMD and Field Survey

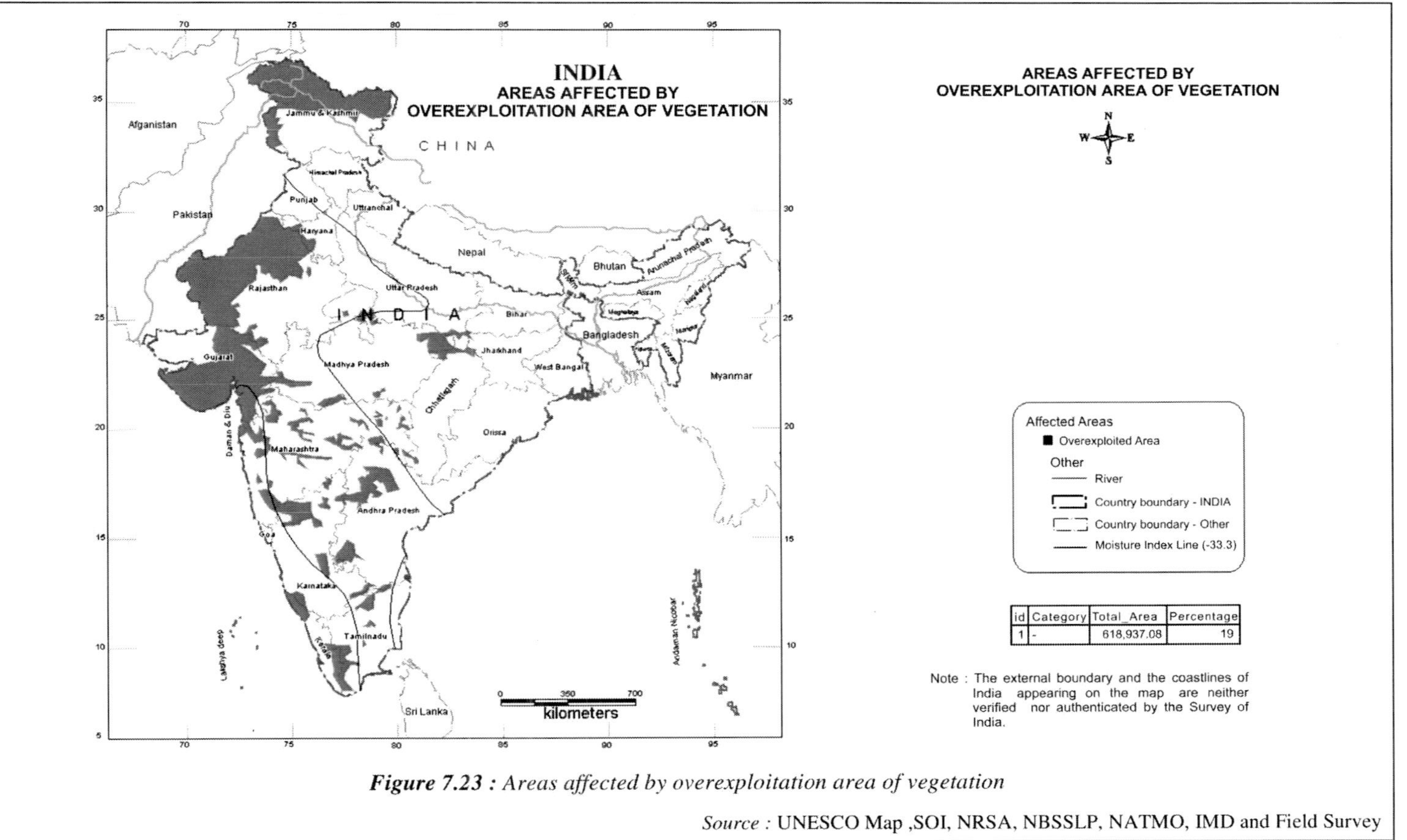

id	Category	Total_Area	Percentage
1	-	618,937.08	19

Figure 7.23 : *Areas affected by overexploitation area of vegetation*

Source : UNESCO Map ,SOI, NRSA, NBSSLP, NATMO, IMD and Field Survey

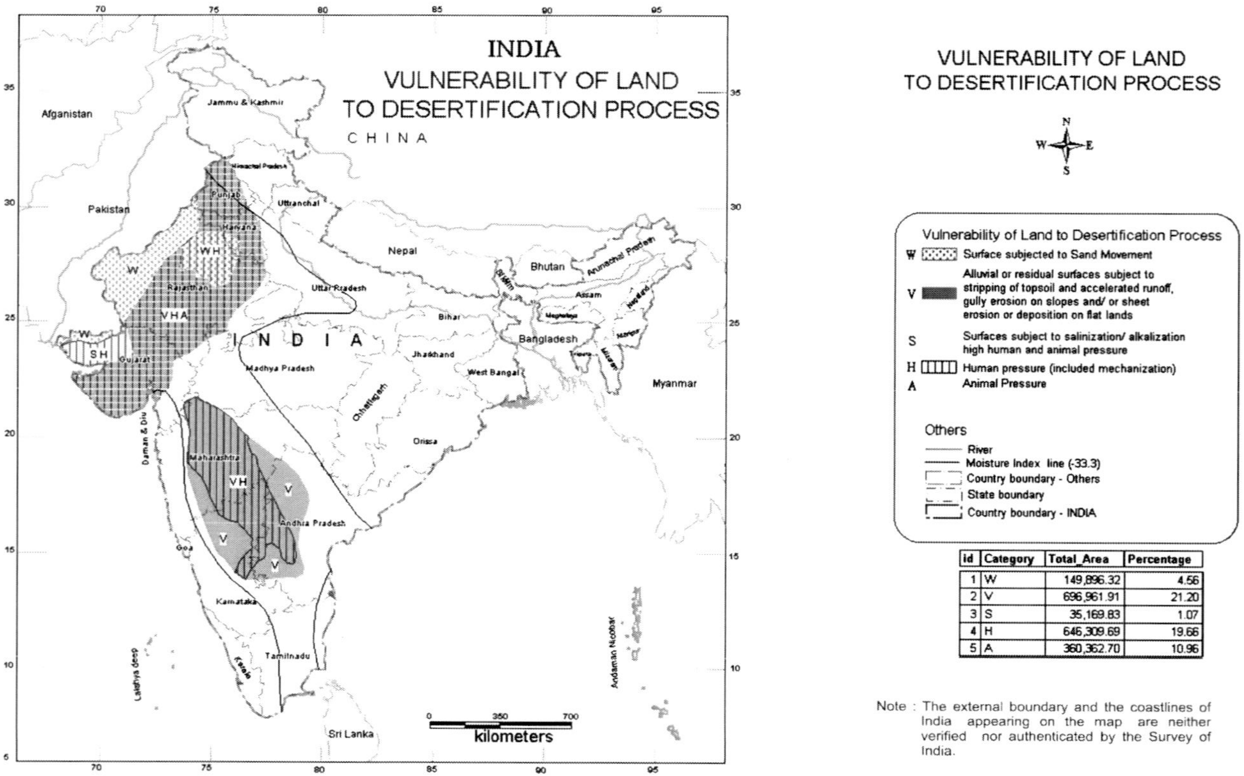

id	Category	Total_Area	Percentage
1	W	149,896.32	4.56
2	V	696,961.91	21.20
3	S	35,169.83	1.07
4	H	646,309.69	19.66
5	A	360,362.70	10.96

Note : The external boundary and the coastlines of India appearing on the map are neither verified nor authenticated by the Survey of India.

Source : UNESCO Map and IMD Pune

Figure 7.24 : *Vulnerability of land to desertification process*

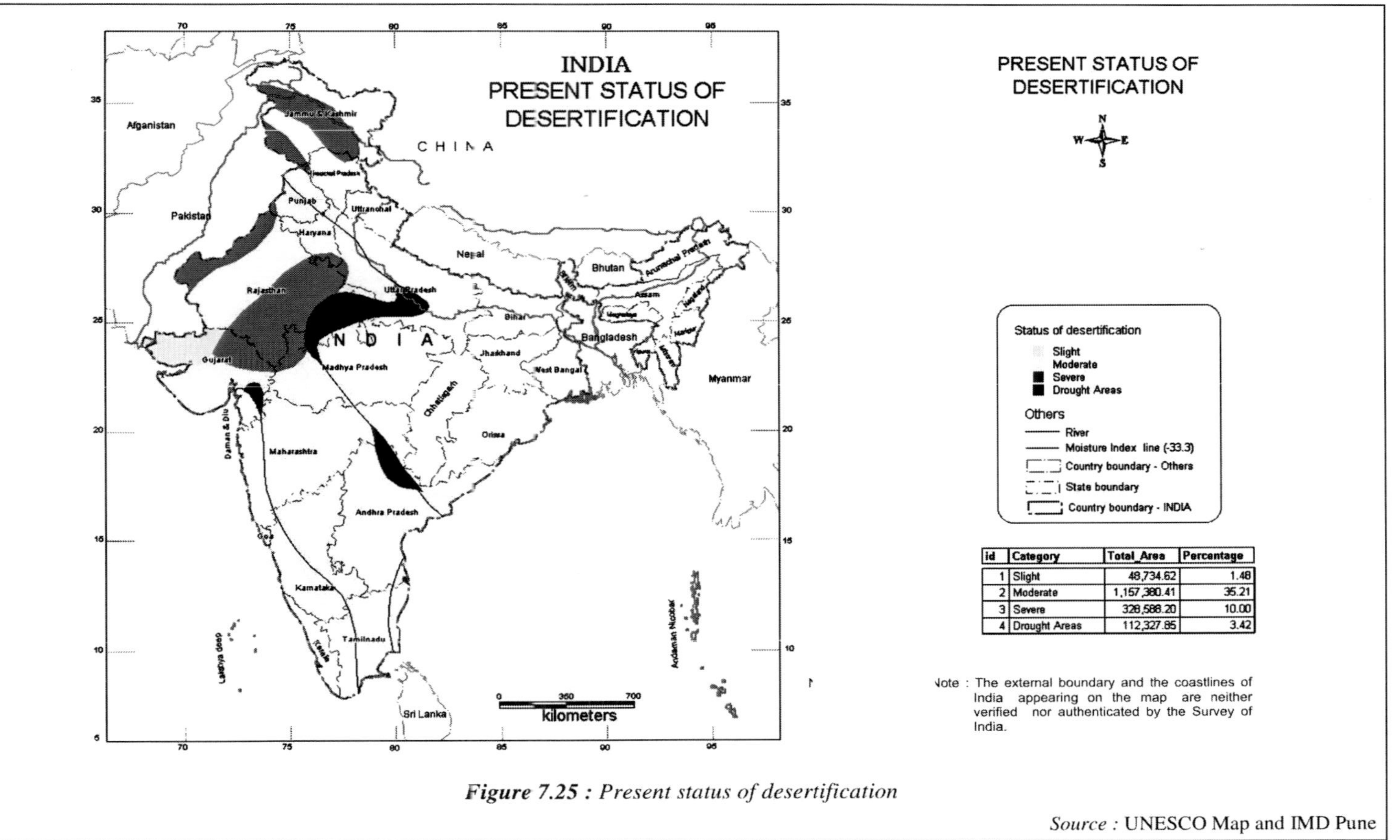

id	Category	Total_Area	Percentage
1	Slight	48,734.62	1.48
2	Moderate	1,157,380.41	35.21
3	Severe	328,588.20	10.00
4	Drought Areas	112,327.85	3.42

Figure 7.25 : Present status of desertification

Source : UNESCO Map and IMD Pune

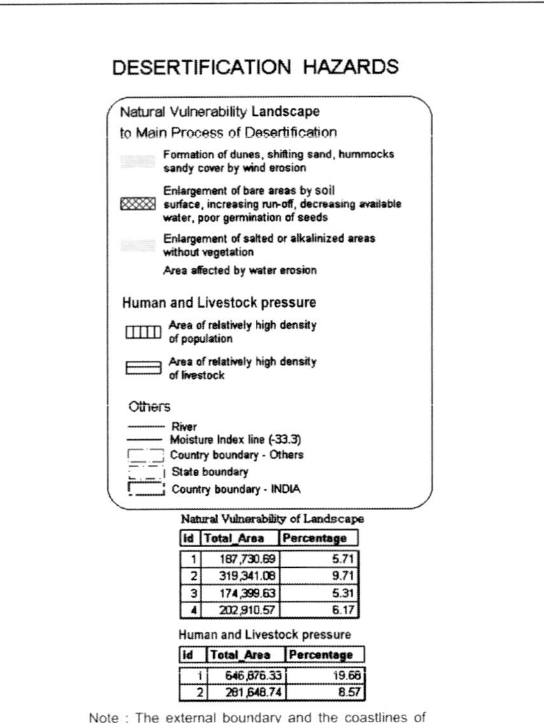

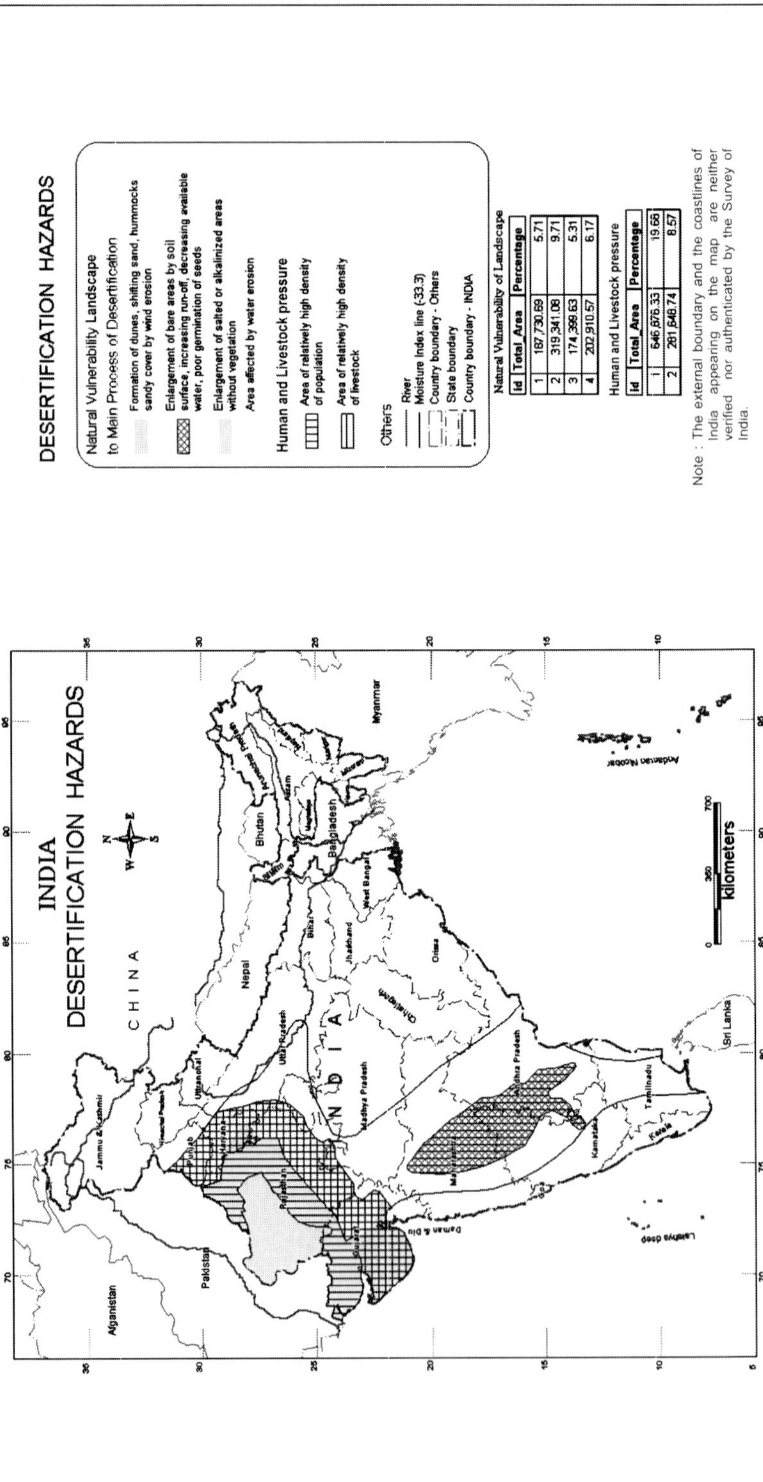

Natural Vulnerability of Landscape

Id	Total_Area	Percentage
1	187,730.69	5.71
2	319,341.08	9.71
3	174,399.63	5.31
4	202,910.57	6.17

Human and Livestock pressure

Id	Total_Area	Percentage
1	646,876.33	19.68
2	281,648.74	8.57

Note : The external boundary and the coastlines of India appearing on the map are neither verified nor authenticated by the Survey of India.

Figure 7.26 : Desertification hazards

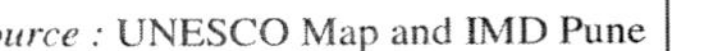

Source : UNESCO Map and IMD Pune

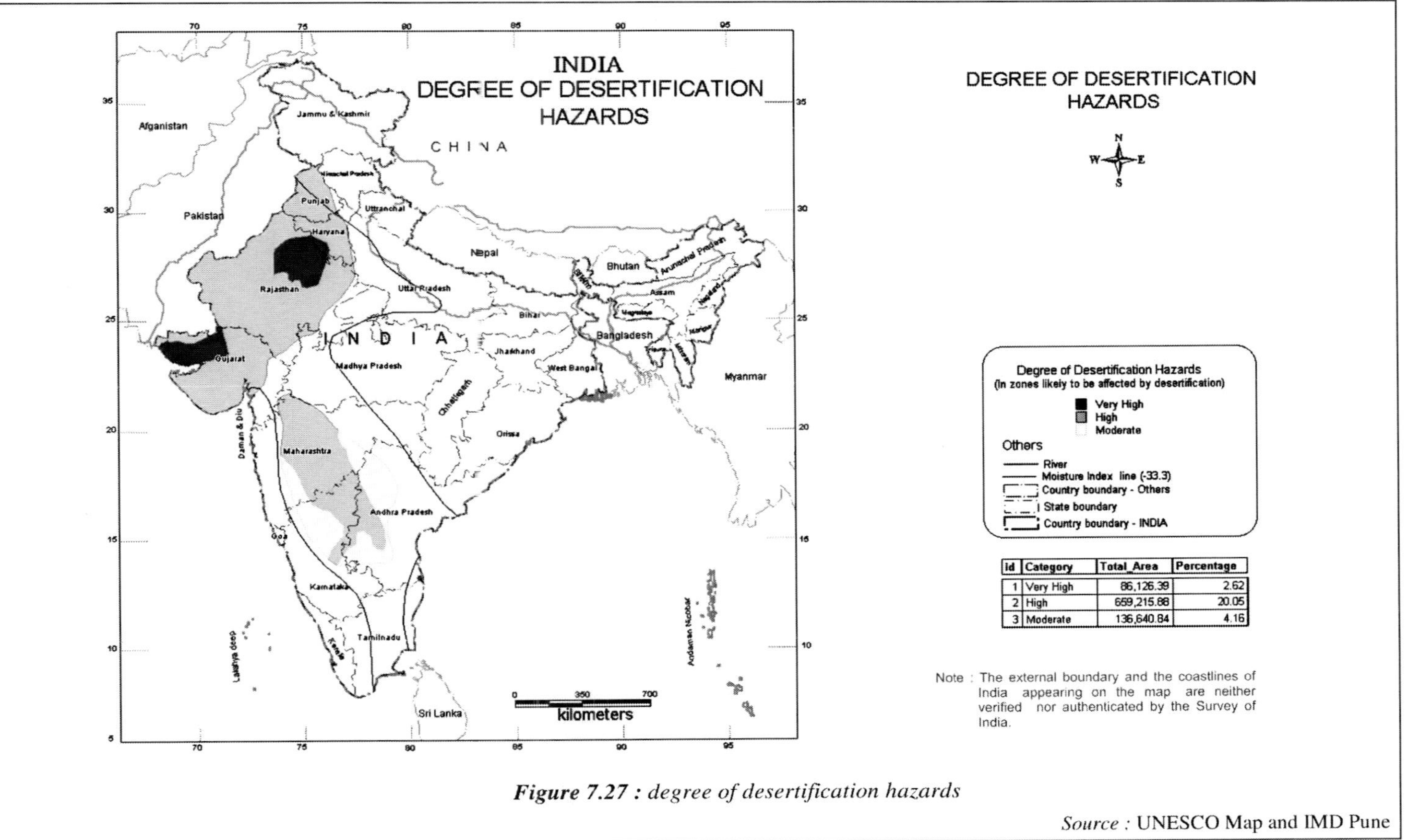

id	Category	Total_Area	Percentage
1	Very High	86,126.39	2.62
2	High	659,215.88	20.05
3	Moderate	136,640.84	4.16

Figure 7.27 : degree of desertification hazards

Source : UNESCO Map and IMD Pune

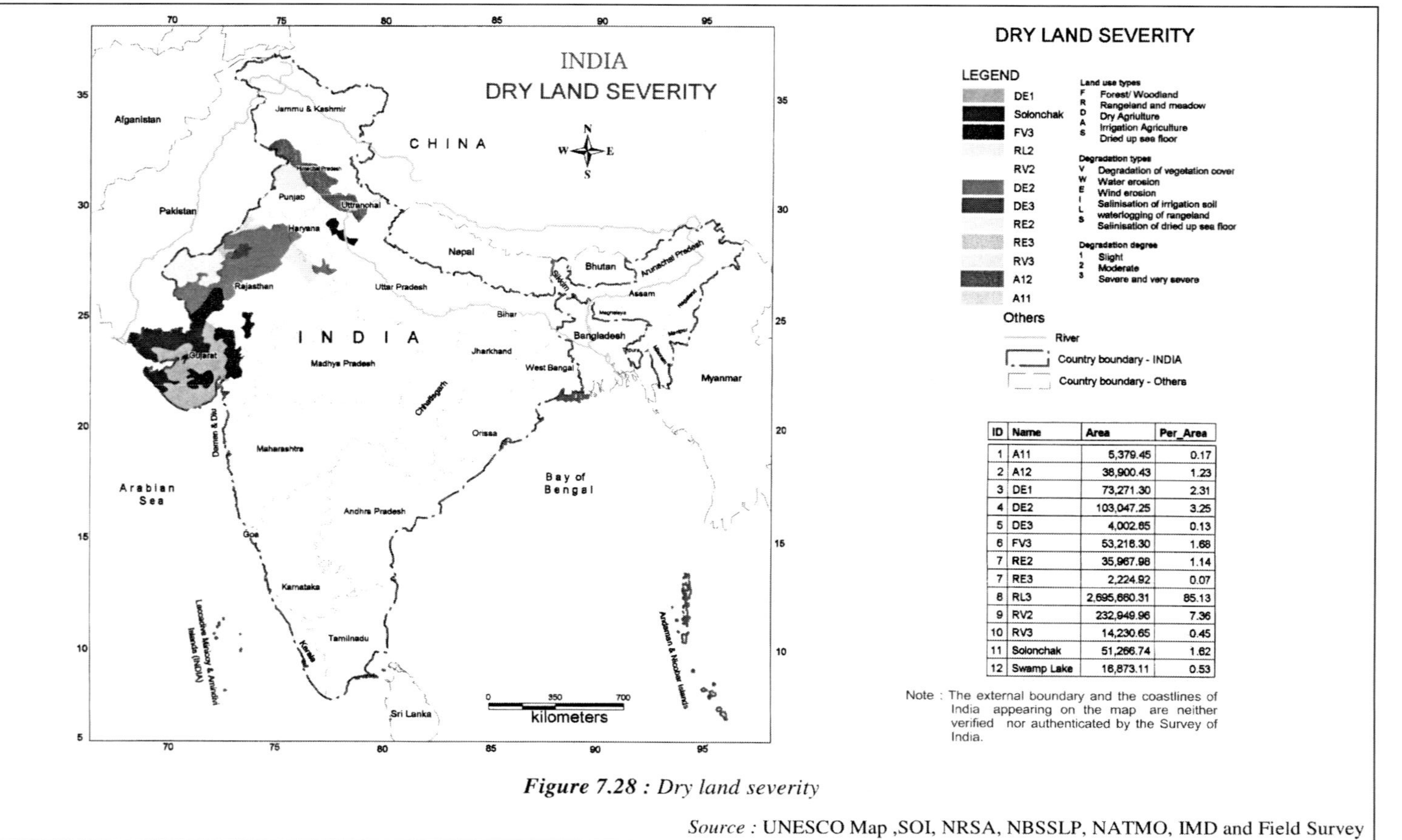

ID	Name	Area	Per_Area
1	A11	5,379.45	0.17
2	A12	38,900.43	1.23
3	DE1	73,271.30	2.31
4	DE2	103,047.25	3.25
5	DE3	4,002.65	0.13
6	FV3	53,216.30	1.68
7	RE2	35,967.98	1.14
7	RE3	2,224.92	0.07
8	RL3	2,695,660.31	85.13
9	RV2	232,949.96	7.36
10	RV3	14,230.65	0.45
11	Solonchak	51,266.74	1.62
12	Swamp Lake	16,873.11	0.53

Figure 7.28 : Dry land severity

Source : UNESCO Map ,SOI, NRSA, NBSSLP, NATMO, IMD and Field Survey

PART–C

Desertification and Human Response - Indian Experience

Desertification has been recognized as major environmental problem affecting living conditions of people in affected regions. Desertification has been defined in international Conference in 1977 in Nairobi as "Diminution or destruction of the biological potential of the land, which leads ultimately to desert like conditions and is an aspect of the widespread deteriorations of ecosystem under the combined pressures of adverse and fluctuating climate and excessive exploitation." Dregne has defined desertification in United Nations Conference on Environment and development in 1991 as "land degradation in arid, semi-arid and dry sub-humid areas resulting from various factors, including climate variations and human activities." For land degradation purpose, aridity is categorized by ratio of P= Mean annual precipitation to PE= Mean annual potential evapo-transpiration, using modified Thornwaite formula and world regions have been classified on the basis of aridity index.

Table 7.19 : Aridity Index of World Dryland Regions

Climatic Zone	P/ PE Ratio	% of world covered
Hyper -arid	<0.05	7.5
Arid	0.05 to 0.20	12.5
Semi-arid	0.21 to 0.50	17.5
Dry-sub-humid	0.51-0.65	9.9
Humid	>0.65	39.2
Cold	>0.65	13.6

The drylands are defined as regions where ratio of mean annual precipitation to mean annual evapo-transpiration is in the range of 0.05-0.65 where arid, semi-arid and dry sub-humid regions are considered as dryland, but excludes hyper-arid region, where P/PE ratio is less than 0.05 from the ambit. Similarly moist sub-humid, humid and per-humid zones are also excluded. In addition to global interpretation, country and regional level the desertification is termed in view of problems faced most due to land degradation impact. Desertification is very slow process of land deterioration and could not get due attention like natural disasters but the slow process is understood by local population in general and cultivators in particular who are directly affected. As environment problem, the impact is visible in the form of drought and rainfall pattern.

Desertification of environment is generally related to human interface to earn living in drylands. Usually, desertification is most critical problem in drylands as these regions are severely affected with the impact and consequences. Human interface and natural factors are responsible for desertification and role of natural factors becomes critical when the ecology is disturbed by human acts. In brief, human acts are reflected in the form of uncontrolled grazing of livestock, large scale felling of trees, squeezing the vegetative cover and over-exploitation of ground water resources. Every given factor is cause and consequences of environment degradation and cycle of land degradation is framed, where most critical impact lies on productivity of the land, which is survival base of many people, who are completely depended on land resources.

Term 'desertification' was employed in 1949 by French forester Aubreville, who used this term to refer displacement of tropical rainforest by secondary savannah and scrub in those parts of Africa, where forest was being cleared and burned to provide land for cultivation. Aubreville concluded that process was specially active in sub-humid tropics of Africa and was akin to creation of deserts in formerly forested areas. The extent of accelerated soil erosion induced by indiscriminate felling and burning of forest and woodland in Africa and changes in the soil-water budget and hydrological cycle were understood as some of factors leading to land degradation. There was also a growing recognition of the part played by human activities and climatic changes in the form of prolonged and frequent droughts aggravating land degradation. The voice of Aubreville was heard by the world community and there was long debate over the problem raised by a forester.

In 1997, United Nations Conference on Desertification (UNCOD) was convened in Nairobi, Kenya to produce an effective, comprehensive and coordinated programme for addressing the problem of land degradation. UNCOD was an outcome of extensive studies and consultations undertaken at global, regional and local level involving scientists, policy and decision makers and experts from research and development institutions and other organization from all over the world. UNCOD recommended the United Nations Plan of Action to Combat Desertification (PACD), which was hampered due to lack of resources. Assessments made by UNEP in 1984, 1987 and 1989 revealed that desertification was spreading continuously. The Un Commission for Sustainable Development Report, 1988 observed that desertification had become one of the most serious environmental and socio-economic problems of the world.

In 1992, UNEP produced a World Atlas of Desertification, which revealed that over preceding 20 years, problem of land degradation continued to worsen. In the past, drylands recovered easily following long droughts and dry periods, but under modern conditions, they tend to lose their biological and economical productivity quickly, unless they are managed in sustainable manner. Such studies further indicated that over-cultivation, over-grazing, deforestation and poor irrigation practices are degrading drylands in every continent where major factors are human and livestock population pressure, inappropriate land use and agricultural practices, social conflicts and drought. Desertification has been recognized as a major global concern affecting over 300 million people directly and over one billion at risk, changing traditional lifestyle, culture and composition of rural societies.

The UN Conference on Environment and Development, also known as Earth Summit, held in Rio de Janeiro, Brazil in June 1992 provided a platform for addressing various major global environmental concerns like climate change, bio-diversity, deforestation etc. The Rio Summit highlighted problem of desertification and recommended that UN General Assembly to establish Inter-governmental Negotiating Committee to prepare Convention to Combat Desertification for countries experiencing severe drought and desertification, particularly in Africa. The Committee established in 1993 adopted convention in Paris and 172 countries acceded or ratified the Convention. Central Arid Zone Research Institute (CAZRI) scientists have done commendable work in the field of combating desertification

7.11 EXTENT AND CAUSES OF DESERTIFICATION

Desertification is slow and continuous process of land degradation, loss of vegetation cover over earth surface, prevalence of dryness in atmosphere, environment deterioration and serious impact on ecology of area. Desertification is the characterized by sparse and highly variable precipitation and high evaporation resulting into severe droughts. Vegetation is seriously affected due to limited availability of water. Population is very limited due to scanty sources of livelihood. Desert regions of the country are most affected due to limited water resources and low rainfall zones. Though desert regions are vulnerable area possessing limited potentials of development but deterioration process of all the positive factors create serious problems. With this point, extent of desertification is necessary to be assessed as causes of desertification are closely associated.

7.12 EXTENT OF DESERTIFICATION

India is vast country having total geographical area of 3287263 sq kms with varied geophysical conditions, soil structures, temperatures, rainfall and vegetation. Desertification is serious phenomena of hyper arid, arid and semi arid regions of India. Hot and cold desert regions combined are spread over 37 districts of seven states but cold desert is limited to Ladakh district of Jammu and Kashmir. Hot desert regions cover six states viz. Rajasthan, Punjab, Haryana, Gujarat, Andhra Pradesh and Karnataka. Total geographical area of all the 37 desert districts is 513,911 sq kms which is 15.63 percent of country geographical area. Extent of desertification in India is given in Table 7.20.

Table 7.20 : Extent of Desertification in India

S. No.	State/ District	Total Geog. Area (Sq. Km)	Desert Area (Sq. Km)	% to Total Geographical Area of district	% to total Desert Area
	I. RAJASTHAN				
1.	Barmer	28387	28387	100.00	6.42
2.	Bikaner	27244	27244	100.00	6.16
3.	Churu	16830	16830	100.00	3.80
4.	Ganganagar	10978	10978	100.00	2.48
5.	Hanumangarh	9656	9656	100.00	2.18
6.	Jaisalmer	38401	38401	100.00	8.68
7.	Jalore	10640	10640	100.00	2.41
8.	Jhunjhunu	5928	5928	100.00	1.34
9.	Jodhpur	22850	22850	100.00	5.17
10.	Nagaur	17718	17718	100.00	4.01
11.	Pali	12387	12387	100.00	2.80
12.	Sikar	7732	7732	100.00	1.75
	Total - Rajasthan	**208751**	**208751**	**100.00**	**47.20**
	II. ANDHRA PRADESH				
1.	Anantapur	19130	19130	100.00	4.33
2.	Kadapa	15359	3856	25.11	0.87
3.	Kurnool	17658	4517	25.58	1.02
	Total - Andhra Pradesh	**52147**	**27503**	**52.74**	**6.22**
III	GUJARAT				
1.	Banas Kantha	10757	10757	100.00	2.43
2.	Jam Nagar	14125	14125	100.00	3.19
3.	Kachchh	45652	45652	100.00	10.32
4.	Mehsana	4384	2076	47.35	0.47
5.	Rajkot	11203	5568	49.70	1.26
6.	Surendra Nagar	10489	5417	51.64	1.23
	Total - Gujarat	**96610**	**83595**	**86.53**	**18.90**
	IV. HARYANA				
1.	Bhiwani	4778	4778	100.00	1.08
2.	Fatehabad	2538	1257	49.53	0.28
3.	Hissar	3983	3983	100.00	0.90
4.	Mahendragarh	1899	947	49.87	0.22
5.	Sirsa	4277	4277	100.00	0.97
	Total - Haryana	**17475**	**15242**	**87.22**	**3.45**
	V. JAMMU AND KASHMIR				
1.	Ladakh*	82665	70300	85.04	15.89

* Includes unsurveyed area of 37555 Sq. Kms under unauthorized occupation of China.

S. No.	State/ District	Total Geog. Area (Sq. Km)	Desert Area (Sq. Km)	% to Total Geographical Area of district	% to total Desert Area
			VI. KARNATAKA		
1	Bellary	8419	8419	100.00	1.91
2.	Bijapur	10494	7897	75.25	1.79
3.	Chitra Durga	8440	2316	27.44	0.52
4.	Dharwad	4230	1082	25.58	0.24
5.	Raichur	6828	3546	51.93	0.80
	Total - Karnataka	**38411**	**23260**	**60.56**	**5.26**
			VII. PUNJAB		
1.	Ferojpur	5303	5303	100.00	1.20
2.	Faridkot	1469	1469	100.00	0.33
3.	Bhatinda	3385	3385	100.00	0.77
4.	Nawan Shahr	1267	1267	100.00	0.28
5.	Sangrur	5020	2214	44.10	0.50
	Total - Punjab	**16444**	**13638**	**82.94**	**3.08**
	GRAND TOTAL	**512503**	**442289**	**86.30**	**100.00**

Thus total area in hyper arid, arid and semi-arid regions is 442,289 sq. kms desert of varied categories, which is 13.45 percent of country geographical area. Thus all the 37 districts of the country have 86.45 percent desert area of their respective geographical area. Rajasthan is single largest desert region of the country having 208,751 sq. kms of area spread over in 12 western districts of state with 47.20 percent of total desert area. This region is part of the Great Thar Desert. Ecosystem of Indian Desert have mixed situation. Desert regions suffer acute water problem, which is certainly a hindrance of development and ecosystem has been deteriorated to significant extent. The extent and intensity is varied with local conditions, but some salient features are common. Deforestation is common problem of the desert region, which is the prime reason for land degradation. Soil erosion, loss of vegetative cover and human acts are the end results of such happenings. There are various factors, which are inter-related and consequential impact to misbalance the ecosystem. General features of desert regions having impact on the ecosystem of desert regions are discussed below which are impacts of above narration.

Andhra Pradesh has 27503 sq kms area under desertification, which is 6.22 percent of total area under desertification. This part of the state is shadow desert spread over in Anantpur, Kadapa and Kurnool. Area under desertification is spread over in 83595 sq kms or 18.40 percent of total dryland region covering six districts of the state viz Banas Kantha, Jam Nagar, Kachchh, Mehsana, Rajkot and Surendra Nagar district. First three districts have total dryland region and remaining three districts have 47.35, 49.70 and 51.64 percent of their respective geographical area. Haryana state is partly covered with canal irrigation, but five districts, viz. Bhiwani, Fatehabad, Hissar, Mahendragarh and

Sirsa are affected with desertification spread over in 15242 sq kms or 3.45 percent of total area under desertification. Bhiwani, Hissar and Sirsa are 100 percent affected and Fatehabad and Mahendragarh have 49.53 and 49.87 percent of their respective geographical area.

Ladakh district of Jammu and Kashmir state has 70300 sq kms area affected by cold desertification and this is 15.89 percent of total area under desertification. Karnataka state has 23260 sq kms area under shadow desertification where Bellary is fully affected and remaining four districts- Bijapur, Chitra Durga, Dharwad and Raichur have 75.25, 27.44, 25.58 and 51.93 percent of their respective geographical area affected by desertification problem. Punjab state is mostly covered with canal irrigation but four districts- Ferojpur, Faridkot, Bhatinda, Nawan Shahr are fully affected by desertification while Sangrur district has 44.10 percent of geographical area affected with desertification problem. All the 37 districts of India have 86.30 percent area under desertification with varied problems.

7.13 CAUSES OF DESERTIFICATION

Desertification is slow process of land degradation and present condition is the resultant impact of more than 5000 years. These lands were dense forests and fertile for cultivation practices. The present aridity conditions are the consequential impact of various factors responsible for severe degradation. Desertification is cause and impact of various inter-related factors, which play vital role in degradation of environment and ecology of the region. These factors have been classified into two broad categories viz. physical factors and anthropogenic factors. Detailed causes of desertification are elaborated hereunder:

7.14 PHYSICAL FACTORS

Total area affected by desertification mostly fall under 15 to 30 degree north from tropics, which include the Great Indian Desert- the Thar, cold desert of Himalayan region and shadow desert of coastal regions obstructed by mountains. There is significant difference in desert and desertification as various desert regions on earth surface are vegetated lands but desertification is the process of land degradation where vegetative cover of earth surface is completely or largely denuded, surface or ground water is very limited and means of livelihood become quite scarce. Physical factors are not the only land degradation factor but part of long cycle creating adverse conditions through following manner:

1. Variability in Rainfall Pattern

Desertification is the cause in rainfall variability, visible from annual rainfall occurrence trends, which reveal that rainfall significantly vary from normal factor. This variability is severe in dryland regions and variation from normal rainfall is becoming vital in terms of occurrence, time duration, frequency between spells, late occurrence and early departure. The analysis of trends rainfall cycle of five years in last two decades reveal that only one year is in proximity to normal rainfall and another one years gets more rains than the normal, whereas three years are quite critical. Crop cultivation based on monsoon, drinking water arrangements and rural economy are seriously affected with these variations. Even the normal and excess rainfall is not always beneficial to crops due to variability factors. Vegetation is seriously affected causing denuded land surface, loss of trees and bushes, increase in aridity and imbalancing the environment of these regions.

2. Climatic Shifts

Climatic changes caused as impact of desertification reflect in terms of temperature, humidity and moisture, which create hindrances in growth and maintenance of vegetation. Temperature variations are extreme in summer and winter in hyper-arid, arid and semi-arid regions, which reflects in terms of high evaporation, dryness in atmosphere and even trees and bushes find difficult in survival. Upper surface becomes completely dry and hard or cracks occur in hard soil strata. Dryness in atmosphere puts vegetative impact on monsoon crop and crop failure is regular feature. In years of failure of monsoon the situation further worsens. Mean moisture index reveals negative impact affecting most in extreme-arid and arid regions. Semi-arid regions are able to coup up with the situation to some extent in view less negative impact and vegetation is visible in these regions. Extreme arid and arid regions look fully or mostly denuded over the earth surface.

3. Moisture Factor

In hyper arid and arid regions where density of population is low, cultivators can grow crops or plantation through moisture system, which is common in these regions. But the extreme temperatures dry the sub-surface level in years of low rainfall. Moisture retention is necessary in erratic rainfall conditions and one widespread rain can help in cropping in most monsoon session with proper treatment of land. But this process is possible if there is single heavy showers. Usually such rains are possible once in two years. Therefore, moisture retention is also not fully assured system in hyper-arid and arid regions where rainfall is comparatively low.

4. Albedo Factor

If earth surface has no vegetative cover or there is decrease in vegetative cover, the reflectivity of earth surface is seriously affected by dryness. In this situation the process of evaporation is further intensified and even good rainfall is not fully effective for crops and vegetation to desired level. The process of desertification is further intensified as resultant impact. Vegetation cover helps in decreasing dryness in environment but when the vegetative cover is lost for various reasons, the dryness in atmosphere is further intensified and hyper desert regions with least population or biological presence.

5. Carbon-di-oxide Factor

Vegetative cover helps in function of oxygen factor and carbon-di-oxide is converted into oxygen helpful for biological presence. In dryland regions, vegetation presence is diminished for various regions and quantity of carbon in the atmosphere is increased to substantial extent. In crease in carbon in atmosphere has resultant impact of increased temperature, which is harmful for growth and survival of vegetation. Increased temperature helps in increase of evaporation process and the process of desertification is further intensified.

6. Dust Particles Factor

Frequency of dust storms rises with reduced precipitation in preceding year. This is very significant factor in aggravating erosion and desertification by ultimate removal of soil cover by winds. Measurement of dust from the surface and three metre above, varies between 0.40 to 5.20 quintals per hectare through dust storms, which move sand dunes from one place to another. Size and shape of sand dunes continue to change creating various problems like blockage of roads, railway track, disturbing the electricity transmission system, canals and buildings. Vegetation loss helps in sand and silt movement through dust storms. Dust rises to more than 5000 metres over hyper arid and arid regions cause for the midtropospheric divergence that reduces rainfall in the area, despite presence of moisture. Dust particles and rainfall are adversely related. In these regions, lesser presence of dust particles in atmosphere do not allow clouds to form centric position resulting into low rainfall. On the contrary higher presence of dust particles in atmosphere do not create favourable conditions for rainfall.

7. Accumulation and Accretion of Sand and Sand Dunes

Movement of sand in huge quantity during summer accumulate sand over hills, river valleys, saline depressions, salty lakes and even with small blockades of trees or bushes. Such movements form sand dunes of various types as parabolic, barkhans, etc. Even the sand layer is spread over plain areas and cultivated and moisture lands are covered with sand layer causing further desertification conditions.

8. Geological and Geo-morphological Factors

Geological, hydrological and geo-morphological factors contribute to aridity and desertification through varied forms, some of these conditions are narrated hereunder:

(a) Recurrent aridity in geological history of tract with particularly significant phase of wider extent in early quaternary, accompanied by disorganization of drainage, seasonal river basins, creation of inland basins with ephemeral, centripetal drainage, salt lakes, desert landscape dunes, bornhardts and other associated features. As a results of these factors, considerable tracts outside existing limits of arid regions, sensitive to irrational land use practices, well within semi arid belts.

(b) Even after an interlude of ameliorating humid climate from 3000 to 1000 BC, second phase of aridity is still in progress. The terrain did not have enough time to recover from the effects of earlier aridity phase .and further disorganization of erosion of productive features like tafonis, bornhadts, rock pediments etc.

(c) Present configuration effectively precludes the beneficial influence of monsoon to notable extent in arid tract to its west and provides eastern climatic barrier.

(d) Geomorphic setting have several wind gaps and streams crossing various blocks, which are potential passages for wind transported material into outer fringe areas, in the directions of prevailing wind pattern.

(e) In a terrain with evaporation in excess of precipitation, presence of inland basins or depressions with centripetal drainage and lack of flushing system, help in accumulation of saline material in entire catchment region, which is one of major attribute of desertification.

(f) Hydro-geological situation in the tract is one of restricted aquifers often with brackish or saline water table. Recharge facilities are limited due to low precipitation even in fresh water aquifers. These factors seriously hamper exploitation possibilities, recuperation of aquifers and maintenance of ground water regime.

(g) Due to combination of climatic and geomorphic factors, there is hardly any surface water system with integrated drainage and appreciable discharge for greater part of the year, except of few exceptions, which too misfit system, disorganized without any major sustained flow.

(h) Extensive salt Ranns of Kutch into which Luni discharges, also contributes to salinity of tract. Luni system is also erratic as from origin to about half distance, water is pure but later becomes saline creating serious problem of land degradation.

Geological successions in dryland regions of India are given in Table 7.21.

Table 7.21 : Geological succession in dryland regions of India

Group	System	Rajasthan	Gujarat	Punjab and Haryana	Andhra and Karnataka	Jammu and Kashmir
1	**2**	**3**	**4**	**5**	**6**	**7**
Quaternary	Recent Pleistocene	Blown sand and Alluvium Mar and Shumar series	Alluvium Miliolite limestone	Alluvium	Monazite	Alluvium
Tertiary	Miocene Eocene	Bandah and Jogira series Khuila and Palana series	Nari and Gaj series	—	Plagioclase	Granite
	Eocene Cretaceous	—	Deccan traps	—	Daccen Trap	—
Mosozoic	Cretaceous	Abur beds and Parihar beds	Bagh beds	—	Bagh beds	—
	Jurassic	Baisakhi beds, Jaisalmer limestone Umia sst. Chari Est. Katrol Sst. Patcham Lst.	—	—	—	Infra stacean
Paleaozoics	Upper Carboni- ferous	Bap Boulder beds	—	—	Kaladi series	—
	Vindhyan System	Birmania and Bilara limestone Jodhpur, Nagaur and Bhadura sandstones	—	—	Daccan trap	Tanawal series
	Malani suite of igneous rocks	Granite and Rhyolites	—	—	Kornool Series	—
Archaeans	Delhi System	Erinpura Granite Galc-schists Basal Grits, Quartzites	—	—	Dharwar, Gneissic, Dyke rock	Dogra slates
Pre-Cambrians	Rialo series	Marble Mica- Schists	—	—	Gneissic complex	—
	Aravalli System	Quartzites Mica Schists gneises	—	—	Quartz schists	—

Source: GSI

9. Soil Characteristics

Major factors in pedological character of arid region have significant contributions for desertification conditions and lack of adequate vegetal growth. In-adequate organic accumulation in soils due to prevalence of climate unfavourable to vegetal growth, not only during the current arid phase but from the precarious extensive phase of aridity. Even the soils of semi-arid zone manifest limited soil profile development, often on ancient established dunes. The relatively richer soils are restricted to the fluvial margins and flood plains.

Well developed calcium carbonate material formations in solid profile to variable depths of this region with considerable consolidation inhibit normal root and vegetal growth. Apart from these basic pedogenic limitations, the intrinsic climate conditions that promote erosion and wind transport, constantly remove the surface layers of soil cover resulting into impoverishment of plant growth environment. Movement of salinized water in some areas and deep water tables in many other tracts are also deterrents to soil productivity. Agro-ecological sub-regions, zones and sub-zones of hot and cold desert regions of the country are given in Table No. 7.22.

Table 7.22 : Agro-Ecological Sub-regions, Zones and Sub-zones of Hot and Cold Desert Regions of India

Agro-ecological Sub-region/ Zones / Sub-zones	Area million hectares and %	Rain fall (mm)	Soils	Land use/ major crops	State/ Districts
A. WESTERN SANDY PLAIN SUB-REGION					
1. Hyper arid zone:					
1.1 Hyper arid	0.86 (1.94)	<100	Torripsamments (D)	Sandy waste	Rajasthan- Jaisalmer
2. Dune complex with scrub and grasses:					
2.1 Western dune complex with Lasiurus Sindicus	3.33 (7.49)	100- 150	Torripsamments (D)	Open grazing	Rajasthan- Jaisalmer, Barmer, Bikaner
2.2 Western dune complex with scrub	0.26 (0.59)	100- 150	Torripsamments (D)	Open grazing	Rajasthan- Barmer
3. Hard pan zone					
3.1 Hard pan soil	1.57 (3.53)	150- 250	Petrocalcids lithic cambids Lithic Calcids	Limited kharif cultivation, grazing land	Rajasthan- Jaisalmer, Barmer, Jodhpur, Bikaner
4. Sandy plain with scattered dunes					
4.1 Eastern dune complex with limited cultivation	1.32 (2.97)	250- 350	Torripsamments (D)	Kharif- Bajra, Moong, Moth, Rabi- gram	Rajasthan- Bikaner, Churu, Sikar, Jhunjhunu
4.2 Sandy plain with < 300 mm rainfall	3.68 (8.28)	250- 350	Torripsamments (D) Haplocambids, Haplocalcids	Kharif- Bajra, Moth, Guar	Rajasthan- Barmer, Jodhpur, Nagaur, Bikaner, Ganga-nagar, Churu, Jalore
4.3 Sandy plain with > 300 mm rainfall	0.80 (1.80)	300- 450	Torripsamments (P and D)	Kharif- Bajra, Moong and Til	Ganga-nagar, Churu, Jhunjhunu

Agro-ecological Sub-region/ Zones / Sub-zones	Area million hectares and %	Rain fall (mm)	Soils	Land use/ major crops	State/ Districts
					Haryana- Bhiwani
4.4 Sandy plain with inland drainage	1.25 (2.81)	300- 400	Torripsamments (P) Haplocambids	Kharif- Bajra, Moong, Rabi- Wheat and Mustard	Rajasthan- Nagaur, Sikar, Jhunjhunu
4.5 Western Haryana –Punjab plain	1.21 (2.72)	350- 450	Torripsamments (P) Haplocambids	Kharif- Bajra, Moth, Moong, Til. Rabi- Wheat, Mustard	Bhatinds, Faridkot, Hisar, Sirsa
B. CENTRAL ALLUVIAL PLAIN SUB-REGION					
5. Luni-Banas Basin:					
5.1 Luni basin with coarse loamy soils	2.87 (6.46)	300- 450	Haplocalcids Haplocambids	Kharif- Bajra, Guar, Moong, Til Rabi- Cumin, Mustard, Wheat, Isabgol with Brackish Water Irrigation	Rajasthan- Nagaur, Pali, Jodhpur, Jalore, Barmer, Bujarat- Banas-kantha
5.2 Luni basin with fine loamy soils	1.44 (3.24)	300- 450	Haplocambids, Haplocalcids, Haplosalids	Kharif- Bajra, Jowar, Til, Maize. Rabi- Wheat, Mustard, Cotton	Rajasthan- Pali, Jalore, Nagaur, Jodhpur
5.3 Mendha basin	0.17 (0.38)	400- 450	Torripsamments (P), Torrifluvents, Haplocambids, Haplosalids	Kharif- Cotton. Rabi- Wheat, Mustard, Spices	Rajasthan- Sikar, Nagaur
5.4 Aravalli foot hills	0.76 (1.71)	450- 500	Lithic cambids, Haplocambids	Kharif- Maize, Jowar, Til, Cowpea.	Rajasthan- Jhunjhunu, Sikar, Pali, Jalore, Nagaur
5.5 Aeo-alluvial plains of Banas	0.17 (0.38)	450- 500	Ustipsamments	Kharif- Bajra, Bulses, Castor. Rabi- Wheat,	Gujarat- Banas-kantha

Agro-ecological Sub-region/ Zones / Sub-zones	Area million hectares and %	Rain fall (mm)	Soils	Land use/ major crops	State/ Districts
				Cumin Isabgol	
5.6 Sabarmati plain	0.54 (1.22)	400- 500	Haplocambids, Ustipsamments (P)	Kharif- Maize, Jowar, Ground-nut, Castor. Rabi- Wheat, Mustard, Spices	Gujarat- Banas-kantha, Mehsana
C. NORTHERN CANAL IRRIGATED SUB-REGION					
6. Canal irrigated sandy plain					
6.1 Dune complex with canal irrigation	1.18 (2.66)	150- 250	Torripsamments (D), Petrogysids, Petrocalcids	Sand Dunes- Silvipasture Inter Dune Wheat, Mustard, Groundnut	Rajasthan- Jaisalmer, Bikaner
6.2 Aeolian plain of north-east	0.75 (1.69)	250- 300	Torripsamments (P) Haplogypsids	Kharif- Bajra, Guar, Cotton. Rabi- Wheat, Mustard	Rajasthan- Ganga-nagar, Haryana-Sirsa, Punjab-Faridkot
6.3 Ghaggar flood plain	1.26 (2.84)	250- 300	Torrifluvents, Torripsamments (P), Haplosalids	Kharif- Cotton, Paddy, Sugarcane. Rabi- Wheat	Rajasthan, Ganga-nagar. Punjab-Ferojpur, Faridkot
7. Canal irrigated alluvial plain					
7.1 Satluj plain with coarse loamy soils	0.81 (1.82)	400- 450	Ustipsamments (P)	Rainfed- Bajra, gram Rabi- Wheat, Paddy	Punjab- Ferojpur, Faridkot, Haryana-Bhatinds, Hisar.
7.2. Yamuna plain with loamy soils	0.51 (1.15)	450- 500	Haplocambids, Ustorthents, Ustochrepts	Kharif- Maize, Urad, Jowar, Rabi- Wheat, Paddy, Sugarcane, Gram	Punjab- Sangrur Haryana- Hisar, Bhiwani
7.3 Satluj plain with fine loamy	0.67	400- 500	Torrifluvents, Haplocambids,	Kharif- Maize, Urd,	Punjab- Ferozpur,

Agro-ecological Sub-region/ Zones / Sub-zones	Area million hectares and %	Rain fall (mm)	Soils	Land use/ major crops	State/ Districts
soils	(1.51)		Haplosalids	Jowar, Rabi- Wheat, Paddy, Sugar Cane, Gram	Faridkot Haryana- Bhatinds
D. SOUTHERN COASTAL AND BASALT SUB-REGION					
8. Coastal saline/ rocky uplands:					
8.1 Rann and mudflats	2.07 (4.66)	200- 300	Salids	Saline Waste Salt Pan	Gujarat-Kuchchh, Jamnagar
8.2 Banni eco-system	0.43 (0.97)	250- 300	Torrifluvents	Open Grazing Salt Waste	Gujarat- Kachchh, Jamnagar
8.3 Mangrove eco-system	0.29 (0.65)	150- 200	Unifluvents / Salids	Mangroves	Gujarat- Kachchh, Jamnagar
8.4 Rocky uplands with limited agriculture	0.33 (0.74)	200- 250	Lithic Orthents	Kharif- Bajra, Pulses Limited Cropping	Gujarat- Kachchh
8.5 Coastal saline plain	0.25 (0.56)	200- 300	Salids	Open Grazing Saline Waste	Gujarat- Kachchh, Jamnagar
8.6 Dwarka coastal plain	0.35 (0.79)	150- 200	Lithic Haplocalcids, Haplosalids, Petrocalcids	Rainfed- Groundnut, Castor, Cowpea, Jowar	Gujarat- Jamnagar, Junagarh
9. Kachchh alluvial plain:					
9.1 Central plain with valley fill and rocky upland	0.98 (2.21)	250- 350	Haplocambids, Ustipsamments, Ustertic, Haplocambids, Lithic Calcids	Limited Irrigation- Groundnut, Cotton, Jowar, Pulses, Horticulture	Gujarat- Kachchh
9.2 Coastal alluvial plain with intensive agriculture	0.45 (1.01)	350- 450	Haplocambids, Haplocalcids	Extensively Irrigated Groundnut, Cotton, Pulses	Gujarat- Kachchh
10. Santhalpur plain with saline soils:					

Agro-ecological Sub-region/ Zones / Sub-zones	Area million hectares and %	Rain fall (mm)	Soils	Land use/ major crops	State/ Districts
10.1 Santhalpur plain with coarse loamy soils	0.75 (1.69)	250- 350	Torripsamments (P), Haplocambids	Rainfed- Groundnut, Til, Castor, Saline Waste	Gujarat- Banas-kantha, Mehsana Rajasthan- Jalore
10.2 Rapar plain with fine loamy soils	0.30 (0.67)	250- 350	Palcargids	Rainfed- Jowar, Til, Bajra, Castor. Rabi- Wheat	Gujarat- Kachchh
11. Basaltic piedmont plain:					
11.1 Basaltic gravelly/ rocky upland	0.22 (0.50)	250- 350	Lithic orchents	Limited Rainfed- Groundnut, Open Grazing	Gujarat- Jamnagar
11.2 Basaltic piedmont plain with shallow soils	0.28 (0.63)	400- 500	Lithic Ustorthents	Limited Irrigation- Groundnut, Spices	Gujarat- Jamnagar, Rajkot
11.3 Basaltic piedmont plain with moderately deep soils	0.22 (0.50)	250- 400	Vertic Ustocherpts, Typic Haplusterts	Groundnut, Cotton, Sugarcane, Spices	Gujarat- Jamnagar, Rajkot
E. WESTERN HIMALAYAS					
12.1 Shallow sandy to loamy skeletal soils with low availability of water capacity	7.03 (15.82)	<150	Cryorthents, Cryorthids	Millets and Wheat in Patches	Ladakh
F. DECCAN PLATEAU					
13.1 Shallow and medium loamy to clayey, black soils with low and medium availability of water.	2.75 (6.19)	600-1000	Ustropepts, Ustorthents, Chromusterts	Millets, Cotton, Pusles, Sugarcane under irrigation	Anantpur, Cuddapah, Kurnool of Andhra Pradesh
13.2 Medium red loamy soils with low and medium water availability	2.32 (5.22)	600-1000	Ustropepts, Rhodustalfs, Plinthustalfs, Ustorthents	Millets, Oilseeds, Rice Cotton and Sugarcane under irrigation	Bellary, Bijapur, Chitradurg, Dharwad, Raichur of Karnataka

Source : ICAR/BSS & LUP

10. Recurrence of Droughts

Drought is recurrent phenomena of desert regions, which is consequential impact of failure of monsoon. Analysis of past 103 years from 1901 to 2003 reveals that sixteen years were very severe droughts, 9 years severe droughts, 19 years of moderate droughts, 18 years of light drought and 44 years of normal droughts. Desert areas are worst affected with occurrence of droughts as people of rural areas depending on land are put to very miserable condition. Livestock is also affected, which is most dependable source of livelihood in desert regions. During 53 years from 1950-51 to 2002-03 there were only six years without drought in any part of desert region. In remaining years drought intensity was very severe to moderate In drought conditions, people of desert region suffer most as poor people depend on single cropped area and monsoon failure create food, drinking water, fodder problems for survival of human beings and their livestock. For meeting food and fodder requirement, people start cutting trees to survive them and livestock. Such conditions misbalance development process and ecology of desert regions is worst affected, which take years to restore. Since droughts are recurring process, ecology of the area is disturbed to maximum possible extent.

Monsoon failure does not mean decrease from normal rainfall in any zone of regions, but it affects in many ways like late occurrence of rains specially after a fortnight after sowing, timely rains in early spell but failure in subsequent duration and lacking time dispersal in spells. All these factors are responsible for drought conditions. Intensity of drought is measured in terms of crop failure more than 50 percent. In case of 100 percent crop failure, the intensity of drought is treated as very severe. Consequential impacts are calculated on magnitude of crop failure. Poor people belonging to small and marginal cultivators, landless labour are worst sufferers of drought conditions.

Arid lands laving loose soil structure also suffer from heavy rains, which is measured in terms of time duration in single spell and its velocity. In absence of lack of drainage regular systems, flow water start damaging structures like roads, railways, bridges, buildings, individual houses, cultivated fields etc. Loose sandy soil areas are seriously affected and people lose their belongings. Occurrence of flood in desert regions in once in five years while droughts on three years with varying intensity. Five-year monsoon cycle has one favourable year when crop production is in abundance. Desert zones have comparatively low density of population and operational holdings are higher than non-desert regions. Though most of land is unirrigated but agriculture production is substantial in favourable monsoon conditions.

11. Wind Erosion

Wind erosion is common feature of arid regions, which takes places from fields, sand dunes and drought, accelerates impact. Nature and extent of socio-ecological parameters very with severity of wind erosion due to drought and ecological differentiation. It affects roads, railway network, communication system, houses, cultivated fields and water-bodies. Movement of dunes disrupts human activities, which is serious setback over individual physical assets required for livelihood. Droughts and floods are natural factors of disaster conditions, but human interface is also responsible for creation of such conditions. The root cause of such destructions is deforestation at large scale. Occurrence of monsoon is dependent on vegetative cover and desert ecology has been disturbed completely, droughts and floods are bound to create destruction Desert regions of the country are generally low rainfall areas, where water charge in the ground is lesser than its withdrawal. Decreasing trend of monsoon, occurrence of drought and floods are resultant impact of distortion of ecosystem, where major role has been played by deforestation at large scale.

In view of loose sandy soil, floods create havoc in dryland regions, as the water seepage in desert land is slow. Structures like railways, roads, electric transmission lines, bridges are destroyed severely due to displacement of soil. Areas having lignite or gypsum at a depth below 10 meters are worst affected as water is not possible to be charged inside the soil. There is no permanent drainage system due to movement of dunes and in view of heavy downpour, water is stored in depression areas creating health hazard to the people and vegetation. Such situation arises in five to fifteen years in any part of the desert region. Land forms with associated characteristics are summarized in Table 7.23.

Table 7.23 : Landforms with associated characteristics

Landform	Major shaping force	Soils	Dominant plant	Adaptive character
1. Mountains and Hills	Chemical and Physical	Exposed bed rock	Anogeissun latifolia Acacia Senegal, Compiphora wightii	Regions with rainfall less than 25 cm
2. Upper piedmont zone	Water erosion	Sandy loam	Salvadora oleoides, Capparis deciduas, Acacia Senegal (sparse)	Evergreen shrubs with low diffuse roots
3. Lower piedmont merging with plains	Water erosion	Sandy loam to loamy sand	Acacia arabica, Prosopis cpicigera, Salvadora oleoides, Acacia senegal, Capparis deciduas (rainfall less than 25 cm)	Evergreen with deep tap root Evergreen tree and shrub
4. Depression (saline)	Water deposition	Clay loam	Salvadora persica, Tamarix dioica, Halophytic succulents	No vegetation when salinity is high
5. Water courses	Water erosion	Coarse sand	Acacia arabica, Salvadora olecodes	Deep tap root
6. Depression (non-saline)	Water deposition	Clay loam	Acacia arabica	Deep tap root

Landform	Major shaping force	Soils	Dominant plant	Adaptive character
7. Sandy plains	Wind erosion and deposition	Wind deposited sand	Prosopis cineraria, Capparis decidua	Deep tap root
8. Sand dunes	Wind deposition	Fine sand	Calligonum polugonodes	Deep tap and shallow side roots
9. Inter dunes	Water deposition	Clay loam covered with sand	Capparis Haloxylon	Deep tap root
10. Flood plains (old)	Water deposition	Clay loam	Saloxylon salicornicum and H. recurvum	Deep tap root

7.15 ANTHROPOGENIC FACTORS

Ecology of dryland regions has been disturbed significantly in last fifty years due to anthropogenic factors. These factors direct human interface for degradation of environment, ecology and ecosystem of the earth but it is severe in hyper-arid, arid and semi-arid regions as these areas are already facing land degradation and water scarcity problems. Cutting of forests and clearance of vegetated lands for meeting cultivation requirement of increasing population. Human interface for aggravating the problems in desert regions are multi-faceted. In low density regions, small livestock rearing in large number is one of the serious issue. People keep small livestock like sheep and goat for supplemental livelihood due to erratic monsoon and crop failure. Sheep and goat are migratory in the desert region as there is very little to feed the large number and people move with herd to vegetative areas of surroundings states. The herd of small livestock increases significantly and owners get regular income from their livestock. However, vegetative loss is serious loss to the areas of their migratory routes.

1. Pressure of Human Population:

Density of population in dryland regions is 127 persons per sq kms against country density of 324. State desert region specific density of Rajasthan (108), Andhra Pradesh (187), Gujarat (130), Jammu and Kashmir (3) and Karnataka (245) is below national average, while Haryana and Punjab have higher density of 326 and 369 persons respectively. Districts having density with less than 40 persons are Ladakh (3), Jaisalmer (13) and Kachchh (35) due to tough geographical conditions. In addition pressure of livestock is significant in desert regions as livestock rearing is rather helpful in assured supplemental income but at the cost of environment degradation. Demographic features of dryland regions are depicted in Table 7.24 below.

Table 7.24 : Demography in Indian Desert Regions

States/ Districts	Area	Population (Total)	Population (Rural)	Growth rate 1991-2001	Density 2001	Sex Ratio 2001
I. RAJASTHAN						
1. Barmer	28387	1964835	1819431	36.90	69	892
2. Bikaner	27244	1674271	1079235	38.20	61	890
3. Churu	16830	1923878	1387682	24.70	114	948
4. Ganganagar	10978	1789423	1336066	27.60	163	873
5. Hanumangarh	9656	1518005	1214467	24.40	157	894
6. Jaisalmer	38401	508247	431853	47.50	13	821
7. Jalore	10640	1448940	1338946	26.80	136	964
8. Jhunjhunu	5928	1913689	1518573	20.90	323	946
9. Jodhpur	22850	2886505	1909423	34.00	126	907
10. Nagaur	17718	2775058	2297721	29.40	157	947
11. Pali	12387	1820251	1429364	27.50	147	981
12. Sikar	7732	2287788	1815250	24.10	296	951
Total Rajasthan	**208751**	**22510890**	**17578011**	**27.62**	**108**	**847**
II. ANDHRA PRADESH						
1. Anantapur	19130	3640478	2720915	16.37	191	958
2. Kadapa	15359	2601797	2014044	15.81	169	974
3. Kurnool	17658	3529494	2712030	16.21	199	965
Total A.P.	**52147**	**9771769**	**7446989**	**16.18**	**187**	**965**
III. GUJARAT						
1. Banas Kantha	10757	2524244	2228743	26.38	233	930
2. Jam Nagar	14125	1904278	1068022	21.79	135	941
3. Kachchh	45652	1583225	1108333	25.40	35	942
4. Mehsana	4384	1837892	1426175	12.05	419	927
5. Rajkot	11203	3169881	1544019	26.08	283	930
6. Surendra Nagar	10489	1515148	1112700	25.34	144	924
Total Gujarat	**96610**	**12534668**	**8487992**	**22.84**	**130**	**932**
IV. HARYANA						
1. Bhiwani	4778	1425022	1154629	22.49	298	879
2. Fatehabad	2538	806158	664001	24.76	318	884
3. Hissar	3983	1537117	1138999	27.11	386	851
4. Mahendragarh	1899	812521	702885	19.16	428	918
5. Sirsa	4277	1116649	823184	23.59	261	882
Total Haryana	**17475**	**5697467**	**4483698**	**23.42**	**326**	**883**
V. JAMMU AND KASHMIR						
1. Ladakh*	82655	117232	88593	29.97	3	823

* Includes unsurveyed area of 37555 Sq. Kms under unauthorized occupation of China.

States/ Districts	Area	Population (Total)	Population (Rural)	Growth rate 1991-2001	Density 2001	Sex Ratio 2001
			VI. KARNATAKA			
1. Bellary	8419	2027140	1587871	18.06	241	952
2. Bijapur	10494	1806918	1282911	17.98	172	937
3. Chitra Durga	8440	1517896	1147348	18.66	180	948
4. Dharwad	4230	1604253	1187147	19.34	379	956
5. Raichur	6828	1669762	1269019	17.36	244	967
Total Karnataka	**38411**	**8625969**	**6774296**	**18.28**	**245**	**952**
			VII. PUNJAB			
1. Ferojpur	5303	1746107	1295382	20.51	329	885
2. Faridkot	1469	550892	357321	21.07	375	883
3. Bhatinda	3385	1183295	831541	20.09	350	870
4. Nawan Shahr	1267	587468	506402	10.58	464	914
5. Sangrur	5020	2000173	1415354	18.67	399	871
Total Punjab	16444	6067935	4406000	18.84	369	880
GRAND TOTAL	**512503**	**65325930**	**49265579**	**22.84**	**127**	**907**

Source: Census of India

Total density of population of dryland regions is 6.53 crores varied from 3 persons per sq kms in Ladakh to 464 persons per sq kms in Nawan Shahr of Punjab. The population growth rate in 1991-2001 decade remained of 22.84 percent for dryland regions as a whole varying from 10.58 percent in Nawan Shahr of Punjab to 47.50 percent in Jaisalmer district of Rajasthan. Rural population of these district is 75.41 percent of total population. Sex ratio is 907 female for 1000 but varied from 981 in Pali district of Rajasthan to 823 in Ladakh of J and K state.

In dryland region, 75.41 percent of total population resides in rural areas, where means of cooking are wood, agriculture residue and cattle dung, which is easily available without any cost. Out of 49 million of dryland population, 32 million are cultivators and 11 million agricultural labourers dependent of agriculture for their livelihood. With increase in population, size of operation holdings is decreasing constantly causing small and fragmented lands, which decrease productivity with small holdings. Though operational holdings in dryland regions are comparatively larger, but these are unable to provide sufficient return due to scarcity of water from all available sources. Encroachment on forest lands, common property resources, grazing and pasture lands for meeting cultivation requirement has disturbed ecological balance. Use of marginal lands under cultivation has resulted into uneconomic holdings and cost of cultivation is increasing substantially. Half of operational holdings are small and marginal where cultivators could not get sufficient yield for their livelihood. Such people lack in technology to increase production from available land resources and conventional practices of cultivation are still prevailing in small operational holding less than two hectares of cultivated land.

Livestock population in dryland regions is equal to human population, engaged in uncontrolled grazing causing serious impact of soil structure and vegetation loss is vital. Livestock is kept for assured and supplemental income but at the cost of environmental deterioration. In drought conditions, trees are cut for survival of livestock This has lost vegetated cover of earth surface and plantation activities are seriously affected. There are very few protected areas where trees and vegetation cover is intact, but remaining areas are completely denuded. In hyper-arid and arid regions, trees and vegetation are necessary for moisture conservation and decreasing dryness in the environment. Present conditions dryness and denudation of land are the resultant of human interface. Even forest areas have been encroached for cultivation practices.

2. Changes in Land use Pattern

Present land use pattern of total geographical areas of dryland regions has been given in Table 7.25 which reveals that 7.81 percent area remained as forests, 9.71 percent area used in various habitations, infrastructure activities, urbanization and buildings, 8.24 percent area is barren land, 2.85 percent area under permanent pastures and tree groves, 21.82 percent area is wasteland, 18.96 percent area is fallow land and 30.61 percent land is put under cultivation practices. The cultivated area left as fallow land is mostly seriously degraded where cultivation practices are possible in good rainfall conditions. Out of present cultivated area of 155.5 sq kms only 34 percent is irrigated, where single irrigated crop is possible.

Forest land was about 30 percent of total geographical area about one hundred years earlier, but squeezed to merely 7.81 percent and remaining area was continuously occupied for cultivation practices. Barren land and wasteland are product of deteriorated lands due to loss of soil moisture. With present land use pattern, plantation area can be significantly increased to 40.72 percent by integrating forest land, barren land, permanent pasture and tree groves and waste land and cultivable land of 49.57 percent can be treated without changing the land use pattern having 19.13 percent irrigated land and 30.44 percent is dependent on the mercy of monsoon. In irrigated regions, intensity is quite low and in 10 percent of irrigated lands, only single crop is possible with ground water. Canal irrigated lands are capable for double crop but most of the area is affected with water logging problem for mismanagement of water.

Table 7.25 : Land use pattern in dryland regions of India 2002-03

(Area in Sq. Kms)

States/ Districts of Indian Desert	Geo. Area (Village papers)	Forests	Non-Agriculture use	Barren land	Pastures and Tree groves	Waste land	Fallow land	Net Culti. Area
I. RAJASTHAN								
1. Barmer	28387	310	737	1270	2029	2665	10773	10603
2. Bikaner	27244	825	2578	359	505	8238	12136	2603
3. Churu	16830	64	633	7	384	121	10175	5446
4. Ganganagar	10978	593	683	28	20	385	3866	5403
5. Hanumangarh	9656	150	586	7	41	93	2031	6748
6. Jaisalmer	38401	233	1184	3691	1035	25941	5538	779
7. Jalore	10640	201	403	834	474	255	3105	5368
8. Jhunjhunu	5928	396	202	154	403	65	1450	3258
9. Jodhpur	22850	70	789	1459	1226	744	16661	1901
10. Nagaur	17718	180	855	580	733	148	7849	7373
11. Pali	12387	802	581	1459	908	493	4605	3539
12. Sikar	7732	610	343	173	409	106	1871	4220
Total Rajasthan	**208751**	**4434**	**9574**	**10021**	**8167**	**39254**	**80060**	**57241**
II. ANDHRA PRADESH								
1. Anantapur	19130	1968	1618	1968	724	518	1622	10712
2. Kadapa	15359	5055	1733	2382	238	690	1069	4192
3. Kurnool	17658	3182	1342	994	61	784	2504	8791
Total Andhra Pradesh	**52147**	**10205**	**4693**	**5344**	**1023**	**1992**	**5195**	**23695**

States/ Districts of Indian Desert	Geo. Area (Village papers)	Forests	Non-Agriculture use	Barren land	Pastures and Tree groves	Waste land	Fallow land	Net Culti. Area
III. GUJARAT								
1. Banas Kantha	10449	1052	529	341	651	186	326	7364
2. Jam Nagar	10203	482	650	1527	763	321	463	5997
3. Kachchh	45652	2883	736	17045	701	16812	786	6689
4. Mehsana	4392	72	315	52	268	49	172	3464
5. Rajkot	11074	359	702	1053	891	140	551	7378
6. Surendra Nagar	10458	502	547	1270	460	158	659	6862
Total Gujarat	**92228**	**5350**	**3479**	**21288**	**3734**	**17666**	**2957**	**37754**
IV. HARYANA								
1. Bhiwani	4778	33	247	205	—	—	164	4129
2. Fatehabad	2538	5	217	26	—	—	34	2256
3. Hissar	3983	2	332	61	—	—	604	2984
4. Mahendra-garh	1899	23	226	148	—	24	262	1216
5. Sirsa	4277	17	211	5	—	—	183	3861
Total Haryana	**17475**	**80**	**1233**	**445**	**—**	**24**	**1247**	**14446**
V. JAMMU AND KASHMIR								
1. Ladakh	82665	13047	20977	252	22	48245	3	119
VI. KARNATAKA								
1. Bellary	8419	1431	1729	1168	214	1091	1365	1421
2. Bijapur	10494	1679	3112	1279	156	330	1413	2525
3. Chitra Durga	8440	1266	1722	1146	317	1191	1376	1422

States/ Districts of Indian Desert	Geo. Area (Village papers)	Forests	Non-Agriculture use	Barren land	Pastures and Tree groves	Waste land	Fallow land	Net Culti. Area
4. Dharwad	4230	634	836	108	104	145	1189	1214
5. Raichur	6828	1092	885	756	736	745	1287	1327
Total Karnataka	**38411**	**6102**	**8284**	**4457**	**1527**	**3502**	**6630**	**7909**
VII. PUNJAB								
1. Ferozpur	5303	121	383	14	—	—	31	4754
2. Faridkot	1469	22	117	6	—	35	25	1264
3. Bhatinda	3385	83	121	—	—	72	133	2976
4. Nawan Shahr	1267	169	71	42	—	5	34	946
5. Sangrur	5020	70	402	16	5	83	36	4408
Total Punjab	**16444**	**465**	**1094**	**78**	**5**	**195**	**259**	**14348**
GRAND TOTAL	**508121**	**39683**	**49334**	**41885**	**14478**	**110878**	**96351**	**155512**

Source: Agricultural Department Govt. of India

Above information of land use pattern is based on actual revenue set up based on village land records, which depicts the position of each category of land put to various uses. This information lacks in land degradation factor under each category. Wastelands of various nature have also been identified with the help of space technology and given in Table No. 7.26, which reveals that total 32.8 percent of geographical area of dryland regions is seriously affected by land degradation. Categories of wasteland are given in concluding part of the table to understand the nature of degradation among respective categories.

Table 7.26 : Combined position of Wastelands in dryland states

Category	Rajasthan	Gujarat	Punjab	Haryana	AP	J and K	Karnataka	Total
1	52	53	54	55	56	57	58	59
1	265.99	131.23	—	—	8.17	10.79	1.76	417.94
2	121.45	207.66	—	—	14.36	—	13.71	357.18
3	4.29	—	—	—	22.17	—	0.04	26.50
4	17638.08	7596.34	47.17	288.91	4092.72	—	1033.75	30696.97
5	5110.29	3977.54	9.00	21.17	335.11	—	251.27	9704.38
6	68.20	—	125.89	4.74	1.69	—	—	200.52
7	143.30	8.59	0.23	9.22	27.78	—	—	189.12
8	478.99	—	—	0.23	84.86	—	0.30	564.38
9	404.66	874.02	—	2.62	24.87	—	106.12	1412.29
10	387.39	—	0.10	0.98	4.73	—	10.99	404.19
11	—	—	—	—	—	—	—	—
12	—	—	—	—	—	—	—	—
13	1119.39	144.83	—	7.01	5456.55	—	1557.50	8285.28
14	16.56	—	—	—	152.76	—	39.20	208.52
15	6083.44	42.89	—	382.25	0.45	61.58	19.83	6590.44
16	47.45	—	267.81	105.37	0.96	—	—	421.59
17	92.97	—	19.75	0.18	—	10.87	18.81	142.58
18	—	—	—	—	—	—	—	—
19	—	16.29	—	—	—	—	—	16.29
20	2672.88	—	—	—	—	—	—	2672.88
21	16380.70	—	—	—	—	—	—	16380.70

Category	Rajasthan	Gujarat	Punjab	Haryana	AP	J and K	Karnataka	Total
22	8288.24	—	—	127.20	—	1565.17	—	9980.61
23	1735.57	—	—	0.71	—	—	—	1736.28
24	112.28	34.58	0.23	0.45	92.61	—	70.84	310.99
25	14.95	—	—	—	2.35	—	1.07	18.37
26	2391.64	34.14	—	50.29	1154.24	23565.93	377.45	27573.69
27	—	10.41	—	—	103.16	1893.62	—	2007.19
28	—	—	—	—	—	48245.28	—	48245.28
Total	**63578.71**	**13078.52**	**470.18**	**1001.33**	**11579.54**	**75353.24**	**3502.64**	**168564.16**
TGA	208751.00	97515.00	16444.00	17475.00	52147.00	82665.00	38411.00	513911.00
% TGA	**30.46**	**13.41**	**2.86**	**5.73**	**22.21**	**91.15**	**9.12**	**32.80**

Source: NRSA

Detailed Categories of Wastelands

01 : Gullied and/ or Ravinous Land (Shallow- less than 2.5 m depth)
02 : Gullied and/ or Ravinous Land (Medium More than 2.5 m and up to 5 m deep)
03 : Gullied and/ or Ravinous Land (Deep- more than 5 m)
04 : Land with Scrub
05 : Land without Scrub
06 : Waterlogged or Marshy- Permanent
07 : Waterlogged or Marshy- Seasonal
08 : Saline / Alkaline- Strong
09 : Saline/ Alkaline Moderate
10 : Saline/ Alkaline- Slight
11 : Shifting Cultivation: Abandoned
12 : Shifting Cultivation- Current
13 : Degraded Forests- Scrub dominated
14 : Agriculture Land inside Notified Forests
15 : Degraded Pasture / Grazing land
16 : Degraded Land under Plantation Crops
17 : Sands- Flood Plain
18 : Sands- Levees
19 : Sands- Coastal Sand
20 : Sands- Semi Stabilized to Stabilized (> 40 m)
21 : Sands- Semi Stabilized to Stabilized- Moderate (15-40m)
22 : Sands- Semi Stabilized to Stabilized- Low (< 15 m)
23 : Sands- Closely spaced Inter-Dunal Area
24 : Mining Wastelands
25 : Industrial Wastelands
26 : Barren Rocky/ Stony Waste Areas
27 : Steep Sloping Areas
28 : Snow Covered and / or Glacial Areas

3. Deforestation

Afforestation, vegetative cover and agro-forestry land use system hold serious social and economic implications in fragile ecosystem of desert regions. Ecological deterioration and accentuation of natural hazards affecting balanced land use system has been occurring in region due to human and livestock pressure. Tree cover in desert region is helpful for stabilization of desert sand and deforestation is responsible for various problems related to movement of sand dunes, drought and flood condition. In real sense, land degradation starts from deforestation and other problems are consequential impact of this very situation. Wind erosion is causatic impact of soil erosion and deforestation remove all barriers for creating further disturbance.

4. Land Degradation and Desertification

There exist significant wasteland in desert regions, which is not fit for useful land activities and cultivated land is also losing its productivity to due various human factors. Conventional practices for management of land are regularly diminishing and survival of people in desert regions is becoming quite difficult. Existence of wasteland and habitation of people in desert lands are adversely correlated. It has been seen that concentration of population in desert regions is possible in areas where wasteland is minimum. Ladakh of Jammu and Kashmir state, Kachchh of Gujarat state and Jaisalmer of Rajasthan state have very low density of population as there exist very meagre means of livelihood for survival of people. Desert districts of Punjab, Haryana and Ganga Nagar and Hanumangarh districts of Rajasthan have greater concentration of population as there are very low wastelands. Existing wastelands cannot be used for any economic purpose without treatment of wasteland conditions. It is true that all wastelands are not treatable due to financial constraints as cost of treatment of higher than possible returns but some category of wastelands can be put to economic use with little investment and such areas can be helpful in regaining their ecological balance.

Cultivated lands have different problems of land degradation and desertification. These are basically related to excessive land use, excess pressure of human and livestock population, use of marginal lands for cultivation activities without treatment of their existing problems. In addition, decrease in better lands for urban population, construction of roads, railways, bridges, dams, electricity transmission lines etc. are the areas where productive land is used for provision of these facilities. Overall impact of all these activities fell on the people residing in desert regions.

5. Water Management Problems

Water is most essential and scarce factor in desert regions. Desert regions having water availability through any possible source change the desert land in most efficient manner as desert regions like Punjab, Haryana and some areas of Rajasthan have gained prosperity due to availability of water. Deficiency of water in desert regions is helping in deterioration of ecological balance. Water management in desert regions is most essential component for rationale use of available land resources, which is completely lacking and various ecological problems have been cropped up as resultant impact. Areas with canal irrigation system are facing serious problem of water-logging due to seepage of water and cultivated areas are seriously affected and loosing productivity. This problem is prevalent in all desert regions as sandy soil has no capacity to absorb excess water.

Areas served with ground water are also affected with water problems due to excessive drawl of ground water than charged through monsoon system. Most of the desert regions have reached at critical stage due to over drawl of ground water and cultivated lands are severely affected by alkalinity and salinity problems. In desert areas occurrence of monsoon is not sufficient to cultivate all lands. With increasing human and livestock pressure on land, cultivation is essential and ground water is only source to get two or more crops to meet growing requirements. Cultivators try to draw maximum water from their sources to get good crops, which has seriously affected water sources and ground water depletion in resultant impact. People need to dug tube-wells for irrigation and drinking water purposes. Such sources are sunk at large scale and only solution is to further drill to get water. Water depletion rate in desert regions is ranging between 3 to 10 meters per year. Status of Ground water availability and its exploitation in desert districts has been given in Table 7.27 and 7.28.

Sub-surface hydrograph follows definite trend with peak followed by recession limb, post-recharge period characterized by distinct steep and gentle slopes. Utilization recharge is estimated on pre and post monsoon water level fluctuations for specific areas. Rainfall infiltration depends on loose and hard soil strata. Annual recharge water includes monsoon and non-monsoon rainfall, seepage from canal, return flow from irrigation, inflow from influent rivers, recharge from sub-merged lands and lakes. Taking water charge and drawl factors, stage of ground water development is drawn for adjudging ground water potential. With this process areas up to 65 percent are graded as white zone, areas more than 65 percent but less than 85 percent are treated as grey zones, areas with more than 85 percent but less than 100 percent are termed as dark zone and areas more than 100 percent level are categorized as over-exploited regions. Information of ground water development level of 1994 and 1998 and trend decrease of water till date reveals that most of desert regions have reached at least dark zone and few are under over exploited regions. This situation is most dangerous from availability of sub-surface water for drinking water and irrigation requirements. It is also feared that extinction of ground water would be harmful for life over earth.

Table 7.27 : Ground water resources and Irrigation potential due to Natural Recharge (1994)

(Figures in Thousand hectare metre/ year)

States/ Districts of Indian Desert	Total Replenish-able GW resources from natural recharge	Provision for domestic, industrial and other uses	Available GW resources for irriation in net terms	Utilizable GW Resources for irrigation in net terms	Gross draft from natural recharge estimation on prorata basis	Net draft from natural recharge	Balance GW resources for future use	Level GW develop-ment (%)
I. RAJASTHAN								
1. Barmer	30.75	5.57	25.18	22.66	9.30	6.51	18.67	25.85
2. Bikaner	12.50	2.03	10.47	9.42	0.00	0.00	10.47	0.00
3. Churu	15.26	3.06	12.20	10.98	0.54	0.38	11.82	3.11
4. Ganganagar	32.94	4.94	28.00	25.20	12.86	9.00	19.00	32.14
5. Hanumangarh	32.94	4.94	28.00	25.20	12.86	9.00	19.00	32.14
6. Jaisalmer	10.09	1.55	8.64	7.78	0.07	0.05	8.59	0.58
7. Jalore	29.47	4.51	24.86	22.37	50.57	35.40	-10.54	142.40
8. Jhunjhunu	21.57	4.32	17.25	15.52	21.81	15.27	1.98	88.52
9. Jodhpur	31.82	6.82	25.00	22.50	28.31	19.82	5.18	79.28
10. Nagaur	57.83	9.40	48.43	43.59	23.69	16.58	31.85	34.23
11. Pali	42.96	6.44	36.52	32.87	25.54	17.88	18.64	48.96
12. Sikar	33.60	5.09	28.51	25.66	33.86	23.70	4.81	63.13
II. ANDHRA PRADESH								
1. Anantapur	91.64	13 75	77.89	70.10	41.00	28.70	49.19	36.85
2. Kadapa	88.89	13 33	75.56	68.00	37.81	26.47	49.09	35.03
3. Kurnool	74.71	11 21	63.50	57.15	15.91	11.14	52.36	17.54

States/ Districts of Indian Desert	Total Replenish-able GW resources from natural recharge	Provision for domestic, industrial and other uses	Available GW resources for irriation in net terms	Utilizable GW Resources for irrigation in net terms	Gross draft from natural recharge estimation on prorata basis	Net draft from natural recharge	Balance GW resources for future use	Level GW develop-ment (%)
III. GUJARAT								
1. Banas Kantha	128.78	19.32	109.46	98.51	99.47	69.63	39.38	63.61
2. Jam Nagar	87.15	13.07	74.08	66.67	35.16	24.61	49.47	33.22
3. Kachchh	66.45	9.97	56.48	50.83	28.69	20.08	36.40	35.55
4. Mehsana	55.98	8.40	47.58	42.82	84.37	59.06	-11.48	124.13
5. Rajkot	111.74	16.76	94.98	85.48	50.00	35.00	59.98	36.85
6. Surendra Nagar	73.81	11.07	62.74	56.47	32.51	22.76	39.98	36.28
IV. HARYANA								
1. Bhiwani	24.27	3.64	20.63	18.57	10.67	7.47	13.16	36.21
2. Fatehabad	22.97	3.45	19.52	17.57	20.29	14.20	5.32	72.75
3. Hissar	34.40	5.16	29.24	26.32	13.21	9.25	19.99	31.63
4. Mahendragarh	13.24	1.99	11.25	10.13	26.89	18.82	-7.57	167.29
5. Sirsa	20.34	3.05	17.29	15.56	10.46	7.32	9.97	42.34
V. JAMMU AND KASHMIR								
1. Ladakh	Not Surve-yed							
VI. KARNATAKA								
1. Bellary	38.73	5.81	32.92	29.63	11.44	8.01	24.91	24.33
2. Bijapur	101.57	16.14	91.43	82.29	56.82	39.78	51.65	43.51

States/ Districts of Indian Desert	Total Replenish-able GW resources from natural recharge	Provision for domestic, industrial and other uses	Available GW resources for irriation in net terms	Utilizable GW Resources for irrigation in net terms	Gross draft from natural recharge estimation on prorata basis	Net draft from natural recharge	Balance GW resources for future use	Level GW develop-ment (%)
3. Chitra Durga	59.76	8.96	50.80	45.72	31.60	22.12	28.68	43.54
4. Dharwad	98.00	14.70	83.30	74.79	23.86	16.70	66.60	20.05
5. Raichur	71.40	10.71	60.69	54.62	12.66	8.86	51.83	14.60
VII. PUNJAB								
1. Ferojpur	111.82	11.18	100.64	90.58	92.50	64.75	35.89	64.34
2. Faridkot	135.93	13.59	122.34	110.11	113.54	79.48	42.86	64.97
3. Bhatinda	83.35	8.33	75.02	67.52	46.21	32.35	42.67	43.12
4. Nawan Shahr	96.71	9.67	87.04	78.34	83.51	58.46	28.58	67.16
5. Sangrur	62.52	6.29	56.63	50.97	121.16	84.81	-28.18	149.76

Sorce: CGWB

Table 7.28 : Ground Water resources and Irrigation potential due to natural recharge (1998)

(Figures in Thousand Hectare Metre/ Year)

States/ Districts of Indian Desert	Total Replenish-able GW resources from natural recharge	Provision for domestic, industrial and other uses	Available GW resources for irriation in net terms	Utilizable GW Resources for irrigation in net terms	Gross draft from natural recharge estimation on prorata basis	Net draft from natural recharge	Balance GW resources for future use	Level GW develop-ment (%)
I. RAJASTHAN								
1. Barmer	21.14	5.14	16.00	14.40	21.16	14.81	1.19	92.56
2. Bikaner	19.49	0.71	18.78	16.90	4.90	3.43	15.35	18.26
3. Churu	14.18	3.91	10.27	9.24	6.15	4.30	5.96	41.91
4. Ganganagar	33.83	7.42	26.41	23.77	12.91	9.04	17.38	34.21
5. Hanumangarh	26.56	6.28	20.28	18.26	16.15	11.31	8.98	55.75
6. Jaisalmer	4.74	1.00	3.74	3.37	1.05	0.74	3.01	19.68
7. Jalore	39.31	5.90	33.41	30.07	98.95	69.27	-35.86	207.33
8. Jhunjhunu	22.03	5.70	16.33	14.69	33.52	23.47	-7.14	143.72
9. Jodhpur	31.41	8.04	23.37	21.03	47.88	33.51	-10.14	143.40
10. Nagaur	49.48	11.19	38.29	34.46	63.51	44.45	-6.17	116.11
11. Pali	41.84	6.28	35.56	32.00	39.30	27.51	8.05	77.37
12. Sikar	32.28	4.87	27.41	24.67	38.66	27.06	0.35	98.74
II. ANDHRA PRADESH								
1. Anantapur	124.91	18.74	106.17	95.55	61.70	43.19	62.98	41.00
2. Kadapa	105.72	15.86	89.86	80.88	49.66	34.76	55.10	38.68
3. Kurnool	115.34	17.30	98.04	88.24	27.13	18.99	79.05	19.00

States/ Districts of Indian Desert	Total Replenish-able GW resources from natural recharge	Provision for domestic, industrial and other uses	Available GW resources for irriation in net terms	Utilizable GW Resources for irrigation in net terms	Gross draft from natural recharge estimation on prorata basis	Net draft from natural recharge	Balance GW resources for future use	Level GW develop-ment (%)
III. GUJARAT								
1. Banas Kantha	145.02	21.76	123.26	110.93	133.31	93.32	29.94	75.71
2. Jam Nagar	99.36	1.49	74.45	76.01	47.71	3.34	51.05	39.55
3. Kachchh	80.29	12.04	68.24	61.41	41.25	28.88	39.37	42.31
4. Mehsana	86.43	12.97	73.46	66.11	155.01	108.51	-35.05	147.71
5. Rajkot	136.17	20.42	115.75	104.17	7.25	50.75	65.00	43.84
6. Surendra Nagar	86.72	13.01	73.71	66.34	45.44	31.81	41.90	43.15
IV. HARYANA								
1. Bhiwani	42.87	6.43	36.44	32.79	39.56	27.69	8.74	76.00
2. Fatehabad	90.09	13.51	76.58	68.92	54.05	37.83	38.74	49.41
3. Hissar	153.21	22.98	130.23	117.21	100.23	70.16	60.07	53.88
4. Mahendragarh	19.99	3.00	16.99	15.29	35.72	25.01	-8.02	147.20
5. Sirsa	78.13	11.72	66.41	59.77	44.03	30.82	35.59	46.41
V. JAMMU and KASHMIR								
1. Ladakh	Not Surveyed							
VI. KARNATAKA								
1. Bellary	70.11	10.52	59.59	53.63	23.14	16.20	43.39	27.19
2. Bijapur	114.11	17.60	96.51	87.20	67.43	47.20	49.30	48.91

States/ Districts of Indian Desert	Total Replenish-able GW resources from natural recharge	Provision for domestic, industrial and other uses	Available GW resources for irriation in net terms	Utilizable GW Resources for irrigation in net terms	Gross draft from natural recharge estimation on prorata basis	Net draft from natural recharge	Balance GW resources for future use	Level GW develop-ment (%)
3. Chitra Durga	74.60	11.19	63.41	57.07	37.16	26.01	37.40	41.02
4. Dharwad	103.80	15.57	88.23	79.50	28.33	19.83	68.40	22.48
5. Raichur	107.30	16.30	91.00	81.80	21.46	15.02	75.98	16.51
VII. PUNJAB								
1. Ferojpur	302.38	30.24	272.14	244.93	249.55	174.68	97.46	64.19
2. Faridkot	58.97	5.90	53.07	47.77	37.91	26.54	26.54	50.00
3. Bhatinda	55.78	5.58	50.20	45.18	51.01	35.71	14.49	71.13
4. Nawan Shahr	44.06	4.41	39.65	35.68	54.97	38.48	1.17	97.04
5. Sangrur	151.82	15.18	136.64	122.97	302.96	212.08	-75.44	155.21

Source: CGWB

6. Use of Chemical Fertilizers and Chemical Pesticides

Use of chemical fertilizers and chemical pesticides is only alternative with cultivators to maintain productivity level. These measures are adversely affecting human health, livestock, environment and ecology of area and natural fertility of soil has been seriously affected as increase of doses is necessary to get production from land. Chemical fertilizers and pesticides require more water for survival of plant and this has become unending process. Use of chemical fertilizers and chemical pesticides has been banned in developed countries due to adverse impact on human beings but it is still flourishing in India. In addition to disturbing ecology of area, most of prevalent diseases in human and livestock is the impact of these elements.

7. Deterioration in Ethnic Values

Ethnic values helping environment and ecology are at the verge of extinction as people are trying to put maximum land for cultivation requirements. In recent past people grow trees as per their ethnic values in larger areas which was generally devoted to some religious these. Oran was an example of such ethnic values where people raise trees where no one was permitted to cut live trees. Even entrance in such areas was permitted for collection of dry tree branches for fuel wood requirements. Collection of medicinal produce was also permitted without damaging trees or plants. Oran of 40,000 hectares of area was existing in Barmer in recent past. Now all such lands are at the verge of extinction and being converted into cultivated fields. Orans were shelters of wild life and no one was allowed to shoot wild animals due to vigilance of local residents. Such orans were attractive sites in desert regions and people were helping in maintaining ecological balance with such efforts. Bishnoi community was famous for preservation and protection of such orans.

8. Flora and Fauna

Due to water scarcity, no significant vegetation can survive in desert regions but there are various species grow automatically with occurrence of monsoon and extinct before summer. Similar wild life is rare in desert regions. However there are some pockets where wild life is found are chinkara, desert fox, jackal, black deer, black buck, great Indian bustard, Indian fox, tailor etc. Some efforts have been made to establish national parks and game sanctuaries but human interface is main hindrance in survival of trees and wild life.

7.16 RESULTS AND DISTRIBUTION OF DESERTIFICATION IN INDIA

For past centuries, Indian desert has been a unique human ecosystem. All life forms like human, animal and vegetation have survived by evolving delicate and precarious relationship with fragile ecosystem, which was by way of working out symbiotic relationship between people and environment. most marvelous inheritance of this desert civilization is its great physio-cultural institutions and biotechnological capacities to sustain a variety of life forms. This region as compared to any other world desert has highest biotic concentration of human and livestock population. Over time, unique socio-ecological organism, multi-ethnic, multi-caste hierarchical society interwoven in symbiotic relationship, was born.

In present era, general problem of Indian desert region environment is essentially one of human ecology. Man has been one of most important and active agent of biotic interference in fragile ecosystem. Over-exploitation of natural resources by men has disturbed delicate ecological balance. Overgrazed lands, shrinking forests and eroded agricultural fields show imprints of human activities on its environment. Life of people of this region and their sources of livelihood have chiefly been conditioned by disturbances in ecological balance due to severity of desert climate, which led to degradation of vegetation and diminution of crop potential increasing productivity per unit area. These characteristics of land degradation have certainly human significance. Desert are natural phenomena while desertification or land degradation is human process, which poses potential threat to resources and ultimately lives and livelihood of inhabitants.

Assessment of desert ecosystem requires analyzing problems and prospects for ameliorating their impact by human interface and people's traditional wisdom related to socio-ecological aspects. Assessment of ecological degradation through human interface is necessary in terms of intensity and magnitude of problem. In addition to their susceptibility to natural hazards, desert regions have conditions, which quickly deteriorate through mismanagement of their fragile resources. Environmental degradation arising from overgrazing has been widespread in many desert regions. Felling of trees and forest clearance for agriculture and fuel have contributed significantly to spread of deserts. In such areas human activities for utilization of natural resources need to have equilibrium with carrying capacity of those resources.

Occurrences of desert are attributed to natural phenomena and human activities degrade land through mismanagement of basic factors of land use system, including soil, water and vegetation resources. These human activities are ecologically destructive in nature, over-exploitation of natural resources to satisfy increasing needs of growing human and livestock population. Cumulative impact of these factors coupled with occurrence of drought resulting in land degradation and decreasing vegetative cover is very critical issue as people surviving in desert regions, prefer to understand land degradation and disturbing desert conditions to significant extent. Droughts occurred during 1980s and earlier were most devastating for human and livestock at large scale.

Natural constraints prevail in all parts of world and people have to adjust themselves according to the environment and ecology of the area. This also hold to desert regions of India where people residing for centuries are accustomed with the local conditions and make their lifestyle accordingly. Basic problem with the desert regions of India is the judicious use of available resources for keeping them intact in same form provided by their ancestors. Problems arising due to human interface are quite serious and need retrieval for regeneration of the scarce resources.

The desert regions of India are situated in different parts having changed climatic conditions, ecology and environment. Even conditions vary in different states where desert lands are located. Therefore it has been deemed proper to narrate the problems and consequences prevailing in various states of Indian Union and their intensity and magnitude in terms of degradation of the ecosystem. Since reasons also vary from area to area, these issues are relevant due to geographical locations. Desert in broad terms is an area containing sandy and dunes but other ingredients are rather effective in their role, which makes it different in terms of livelihood, density, life conditions and atmospheric events.

Creation of irrigation facilities has changed geophysical conditions of the desert regions to reasonable extent. Canal irrigation in some parts has yielded success in establishing survival base of people facing severe problems. Desert regions connected with canal system have developed tremendously visible from crop yield and increase in density. Perennial irrigation facilities have proved that desert regions are also productive regions of country and can produce all major crops with substantial yield. Cultivated area has been increased substantially creating ecological problems. Application of chemical fertilizers and pesticides water requirement has been increased in canal regions which has helped in increasing of ground water and creating water logging problems. Chemical fertilizers are hazardous for human health and environment and for maintaining the productivity level, increase in chemical fertilizers has become necessity of cultivators resulting in extinction of natural fertility of soil. This cycle has created serious problem and there seems no end as cultivators use these inputs without bothering consequential impact. However, canal irrigation facilities have their limitations and even after significant efforts, it is not possible to provide surface irrigation facilities to all the desert regions.

Water logging is resultant impact of canal irrigation system and most intensive in upper and middle reaches of canal, where water is available in abundance while tail end cultivators suffer from non-availability of water. Water logging is caused due to increase of water level of ground and there comes a situation when capacity of earth is unable to receive further water. Such situation arises after 10-20 years depending on absorption capability of soil. Cultivators bother for increase in productivity by using more and more water, which create water logging and decrease in productivity start witnessing its impact. In long run the cultivated fields are become waterlogged or marshy lands and cultivation practices are stopped completely.

Water deficient areas have been provided with tube-well and well irrigation through investigation of groundwater availability, which has helped in creating irrigation facilities as per availability of groundwater availability and its recharge capacity. With increase of cultivated area and addition of irrigation facilities, ground water depletion has reached to great extent. Ground water depletion has reached in most part of desert regions at most critical condition and most of ground water irrigated zones have categorized as critical sensitive zones where further depletion would abandon cultivation practices and most part of such areas are facing salinity and alkalinity problem. Thus human efforts have deteriorated desert ecosystem to most severe extent, which is common problem of ground water irrigated areas.

Desert regions mainly develop due to erratic monsoon, deforestation and human interface with ecology of region. Natural factors are beyond capability of management but best attempts can be made for getting forecast to maximum accuracy, which is still lacking in India and land based activities suffer most. Cultivation practices are backbone of Indian economy and crop failure creates problem to economy of people and country as total economy is primarily dependent of efficient cultivation practices. Occurrence of drought are common feature worldwide but its severity is highest in desert regions as most of cultivation practices are single cropped and survival base of poor, who are unable to manage irrigation facilities from ground water. In water deficient areas, big cultivators are able to manage monsoon failure by irrigating crops for minimizing losses.

7.17 COMBATING MEASURES

Combating measures for desertification have remained focus of the development process. Government of India introduced Desert Development Scheme for hyper arid, arid and semi arid regions, where focus remained on sand dune stabilization and watershed practices. These efforts were mainly for non-arable land. Though forest areas were clearly identified under reserve forests, protected forests and unclassified forests and use of forest land for other activities was completely banned and change of its status was possible with the regulatory measures, where equal area need to be earmarked for forestry activities. Even with strong regulatory measures, encroachment in forest lands could not be controlled and uncontrolled grazing continued in reserve and protected forest. Presently most of the forest areas are mostly denuded with human interface. Uncontrolled felling of trees lost the moisture balance on earth surface and dryness prevailed in total atmosphere of the dryland regions. No integrated measures for combating desertification could be taken on the analogy that cultivators land is private property and no measures for treatment of such areas could be deemed proper.

Desertification is serious integrated land degradation problem, where climate, temperature, environment and ecology have inter-related effects and problem persists in total area. Therefore, combating desertification requires integrated efforts and that can be helpful in treatment of all land degradation problems in keeping total land into consideration.

7.18 CONCLUSIONS AND SUGGESTIONS

Environment of earth surface. All the problems are not solely for human acts as natural factors also play destructive role particularly in desert regions, but cursing all the situations is not the solution of problems. Keeping aside natural factors, there is immediate necessity to take suitable steps for changing situation of desert regions. So far as natural factors are concerned, those are also resultant impact of disturbances in ecosystem and when corrective measures within control of man are taken, natural factors will also change in due course of time or at least severity of disastrous impacts would be minimized to great extent. With this strategy Desert ecosystem of India is completely disturbed as all the factors responsible for development and maintenance of ecosystem of desert regions are managed in very uncontrolled manner. This condition is not only for desert regions of country but even other regions are facing similar problems with variation in intensity and magnitude. Vegetation over earth surface has been suggested at least thirty percent of total land area which includes forests, pastures, tree groves etc. which are helpful in attracting monsoon, decreasing dryness in environment and water evaporation, increase in humidity and availability of water for survival of human, livestock, plants, animals and, some suggestions are given for consideration of country government, political institutions, policy makers, geographers, administrators and man whose existence is at stake due to serious environmental and ecological conditions. Further corrective measures need to be practical, cost effective, acceptable to people of area and keeping in view human requirements. It is certainly true that uncontrolled activities or destructive acts need to be checked effectively to derive desired results. Forests are cleared for cultivation requirement but that is not proper solution of the problem. Similarly cutting live trees is not solution of fuel wood, fodder and industrial requirements. Therefore for sustainable development and maintenance of ecological balance, all due requirements need to be kept in view as development and destruction cannot act together. Political support for destructive acts is pre-requisite for success of these measures. Therefore, awareness generation is process for curbing the human acts and this is essential tool to practice the suggestions. Peoples participation in total decision making process is essential as there is need to develop ownership feeling in people who will be ultimate beneficiary through proposed measures. Unemployment problem is also grave specially in rural areas and peoples involvement must provide them substantial earning from created assets. With this background following suggestions are made for revival of desert ecosystem of India.

1. Management of Land Resources

Present status of land use under various categories reveals that 7.81 percent area has been left under forests, 30.61 percent under cultivation, 18.96 percent fallow lands and 9.71 percent area is put to other uses. For revival of ecosystem, it is necessary that at least 30 percent area need to be earmarked under forests at all levels. This process can be done by using present forest lands, barren lands, pastures and half of wastelands. Similarly cultivated land, fallow land and half of waste land need to be put under cultivation for meeting all human requirements. For this purpose, treatment of all above category lands is necessary by meeting all deficiencies. Total 30 percent forest land need to be classified into reserve, protected and unclassified forests. Areas put under reserve and protected forest category need to be fensed for growth of trees, shrubs and wild life. These lands can be managed by forest department or by any agency as per suitability for result oriented task. Unclassified lands need to be treated and used by willing local people for horticulture, fuel wood, fodder and industrial needs. People associated with task need to be allowed to take benefit of produce but without harming trees. If one hectare area is allotted for each willing person for sharing benefits of produce and asked to treat land, that would be managed without any problem and five million people can be put on direct employment through this process.

Added area under cultivation will be helpful for meeting foodgrain, fodder and fuelwood requirement of people but uncontrolled grazing and cutting live trees need to be checked completely. This arrangement will provide ownership to all land put to productive use. Safety of government and community land is problem and ownership of land even for getting benefit of produce would curb similar destruction completely.

One serious problem of land is diminishing natural fertility of soil due to excessive use of chemical fertilizers and pesticides, which is harmful to people, livestock, wild life and environment. For maintaining crop production level, use of chemicals is increased which has completely lost natural fertility of soil and also harmful to great extent. Half of human diseases, danger to wild life, environment degradation are serious issues which can be checked completely through shifting to biofertilizers. With maintenance of natural fertility of soil, ten percent production would increase from same area of land and also save 15 to 30 percent water for irrigation. Biofertilizers and persticides can be prepared by cultivators with local resources and help in decreasing cost of cultivation.

With proper land management, most of problems of ecosystem would be solved as people would welcome this approach. Preparation of biofertilizers would generate employment in rural sector and people would start understanding value of land and its treatment. Land treatment is to be done through demonstration approach on few lands and people would get their land treated in their own interest. Demonstration of land treatment is necessary for managing every category of deficiency as people need to understand process. Treatment of alkalinity and salinity need different modes and every land differ in deficiency level and suggestion for every category of land is to be given at village level.

2. Management of Water Resources

Water is most needed factor for all human requirements but its judicious use is necessary. Water management for availability and use is necessary which is not a problematic issue as people of desert region are well aware of such practices which are stopped with availability of drinking water facilities and easy access to water for irrigation. For availability of water, collection of rainwater and use of ground water need to be managed in logical manner and ground water need to be kept as reserve source to be used in bare necessity. For agriculture and plantation activities more water would not be required but total requirement can be managed with same quantum of water through water saving devices. Rain water harvesting can solve water problems to significant extent. Wimilarly water saving through use of biofertilizers and biopesticides can solve ground water depletion to significant extent. Water saving devices need to be demonstrated to users of various categories and change in habit of people can work in effective vegetation. If people are made aware to match availability and use of water for all requirements, water management is possible even in most critical areas as quantum of wastage of water in desert regions is also quite high. Use of water for human requirement in desert regions need to be checked for vegetation cover.

3. Human Resource Management

Human and livestock population is increasing very fast and causing serious vegetation loss which is necessary to be checked. For meeting human food requirement, productivity of land need to be increased and simultaneously population control measures need to be effective where suggestive and control measures are essential. Desert regions are not in position to bear further burden of man and livestock at the cost of deteriorating ecosystem. With proper land and water management at least 10 million people can be provided employment on sustainable manner. For achieving higher gross domestic product, contribution of land based activities is essential as this is necessary for survival and other productive activities where land produce is used as input.

4. Controlling Human Interface

For reversal of ecosystem of desert regions, control over human acts is essential. Present level of degradation of ecosystem is mostly due to destructive acts and for sustainable ecosystem; human acts need to be made pro-environment. If land and water related suggestions are implemented, most of human interface issues will be curbed substantially. Political support to destructive acts and large scale deforestation can only be checked with awareness generation in people who will be main sufferer of such degradation.

Though ecosystem has been deteriorated to dangerous level, but it is not impossible to change this situation. It is essential to initiate effective steps, as every one is aware of problem and future consequences but trying to take lead by others.

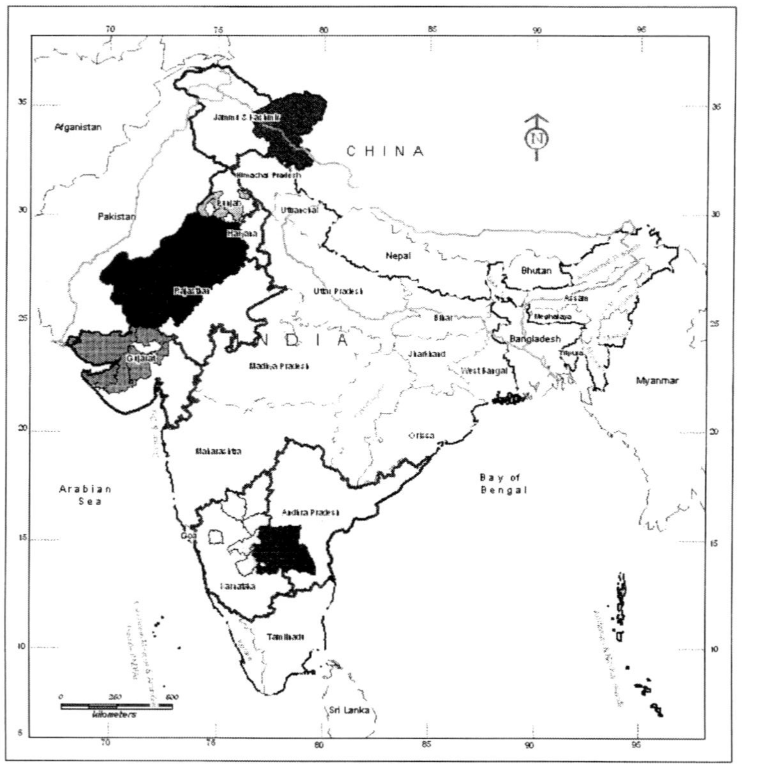

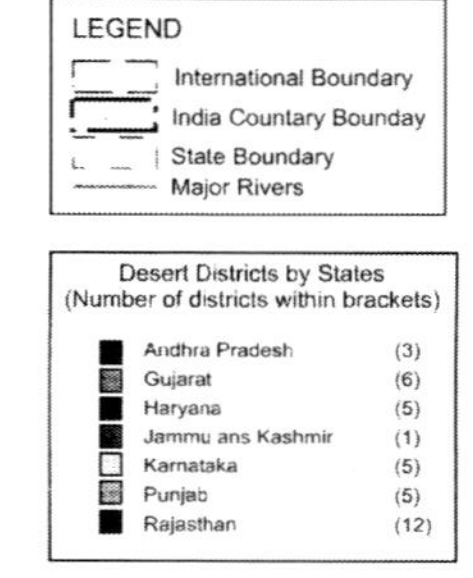

Note : The external boundary and the coastlines of India appearing on the map are neither verified nor authenticated by the Survey of India.

Figure 7.29 : Desert districts by states

Source : IMD, Pune

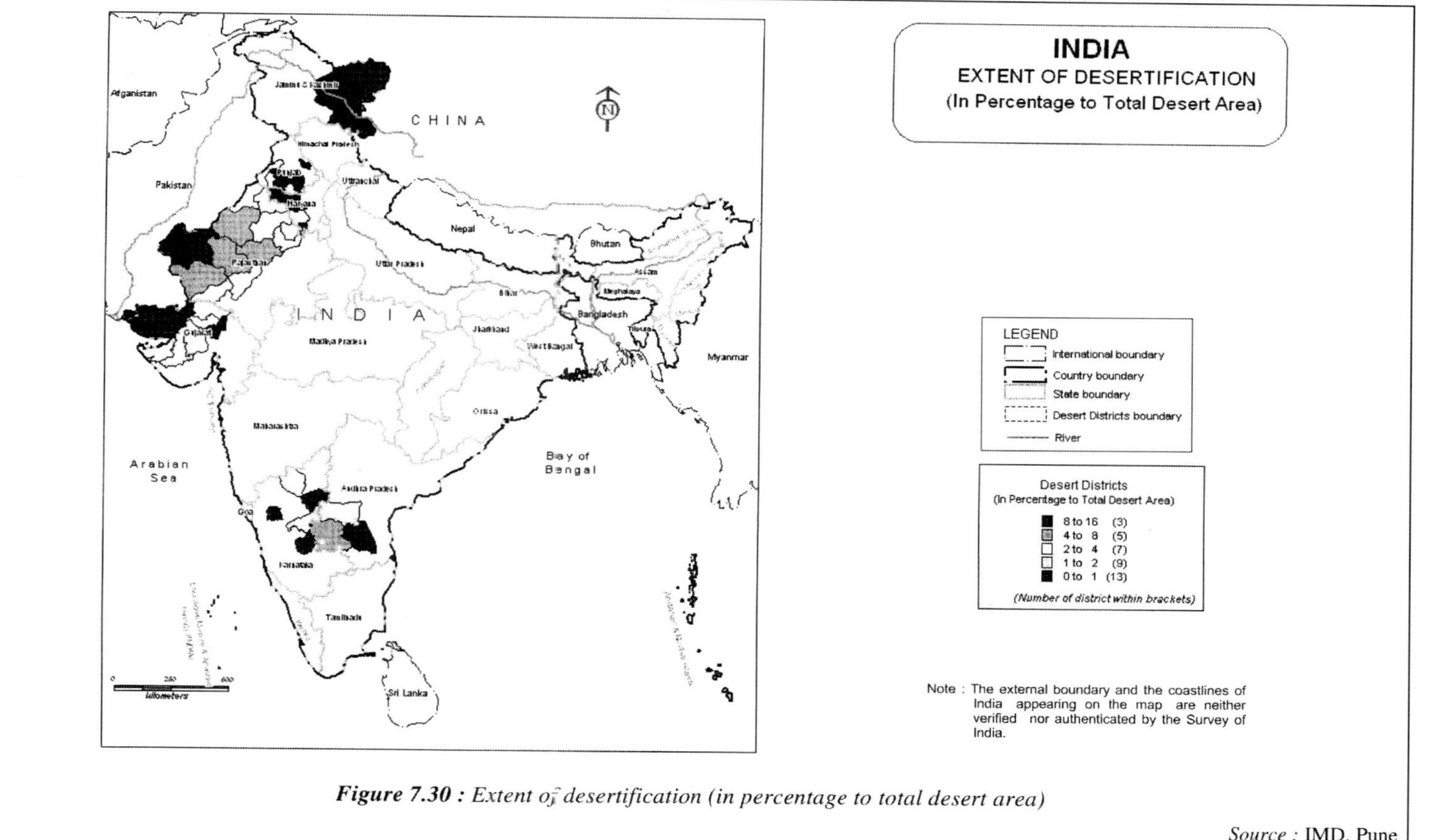

Figure 7.30 : *Extent of desertification (in percentage to total desert area)*

Source : IMD, Pune

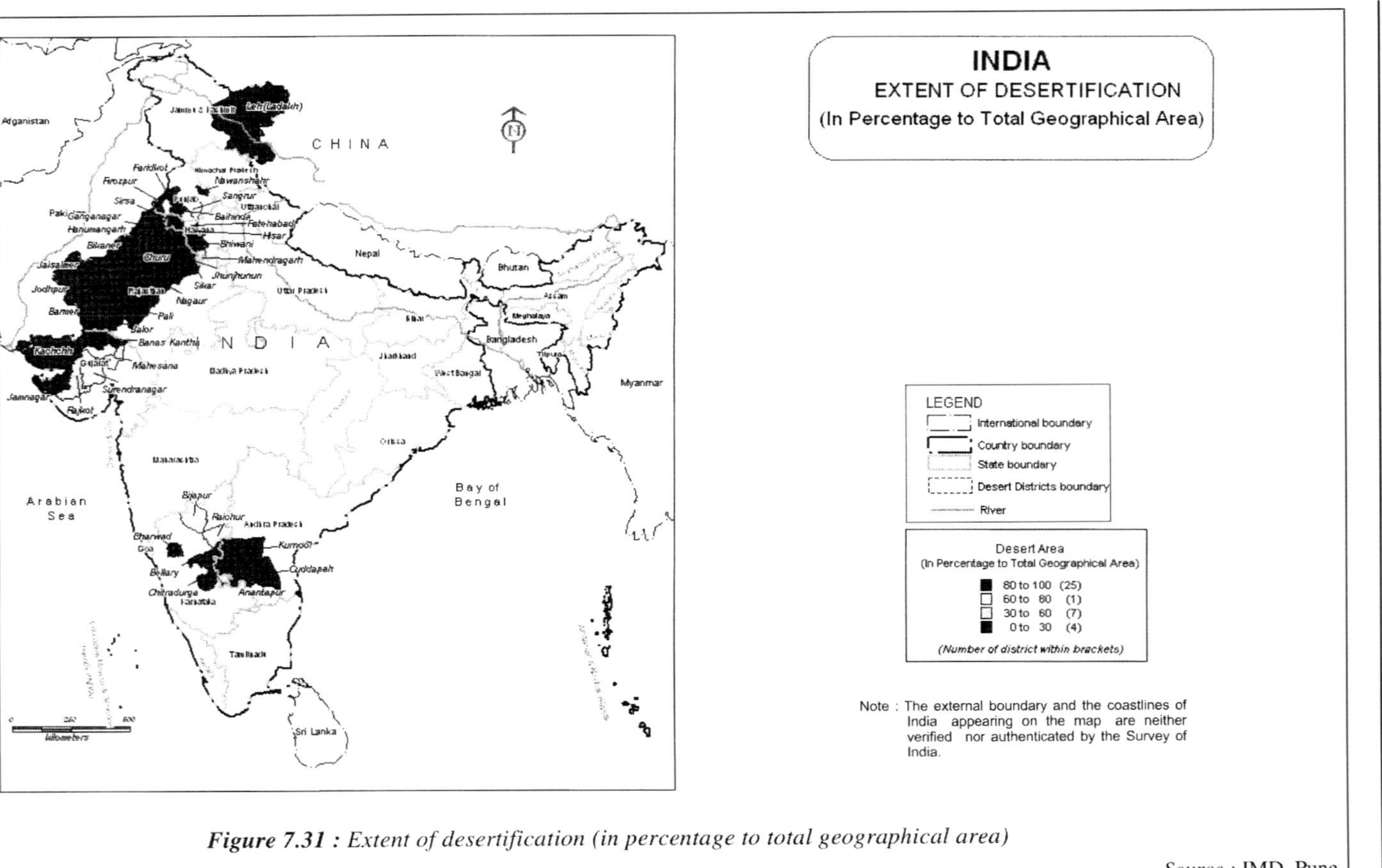

Figure 7.31 : Extent of desertification (in percentage to total geographical area)

Source : IMD, Pune

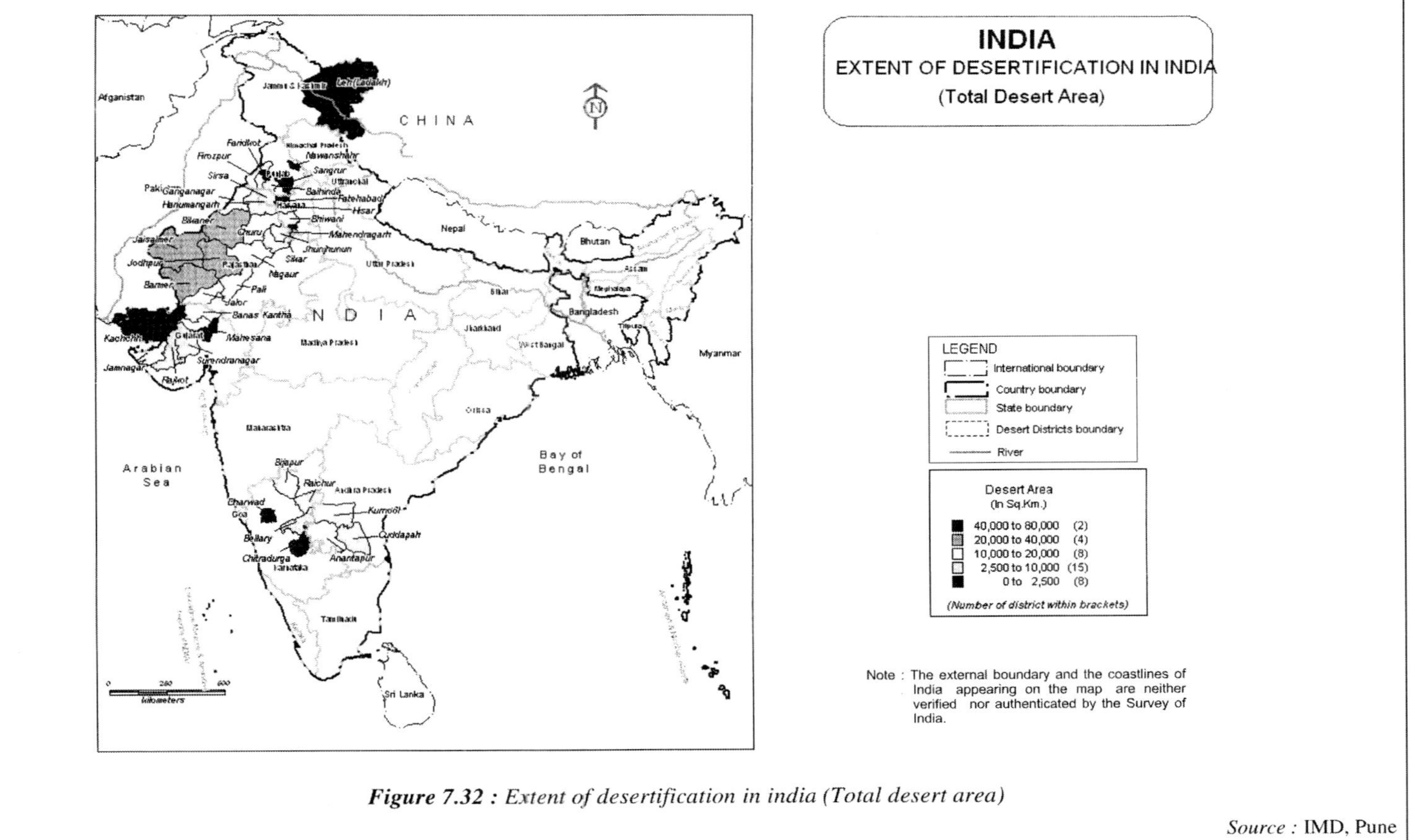

***Figure 7.32 :** Extent of desertification in india (Total desert area)*

Source : IMD, Pune

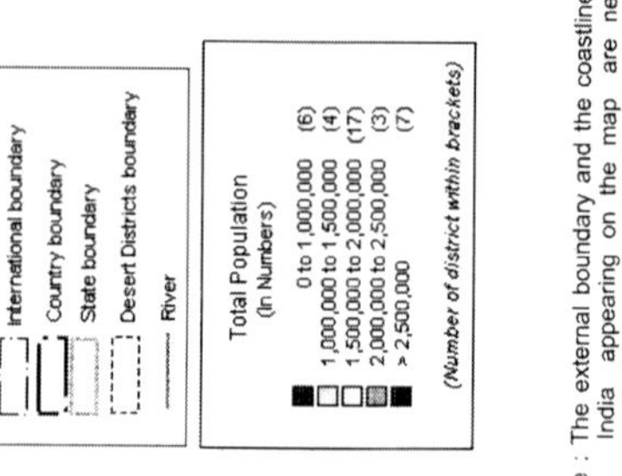

Note : The external boundary and the coastlines of India appearing on the map are neither verified nor authenticated by the Survey of India.

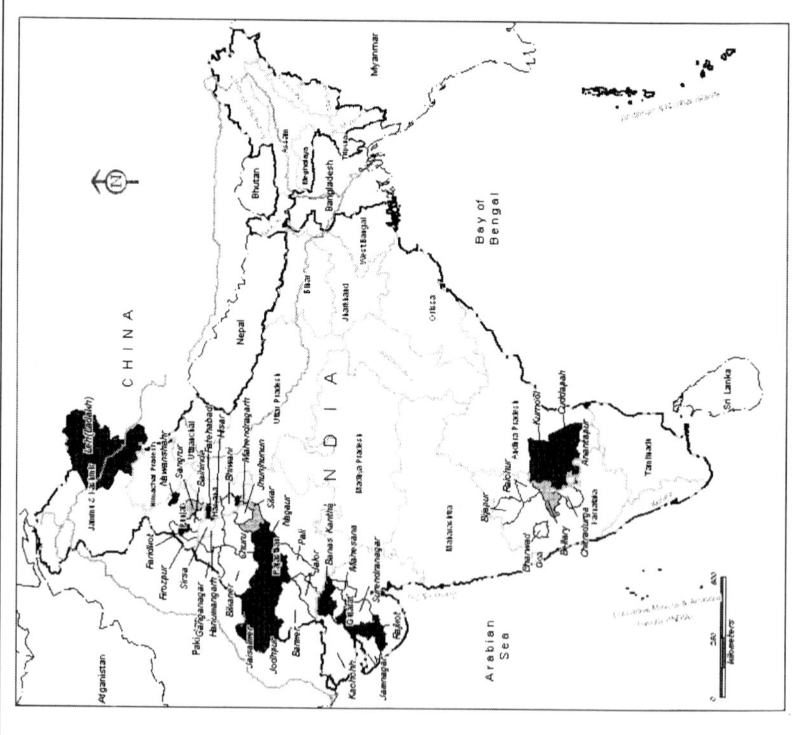

Figure 7.33 : INDIA Total population

Source: Census of India-2001

INDIA

RURAL POPULATION

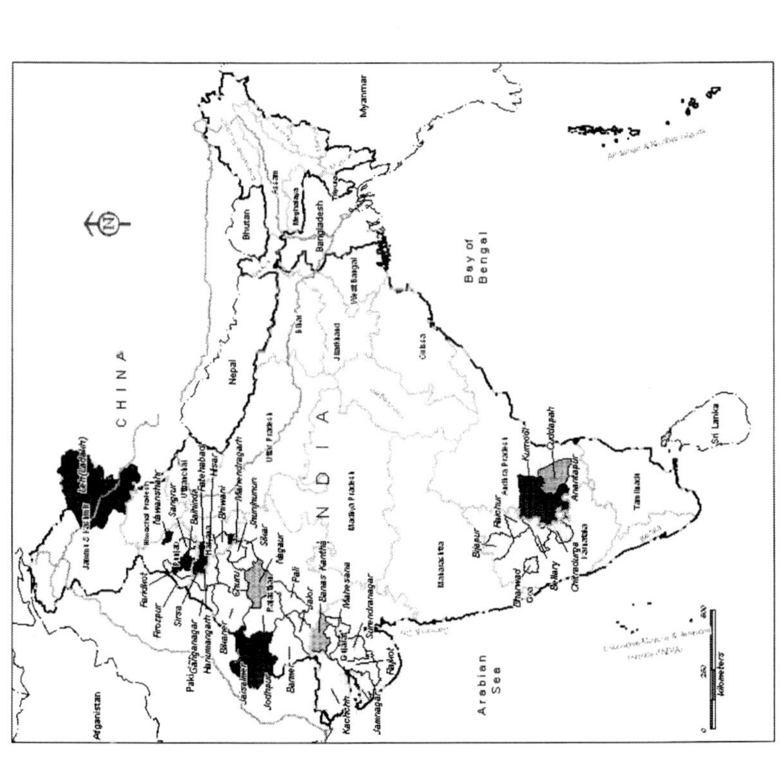

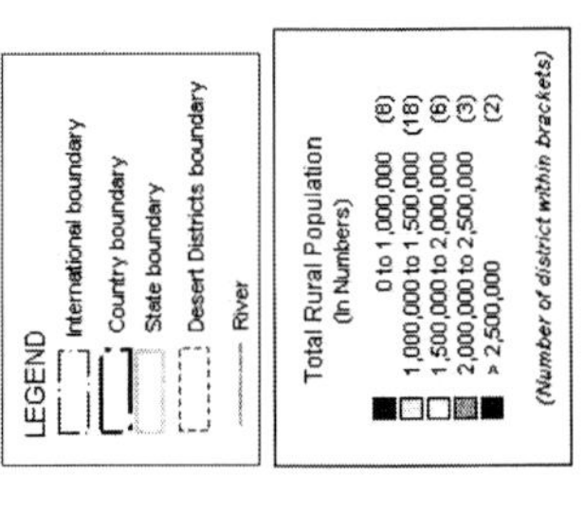

Note : The external boundary and the coastlines of India appearing on the map are neither verified nor authenticated by the Survey of India.

Figure 7.34 : INDIA (Rural population)

Source : Census of India-2001

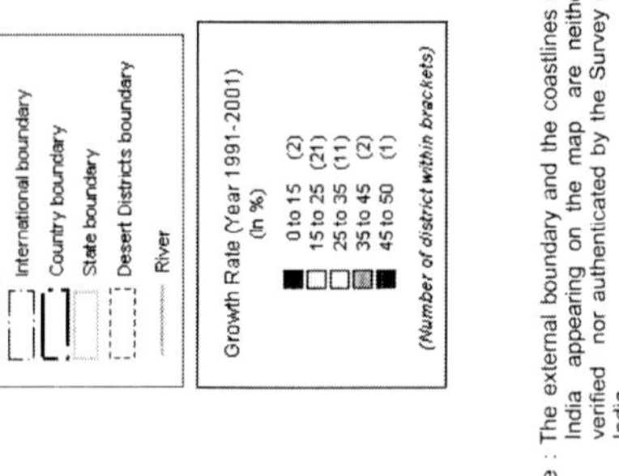

Note : The external boundary and the coastlines of India appearing on the map are neither verified nor authenticated by the Survey of India.

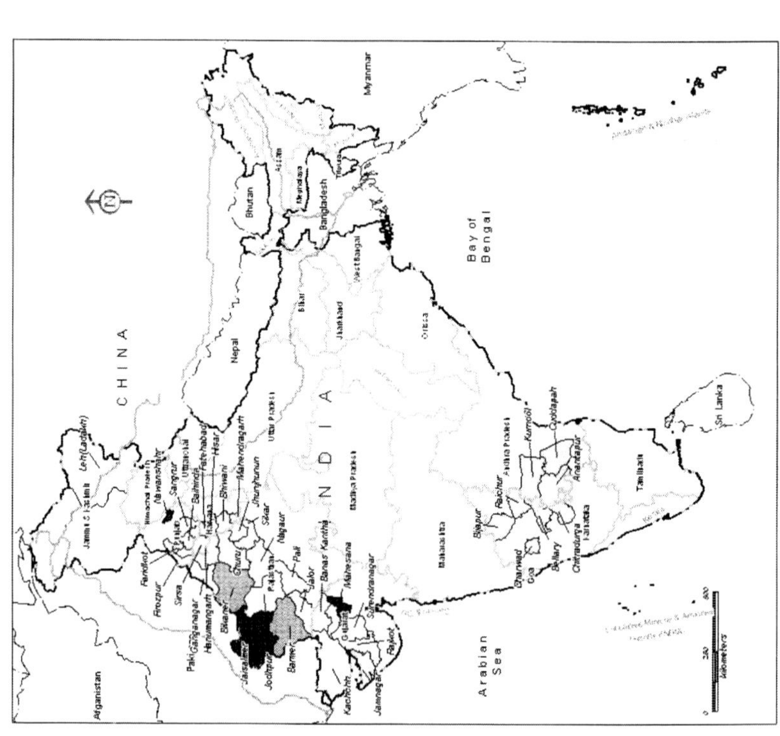

Figure 7.35 : Growth rate of population

Source : Census of India-2001

INDIA
POPULATION DENSITY

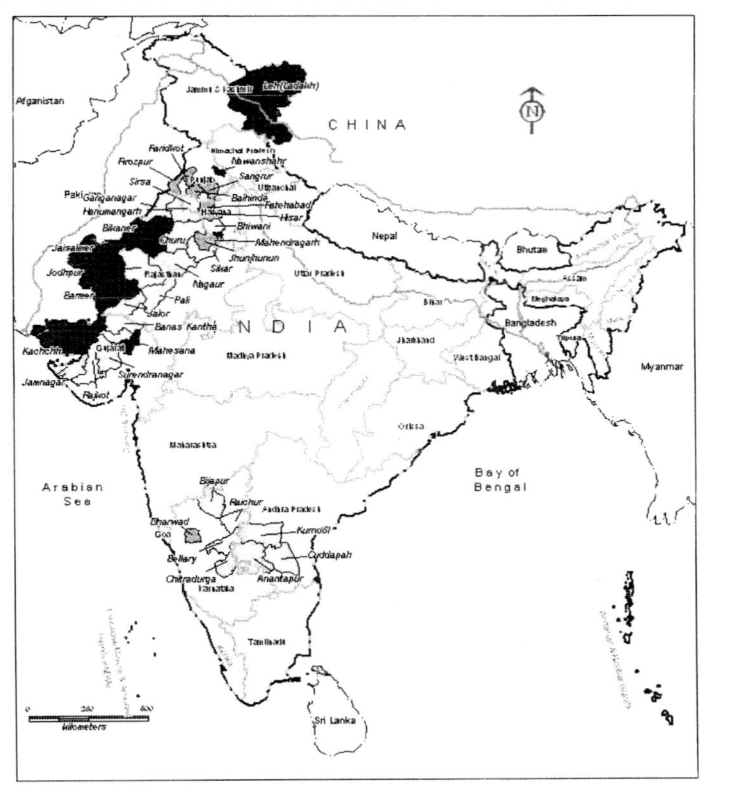

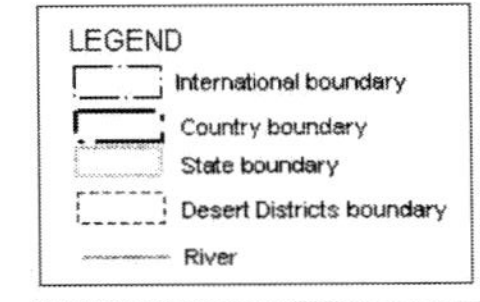

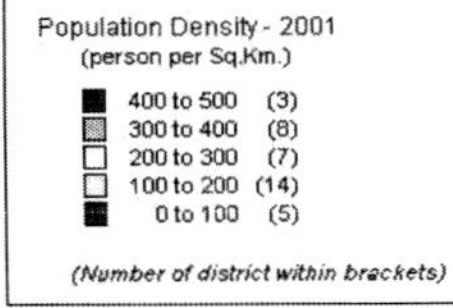

Note : The external boundary and the coastlines of India appearing on the map are neither verified nor authenticated by the Survey of India.

***Figure 7.36 :** Population density*

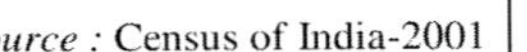

Source : Census of India-2001

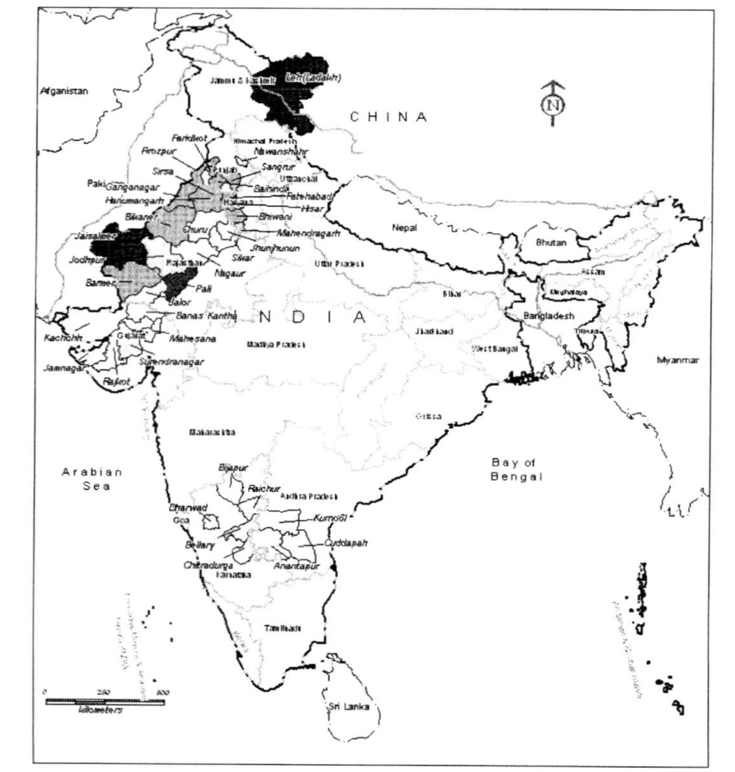

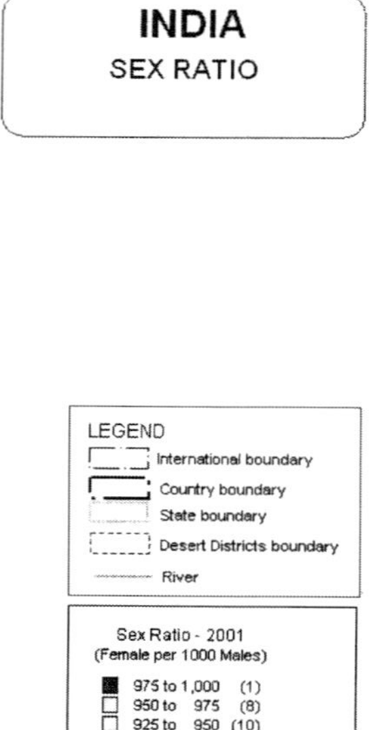

Note : The external boundary and the coastlines of India appearing on the map are neither verified nor authenticated by the Survey of India.

Figure 7.37 : *INDIA Sex ratio*

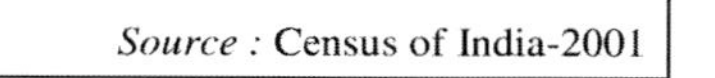
Source : Census of India-2001

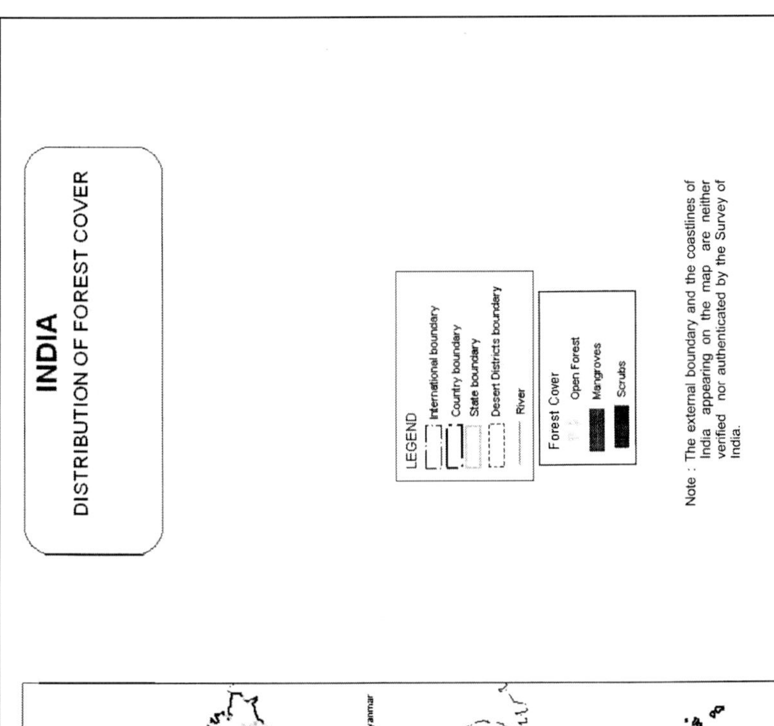

Source : MOEF

Figure 7.38 : INDIA Distribution of forest cover

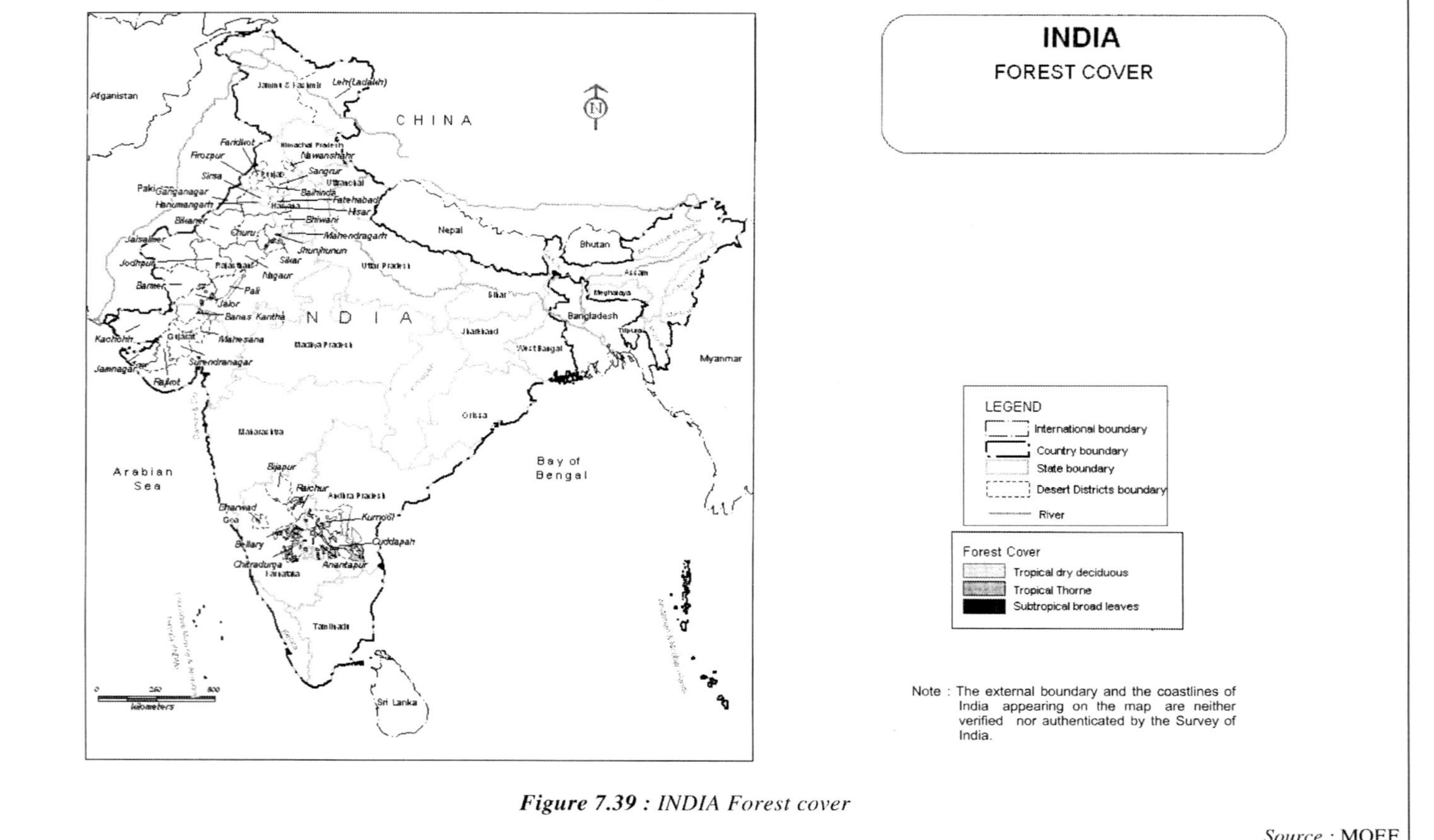

Figure 7.39 : INDIA Forest cover

Source : MOEF

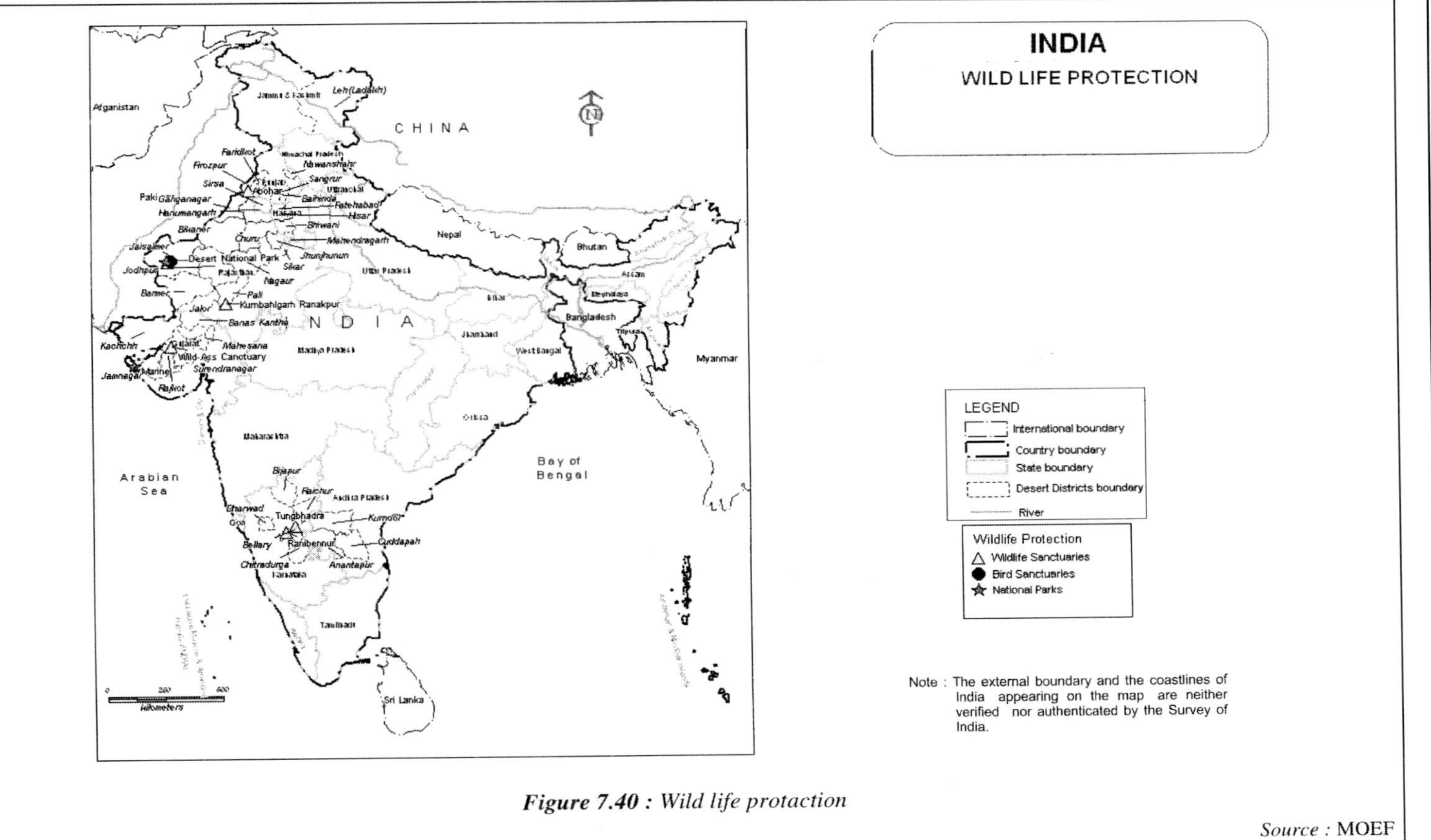

Figure 7.40 : *Wild life protaction*

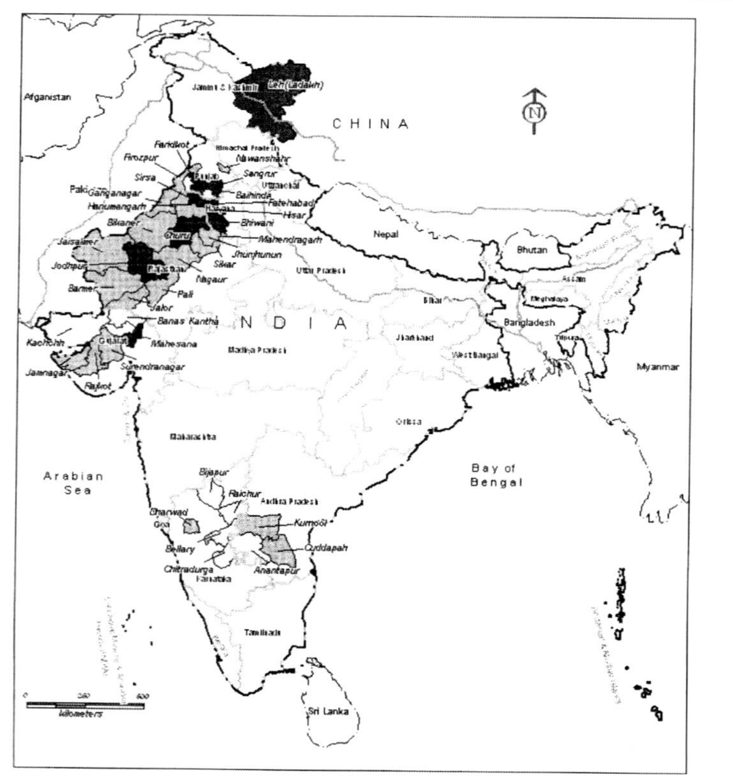

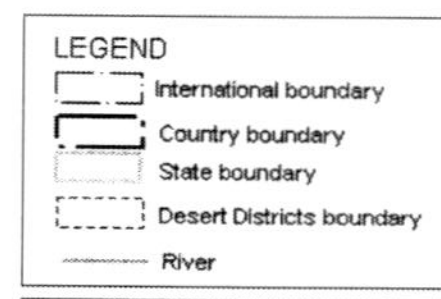

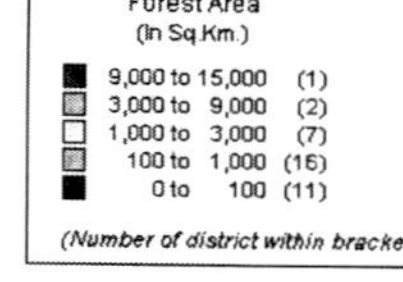

Note : The external boundary and the coastlines of India appearing on the map are neither verified nor authenticated by the Survey of India.

Figure 7.41 : *INDIA forest area*

Source : Satellite Imagery

INDIA
NON-AGRICULTURAL LAND USE

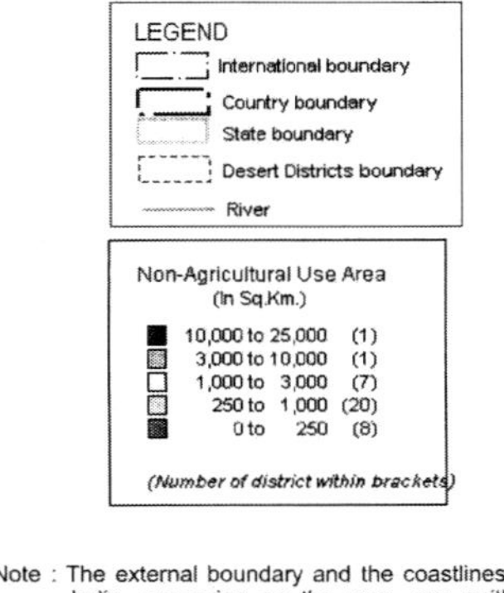

Figure 7.42 : *INDIA non-agriculture land use*

Source : Satellite Imagery

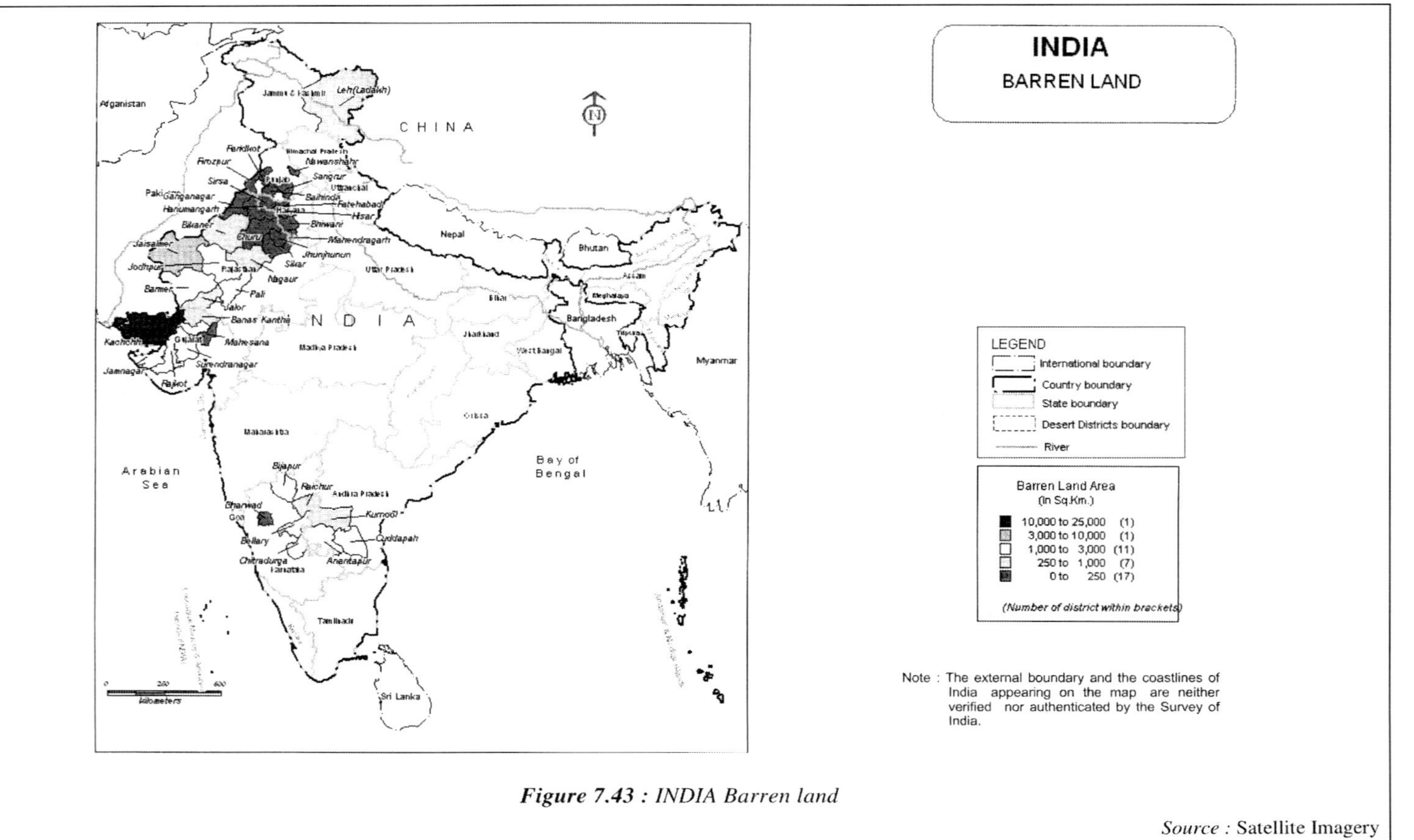

Figure 7.43 : INDIA Barren land

Source : Satellite Imagery

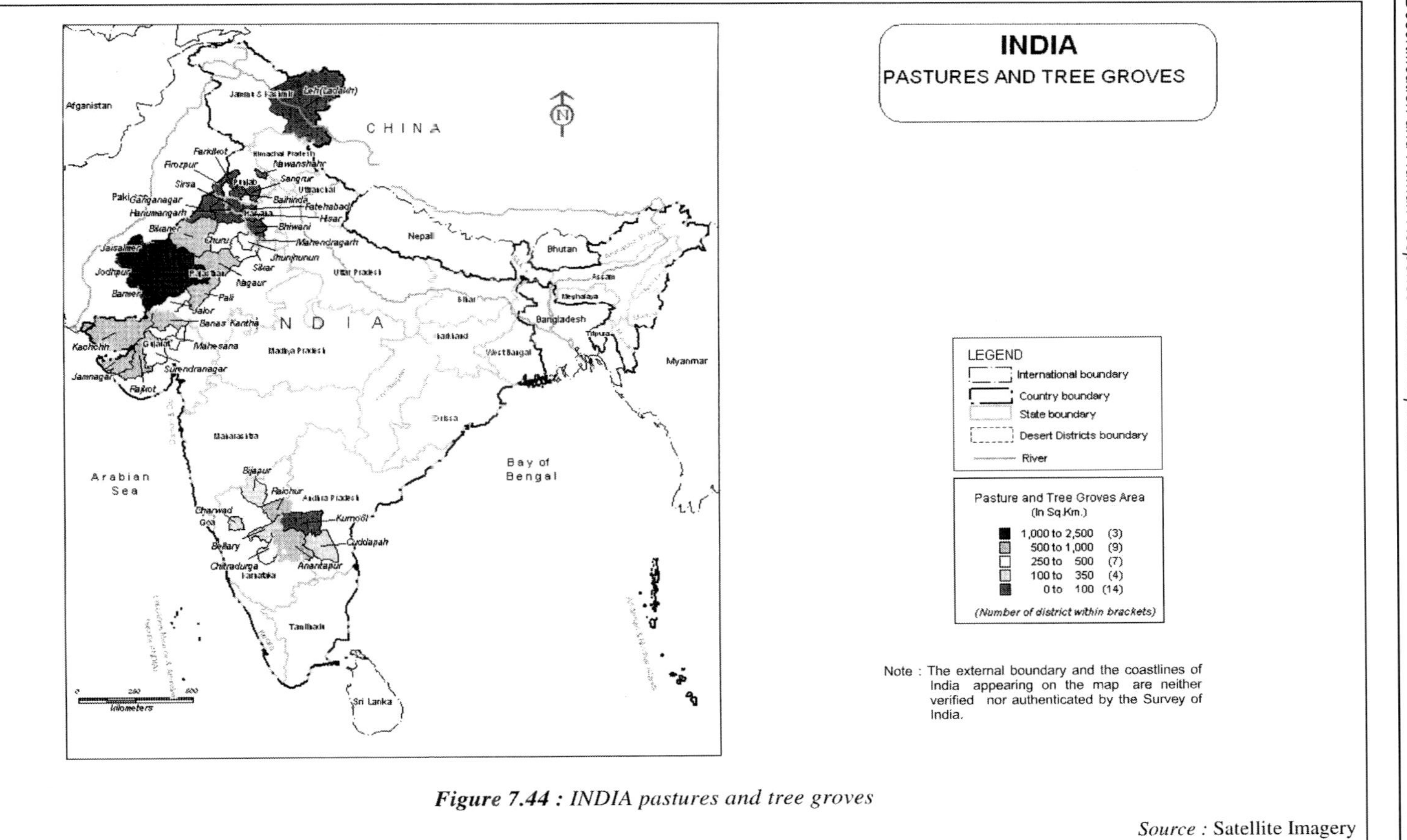

Figure 7.44 : INDIA pastures and tree groves

Source : Satellite Imagery

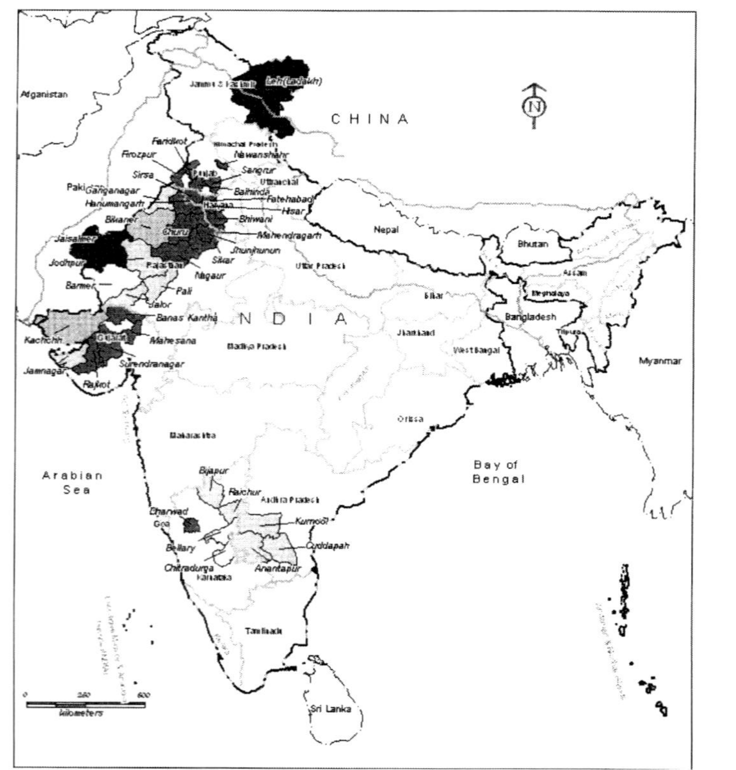

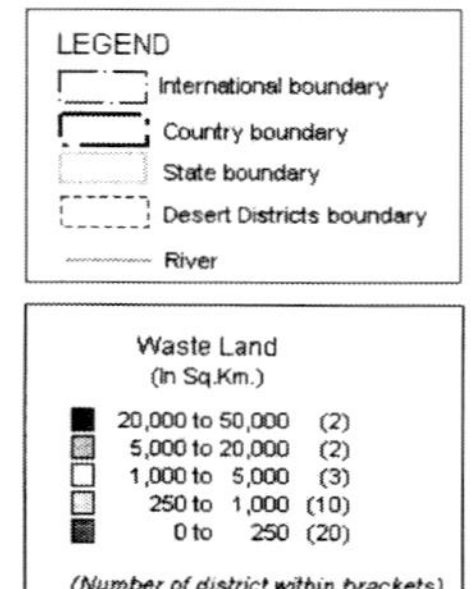

Note : The external boundary and the coastlines of India appearing on the map are neither verified nor authenticated by the Survey of India.

Figure 7.45 : *INDIA waste land*

Source : Satellite Imagery

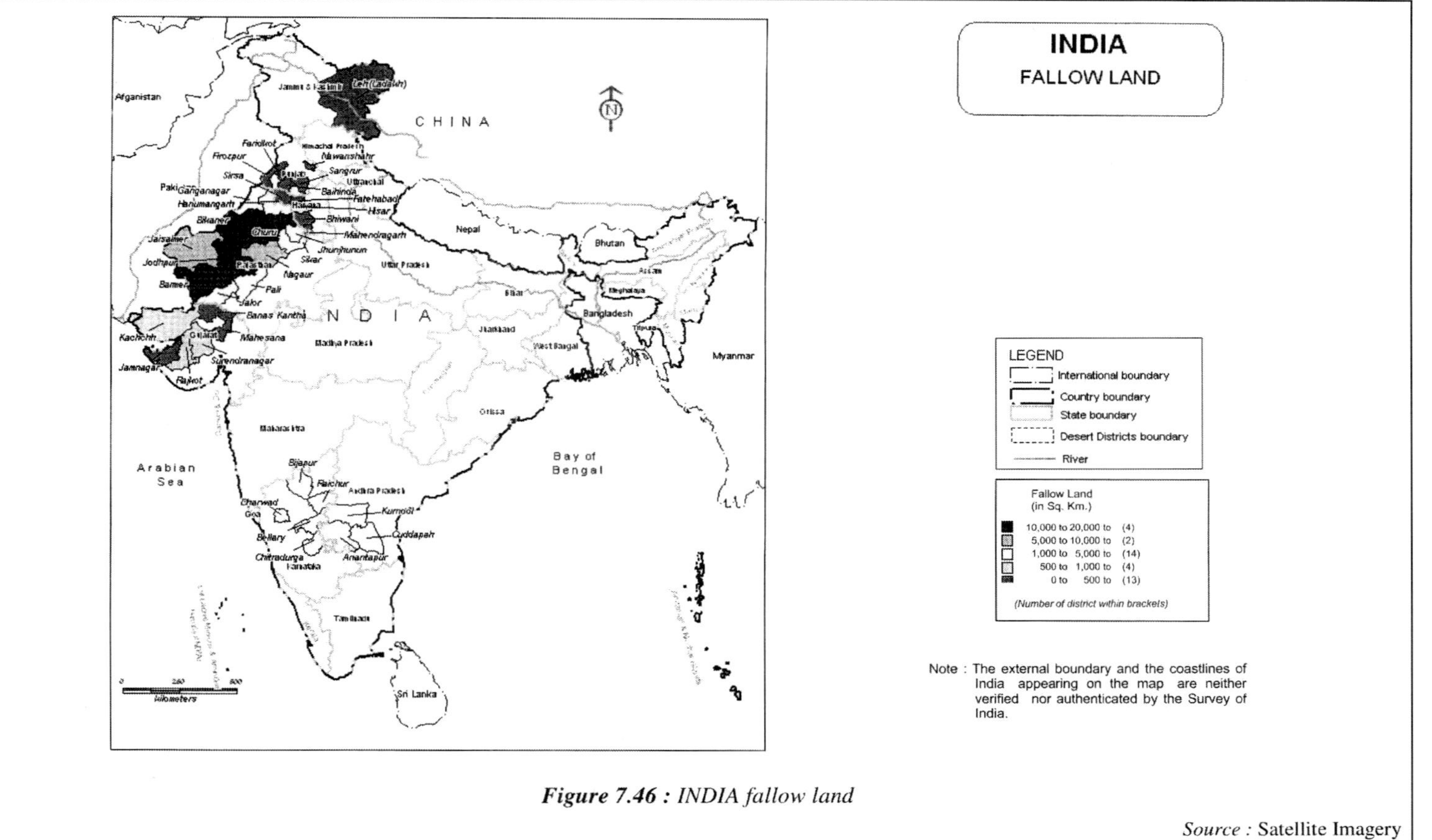

Figure 7.46 : INDIA fallow land

Source : Satellite Imagery

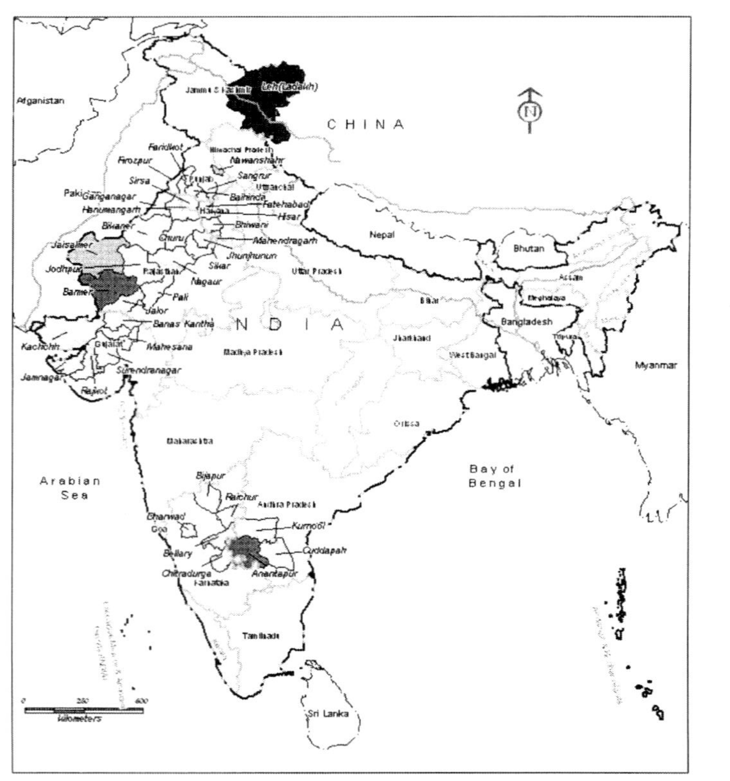

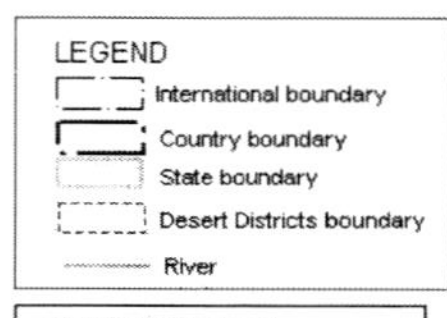

Note : The external boundary and the coastlines of India appearing on the map are neither verified nor authenticated by the Survey of India.

Figure 7.47 : *INDIA net cultivable area*

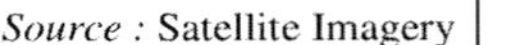

Source : Satellite Imagery

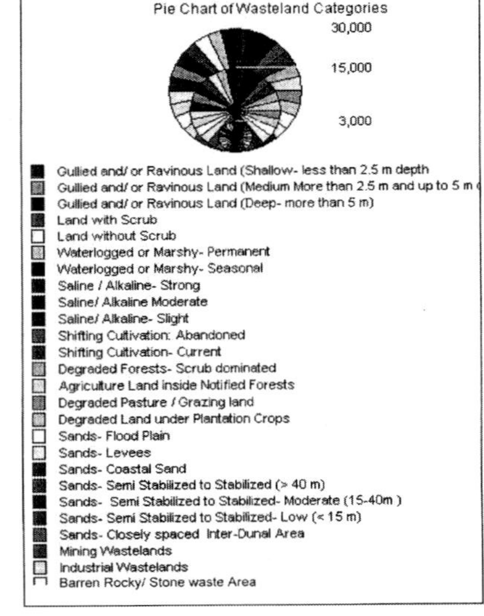

Figure 7.48 : INDIA waste-lands in dryland states, combined psitions

Source : Satellite Imagery

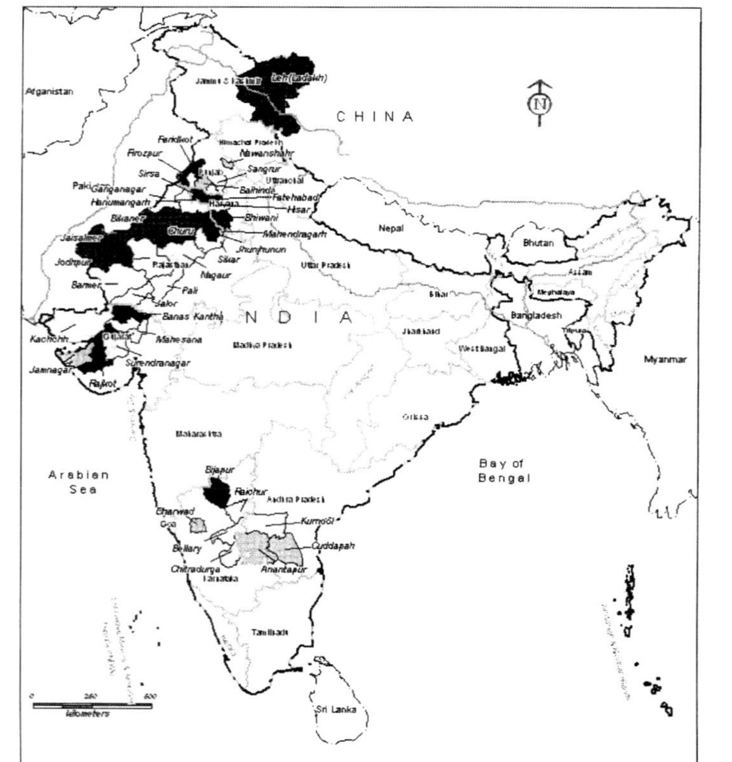

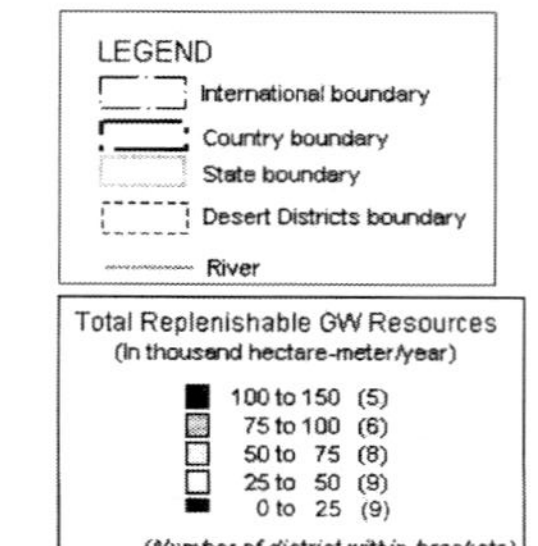

Note : The external boundary and the coastlines of India appearing on the map are neither verified nor authenticated by the Survey of India.

***Figure 7.49 :** INDIA ground-water resources and irrigation potential, Total replenishable GW resources*

Source : CGWB

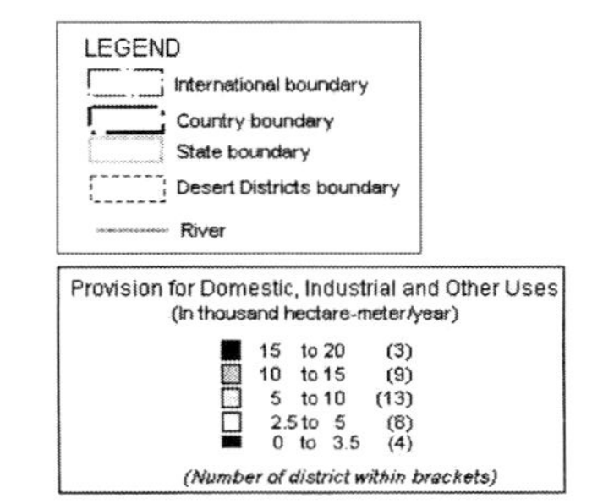

Note : The external boundary and the coastlines of India appearing on the map are neither verified nor authenticated by the Survey of India.

Figure 7.50 : *INDIA ground-water resources and irrigation potential, Domestic industrial and other uses*

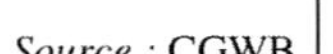

Source : CGWB

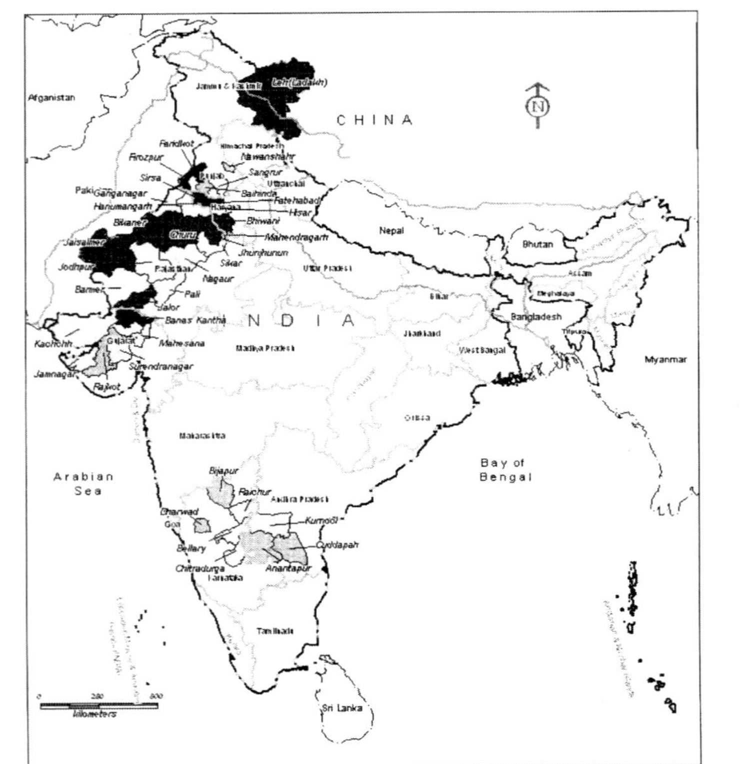

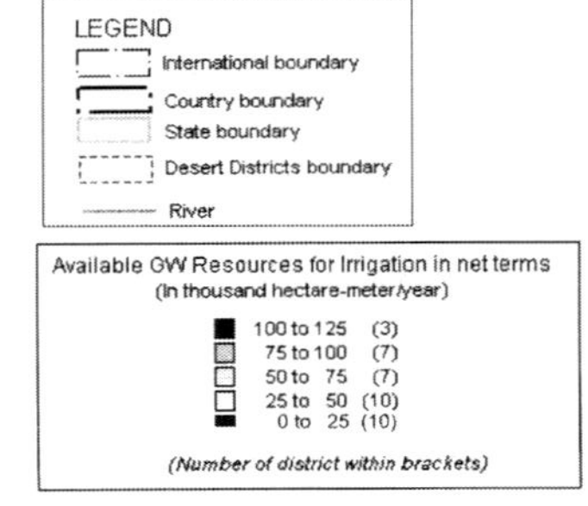

Note : The external boundary and the coastlines of India appearing on the map are neither verified nor authenticated by the Survey of India.

Figure 7.51 : *INDIA ground-water resources and irrigation potential, Available GW Resources for irrigation*

Source : CGWB

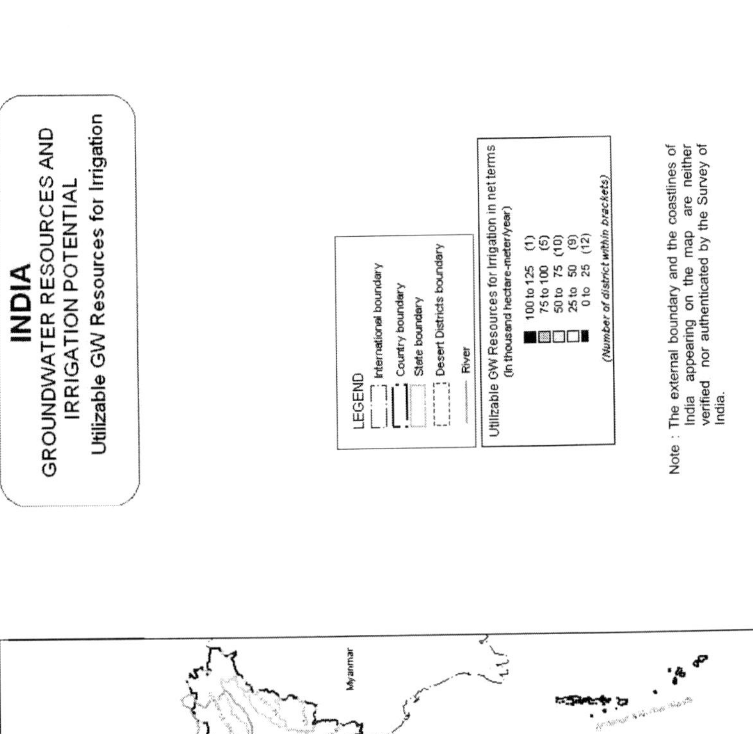

Source : CGWB

Figure 7.52 : *INDIA ground-water resources and irrigation potential, utilizable Gw resources for irrigation*

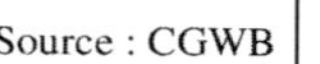

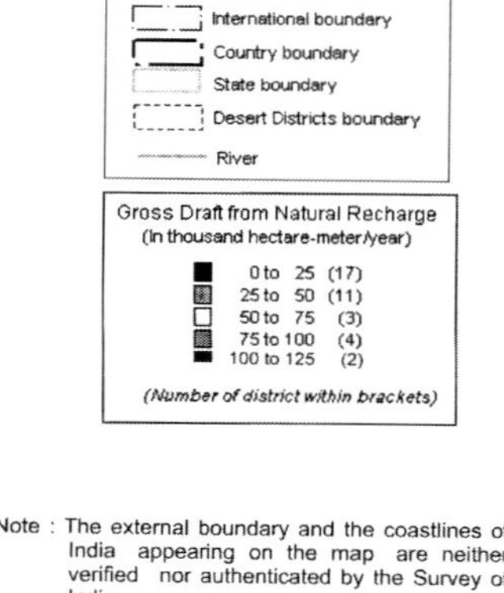

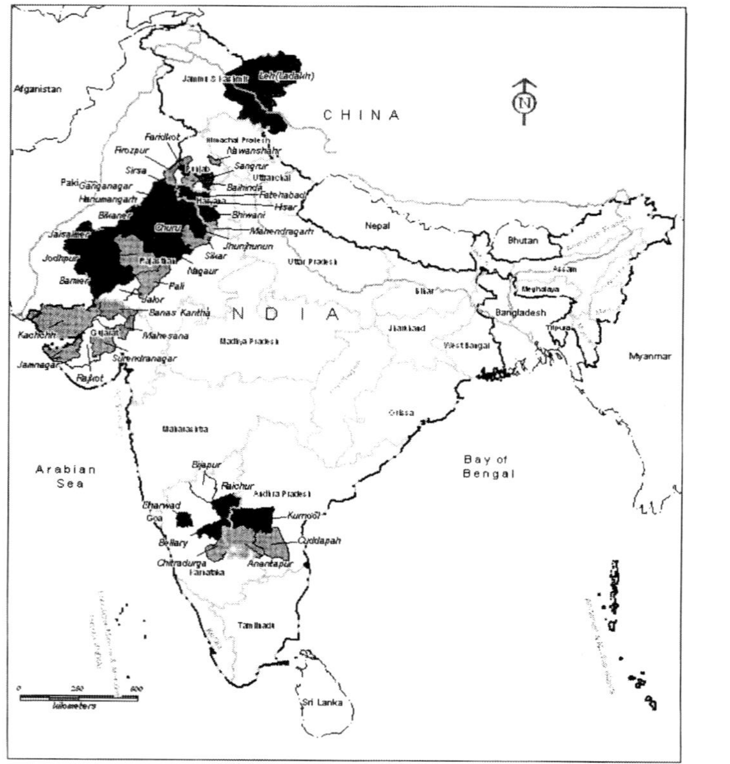

Figure 7.53 : INDIA ground-water resources and irrigation potential, Gross draft from natural recharge

Source : CGWB

INDIA
GROUNDWATER RESOURCES AND IRRIGATION POTENTIAL
Net Draft from Natural Recharge

LEGEND
International boundary
Country boundary
State boundary
Desert Districts boundary
River

Net Draft from Natural Recharge
(In thousand hectare-meter/year)

0 to 10	(12)
10 to 25	(13)
25 to 50	(6)
50 to 75	(4)
75 to 100	(2)

(Number of district within brackets)

Note : The external boundary and the coastlines of India appearing on the map are neither verified nor authenticated by the Survey of India.

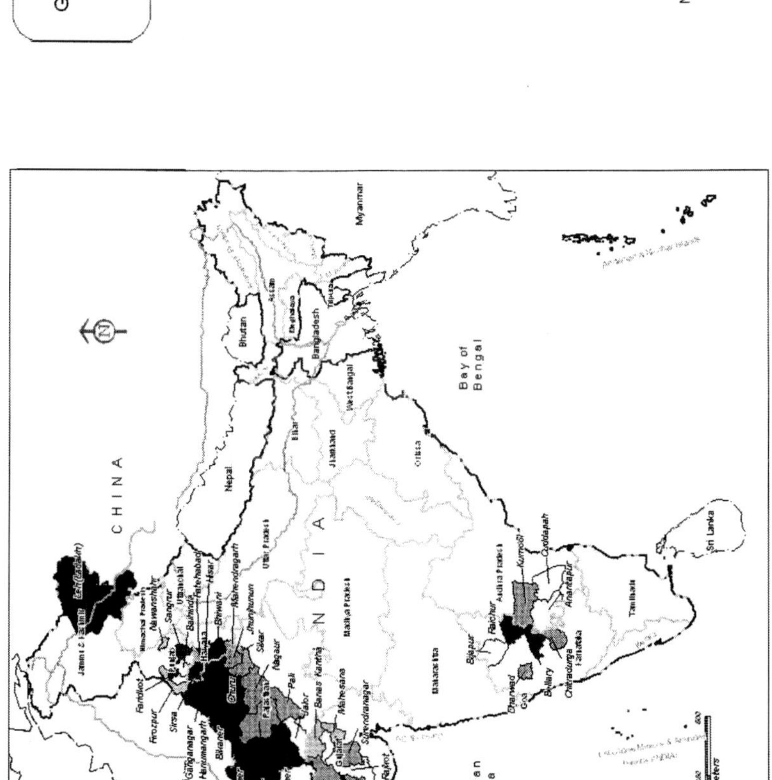

Source : CGWB

***Figure 7.54 :** INDIA ground-water resources and irrigation potential, Net draft from natural recharge*

INDIA
GROUNDWATER RESOURCES AND IRRIGATION POTENTIAL
Balance Groundwater Resources

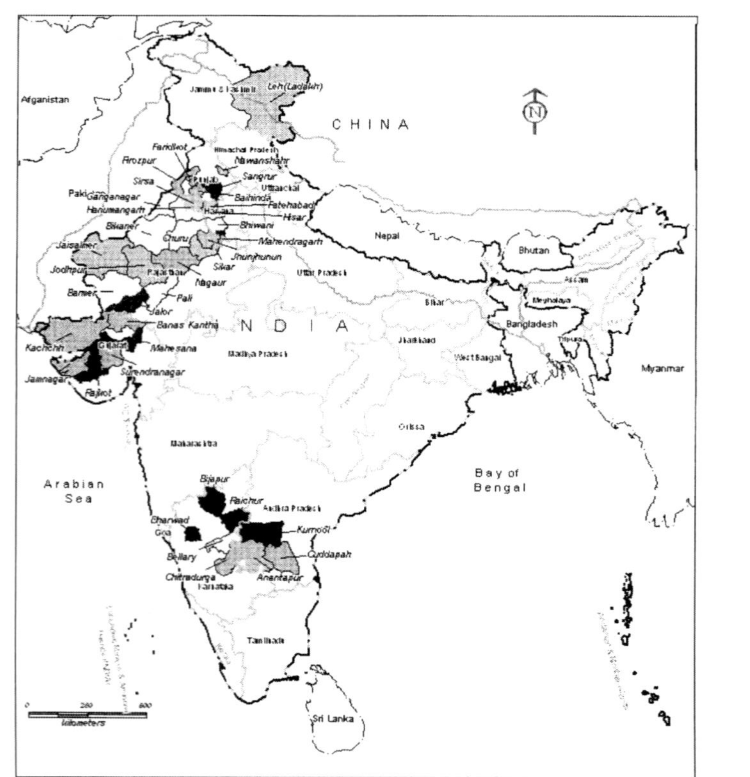

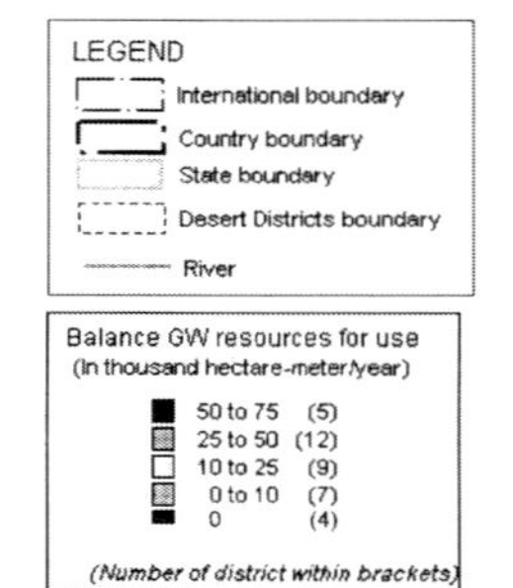

Note : The external boundary and the coastlines of India appearing on the map are neither verified nor authenticated by the Survey of India.

***Figure 7.55 :** INDIA ground-water resources and irrigation potential, Balance ground-water resources*

Source : CGWB

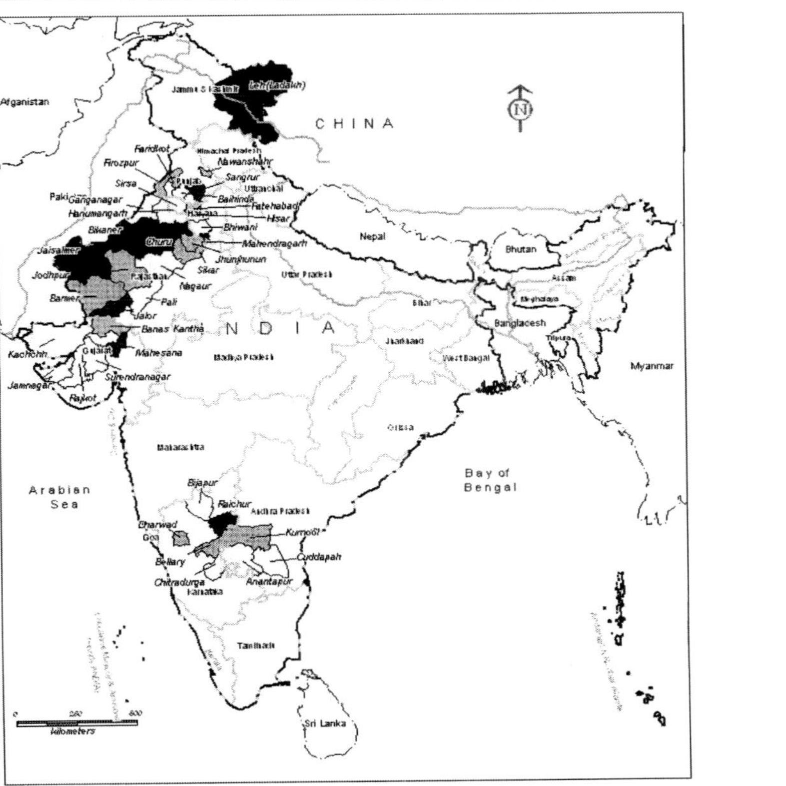

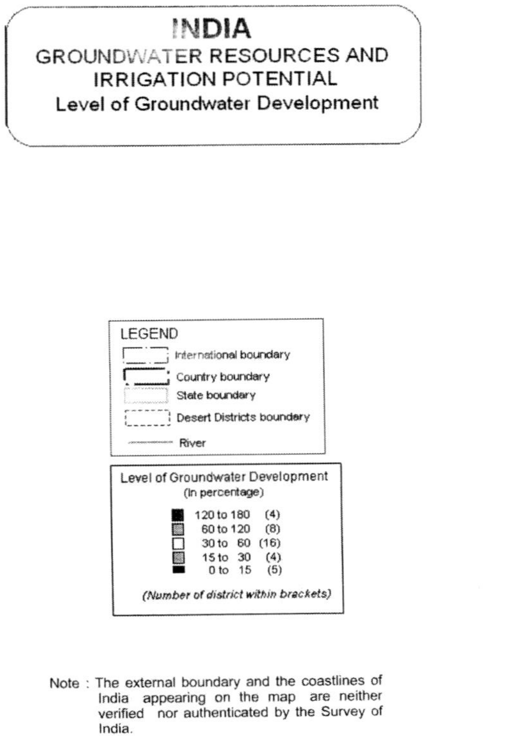

Note : The external boundary and the coastlines of India appearing on the map are neither verified nor authenticated by the Survey of India.

Figure 7.56 : ***INDIA ground-water resources and irrigation potential, Level of ground-water development***

Source : CGWB

INDIA
GROUNDWATER RESOURCES AND IRRIGATION POTENTIAL
Total Replenishable GW Resources

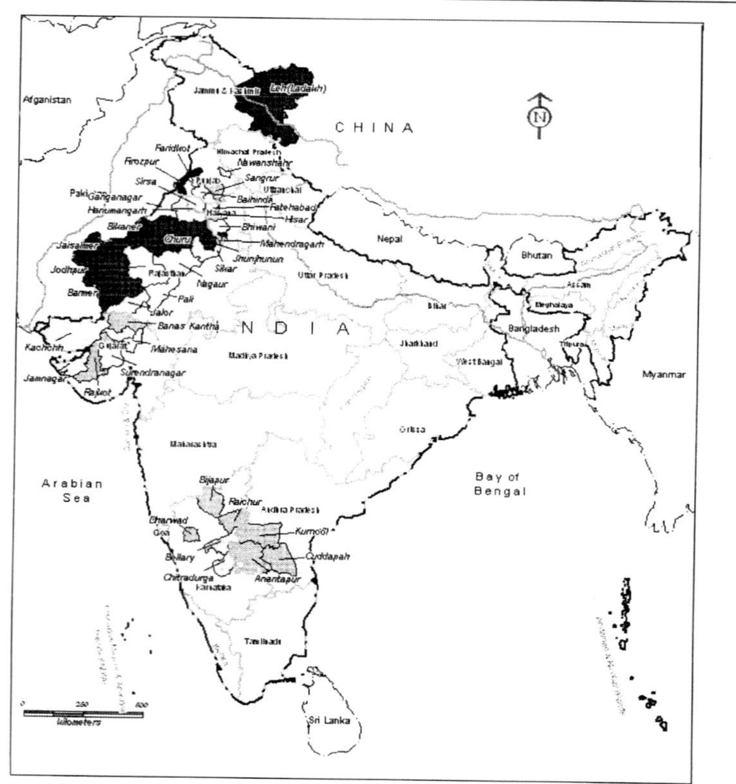

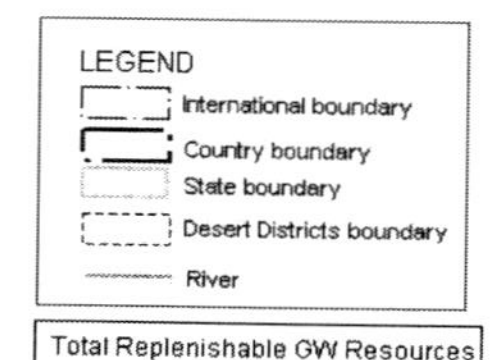

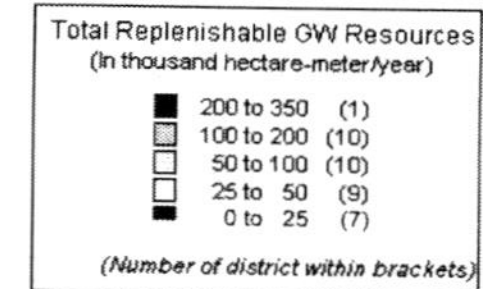

Note : The external boundary and the coastlines of India appearing on the map are neither verified nor authenticated by the Survey of India.

Figure 7.57 : *INDIA ground-water resources and irrigation potential, Total replenishable GW resources*

Source : CGWB

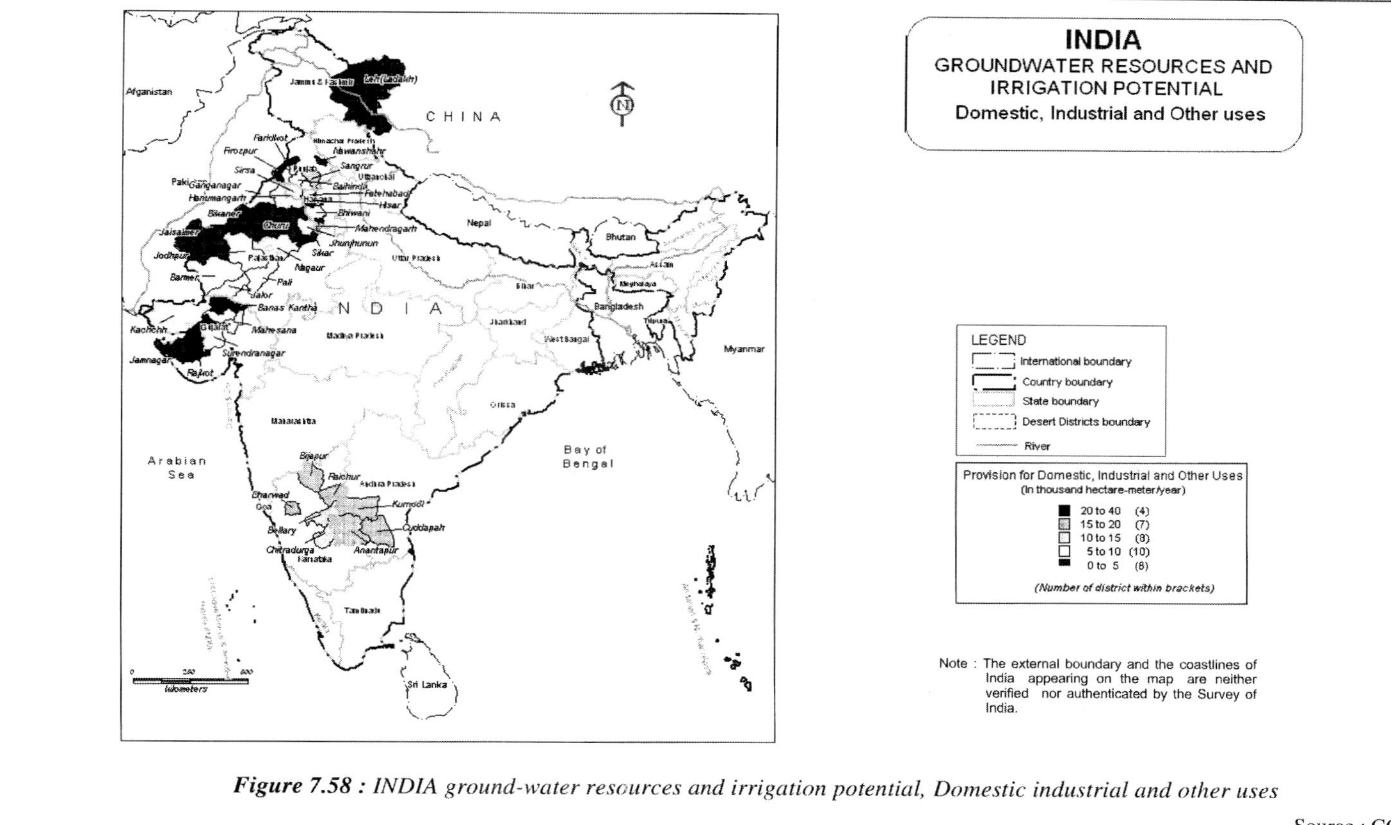

Figure 7.58 : INDIA ground-water resources and irrigation potential, Domestic industrial and other uses

Source : CGWB

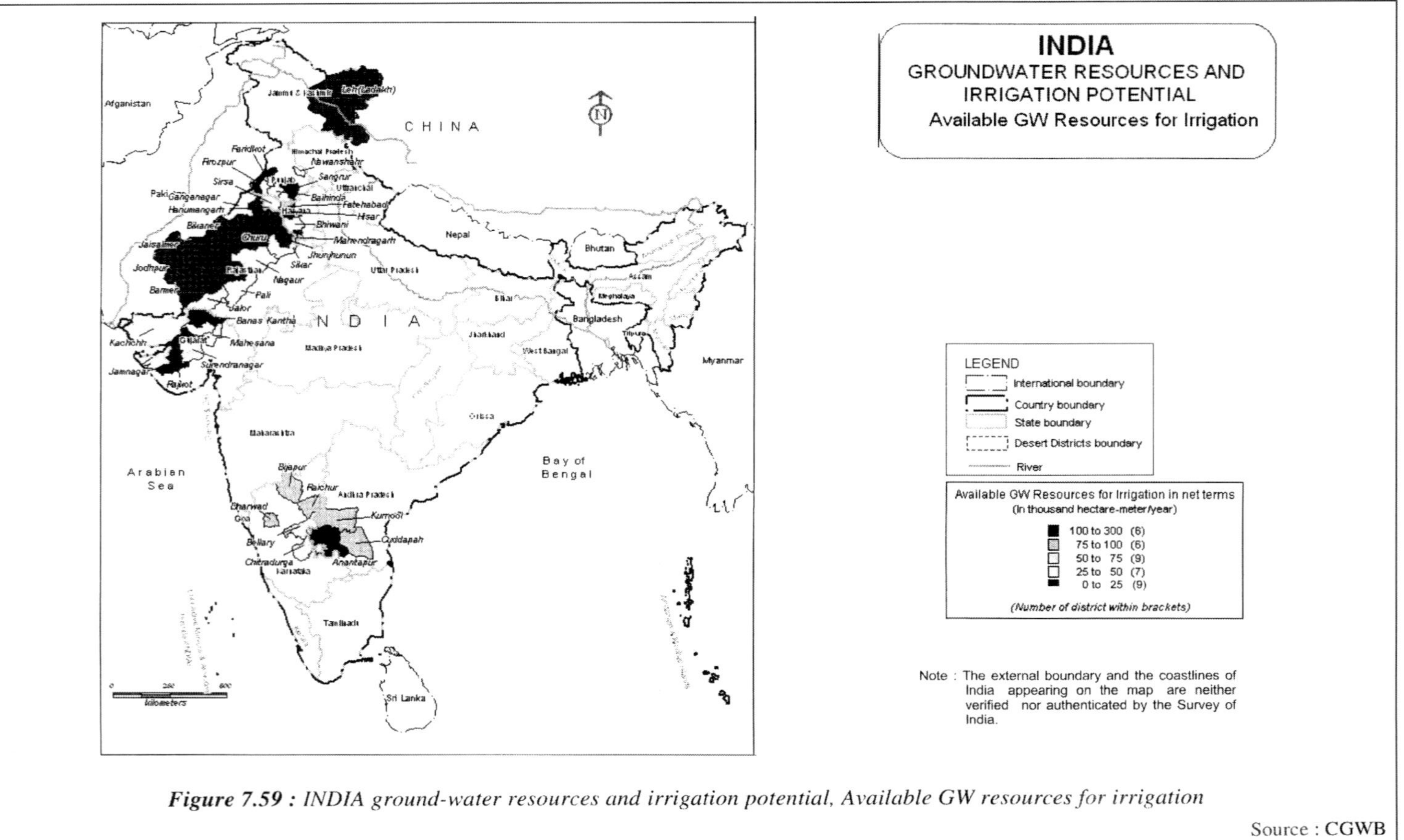

Figure 7.59 : INDIA ground-water resources and irrigation potential, Available GW resources for irrigation

Source : CGWB

INDIA
GROUNDWATER RESOURCES AND IRRIGATION POTENTIAL
Utilizable GW Resources for Irrigation

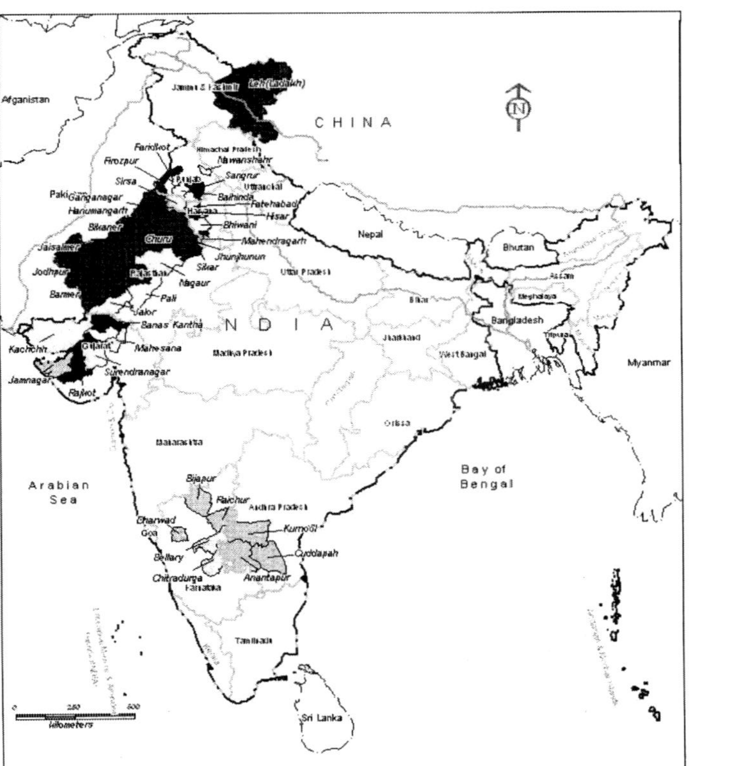

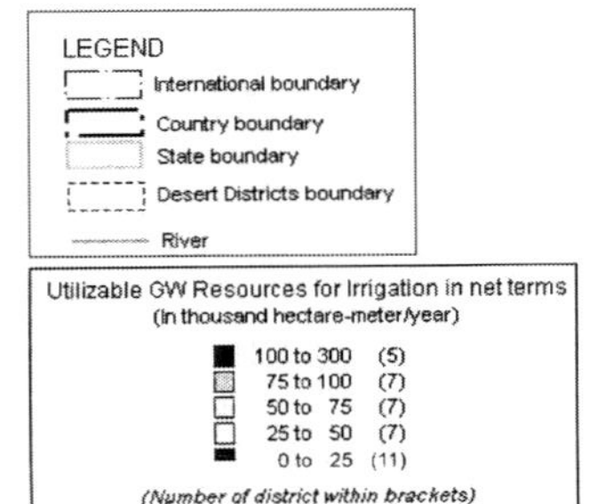

Note : The external boundary and the coastlines of India appearing on the map are neither verified nor authenticated by the Survey of India.

***Figure 7.60 :** INDIA ground-water resources and irrigation potential, Utilizable gw resources for irrigation*

Source : CGWB

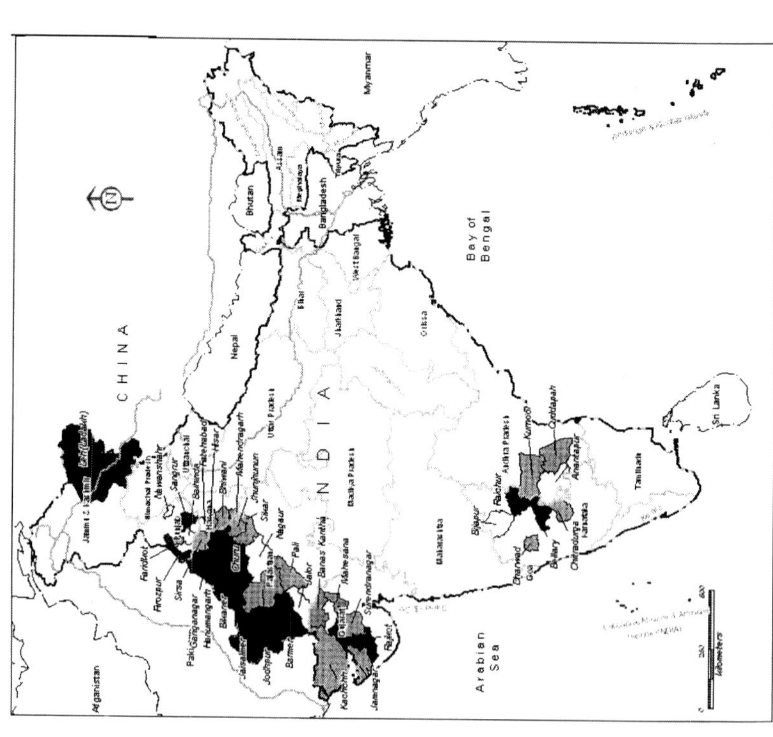

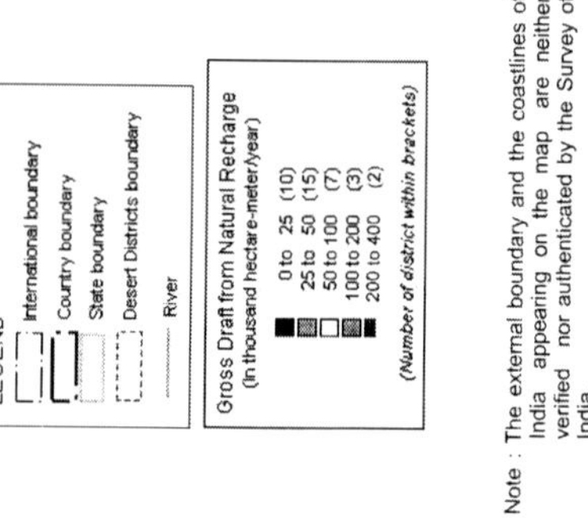

Note : The external boundary and the coastlines of India appearing on the map are neither verified nor authenticated by the Survey of India.

Source : CGWB

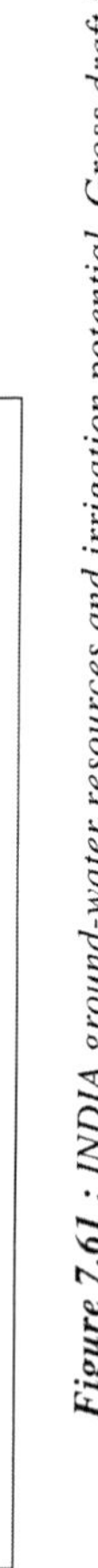
Figure 7.61 : INDIA ground-water resources and irrigation potential, Gross draft from natural recharge

INDIA
GROUNDWATER RESOURCES AND IRRIGATION POTENTIAL
Net Draft from Natural Recharge

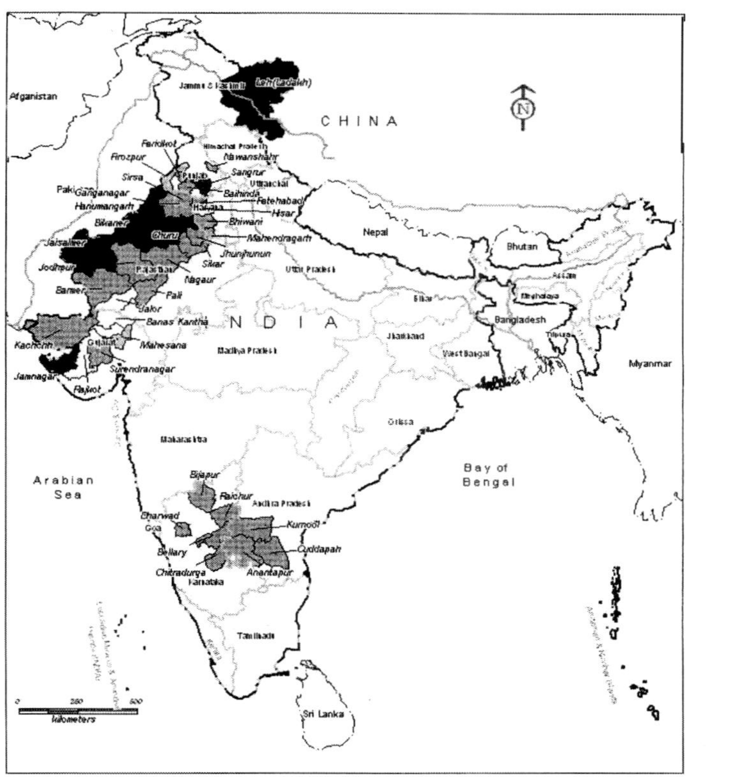

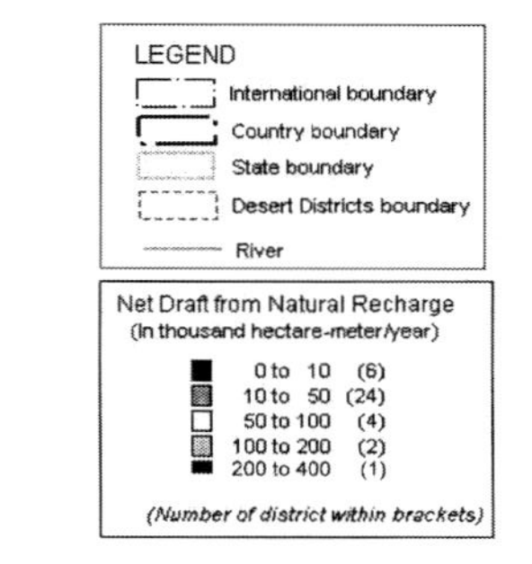

Note : The external boundary and the coastlines of India appearing on the map are neither verified nor authenticated by the Survey of India.

Figure 7.62 : INDIA ground-water resources and irrigation potential, Net draft from natural recharge

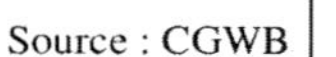
Source : CGWB

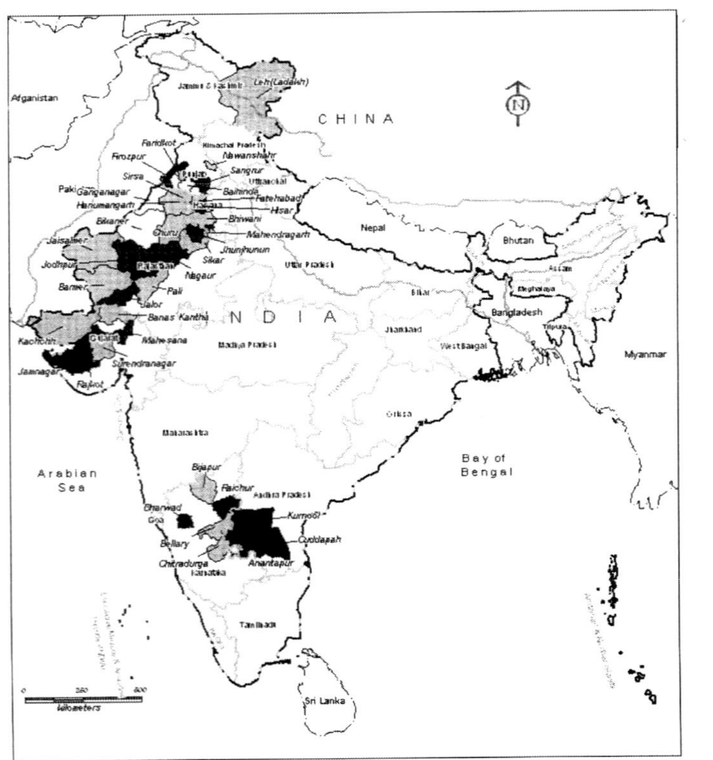

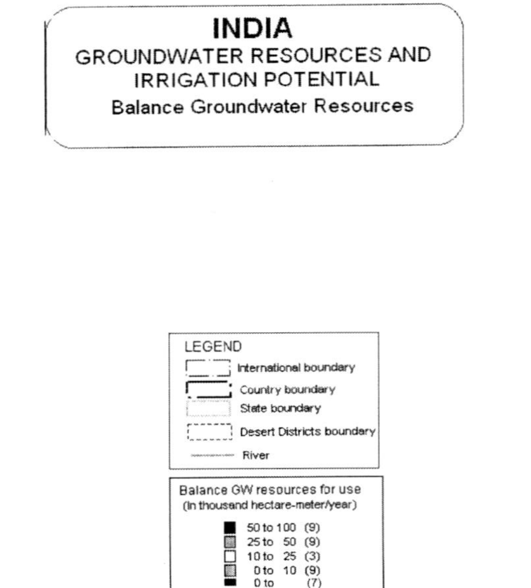

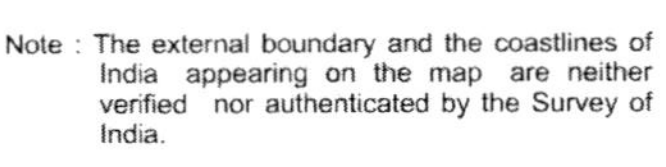

Figure 7.63 : *INDIA ground-water resources and irrigation potential, Balance ground-water resources*

Source : CGWB

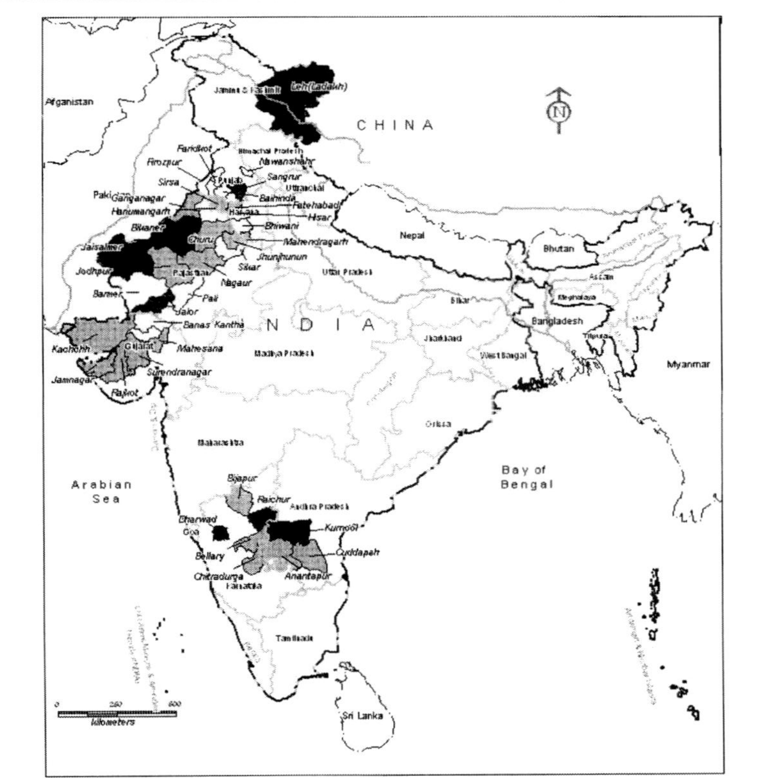

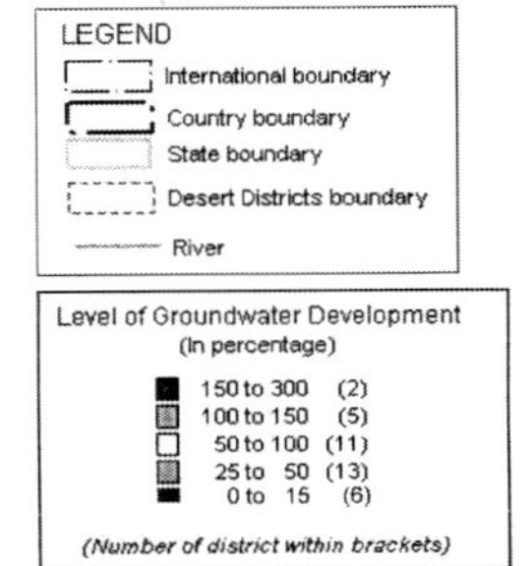

LEGEND

- International boundary
- Country boundary
- State boundary
- Desert Districts boundary
- River

Level of Groundwater Development (In percentage)

- 150 to 300 (2)
- 100 to 150 (5)
- 50 to 100 (11)
- 25 to 50 (13)
- 0 to 15 (6)

(Number of district within brackets)

Note : The external boundary and the coastlines of India appearing on the map are neither verified nor authenticated by the Survey of India.

Figure 7.64 : *INDIA ground-water resources and irrigation potential, Level of ground-water development*

Source : CGWB

Chapter 8

Bio-Diversity in Susceptible Drylands

The biological diversity of susceptible drylands may be less publicized than that of the tropical rain forest but many important crops and animals originate in the drylands and humans rely significantly on the biodiversity of these areas. Wheat, barley, sorghum, millets, pulses and cotton, as well as animals that have been closely linked with the development of human civilizations originated in the drylands. In the face of burgeoning human populations, the challenge is to maintain the species richness of the susceptible drylands within the context of sustainable environment management.

Some areas of the susceptible drylands are in fact wetlands e.g. the Aral Sea. Wetlands play key role in local biophysical conditions and also in global migratory patterns of birds. They have significance far beyond their local expression. They also have unique and highly susceptible biological characteristics of their own. Large parts of many critical wetland habitats, including estuaries, mangroves, open coasts, flood plains and swamps and the species they support, have been lost due to a wide range of human activities (UNEP 1995).

The biota of drylands consists of a series of dynamic and delicate ecological balances, Dominant among these are the balances that have been established between herbivores, pollinators and vegetation. Vegetation requires nutrients, predators require prey; and herbivores need vegetation.

Productivity of the soil depends on a balance being maintained between vertebrates, invertebrates, microbial activity and nutrient fluxes related to seasonal changes, moisture and solar energy availability. The balances that have been established are not only bilateral ones, but complex interactions that fluctuate with seasons and vary from place to place. These have been referred to as 'symphonic guilds' of interacting organisms (UNEP 1995: 350). They are part of complex web of life in the susceptible dryland ecosystems.

Biological diversity in the drylands is characterized by a high degree of within-species variation representing adaptation to the variable and sometimes extreme conditions. There is a moderately high diversity of species in the drylands, except in the driest areas where specially adapted species predominate. Dryland species are highly adapted to environmental stress. This makes them a vital source of genetic material for improving crop varieties by increasing their resistance to drought and disease.

Predominant among the species in the dryland ecosystems are the micro-organisms, annual plants and the ground-living creatures such as predatory arthropods, ants and termites, grasshoppers, snakes, lizards and rodents (UNEP 1995). There is also considerable variation in species richness among the drylands of different continental areas.

Some of the dryland species such as large game animals are well known but other species are known only to specialists. The role of fungi, for instance, in the maintenance of the productivity of the soils is indispensable. The sub-surface life in the drylands vastly exceeds that above ground. Dryland plant species are also important sources of commercial and industrial products.

Efforts have been made by country governments in the past few decades to conserve wild life species in a systematic manner. Establishment of wild life sanctuaries and national parks has been accepted in view of tourist attraction and every government of developing world is eager to attract tourists for acceleration of the economy. This has significantly helped in safety and security of wild life. The areas demarcated for wild life have been protected from destruction and regulatory legislation measures for their survival and growth are taken to keep the ecology balance.

Vulture is the key protector of ecology was at the stage of destruction in past decades has been survived with suitable protective measures and every effort is made for its growth and prosperity. Vulture helps in finish the dead animals and protects from environmental deterioration through spread of diseases through surviving insects over the dead animal. Other birds and animals join in their efforts once vultures remove the upper skin. Establishment of carcass plants by governments has further helped in environmental upgradation and utilization of ingredients for various use. So far urban areas are concerned, sufficient measures have been introduced or are under process of introduction. But the vultures and other birds and animals tackle the rural areas problems. Their protection has helped in maintenance of ecological balance.

Suitable measures have been taken for protection of forest areas by demarcation into reserved and protected forests. The deforestation done in past decades is being set right through government efforts with people's cooperation. Awareness generation and involvement of the people has helped significantly. Other areas deforested earlier are being reclaimed under national programme of Afforestation, reforestation and protection measures for safety and security by forest departments and the peoples' support.

For protection of cultivated lands, vegetative barriers are being used for meeting the growing needs of population and commercial use. Mixed cropping system has helped in maintaining the fertility of the land. Farm forestry and silvi-pastoral cultivation is another example of awareness generation, which was not being applied in the past. The growing needs of population and value of land has helped in protective use of land through bio-diversity. Though the concept has been introduced in recent past but the acceptance is widespread.

8.1 HUMAN IMPACT ON BIODEVERSITY

While the human use for commercial purposes of some dryland species has resulted in the loss of some and significant extensions in the range of others, the most widespread human-induced impact on dryland biodiversity stem from the introduction of fire for hunting, clearance of natural habitats for cultivation and soil fertility improvement and excessive grazing of livestock, introduction of domestic and game animals have been accompanied by the creation of water points, encouraging concentration of livestock. The introduction of non-native plant species, the removal of non-palatable plants and removal of predators and of burrowing and herbivorous animals have all hand impacts on the natural biology of the drylands. Construction of roads, canals, fences, dams and irrigation schemes has fragmented and even obliterated natural areas, replaced established species and interfered with or blocked natural migratory routes.

Among the impacts on species diversity are the following:

- Disturbance of 'symphonic guilds' has had direct influences on the structure and functioning of dryland ecosystems. Changes in plant root systems affect the underground decomposers. Termites and nesting or burrowing mammals provide micro-environments that enhance decomposition and nutrient cycling. Their removal shows decomposition and increases the chances of loss or transport of nutrients from the surface (Whitford 1991).
- Introduction of hoofed livestock to regions lacking a recent evolutionary history of grazing is often accompanied by reduction of populations of native burrowing herbivores, thus reducing the soil disturbing activities of these animals (UNEP 1995).
- Conversion to agriculture may increase local nitrogen values but native plant species may be displaced and are notoriously difficult to replace (Jackson *et. al.* 1991, UNEP 1995).
- Reductions in soil crusts may reduce nitrogen fixation, increase soil erosion and alter local moisture infiltration rates since microbial crusts play significant roles in nitrogen cycling and in stabilizing the soil surface against erosion (Eldridge and Green 1994).
- Replacement of native plants with non-native plants, to remove forage or as invasive weeds, reduces local plant and animal diversity. Aquatic habitats are especially vulnerable to invasion of non-native plants.
- The loss of large woody species, overused for fuel wood or as carving material, negatively influence the persistence of many typical species of drylands plants (Franco and Nobel 1989, UNEP 1995).

- Introductions of large animals as game or livestock have often established wild or feral populations e.g., horse, camels, Oryx, that affect native vegetation and deplete or compete with native herbivores.
- High fire frequency reduces the number of tree species and the number of tree species and the numbers of individuals per species in most savanna sites (San Jose and Farinas, 1991). Fire frequency has increased considerably under human influence. As a result, tree / grass ratios have change markedly.

1. Ecological Consequences of Forest Destruction

The destruction of forest undermines basic operations of the eco-system and may thus cause irreversible changes. The most serious of these change appear to be due to large scale exposure of natural soil systems, leading to increased soil erosion and, in turn, having an indirect impact on water resources development. Forests have a regulatory role in water flow, the hydrological cycle and in atmospheric and climatic phenomena-biogeochemical cycling of carbon temperature relationships- albedo. They represent a considerable stock of biomass in tropical rain forest up to 650 tonnes per hectare, the highest for all natural eco-systems. The ecological services that forests provide are difficult to assess in terms money. But Postel (1984) writes, "Flooding, droughts and siltation in many parts of the world are more severe because of deforestation. A half billion people in Bangla Desh, India and Pakistan are affected by water run-off from the upper Himalayan watershed. The cost of repairing food damages in India below Himalayan watershed has averaged US $ 250 million per year."

Apart from wood products collected and consumed at home, forest products represent an important export item for many tropical countries, the trade value of forest timber exports is about $ 4.7 billion a year. Other forest products like latexes, gums, essential oils, medicines, nuts and ornaments, contribute to export trade. Depletion of forest is changing the global trade pattern and exports are dramatically subsiding, especially from tropical Africa. Tropical forests are major source of useful genetic materials for agriculture, medicines and other pharmaceutical products, a trade with potential sale values of billion dollars.

2. The Problem of Environmental Degradation

Awareness of increasing pressure on grazing resources and the creeping land degradation has not been a product of the environmental and conservation problems. Schapera drew attention to the problem and Protectorate Government made various attempts to control cattle numbers and spread grazing pressure. Land degradation takes a variety of forms and does not merely affect the vegetation. It include the following:

(i) The extension of completely bare ground, especially around watering points, through excessive grazing and trampling;

(ii) The disappearance of the more palatable grass species and their replacement by less desirable plants, which increasingly include woody and thorny shrubs;

(iii) Loss of top-soil by wind and by sheet and gully erosion.

(iv) Cattle creeping on a large scale in the fragile environment, especially with the low standard of range management, a high risk of more or less irreversible land degradation.

3. Desertification and Land Degradation by Human Impact

Human efforts are the main cause of desertification and land degradation. People of native lands use wood for their fuel requirement and grazing of livestock. Basically the efforts are limited to collect dry wood from the wastelands but their efforts gradually move at cutting of trees, as fuel is their basic requirement. Even the tribals residing in forest areas use dry wood without making any loss to grown trees but in view of increasing population, devastation is started. But timber users make the significant loss and they work like gangs without bothering the government protective machinery and the local people. Such groups have political shelter and there is no check to their efforts of destruction of trees and bushes. Destruction caused in past decades had significant loss to environment and ecology. Even the monsoon cycle was disturbed by loss of vegetative cover. Natural disasters from flood are mainly caused due to loss of trees to protect soil from moving with fast flow of water.

Loss so caused has been experienced by human beings, society and the government. Efforts for conserving the trees from enmass cutting have been reduced to some extent with the awareness generation among the people and protective arrangement by government. Growing and conservation of trees, vegetative cover has been stopped in areas, where people have come forward for protection and security. Some success stories can be seen in all the developing countries particularly in dryland areas.

In a famous temple of southern Rajasthan in India consumption of fuel wood for preparation of temple sweets was 40 quintal a day, which was met through wood from surrounding hills. The District Collector calculated the loss in annual terms of 1460 tonnes and he proposed temple authorities to provide them gas connection for sweet preparation in view huge environmental degradation. The temple authorities while sharing with his concern maintained that temple sweets couldn't be prepared on gas in view of emotional religious issue. The District Collector also agreed with the religious sentiments of the temple authorities and ordered forest department officers to grow Juliflora on ten kilometers surrounding area of the temple. Now the areas has been converted in thick forest and the loss has been checked significantly as the growth rate is faster than its cutting.

The above example of understanding psychology and helplessness of the people dependent on trees for their fuel wood requirements. Only similar measures with awareness generation can help in ultimate addition of vegetative cover and development of environment and ecosystem. Therefore, awareness generation alone cannot yield to success but the alternative measures to minimize the problem need to be done for sustainable development. This also holds good for all types of impacts of desertification and land degradation. In the dryland areas, to minimize the impact of destruction of vegetative cover during droughts can be minimized by adopting suitable arrangement to meet the challenges. The famine conditions of drylands need to be mitigated through drought proofing measures for sustainable development of the areas.

4. Desertification and Land Degradation Through Livestock

The people of drylands keep large number of livestock for their livelihood. Keeping livestock is their necessity as agricultural production is uncertain in dryland areas due to erratic conditions of monsoon. This condition cannot change and the continued dependence on livestock cannot be stopped. Loss of vegetation and land degradation through livestock is the inherent problem. Any measure to solve the problem can be sustainable, if this problem is accepted as their genuine requirement. The solution can be made by growing trees meeting fodder requirements, raising grass through protective measures and grain fodder to meet the total fodder requirement.

Another area of operation is the breed improvement of livestock. The indigenous livestock breed in not so productive to feed them costly fodder by purchasing through community common property resources grass lands or through grain fodder. The economics of the small livestock like goat and sheep require open land free grazing and unless the breed measures for increasing the productivity of livestock the cost based fodder cannot suffice in economic terms. This is the real necessity and any solution of the problem can be sustainable while keeping all these aspects.

5. Drought Proofing Measures

Failure of monsoon in dryland regions is a common phenomena and occur through a cycle based on geophysical conditions in such regions. The crop damage is certain in view of rain-fed cultivation. Occurrence of further desertification and land degradation is bound to happen, as there remains no alternative source of livelihood. Provision of free food and large scale employment is a temporary arrangement but economic waste and the expenditure incurred on such activities are fully unproductive. Politically, this is a suitable solution to gain the sympathy of the people and external assistance on humanitarian ground is not a wise step for the honour of the nation. The measures for permanent solution rest in creation of opportunities to get rid of worst impact of famine can be sought through local approach.

Possible areas of the problems can be sought out of water harvesting measures, initiating employment oriented activities in lean season through permanent works like construction of road, canal, water harvesting, land upgradation, adoption of biotechnology, afforestation or reforestation, social forestry and similar other measures but the selection of such activities need to be done on areas based activities with peoples participation. Dry farming practices are also a dependable alternative where the seed varieties can survive for longer duration without water. Through these measures, minimum level of production is possible even in worst situations.

8.2 VALUING BIODIVERSITY

Biological resources can be valued economically under two headings: direct values and indirect values (Me Neely 1988). Direct values include values for consumptive use e.g., fuel wood, wild animals hunted for food and productive use values e.g., plant based drugs. In many rural drylands, wild foods form an integral part of everyday diets, as well as providing insurance against crop failure, pest attack or drought. The harvesting of wild resources also represents a ready source of income to cash-poor families, which may amount to a significant proportion of total household incomes, particularly where farming is marginal (TIED 1995). The value of all wild resources to local people, for subsistence consumption or sale, was calculated to be more than US$ 120 million in Tanzania in 1988, representing 8% of the total agricultural contribution of GDP that year (Kiss 1990).

Wild resources also have indirect value, which can be categorized into non-consumptive uses e.g., wooded land, which protects against flooding and may help to regulate local climate, option value for some future use e.g., wild relatives of crops represent a storehouse of genetic diversity for future cross breeding as given in Table No. 8.1. , and the aesthetic value of a species by its very existence. Assigning realistic economic values to some of these indirect benefits can be particularly difficult. Although economic valuation is a useful way of making wild resources financially visible, it must be recognized that not all of these resources can be given a price tag. Nevertheless, such resources remain essential, at local, national, or both scales (IIED 1995).

Table 8.1 : Examples of the productivity contributions of wild relatives of crops

Crop	Found in	Effect on production
Wheat	Turkey	Genetic resistance to disease; valued at US$ 50 million per year
Barley	Ethiopia	Protects California's US$ 160 million per year crop from yellow dwarf virus.
Beans	Mexico	Genes from the Mexican wild bean used to improve resistance to the Mexican bean weevil, which destroys up to 25% of stored beans in Africa and 15% in South America.
Grapes	Texas, USA	Texas rootstock used to revitalize the European wine industry in the 1860s after louse infection.

Source: After WCMC (1992).

The Global Biodiversity Assessment (UNEP 1995; 890) tackled the problem of the assessment of the economic value of biodiversity in depth and concluded; "Effective biodiversity conservation depends on the valuation of all goods and services provided by ecosystems in order to determine the configuration and combination of uses that best satisfy the interests of society." A recent IIED study (IIED 1995; 14) found:

"Combining economic concepts with participatory research allows for a more comprehensive valuation of wild resources, recognizing not only the financial value. The local valuation methodologies being developed should be made accessible to local communities to enable them to negotiate with powerful external interests and to enhance local natural resource management structures."

1. Economics of Biodiversity

Nature has kept balance in the environment and ecology of the earth and the gifts of nature provided to mankind are most valuable not possible to quantify in money terms. The gifts of nature in terms of river, stream, lake, wild life for maintenance of environment and ecology are perennial and cannot be quantified in monetary terms, but the benefits to the area and human beings are the well being and prosperity. The lack of such facilities in dryland areas is hardships and natural degradation. Human impact further adds to desertification and land degradation, which also cannot be quantified in money terms. The natural gifts are not equally distributed and therefore all the areas are not so prosperous. The history of mankind tells about the use of land started in the riverside where cultivation was rather easy. The pressure of increasing population compelled to use marginal lands and in due course of time use of further degraded land changed the definition of marginal land. Intensive use of land deteriorated its fertility and the process of continuing regularly.

In the present context, where the population of the world has increased significantly, all out efforts are being made to get maximum possible from the land to feed the population and the variation of population density and pressure on land is the ultimate cause of desertification and land degradation. Since no concrete efforts are made to check population growth, the problem of further deterioration of land is the ultimate situation and therefore, the geologists, environmentalists, social scientists and other specialists are regularly warning for initiating remedial measures to mitigate the problem of desertification and land degradation.

Undoubtedly, every government of the developing world is equally worried for the situation and intended to apply sincere efforts to check the process of further deterioration but this is a tedious, where no single solution is possible. Every degraded aspect require huge money for mitigation and upgradation and the political will to carry out such tasks. One more aspect of present situation is linked with the reality that in case of huge devastation due to natural or human efforts alarms the political system of the government to take immediate remedial measures but the degradation and loss has no similar alarms the those politicians, as the problem does not look so serious.

The problems of desertification and land degradation are not common in all the dryland areas. In areas where irrigation facilities have been provided, the problems o water logging and salinity have been caused by mismanagement of water resources. In rain-fed regions the problem of lesser availability of water is responsible of recurrent droughts. Therefore, at all the levels, mismanagement of the land management system is responsible for land degradation and desertification. The economic aspect of all the issues of land deterioration is unquantifiable in money terms.

In all the cases, desertification and land degradation has been caused due to wrong planning efforts of the country governments to mitigate the problem. The past experiences of execution of larger projects without involving the people in decision-making process was in early years and also prevalent in the present. This is evident that the country governments of developing world are not so serious for solution of the problems. The target based to tackle the problem of total country areas is still prevailing without involvement of the people and without considering local situation. This level of non-serious effort cannot mitigate the problem.

8.3 POLICY RESPONSES

The drylands have been the bread-basket of the world for centuries and many successful efforts have been made to avoid degradation at the same time as increasing productivity. In the last few decades, however, grand development schemes and mechanized agriculture have swept away the hedgerows and windbreaks and the small mixed farms and have fenced pastures and settled many nomads. Now, it is realized that this trend has to be arrested and perhaps reversed.

Conservation of biodiversity is not necessarily incompatible with sustainable agriculture, since ‘much of the generic diversity on which the improvement and future sustainability of agriculture depend is found in and around farmers’ fields, in village woodlands and in grazing lands’ (IIED 1995; 11). Good practices can and do increase sustainable productivity, reduce risk, maintain soil productivity, increase carbon storage, protect or restore biodiversity, prevent desertification and even restore areas, which have become degraded.

Among the prime causes of biodiversity loss in drylands, as in other major world biomass, has been the destruction of habitats and trade in certain dryland plants and animals or parts thereof. International trade has been curbed with legislation under the Convention on International Trade in Endangered Species of Wild Fauna and Flora (CITES), while the conservation response has long been at the forefront of strategies to combat the threatened loss of biodiversity by habitat destruction. Conservation efforts have ranged from the lobbying of conscientious activities, through the designation of certain areas as protected areas, right through to draconian enforcement of Scientific Reserves / Strict Nature Reserves where people were excluded, sometimes after being displaced and nature protected by armed guards or the military.

The conservation approach has achieved much in certain areas and has led to increased understanding of the importance of biodiversity. In the drylands, where 1.8 billion people live and where more than a billion of these people are directly dependent on agriculture and the land for their survival, conservation efforts have faced major economic and livelihood hurdles. So it has been realized that exclusion of indigenous people from the conservation areas is neither productive nor sustainable. Unless, the people also benefit in economic terms, effective conservation or sustainable use of biodiversity in unlikely to become a reality.

As in the other cases, where emerging issues are identified, high quality, reliable assessments of the problem, monitoring of its progress and the success of efforts to overcome it are essential. A first Global Biodiversity Assessment was carried out by UNEP and partners and is drawn on extensively here, while the Survey of Economic Plants for Arid and Semi-Arid Lands (SEPASAL) has taken this work further for dryland flora. But much more needs to be done in the drylands, which have neither yet attracted the attention of the global conscience leaders nor of the financiers.

1. Policy Measures to Mitigate Land Degradation and Desertification

Policy measures adopted by the governments so far have mix results. The efforts made through external assistance in all the developing countries could not yield much success as the ultimate aim of the government remained to use the assistance without peoples' cooperation and through the approach not relevant to local requirements. Involvement of the local people in decision making process has never been considered as necessary. The adaptability of the species grown in dryland areas without knowing its relevance was one of the reasons for failure of the projects.

Degradation of forests is continuing without any check and reforestation activities could not yield much success. Social forestry and upgradation of land as common property resources had never accepted by the people of the area in view of faulty concepts. Requirement of the local population from land resources and the activities to implement the policies could not be matched for various reasons and people never thought of them useful in view of lack of awareness or their active participation.

Country policies for growing cash crops and fulfilment of local requirement of food was never thought while preparing policies for earning foreign exchange. Change in food habits created further problems of regular import of food grain for local requirement. The price of the agriculture produce never considered helpful instrument for providing incentive for growing more food. The input cost of cultivation and farm harvest price never matched to each other.

The watershed approach was promoted by international agencies including the World Bank and huge assistance was given for watershed development. But today no success story of this programme could be maintained and the resultant impaction of replication as envisaged could not be made. Community lift schemes also never accepted by the community of beneficiaries as there was no full proof system for maintenance of created assets. The social system of individual approach could not match with joint sector beneficiary approach.

Individual beneficiary programme of land upgrdation were never considered essential. Command area development approach of irrigation management was one of the unsuccessful events as there was no proper management of sharing of water at all levels. Water logging and salinity could not be foreseen as resultant impact of the problem and when the problem became acute, the attention of the government was drawn at the burning problem.

In groundwater irrigated areas, there became an competition for drawl of as much water from the source to irrigate the crop, resulted in depletion of ground water level and salinity and alkalinity of cultivated land. The depletion level grew to the alarming extent, where the source could receive water sufficient to be drawn within one or two hours. This ultimately resulted in competition to drill the pumps deeper to get maximum possible water. The government carried out survey of underground water level in view of deteriorating situation and half of the areas of total drylands were declared dark zone where further drilling of wells and tube wells was restricted. The bio-diversity was affected by all the measures adopted by the cultivators and the country government was happy as the food problem for increasing population were met by local efforts. The loss of such efforts in land degradation resulted in decreasing the productivity of the soil.

2. Livestock Breed Improvement Programme

In the past efforts were made by import of exotic breed livestock as a breed improvement measure but no success could be achieved due to non-suitability of the area and the results of cross breeding could not yield desired results. Later, efforts were made to improve the local breed through artificial insemination and natural breeding process. Some success was achieved in the efforts but there are different opinions between the animal husbandry experts about the appropriate measure out of natural breeding and artificial insemination. Embryo-transplantation was also tried but could not succeed much.

An imbalance has been created due to non-requirement of male bull of cow, buffalo and camel, which was earlier utilized for ploughing the cultivated lands has been replaced by mechanical system and the male livestock of these species became useless. Famous varieties of bulls in India like Kankrej, Tharparkar, Lathi, etc. are on the verge of extinction in view of their neglect for cultivation use. The policy measures adopted to decrease the number of livestock by improved breed could not yield much success as the migratory livestock was not available for improved breeding and the herdsmen used their own system of insemination through available indigenous male. Collection of semen in far-flung areas could not succeed for want of proper freezing measures. The present level of livestock is equal to human population in dryland areas and no significant improvement in the breed improvement programme was succeeded.

3. Bio- Diversity Imbalance Through Chemical Fertilizers and Pesticides

Excessive use of chemical fertilizers and pesticides has resulted in land degradation. Undoubtedly, use of chemical fertilizers has succeeded in increasing the production and productivity of land to cater the growing population, but it had adversely affected the environment, vegetation, soil fertility and health of human beings and livestock. Application of chemical fertilizers has increased in past four decades and to maintain the fertility level of land increased use is necessary. The developing countries generally import chemical fertilizers in view of increased country demand.

Use of chemical fertilizers is helpful in growth of plants and for maintaining the production level but the quantum of chemical fertilizers need to be increased for retaining the fertility of soil. More water is required for prevention of heat of the chemicals and the cultivators have to use excess water, which can be helpful in addition of additional irrigated area. The adverse impact on the health of the people is increasing due to consumption of infested food with chemical fertilizers. The use of chemical fertilizers has been banned in the developed countries, but the companies producing chemical fertilizers in developed countries are still engaged for meeting the demand of developed countries. In addition to human beings, the adverse impact of livestock and birds has decreased their number.

Use of pesticides in food production, vegetable production, ripening the fruits has developed poisonous effect and the impact on human beings, livestock, birds, environment and ecology is significantly adverse. Chemical pesticides which are harmful even by smelling, are used in food items and the spread of diseases is widespread. In dryland areas where water shortage is quite acute, use of chemical fertilizers and pesticides has deteriorated the environment and bio-diversity to great extent. They are used for growth and protection from insect at the cost of significant hazard. No country in the developing world has bothered to ban these items in view of feared shortage of food items.

The solution of the problem is use of bio-fertilizers and bio-pesticides which can be prepared in the rural areas with local resources, but its application is slow in view of feared decrease of the productivity. In the cultivated fields through chemical fertilizers and pesticides, the fertility of the soil has been deteriorated significantly and replacement of chemical fertilizers and pesticides through bio-fertilizers and pesticides need a cautious approach of replacement in phased manner. In the first year, 25% of bio-fertilizers and 75% of chemical fertilizers, second year on 50-50% basis, third year 75 and 25 basis and fourth year 100 percent bio-fertilizer can safely shift the process without any significant loss of production and productivity.

With this process, the fertility of the soil will sustainably maintained and further increase productivity in longer duration. This would require 30% less water for irrigating the crops. Environment and ecology will be protected from feared deterioration and any adverse impact on human body will be completely removed. Survival of livestock, wild animals and birds will be ensured and the change will be helpful. Presently application of bio-fertilizers and pesticides has been introduced as awareness generation and demonstration for its use in larger area. The countries of developing world will be relieved from import of chemical fertilizers and pesticides and the scarce foreign exchange being used for their procurement would be saved or used for import of any other important item other agriculture inputs.

8.4 CONCLUSIONS

Land degradation and desertification are intricately linked with people s utilization of productive natural resources and the only viable key to the successful maintenance of dryland biodiversity in the long term is to modify the human use of the environments. As the IIED study concluded:

"Incorporation of indigenous crops and other native plant germplasm in the design of self sustaining agro-ecosystems should ensure the maintenance of local genetic diversity available to farmers. In situ management of wild genetic resources is likely to be the most effective conservation method in the long term." (IIED 1995; 11).

Since the key to changing policy is so often the demonstration of the economic valuation of the policy options, their costs as well as their benefits, the Global Biodiversity Assessment concluded:

- The allocation of biological resources on the basis of current market signals is inefficient and inequitable. It leads to losses in social welfare. Moreover, these losses are distributed in a way that bears most heavily on both the poorer members of this generation and all members of future generations. An appropriate strategy for biodiversity conservation should address both the efficiency and equity issues. Valuation is an essential element of this process.
- Efficiency and equity both require the development of effective institutions and incentives that will (a) confront resources users with the full social cost of their behaviour, and (b) enable those, who invest in conservation to appropriate the benefits.
- Traditional approaches to protected areas have sometimes worked against the interest of local populations by removing traditional rights of access to the resources in protected areas and the benefits of that access, thereby taking responsibility for management of the area away from local people.
- Preserving biological resources through nature reserves and other protected areas may be an important short-term step, where biological resources are under immediate threat. But it is not feasible to protect critical ecological systems by excluding human users of wild resources.

The conclusions, recommendations and options for maintaining biodiversity in the susceptible drylands are thus very similar to those for successfully combating desertification in general. As each element involved in effective sustainable environment management is studied, the commonality of need and approach becomes clearer. In order to combat desertification and conserve bio-diversity effectively, sustainable environmental management must be achieved.

Chapter 9

Dryland Plants and their uses

The climatic characteristics of drylands contribute to the survival strategies of plants. Despite moisture deficits at a range of temporal scales and the harsh temperature regimes and high variabilities, there are normally plants of some form in most dryland settings, invisible but present nonetheless as seeds or as dormant tissue below ground. Even the Rub' at Khali, the Empty Quarter of Arabia and one of the most inhospitable places on Earth, has as many as 37 species of flowering plants. The focus here is on the plants of tropical and sub-tropical drylands, excluding anywhere with a mean temperature of < 0 °C in the coldest month, such as much of Central Asia and Central North America.

Dryland plants have their own communities, structures and characteristics, some of which are far-ranging whilst others are more closely associated with particular areas like cacti in the Americas, acacias in Sub-Saharan Africa and the white-barked Ghost Gum (Eucalyptus papuana) in Central Australia. Many dryland peoples, especially in developing countries, rely significantly on indigenous plants for food and livelihoods, but the resource base may be far smaller than in humid areas. While many dryland plants are harvested and used locally, some also enter the market economy at some level, whilst a few even reach international trade.

In drylands, as in other areas, there are pressures affecting the relationships between people and plants. Political measures constrict the movements of nomadic people; increasing human and livestock populations diminish plant resources; dryland rainfall regimes change, because of human activity elsewhere and traditional knowledge about plants and ecosystem management gradually disappears in the face of external pressures. The special needs of drylands are recognized in the United nations Convention to Combat Desertification (CCD). Links between people and plants clearly underlie one of CCD's two objectives: "improved productivity of land and rehabilitation, conservation and sustainable management of land and water resources, leading to improved living conditions, in particular at local level".

9.1 PLANT COMMUNITIES AND COPING STRATEGIES

The least known dryland plants are small ones that live in soils, rocks and even under the light-transmitting stones. They comprise cyanobacteria, algae, lichens (associations of algae and fungi) and bryophytes (mosses and liverworts) and together they commonly form crusts over extensive areas of drylands (fungi too are present, but are not plants). Along the coastal strip of the Namib Desert, lichens are so extensive and dominant a life-form that they can be detected by remote sensing. All these micro-organisms have a great impact on both physical and ecological processes, such as soil nutrition, erosion control, nitrogen fixation and the establishment of the seeds of higher plants. Indeed the crusts that they form on the soil surface are increasingly believed to be indicators of healthy land conditions.

Higher plants, including ferns, gymnosperms, such as conifers and flowering plants and their life cycles are better known. The conservation of water is not a problem for annual plants in drylands, which may complete their entire life cycle within a few weeks and can offer spectacular displays of flower colour, for example in Namaqualand in South Africa. Perennial plants have a number of quite different means of surviving the season of drought. The Creosote Bush (*Larrea tridentata*) of North America and the Mesquite (*Prosopis Juliflora*), a widespread tree native to Central and south America, both have deep roods tapping underground water sources. Succulent plants store water in specialized tissues and often lack or have only rudimentary leaves. Shedding all leaves at times of water stress is a very common adaptation. Many species have spines, which reduce incipient radiation as well as offering protection against browsing animals.

Stomata, through which plants lose water by evapotranspiration, may lie in grooves or furled up leaf surfaces or may even closed by day and open by night in those species, which photosynthesize by fixing carbon at night. The living stone plants (*Lithops*) of Southern Africa reduce exposure by being almost entirely underground and have translucent, flattered leaf tips, which allow light to reach the subterranean parts of the leaves. In coastal Peruvian Desert, bromeliads (*Tillandsia* spp.) dispense with roots altogether and obtain all their moisture from fogs. The fogs of coastal Namibia, where mean annual rainfall is less than 50 mm, are the main source of water for plants such as Trianthema hereroensis and Stipagrostis sabulicola, but the uptake method is quite different for each of them, respectively by direct foliar absorption and by a lattice work of shallow roots exploiting moisture condensed on the surface of the dunes, where they grow (Louw and Seely 1982). One adaptation common to virtually all dryland plants is the ability to their seeds to withstand desiccation survives from one growing season to the next.

1. Plant Communities in Various Drylands

Vast areas that used to be vegetated by dryland Acacia-Cominphora bush in the Sudano-Sahelian region and longer ago are being degraded by action of man, aided by drought. The weak soils are blown and washed away due to poor land management and population pressure leaving desert. For reversal of the trend. The thorn bush, previously thought almost worthless has been converted as natural resource gold mines. The Acacia, Comminphora, Boswellia, Sterculia and many other trees and shrubs in drylands produce gums, oils and other extracts that are currently in demand in many industries. There is great potential to develop more products that could make of these renewable resources. The benefits could be enormous; the creation of employment, new industries, foreign exchange, land conservation and rehabilitation.

In Africa Acacia Comminphora zone stretching more than 5000 km, in view of its commercial demand. In Kenya it is called Nyika and was considered as almost useless bush land while used by nomadic pastoralists and in view of mounting population pressure this specie occupies some 70% of Kenya's land surface, although, the quality of the product is comparatively inferior.

2. Coping Strategies

Many of the indigenous tree and shrub species of this type vegetation hold potential promise as producers of economically valuable products, principally resins, gums, oils and extracts. Gum Arabic, frankincense, myrrh, henna and aloe are some of the better known, but greater potential lies in developments to be made through research in the pharmaceutical, fragrance and flavour, food technology epoxy resin, plastics and industrial coatings industries. These are each multi billion dollar industries, which make use of and are constantly searching for, natural products from trees and shrubs.

The Nyika of eastern Africa is going in the same way in Senegal, Mali, Burkina Faso, Niger, Chad etc. have sparse remnants of this vegetation left, most of it having been destroyed for use of fire-wood, charcoal, livestock enclosures, building materials and intentionally cleared off to make way for grazing land or cultivation. This type of vegetation usually develops under an average annual rainfall regime of between 100 to 600 mm and under high temperatures, thus soils associated with it are generally not well developed. When the trees destroyed and ground cover removed by over grazing or cultivation, the fragile soils became susceptible to erosion by wind and water.

In order to sort out the problem in terms of economic viability of such products and demand of the land for feeding the growing population of various study areas, Deniel Stiles (1988) of UNEP had identified following strategy for economic use of the dryland products for the benefit of country population:

(a) Taxonomic identification of the trees and shrubs of interest down to the sub- species and varieties level and distribution and densities of these plants, noting the environmental parameters, under which they occur, i.e., soils, porosity, slops, altitude, rainfall and associated vegetation community.

(b) Tapping and collecting trials of the gums and resins produced by the trees to gain estimates of yield and potential supply to industries. Growing trials of shrubs need to be carried out and tests conducted.

(c) Chemical analysis of the gums, resins and oils to assess the quality and to research possible new products that can be derived from them.

(d) A detailed study of potential markets for the products to be done which include contacting buyers and end-users to ascertain market volume, quality requirements, product variability to assess potential for development of agro- processing industries in the country, source of availability before earmarking on any economic development project.

(e) Socio-economic study to ensure proper management of the agro-industry in different contexts e.g., (i) wild bush on community land under pastoral land use, (ii) Community land primarily under cultivation land use (iii) small holder private land, (iv)large private lands or individual ranches. The land tenure and land rights system need to be ascertained along with decision making mechanisms at local levels. These must be integrated with government policies and planned at the apex level.

(f) Growing trails eventually need to be carried out to test and propagate the desired products. Different spacing and species combinations need to be tried under various conditions to develop best agro system.

Trees, shrubs and products of some dryland plant communities were identified which were grown and possessing potential uses by agro industries and end users:

Table 9.1

Name of Tree/ shrub	**Product**	**Uses**
Acacia Senegal	Gum Arabic	Emulsifier, stabilizer in processed foods and beverages, ink, paper, textiles, pharmaceuticals, glues, etc.
A. Mellifera	Gum	Emulsifier, stabilizer in processed foods and beverages, ink, paper, textiles, pharmaceuticals, glues, etc.
Astragalus sp.	Gum	Emulsifier, stabilizer in processed foods and beverages, ink, paper, textiles, pharmaceuticals, glues, etc, tragacanth, processed food - very valuable
Boswellia sp.	Frank insense, olibanum	Incense, perfumes, food and beverage flavouring
Commiphora myrrha	Myrrh	Incense, pharmaceuticals, perfumes and flavouring.
Commiphora sp.	Opopanax, opopanom.	Perfumes and flavouring.
Lawsonia sp.	Henna	Shampoos, soaps and cosmetics
Steculia sp.	Gum karaya	Pharmaceuticals, paper and processed food.

Vermonia galamensis	Oil	Epoxies for adhesives, plasticizers, industrial coating, varnishes and paints. Meets EPA regulations to replace solvents as a reactive diluent in resin systems.

Source: Desertification Control Bulletin No. 17, 1988.

9.2 CENTRES OF DRYLAND PLANT DIVERSITY

The Centres of Plant Diversity (CPD) project (WWF and IUCN 1994-97) has set out a basis for conserving plant communities and biodiversity. Worldwide, areas have been identified which, if conserved, would safeguard the greatest number of plant species. Some 234 'first order' sites are included, selected on the basis of the total number of native and / or endemic higher plant species. Other criteria for site selection included the presence of important gene pools of useful plants, the range of habitats represented and the degree threat on the site. As the criteria for selection focused on floristic richness, the vast majority of CPD sites are in the wet tropics. However, a number of dryland areas qualify as given in Table 9.2.

Table 9.2 : Centres of Plant Diversity (CPD) sites containing significant areas of tropical drylands

(Excluding sub-humid areas)

Code (a)	Site	Area (Sq. km) (b)	Flora (c)	Endemism (d)	Types of dryland vegetation
Madagascar IO 1	Madagascar (Whole island treated as site)	587000	9345	Dry southern Domain has highest level of endemism in the island's flora	Dry deciduous forest and deciduous thicket in the south-west including Adansonia, Pachyodium and Didiereaceae (endemic family
Namibia /Angola Af50	Kaokoveld	70000	952*	At least 116 (12%)	Desert vegetation, including Welwitschia, Savanna.
Somalia Af42	Cal Madow	9600	1000	High species endemism	Lowland semi-desert and bush land, to dry montane forest, with Boswellia (frankincense) and Commiphora (myrrh)
Somalia Af44	Hobyo	3000	<1000	High species endemism	Dune vegetation, deciduous bush land.
South Africa/ Namibia Af51	Western Cape Domian (Succulent Karoo)	111212	5000	1750 (35%)	Mostly succulent shrub-land with associated annuals, extremely rich in Aizoaceae, Crassulaceae and Euphorbiaceae; high succulent diversity, unparalleled world-wide.
Australia Au3	Central Australian Mountain Ranges (and surrounding area)	168000	1300	120 (9%)	Grasslands, shrub-lands, wood-lands, rock and cliff communities; the site contains 65% of the known flora of Central Australia
Australia Au 7	South-west Botanical Province	309840	5500	2472 (45%)	Semi-arid scrub in north and east of province.
Mexico MA 12 11	Central regions of Baja California peninsula	36000	>500	c.100 (20%)	Desert vegetation, including Cactaceae and 10 endemic genera.

Code (a)	Site	Area (Sq. km) (b)	Flora (c)	Endemism (d)	Types of dryland vegetation
Mexicc/ USA MA	Apachian/ Madrean region	180000 (whole region)	4000	—	Thorn scrub and other desert vegetation, including Cactaceae, in Chibuahuan and Sonoran deserts
Ecuadcr PO7	Galapagos Islands	7900	541*	229 (42%)	Xerophytic scrub, including endemic Cactaceae, in lowlands.
Brazil SA 19	Caatinga of north-eastern Brazil*	1000000		6 endemic genera	Xerophytic deciduous forest to sparse scrub, savanna, cerrado and grassland.
Chile, Peru SA 42,43	Lomas formations (including those of the Atacama Desert)	>7000	>930	>60% endemism in Chile; 42% in Peru	Desert vegetation of annual, short-lived perennials and woody scrub
Iran SWA 10	Touran Protected Area Biosphere Reserve	18604	1000	Contains most of c. 150 endemics of the Central Iranian deserts	Semi-desert, xerophytic and halophytic communities.
Oman/ Yemen SWA1	Dhofar Fog Oasis	30000	900	c.60 endemics; two dry endemic genera; Dhofaria and Cibirhiza	'drydeciduous woodland, shrubland, semi-deciduous thicket and grassland to desert shrubland and open desert
Yemen SWA 4	Socotra	3625	815*	230-60 (28-32%)	Semi-desert and dry deciduous shrubland, including Boswellia, Commiphona and Dendrosicyos

Notes : (a) Site codes follow WWF and IUCN (1994-97).

(b) Refers to the total area of a CPD site. For large regions this is sometimes an estimate to the nearest 100 Sq. km. or 1000 Sq. km.

(c) Estimates for the number of indigenous higher plant species for the whole CPD site based on current botanical knowledge. An asterisk (*) denotes exact number of indigenous higher plant species so far recorded from the site.

(d) Refers to the number of species whose distribution is restricted to the CPD site. Where a site contains a range of vegetation types, it has not been possible to determine endemism for the arid and/ or semiarid component.

(e) The area covered by caatinga vegetation itself may in fact be closer to 600,000 Sq. km. and comprise c. 10,000 species.

Source : WWF and IUCN (1994-97).

It is worthwhile to mention that many of the sites contain a range of different habitats, including humid areas within dryland areas. The next challenge is to ensure the floras of these sites are adequately conserved. Some, such as Hobyo in Somalia, are reasonably well safeguarded by their isolation and low human population density, whereas others like Madagascar are under severe pressures. Conservation prescriptions will vary from site to site and will need to be formulated with local people's input and with respect to traditional land management practices and resource rights.

Unless political boundaries coincide with the bio-climatic boundaries of plants and plant communities or there has been a specific project to find out, reliable numbers for species are generally difficult to obtain for most dryland areas. Over 1900 native higher plants are found in Australian dryland areas (Jessop 1985). About 1200 species are found in the Sahara region from the Atlantic to the Red Sea (Ozenda 1977, Takhtajan 1985). Niger and Oman are entirely within drylands and coincidentally both support 1180 species (Davis *et. al.* 1986, Ghazanfar 1992). Despite the apparently limited data for tropical drylands, put into context, their broad floristic composition is far better known than for tropical rain forest areas. Details of dryland plant diversities are given for some important species:

(a) *Acacia Senegal:* Produces gum Arabic, used as an emulsifier in foods, diet, drinks, ink and used in textiles, paper, adhesives and paints. 80% of the world's production comes from the Sudan. Production had dropped because of desertification but in view of high demand large scale restocking programme was undertaken in Sudan, Mali, Senegal, Burkina Faso. The countries did not provide much support in terms of costs and weak infrastructure and marketing network. The international agencies came forward to support the cultivators and presently the species are grown in large quantity. A. Mellifera produces a similar gum.

(b) *Commiphora Myrrh:* The specie is used for producing resin myrrh, used as an incense, in perfume, a flavouring additive in soft drinks, sweets and has potential in Pharmaceuticals. It is being exported from Kenya mainly by Somalis. There is also possibility of distilling myrrh into an essential oil or a resinoid and its marketing is developing very fast.

(c) *Commiphora Hortziana:* It produces opopanax, a "bisabol myrrh, similar to myrrh but of lower value than true myrrh, but since the true myrrh is not in sufficient quantity to meet the international demand, it is worth export

(d) *Commiphora* sp.: There are several other uses than perfume and flavouring. In addition to Africa, Asian countries have started growing the species for meeting their country as well as export demands. Pharmaceutical companies also use for preparation of medicines.

(e) *Boswellia Neglecta:* Produces frank incense, also known as olibanum. Used as a scenting agent in perfumes, lotions etc. Since the demand of the product is increasing fast, for indigenous requirements as well for other countries, various dryland regions grow in water deficient areas.

(f) *Astragalus* sp.: Produces gum tragacanth, used in processed food, of high value in local and international market. The commercial tragacanth comes mainly from Turkey, Iran and Syria. Little is in literature about its propagation. There are several Astragalus species in Kenya, which are used for gum production and therefore, meeting the international demand for industries.

(g) *Sesbania Sesban:* It produces gum. similar to guar, used in the food industry. In the past decades, guar has been used as gum extracting specie and its price has been increased similar to this variety.

(h) *Sterculia* sp.: It is also used for gum production and its production has been grown significantly as other species are unable to meet the demand. S. urens gum karaya from India is used as a substitute for A.Senegal gum Arabic and gum tragacanth. In view of economic importance, its production has been geared up by the cultivators of dryland areas.

(i) *Vernonia Galamensis:* Produces an oil which is used in the manufacturing of Polyvinyl Chloride (PVC), adhesives, plasticizers, industrial coatings, varnishes and paints. After adjudging it's very high potential for commercial exploitation, the growth has been increased significantly.

9.3 USE OF DRYLAND PLANTS

Plants have long had greater value to societies subsisting in drylands. There are many dryland plants that transgress geographical boundaries and occur in different regions and continents and some have attained great utility for the populations of different areas, though precise uses vary from place to place. A good example of all this is Dodonaea Viscosa, a member of the largely Australian genus. D. Viscosa is found in many areas, including areas throughout the Pecific, South America, the Sahel, the Karoo, Arabia, east Asia and China. Depending on location, even within Australia it can vary from a multi-stemmed shrub 1.5 m high (in drier areas like Great Victoria Desert) to a tree of 6 m in wet sclerophyll forests (West 1984). Its uses are very diverse. Leaves are chewed like Coca (Erythroxylum coca) in Peru and it is known as Native Hop in Australia for its former uses as a substitute in beer making. In Botswana, it is an important bee plant. In many places its qualities as a good firewood, igniting readily, have been recognized, as has its potential for carrying and as a hedging material. From Gabon and Nigeria, to the Sudan, India and the Philippines, D. Viscosa is a well-known febrifuge, whilst an infusion of the roots has been claimed as one of South Africa's oldest remedies for the common cold. It is also an important fodder plant in Australia, but geographical variability may account for its apparently low palatability to livestock elsewhere.

The Argan (Argania spinosa) is a dryland plant whose ecological, economic and cultural importance is closely interlinked (Prendergast and Walker 1992) and it is the only species of its family north of the Sahara. To local people, it is important in a number of ways. Its timber is dense and strong and makes charcoal more highly prized than less traditional sources such as Australian Eucalyptus species. Despite an armament of thorns, Argan trees are heavily browsed by goats. The seeds provide highly nutritious cooking oil, which is three times the price of olive oil. The Argan is not threatened at the species level but its populations are as a function of clearance and land degradation pressures.

Some dryland plant species have assumed national or regional importance as trading commodities. Two examples show the relative novelty of such trade and how it is being threatened. In the Sonoran Desert of Mexico and USA, the ironwood (Olneya tesota) is specie of both ecological and economic importance. Ecologically, it acts as a nurse plant for other species, including cacti, by creating favourable micro-habitats for germination and seedling establishment and it is important too for the survival of many animals. Local people have traditionally carved Ironwood to make implements needed for their hunter-gatherer life-style. From the 1960s, the demand for Ironwood trees has increased, as carving for the tourist trade has become an important economic activity. Now up to 3000 craftsmen are involved, using an estimated 5000 tonnes per year of Ironwood. However, land clearance for agriculture and charcoal production is threatened Ironwood availability.

In southern Africa, the palm Hyphaene petersiana is important specie with a diverse range of uses. Its large strong leaves are used in basket making, its fruits are edible and its timber used for building and fencing. In north-central Namibia, baskets are almost exclusively destined for local or regional markets (Konstant *et. al.* 1995). In early 1990s, the annual offtake of leaves amounted to 10%, within sustainable levels estimated at 30% per annum, but now severe browsing livestock is a threat to the palm's regional survival. In Botswana, subsistence basket making was turned in the 1970s into a commercial venture for the tourist and export market (Cunningham and Milton 1987). The effects of a seven-fold increase in the value of the industry from 1976-82 reduced both leaf availability, especially near settlements and the possibility for the species to reproduce through seed.

Any destructive harvesting of dryland plants has to be closely reviewed given the low natural productivity and long gestation time. In Western Australia, the highest quality incense oils from Sandalwood Santalum spicatum are produced from trees in semiarid areas (Loneragan 1990). Exports worth US$ 10 million per year are made to the Far East. Whole trees, including their root systems, are harvested but it takes from 50-90 years in these areas for them to reach a commercial size. The sustainability of the trade depends on controlled exploitation and eventually, cultivation.

For the quarter of a million or so, species of higher plants, though relatively few, have become so valued that they are traded globally. Among dryland plants, one of the most important is Frankincense (Boswellia sacra). The Ancient Egyptians were using incense (the tree's resin) by the 5th dynasty i.e., 28oo BC; Greek and Roman physicians, as well as those of India, applied it to an impressive array of illnesses; today it is still burned in religious ceremonies and used medicinally. Originally, Frankincense was traded from the hills of Dhfar in southern Oman where, within Islamic times, the local kings controlled the trade and commanded extra-ordinary wealth (Miller and Morris 1988). Although, official trade statistics are now scarce, Somalia and Ethiopia have been the main exporters since 1970s, with China the leading market (> 1000 tonnes in 1984). The 1994 price for top quality incense from another Boswellia species (B. frereana) was US$ 6 per kg. (Coppen 1995).

Another internationally traded dryland tree is resin, but of far higher commercial value, is Gum Arabic, a polysaccharide produced by Acacia Senegal. Its high water solubility and low viscosity confer much prized emulsifying, stabilizing, thickening and suspending properties, which are exploited in many food products, in the pharmaceutical industry for the manufacture of tablets, in the printing industry for treating lithographic plates and in ceramics for strengthening clay. The production of Gum Arabic is greatly affected by the vagaries of climate like severe Sahelian drought of 1973-74, which halved Sudanese exports in 1975 to about 20,000 tonnes (Coppen 1995). In 1991, the European Community, the largest market for Gum Arabic, imported a total of 32100 tonnes, 17100 from the Sudan, whilst the USA imported a further 8300 tonnes. Since top quality Gum Arabic from the Sudan has a market price of about US$ 5000 per tonne, it is a considerable foreign currency earner. The source is still 'wild' although exploitation is now also taking place of the plantings of A. Senegal established for desertification control. To collect the gum, a strip of bark is levered up and off the wood, causing minimal damage and the tears of gum are subsequently picked by hand. Although yields are variable, c. 250 gram per tree per year is cited usually as an average (Coppen 1995). The development of the Gum Arabic industry has focused on establishing new countries, like Kenya as suppliers, but there is still a need for research on a number of agronomic aspects, including assessments of yields in both wild and planted populations. A potential threat to the trade is the manufacture of synthetic alternatives.

The Economic Botany Data Collection Standard (Cook 1995), developed by the International Working Group on Taxonomic Databases for Plant Sciences, provides a system, whereby plant uses in their cultural context are described, using standardized descriptors and terms and attached to taxonomic data sets. The 13 Level, 1 states as given in Table 9.3, cover all uses of plants and are divisible into a variable number of level of 2 states (there are a total of 107). The Survey of Economic Plants for Arid and Semi-Arid Lands (SEPASAL), (Davis *et. al.* 1997) sues this standard scheme. The sorts of plants it deals with, are now being internationally recognized for their importance as traditional resources (D. Posey and Hinchcliffe 95). By early 1997, SEPASAL contained information on about 6100 dryland species worldwide. The top four families represented are the

Leguminosea (1152 species), Gramineae (769), Compositea (352) and Chenopodiaceae (217) and the top four level, 1 uses are animal food (2689 species), food (2306), materials (2173) and environmental uses (1986). Among the latter's 11 levels 2 states, Erosion Control has the most species currently ascribed to it (571) apart from Ornamentals (762), with 158 and 153 species from the Gramineae and Leguminosea respectively. An important feature of SEPASAL is its ability to search for particular keywords and combinations of characters and its application to answering enquiries about dryland plants. As well as data on plant uses, there are other fields like plant descriptions, vernacular and trade names, sources of seeds and details on soils and climate tolerances.

Table 9.3 : Level 1 sates

LEVEL 1 STATES	LEVEL 2 STATES (Environmental uses)
Food	Unspecified environmental uses
Food additives	Erosion Control
Animal food	Shade/Shelter
Bee plants	Revegetators
Invertebrate food	Indicators
Materials	Soil Improvers
Fuels	Ornamentals
Social uses	Boundaries/ Barriers/ supporters
Environmental uses	Agro-forestry
Vertebrate poisons	Firebreaks
Non-vertebrate poisons	Pollution Control
Medicines	
Gene sources	

Source: Cook, 1995

1. Use of Dry-land Plants in Medicinal Preparation

In various countries of the dryland areas, various species are used for preparation of medicines through Indigenous System of Medicines for centuries. Due to introduction of Allopathic System of Medicines, use of these medicines was decreased significantly. In India, more than 1000 dryland species are used for preparation of medicines, but in view of slow pace of treatment, people all over the world diverted to Allopathic System of Medicines. Large populations in the country prefer to get treatment from the medicines produced through indigenous system. The attraction at the allopathic was its instant result in treatment. There are thousand of doctors of indigenous practice using the medicines collected from hills, mountains, wastelands. The benefit of the system is the medicines for curative and preventive uses and health tonics. The practitioners know the treatment of acute and serious diseases not possible to be cured from any other system. In India, the varieties of species of medicinal use are in plenty and are used for preparation of medicines in different ways. The problem with the system was mainly the neglect by country government and the doctors of this system lived in misery even were unable to attain their livelihood from treatment.

(a) Promotional Activities

The indigenous system of medicine has been geared up in last few decades and various pharmacies have been established for promoting the medicinal system. The government of India has established a separate department in the Ministry of Health and Family Welfare and National Medicinal Plant Board for promoting the cultivation of medicinal plants in cultivated farms, wastelands and for collection of medicines from mountains and hills. These efforts are made for enrichment of the medicinal contents in the plants and its systematic use in preparation of medicines. The promotional measures are welcomed by the cultivators as the economic benefits in commercial cultivation are enormous. In addition, marketing network has been established and every state government has been strengthened with similar boards.

During last decade, the shifting trend from allopathic system has been witnessed in view of various side effects from the medicines. The persons using medicines have developed various diseases including losing the physical capability to face the diseases. The shifting trend has been welcomed by the indigenous system and the promotional activities have further helped in growing medicinal plants to meet the increasing demand.

Benefits from the indigenous system of treatment are many. In addition to free from side effects, the plant content help in removal of disease from root, though it takes its own time. Since the medicinal plants grow in areas other than cultivated fields, these are free from impact of chemical fertilizers and pesticides. The medicinal contents are pure to most effective extent. The promotional efforts of the government has cautioned the cultivators to avoid the use of chemical fertilizers and pesticides to make the product marketable internationally. The commercial values of these species are more than the cash crops and wasteland cultivation is also promoted for the dryland cropping.

In addition, herbal and aromatic species are also promoted for making the products demanded in huge quantity. The future prospects of species of medicinal, herbal and aromatic plants are quite significant. The cultivators are also encouraged for adopting biotechnology practices to eliminate the chemical impact. The developed countries have already banned its use in view of serious health impact, but developing countries are in the process of adopting such practices.

Recently jetropha cultivation has also been promoted for using as bio-diesel. Presently 20% of bio-diesel is admissible to be mixed for plying vehicles, which may be enhanced in due course of time. The benefits of such species are their survival and growth in dryland with little water and free from animal grazing. The species presently grown and proposed to be grown are environmental friendly and adaptable to the climatic conditions.

(b) Nutritive Value of Dryland Species

During the monsoon, edible plant species are abundantly available, the forage plants are either consumed by grazing animals or dry up and deteriorate nutritionally. Dried non-edible shrub species of Prosopis Cineraria and Ziziphus Nummularia are used for feeding livestock. Among the animal feeds and feeding stuffs, cereal grains like seeds of bajra (Pennisetum typhodes), barley (Hordeum vulgare), jowar (Sorghum bicolor), protein rich leguminous seeds like guar (Cyamopsis tetragonoloba) and moth (Phaseolus aconitifolius) and oil bearing oil cakes of ground nut, til (Sesamum indicum) and and cotton seed (Gossypium) are rich sources of nutrients.

Dryland ground flora is mainly composed of annual forages viz. grasses and herbs and perennial grasses. The nutritive value of ground forage is considerably high. The nutritional value of perennial grasses, which resprout with the onset of the monsoon, complete their reproductive life cycles within 50-60 days. On an average, desert grasses harvested during the monsoon contain 10-15% crude proteins, 45-60% cell-wall constituents, 20-28% hemicelluloses, 20-32% cellulose and 1.6 to 3.0% lignin. The average in-vitro true digestibility ranged between 70 and 87%.

Paul *et. al.* (1982) reported that the average crude protein content of the two strains of Cenchrus ciliaris at the flowering sage i.e., 45 days after regrowth, was 7.6% which is dropped to 6.0% at the flowering stage i.e., 75 days after the regrowth. After three months of regrowth, the average crude protein content of these grasses was only 3% on dry matter basis. Cell-wall constituents were considerably increased from the pre-flowering to the maturity stage. Attempts have been made to improve the nutritive value of nutritionally poor grasses harvested at the maturity stage, by ensiling with lactobacillus culture (Pancholi and Mali, Personal communication). Good quality silage was obtained by ensiling C.Ciliaris straws for 40 to 90 days with 1% and 2% urea. Desert grasses on an average, contained 0.8% calcium, 0.7 mg. Magnesium per 100 g, 0.6% potassium, 0.17mg sodium per 100 g, 0.18 mg phosphorus per 100 g, 235 ppm iron, 70 ppm manganese, 54 ppm zinc and 38 ppm copper (Dhir 1984).

The perennial grasses and associated annuals- grasses and herbs, harvested at pre-flowering to maturity stage, calcium, manganese and copper showed little fluctuations but the zinc level increased and the sodium, potassium and phosphorus levels decreased with the maturity of the grasses. Associated herb species contained more calcium and less amount of sodium in comparison to the grasses, the desert ground vegetation apparently contained adequate levels of various minerals and can meet the requirements of the animal foraging on these feeds (Joshi and Gupta 1984). Bohra (*et. al.* 1988) studies on micro-mineral status of the blood of desert goats and sheep, foraging on sown pastures and piedmont scrub areas, have indicated that the plasma of these animals contained appreciable quantities of copper, iron and zinc but plasma manganese level of both the types of animals was lower than the critical level of this element reported to be associated with the deficiency symptoms.

(i) **Crop Residues:** Straw and stovers of bajra (Pennissetum typhoides), jowar (Sorghum bicolor) and wheat (Triticum vulgare) are stored and fed to the penned animals. Other agricultural by-products like leaves and stovers of moong (Vigna radiata), moth (phaseolus aconitifolius) and til (Sesamum indicum) are often fed to the animals. The leaves of P. acinitifolius, S. indicum and V. radiata contained 11.4%, 13.0% and 15.4% crude protein, respectively. Straws are generally, low in crude proteins. Their digestive protein level is generally below one percent. The crude protein contents of C. retragonoloba, V. radiata and S. indicum stovers were 8.1%, 9.8% and 3% respectively. Due to high fibre content, most of the energy provided by these feeds to the animals is used up to mastication, digestion and assimilation and the net amount of energy finally available to the animal is low. Plant breeders' programmes have evolved improved varieties of various crops, which had led to high grain yield but in low nutritionally poor straws. Attempts have been made to improve the nutritive value of poor quality straws and stovers by treating these with calcium hydrozide or spraying with sodium hydroxide solution or ensiling with urea and / or molasses and anhydrous ammonia.

(ii) **Desert top feeds:** Feeds, which are obtained from the upper covers of trees e.g., leaves, are commonly known as top feeds. Top feeds are ain important class of feeds, especially for the range foraging desert livestock. Some country-side desert vegetation, which provide top feeds are khejri (Prosopis Cineraria), bordi (Ziziphus nummularia), ber (Z. mauritiana), neem (Azadirachta indica) Israeli babul (Acacia tortilis). Leaves of ardu (Ailanthus excelsa) and papal (Ficus religiosa) growing near settlements also provide considerably quantities of top feeds. Tree leaves are lopped and freshly offered to grazing animals as well as air-dried, stored and offered to penned animals. The most promising and widely growing desert tree, P. cineraria is rightly called the king of desert trees. Naturally growing P. cineraria plants are very slow in growth. CAZRI scientists had evolved two fast growing varieties (PC-4 andPC-5) of P. cineraria tree.

Most of desert feeds are good to fairly palatable to desert livestock. Sheep and goats tend to consume more top feeds (3.3 kg/ 100 kg b.w.) than the perennial grasses (2 to 2.5 kg/ 100 kg b.w.). Sheep were found consuming as high as 3.8 kg top feeds per 100 kg. body weight. Camels, on an average, consumed daily 1.4 kg of body weight. Consumption of Z. nummularia leaves are generally more than those of P. cineraria leaves of both sheep and goats. Consumption of both these feeds by the goat is considerably higher than by the sheep. This is mainly due to the fact that the goat is basically a browser and has more preference for tree leaves than sheep, which is a grazer.

(iii) **Chemical Composition:** Most of the top feeds are rich in crude protein and calcium, but their phosphorus content is relatively low. Moisture content of fresh tree leaves generally varied from 60-70% in fresh weight basis. This preformed water substantially meets the water requirement of the desert livestock. A excelsa leaves contained as high as 19.6% crude protein whereas Z. nummularia and P. cineraria leaves, on average contained 14% crude proteins. P. bengalensis leaves contained as high as 79% total carbohydrates. On average A. indica and A. excelsa leaves and P. cineraria and Z. nummularia leaves contain 14.6% and 17.4% crude fibre respectively. A. lebbek, F. feligiosa and F. bengalensis contain as high as 31.5%, 26.5% and 26.8% crude proteins.

The nitrogen free extract (NFE) of most of the feeds ranged between 50% and 60%, except for A. nilotica leaves, which had as much as 70% of A. lebbek had as low as 44% NFE on dry matter basis. The highest inorganic matter content has been recorded in Salvadora oleoices leaves - 40-8%, followed by S. persica - 31.4%, A. Senegal - 16.4% and A. excelsa leaves 15.5%. Studies by Dhir (1984) indicated that the desert tree leaves on an average contained 3.8% calcium, 4.8 mg magnesium per 100 g, 1.4% potassium, 0.95 mg sodium per 100 g, 2.0 mg phosphorus per 100 g, 380 ppm iron, 52 ppm manganese, 32 ppm zinc and 19 ppm copper. Similarly, leaves of shrubs have been reported to contain 2.7% calcium, 0.56 mg magnesium per 100 g, 0.51% potallium, 0.13 mg sodium per 100 g, 0.13 mg phosphorus per 100 g, 551 ppm iron, 41 ppm manganese, 31 ppm zinc and 20 ppm copper.

(iv) Nutritive Value: The total digestible" nutrients (TDN) of A. excelsa leaves are as high as 65%. The digestible crude protein (DCP) of these leaves was also recorded as the highest (16%) among all desert feeds. TDN values if P. cineraria and Z. nummularia leaves were 45% and 50% respectively, whereas their average DCP value was 5.6% only. The DCP of the relatively less palatable A. indica (9.3%) and A. lebbek (11.3) were relatively high.

The nutritive value of the important desert top feeds has been evaluated by in-vilto technique. The average in-vitro dry matter digestibility (IVDMD) of the tree leaves was of the order of 55.6 + 5.29 %. A. Senegal leaves with 16.8% crude proteins and 3.4 % lignin but devoid of cutin, tannins and silica, exhibited the maximum IVDMD value of 75.0% in comparison to the other tree leaves containing tannins. P. Juliflora leaves, which contained as much as 33.5 CP and almost negligible quantities of tannins has an IVDMD as low as 55.4%, which may be due to some unidentified factors. The above narration is supplemented by Table 9.4.

Table 9.4 : Effect of level of tannins on digestibility of desert tree leaves.

Tree Species	Crude Proteins %DM	Tannins %DM	IVDMD %DM
Acacia nilotica	14.6	8.0	55.2
A. Senegal	16.8	—	75.0
A. tortilis	14.6	6.4	48.3
P. cineraria	14.6	14.8	42.2
P. Juliflora	20.1	—	55.4

Source: Central Arid Zone Institute, Jodhpur (India) 1991

9.4 CONSERVATION ANALYSIS

International trade in certain plants has reached critical levels. Between 20-30,000 species worldwide are now covered by the Convention on International Trade in Endangered Species of Wild Fauna and Flora (CITES, Mathew 1994). Among them are many groups of plants with characteristic adaptations of dryland environments, like succulence. Many of these plants are highly prized and their wild populations often threatened by plant collectors. include all species in the cactus genera Ariocarpus, Discocactus and Turbinicarpus; Agave arizonica and A. parviflora from North America and all 50 or so species of cycads Encephalartos from Sub-Saharan Africa. In by far the bigger for it includes all c. 2500 species of Cactaceae, all 700 or so succulent species of Euuphorbiaceae, all c. 340 species of the largely African genus Aloe except A. Vera (originally from the Canary Inlands and now widely cultivated for its medicinal properties as would-healing) and all six species of Alluaudia which are endemic to the dry thorn forests of south and south-east Madagascar. In addition to CITES, many species are covered by special national or regional legislation. Madagascar has a 1983 decree banning the export of many succulent Euphorbia and rare Pachypodium species and the European Union has its own more rigorous regulations. Guaiacum officinale and G. sanctum , important for timber and resin in the Americas.

9.5 CONCLUSIONS

The loss of local plant knowledge and its close connection with the loss of the plants themselves is a key issue facing drylands. This is acknowledged in the CCD; Article 17- Research and Development, binds parties to support activities that 'protect, integrate, enhance and validate traditional and local knowledge, know-how and practices'. Traditional Land Management Techniques, now attracting the attention of ecologists and land managers, are part of this knowledge, as is the ability to find, harvest and process those wild, but probably partly rended plants needed for foods, medicines, fuels, fibres and other uses on a daily or seasonal basis. Such plants can be vital to the survival of people at particular times, during drought and especially for women, children and the poor. They may also have significant values in local economies and certain gene pools of cultivated dryland plants. Yet social and environmental changes have threatened the continuity of traditional knowledge. If desertification and other changes are to be reversed, such knowledge, from ecosystem level down to species level, will need to be recorded for the benefit of people, plants and the places where they live.

Chapter **10**

Carbon Sequestration in Drylands

Desertification and global warming are the two major environmental issues that have some significant linkages. The global warming contribute to increased stresses in dryland areas and poor land use practices, especially vegetation destruction and soil loss leads to atmospheric warming. Therefore, strategies that lead to the amelioration of both problems are likely to be of great global importance.

The consequences of human induced global warming are just beginning to be understood Idso and Idso 1994). Likely climate changes caused by the enhancement of greenhouse gases, particularly carbon dioxide (CO_2). in the atmosphere include not only increasing temperatures but also changes in total rainfall and its seasonality and systematic increases in rainfall intensity(Solomon *et. al.* 1993). Along with the direct effects of increasing CO_2 concentrations on plants, these climatic influences interact strongly with the grazing impacts and other land-use pressures that are already being placed on drylands (Sombroek *et. al.* 1994) Fortunately, there are potential opportunities for migration of the rate of increase of atmospheric CO_2, through absorption and storage of carbons in plants and soils, called carbon sequestration. Mitigation strategies involve two possibilities; the conservation and/ or protection of existing carbon sinks and efforts to increase the capacity of the land to sequester carbon.

The principal biological sinks for CO_2 have previously been thought to be the forested regions of the world, notably tropical rainforests and much effort has been concentrated there (Trexler 1991), Kinsman and Trexler 1993). It is now clear that drylands are also a prime candidate for major carbon sequestration efforts (Glenn *et. al.* 1993, UNEP 1995). However, any effort to sequester large amounts of atmospheric carbon in drylands involves significant scientific and organizational challenges. Not the least of these challenges is to find enough land. To absorb even 25% of the atmospheric CO2 emissions into dryland soil and vegetation would require an area of about 2 to 5 billion hectares. While management systems in the susceptible drylands could be altered to sequester more carbon (Glenn *et. al.* 1993), significant incentives may be required to get land users and owners to participate in sequestration programmes.

10.1 RATIONALE FOR SEQUESTERING CARBON IN DRYLANDS

At the first glance, the possibility of large-scale carbon sequestration in drylands might seem unlikely. Compared to other biomass, drylands accumulate only a fraction of the hundreds of tones of carbon per hectare that can accrue in temperate and tropical forest systems and at a fraction of the annual rate. But the world's drylands store about 60 times more than is added to the atmosphere annually by fossil fuel burning. Small unit changes in the rate at which carbon is emitted or sequestered in these soils can have relatively large impacts on the atmospheric carbon budget, given the significant proportion of the total global land area that are drylands. Drylands cover 6.1 billion hectares world-wide. 5.2 billion of which are susceptible to desertification. These susceptible drylands cover 40% of the world's total land area, a greater portion than is covered by cropland of 1.4 billion hectares or 8% or closed forest of 4.4 billion hectares or 25%.

Estimates of the potential productivity of world ecosystems are based on the type of native vegetation (Oldeman *et. al.* 1991). Primary productivity is very low in hyper-arid areas, approximately 250 kg. ha^{-1} yr^{-1} of carbon, but increases in the arid and semiarid areas (Whittaker 1970), reaching levels as high as 4000 kg ha-1 yr-1 of carbon, in dry tropical forests as given in Table 10.1. These rates are nearly as high as those in temperate regions. Increasing carbon storage in the drylands involves converting degraded lands, which are performing well below their potential, into land-use systems that perform closer to natural production levels. Obviously, the greatest potential for carbon storage lies in the regions with highest potential productivity, while hyper-arid areas have little potential for increased carbon storage.

In general, it seems that global dryland systems without effective management will become much stronger carbon sources under a doubled CO_2 climate (Ojima *et. al.* 1993). From a global perspective or from a dryland perspective, it is imperative that they become sinks. Over 1 billion people currently live in susceptible drylands and any efforts to restore the productivity of these areas benefit their habitants.

Table 10.1 : Estimates of carbon sequestration potential of some major land-use types with projected annual carbon storage and time frames

Options	**Area (Million ha)**	**Rate ($tC\ ha^{-1}\ yr^{-1}$)**	**Period (yr)**	**Cost (US$ tC)**	**Total ($MtC\ yr^{-1}$)**
Dryland crop management	450	0.3-1	5-20	1-5	135
Halophytes	130	0.5-5	Indefinite if harvested 5 years if not	170 (irrigated and harvested) 20 (dryland not harvested)	65
Bush encroachment	150	0.1-0.5	15-50	10-20	37
Energy crops	20 (5% of dryland crop area)	4-8	indefinite	150	80
Domestic biofuel efficiency	Not applicable	Not applicable	indefinite	2-5	75
Agro-forestry (Arid)	50	0.2	30	2-10	10
Agro-forestry (Semi arid)	75	0.5	20	2-10	38
Agro-forestry (Sub-humid)	150	1.5	15	2-10	225
Improved pasture (Semi arid Asia)	10 (2500 degraded globally)	0.1	30	10	1

Source: UNEP 1995

10.2 THE CARBON STORY

Soils of the world's drylands hold 300-369 × 1015 grammes (or Pg) of Carbon (C) as soil organic carbon (SOC), and 473-546 Pg C soil carbonate carbon (CAC). This corresponds with about 20-25% and 68-79% of the estimated world's total terrestrial reserves (Batjes 1996). A feature of drylands is that they have low SOC contents and high CAC reserves (Grossman *et. al.* 1995, Sampson *et. al.* 1993). The relatively large contribution of CAC to soil carbon reserves of dryland systems is illustrated by the fact that the value is as high as 77% in hot deserts and only 7% in the boreal forest. This difference in the proportion of SOC and CAC has implications for the potential of drylands as a sink for carbon. Although reserves of carbonate carbon globally are large, this source of carbon does not participate in the carbon flux to other carbon systems as rapidly as organic carbon (Batjes 1996).

The SOC stored in the world's soils as fresh organic matter, stable humus or charcoal, is two or three times higher than the carbon stored in the natural vegetation and in standing crops (Ojima *et. al.* 1993). The soils of grasslands and cropped drylands store up to 10 times as much carbon as the plants growing on them (Schimel 1993). Soil carbon storage is one of the largest pools of carbon in he ecosystem (Batjes 1966).

Soil organic carbon cannot be considered as a homogeneous carbon form. Instead, it must be viewed as a mixture of labile and more resistant material. This effects the dynamics of soil carbon relative to climate and land management changes. Soil organic carbon is divided up into three major components, which include active, slow and passive soil carbon (Parton *et. al.* 1993). Active SOC includes live soil microbes plus microbial products, total active pool is ~2 to 3 times the live microbial biomass level, the slow pool includes resistant plant material, lignin derived material and soil stabilized plant and microbial material, while the passive material is very resistant to decomposition and includes physically and chemically stabilized SOC.

The short-term fluxes of carbon from the soils are derived primarily from changes in the labile soil organic matter pool. The size of this pool is relatively small but it accounts for more than 60% of the soil carbons respiration under 'steady-state' conditions (Ojima *et. al.* 1993). Natural ecosystems may have large stores of carbons in soil or vegetative compartments. The net flux of carbon from natural ecosystems at steady-state conditions to the atmosphere are essentially neutral. When these ecosystems are disturbed through natural disturbances, fire, pest outbreaks, storm damage and floods or converted to croplands or a semi-managed state by modifying land use practices large fluxes of carbons to the atmosphere may result. Release of carbon may come from soil stores or from vegetation reservoirs (Schimel 1993, Raich and Schlesinger 1992).

In the short-term, the carbon balance of terrestrial ecosystems is particularly sensitive to the impact of human activities, including deforestation, biomass burning, land use changes and conversions and environmental pollution. During the period 1850-1980, soil carbon pools have decreased by 40 Pg C from the original 1471 Pg C and carbon held in vegetation is down by 80 Pg C and from 672 Pg C in 1850 (Houghton 1995). The global release of carbon to the atmosphere from land use changes during the 1990s has been between 1.1 and 3.6 Pg C each year, as compared to 5.5-6.5 Pg C each year from fossil fuel combustion. Large amounts of carbon are displaced annually from soils due to global soil erosion and land degradation (Lal 1995).

The potential of drylands to store carbon will be limited by degradation of natural vegetation and soils in response to human population growth and agricultural expansion (Biswas 1994, Oldeman *et. al.* 1991). As the twenty-first century approaches, it has become increasingly clear that the drylands of the world will be subjected to even greater land use pressures as a result of the continued growth of population. Soils of drylands are particularly vulnerable to degradation in view of the slow speed of their recovery after a disturbance. Human activities have led to about 1035 million hectares of degraded soils in the susceptible dryland.

10.3 EVIDENCE OF CARBON STORAGE

Only part of the carbon pool in soils, viz., the SOC in topsoils, is dynamic or labile and contributes to CO_2 exchange and nutrient cycling in the soil-water-atmosphere system (Ojima *et. al.* 1993). The SOC status in natural ecosystems is closely related to the type and age of above-ground vegetation. The projected change in global climate is thought to be accompanied by higher soil surface temperatures, increased drought conditions and larger erosion hazards (Neilson and Marks 1994, Gifford *et. al.* 1990).

In the long term, the most important determinants of the rate of decomposition of SOC are the rate of input of litter, soil moisture content and temperature. There is a tight linkage between the rate, with the mechanisms behind the linkage working in both directions (Owensby 1993, Schimel 1995).

The majority of the SOC is stored in more recalcitrant material and these soil carbon pools also have different degrees of reactivity and, as mentioned, have been defined as either the 'slow' or the 'passive' SOC (Parton *et. al.* 1993). Together, the slow and passive pools constitute about 90% of the total soil carbon, with an estimated ratio of 3:2 between the two pools respectively. The turnover rates in various pools are the function of biotic abiotic, climatic and soil edaphic factors. These factors are affected by management, including grazing and fire management. The flows of carbon between these pools are controlled by decomposition rate and microbial respiration loss parameters, both of which may be a function of soil texture (Davidson 1994). The turnover time of these pools varies with the soil abiotic decomposition parameter, which is a function of monthly precipitation and temperature (Parton *et. al.* 1993, Ojima *et. al.* 1993).

In general, there is a gross uneven partitioning of carbon between above and below ground functions in the dryland soils. Ratios of about 4:1 have been reported (Marshall 1977) Soils under shrubby vegetation also show a higher below-ground fraction than those under grasslands (Gibbens *et. al.* 1983, Owensby 1993). It is projected that as climate change proceeds, shrubs invade grasslands. Grass to shrub species transition shifts below-ground litter input from fine toot biomass, distributed in surface horizons, to deeply rooted, structurally resistant woody material (Connin *et. al.* 1997).

These structural changes may enhance soil carbon storage lower in the soil profile, due to carbon allocation to deep roots by woody plants and deep-rooted grasses (Nepstad *et. al.* 1994). In drylands, higher mass-loss rates are often associated with fine root biomass relative to coarse roots and decreased microbial biomass with soil depth suggests that shrub invasion may result in greater long-term carbon storage in soils. But soil erosion and associated loss of soil carbon may balance these inputs (Lal 1995).

When desertification occurs, net changes in the carbon pool reflect a balance between fluxes in above and below-ground biomass and soil organic carbon. In semi-arid rangelands of New Mexico, USA a shift from grassland to shrub land resulted in a slight increase in soil carbon storage over 110 year period, 1853-1963 (Connin *et. al.* 1997). In many dryland regions, there has already been a shift in vegetation composition towards a greater dominance of shrubs and other woody plants due to grazing pressure. This bush encroachment has, in some areas, been designated a 'woody weed problem, in parts of Australia and the south-western USA. Projections of climate change suggest that woody plants come to dominate a larger area of land as CO_2 level rise and water relations alter.

Global climate change, as currently simulated, could result in broad-scale redistribution of vegetation across the planet (Figure 10.1). Vegetation change could occur through drought-induced 'dieback and fire'. It is predicted that grassland and shrub lands- dryland ecosystems, could expand and that the tundra and boreal forests contract most. A net result is that under the new equilibrium conditions, the terrestrial biosphere may store-up to 30% more carbon above ground than it currently does and this act as a negative feedback to climate change (Neilson and Marks 1994, Sombroek 1995).

Changes in plant communities associated with changes in climate affect litter quality. This leads to either positive or negative feedbacks depending on whether loganiaceous perennials or non-loganiaceous annuals take over. In some drylands, there may be a decrease in shrub abundance, an increase in half-shrubs and herbaceous species, an increase in C4 plants, an increase in the carbon: nitrogen ratio in plant tissues and a decrease in succulents under all scenarios of climate change (Owensby 1993).

While a range of possible climatic change predictions for the twenty-first century exist, including their possible impacts on dryland areas (Williams and Balling 1995), it is difficult to predict some of the impacts that affects biotic systems. Different microbial responses to rainfall and temperature changes are reasonably well known, but complex interactions like effects on N-fixation, n-mineralization, denitrification, cation leaching and ratios of trace gas emissions remain difficult to predict (Davidson 1994).

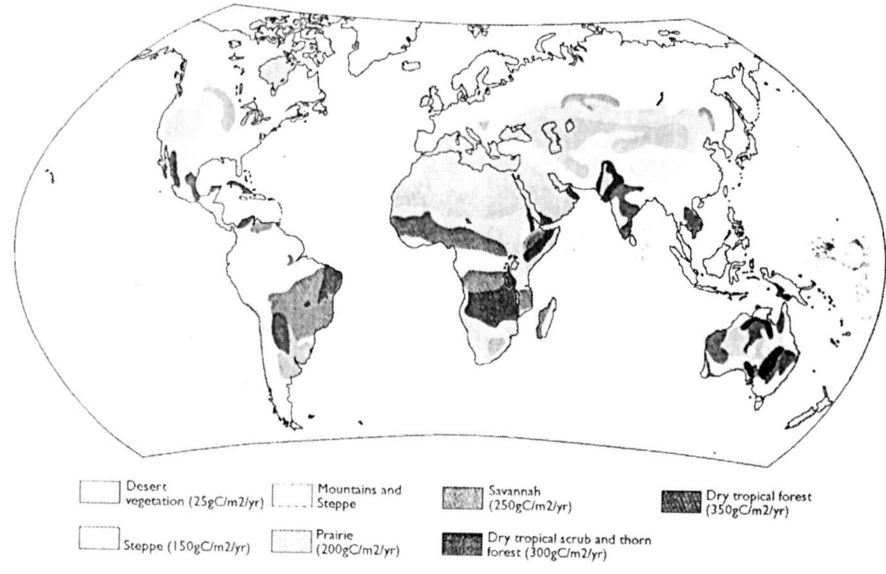

Figure 10.1 *: Potential carbon sequestration by vegetation type (from 1968 and Whittaker 1970)*

10.4 POSSIBLE EFFECTS OF CLIMATE CHANGE ON CARBON STORAGE IN DRYLAND SOILS

Drylands may be among the earliest systems to exhibit vulnerability to the effects of climate changes (OIES 1991), due to their low reserves of water and soil nutrients. Temperature increases and precipitation changes modify processes like evapo-transpiration, decomposition and photosynthesis (Gifford *et. al.* 1990). However, understanding how soil organic matter and nutrient status change in dryland soils in response to climatic change require a knowledge of several biogeochemical input and output processes (Schimel 1993). The effect of global climate change on soil organic matter content, soil organic matter quality and nutrient pools depends on the relative sensitivity of photosynthesis, autotrophic and heterotrophic respiration to climatic changes. There is increasing evidence that water-or temperature-stressed plants are more responsive to CO_2 increase, because higher CO_2 reduces transpiration, than unstressed plants. Rates of carbon gain, decomposition and nutrient cycling are all sensitive to temperature and moisture (Ojima *et. al.* 1993). Projected rises in global temperature and the associated changes in rainfall distribution can have serious implications for the drylands, especially in the tropics (Sombroek *et. al.* 1993).

It appears that the relative growth-enhancing effects of atmospheric CO_2 enrichment is greatest, when resource limitations and environmental stresses are most severe (Idso and Idso 1994, Solomon *et. al.* 1993). It could well be that the percentage growth response of natural ecosystem to atmospheric CO_2 enrichment will be greater than that of managed agricultural systems (Tinker and Ineson 1990).

Since UNCOD in 1977, much attention has been given to the possible climatic effects of rising carbon dioxide (CO_2) concentration in the atmosphere. This rise results primarily from the combustion of coal, oil and natural gas in the industrialized countries, but the CO_2, released is rapidly spread worldwide. Any effects are hence are expected to be global. They may be compounded by rising concentrations of other "greenhouse" gases, notably methane (CH_4), nitrous oxide (N_2O) and various industrial pollutants like chlorofluoromethanes). It is natural to understand the effects of these changes on lands by desertification by following points:

(i) Carbon dioxide, about 343 parts per million by volume of the air, is increased by about 0.4 percent per annum. At probable rates of increase of energy consumption and assuming an important part for fossil fuels in meeting that increase, CO_2 concentration will double in the latter half of the coming years.

(ii) Doubled CO_2, concentration with at equilibrium, probably raise tropical temperatures by 2 or 3 degree Celsius. Other greenhouse gases may increase this by 1 to 2 degree. The effect will probably be unmistakable by the end of the century. Some authorities believe that the global rise of temperature over the past century is the CO_2, effect at work.

Like the albedo hypothesis, the CO_2 effect has been widely tested by GCM experiments. It is reasonable to expect that warmer conditions in low latitudes would increase evaporation from the oceans and hence intensify the whole hydrologic cycle. If evaporation increases, so much rainfall somewhere. Unfortunately, the GCMs are not precise as to location. One experiment (Manabe and Stouffer, 1980) does indeed suggest a slight increase of order 2 to 5 percent, in tropical rainfall for doubled CO_2, but also suggests increased evapotranspiration. There is no clear sign that most arid lands would profit appreciably.

In case of an increase in temperature can only be unwelcome in environments in which heat stress is frequently present, for human as well as for eco-systems. Control of disease might well become more difficult and irrigation costs would probably increase. There is, on this view, no reason to fear that CO_2, effect too greatly. On the other hand, there is little hope that it will be beneficial. A firmer judgement may well have to await confirmation that the effect is actually in progress. As usual, more research is needed; CO_2 is now very high on the research agendas of UNEP and WMO. In a few years there bodies may be in a stronger position to recommend concrete action

10.5 FEASIBILITY OF CARBON STORAGE IN DRYLANDS

On severely degraded lands release of CO_2 due to over-utilization of plant production ensures that drylands are a net source of CO_2 (Sombroek 1995). The benefits of turning this around are obvious. If all the land restoration measures proposed by UNEP (1992) were adopted, the net effect on carbon sequestration would be that over 37 Pg C per year would be sequestered. This represents about 15% of the atmospheric CO_2, emissions (Glenn *et. al.* 1993, Squires and Glenn 1995). This would be a significant contribution to the mitigation of global warming. Some regions will have the potential to sequester more CO_2 than others as indicated in Figure 10.2.

Approx equatorial scale 1:187 million

Figure 10.2 : *Projected changes in leaf area index (LAI) from a double atmpospheic CO_2 scenario. Pink areas have higher LAI, white areas the same or loweer LAI (from Neilson and Marks 1994)*

Dryland soils are low in carbon but there is scope to augment this via better management. Specific sequestration opportunities were calculated for a number of major land use or ecosystem types. Major options considered have been detailed in Table 10.1, along with a time frame and costs.

The option of absorbing excess CO_2 into biomass through global-scale revegetation programmes seems attractive despite problems of cost, availability of land and the difficulty of achieving and monitoring long-term carbon storage (Trexler 1991, Squires and Glenn 1995). Possible carbon offset programmes to remove CO_2 via living vegetation or to store it in the soil have been proposed for every major type of ecosystem. Projected costs of such carbon offset range from US$ 5 to 200 per tonne, which is actually considerably lower than the cost of some carbon source limitation measures (Hour 1991, Trexler 1991). Wisely chosen projects have beneficial social and environmental effects in addition to carbon removal, especially in many parts of the world's dryland, where population pressure is high and quality of life is low. Action to increase biomass on drylands will protect biodiversity, ameliorate the living conditions of local people as well retain carbon (Trexler and Meganck 1993).

1. Treating the Problem

From the measures suggested to assist PACD from the climatological point view, it is highly unlikely that humanity will be able to control climate, but it is very possible that efforts be made to learn to modify it on the local scale. It is also relevant that the affected government can take the concrete efforts successfully. One negative conclusion reached by meteorologists has been that attempts to increase rainfall during drought periods by cloud seeding. Weather modifications as a whole have turned out to be difficult to achieve and still move difficult to confirm. Similarly, attempts to modify climate by flooding desert basin like the Sahara or inland Australia, are most unlikely to repay the effort. The great dry lake of arid zone is flooded naturally from time to time, but re-evaporates. Planting green belts, though desirable as a means of halting soil drilling, is also unlikely to affect large-scale climate. Nevertheless, affected countries can do much to lessen the impact of adverse climate:

(a) The most useful and obvious step is to control land use. It is often not realized that a central purpose of such control is to improve the micro-climate on which natural vegetation and crop depend. Desertification control is amongst the other things, micro-climate control.

(b) In addition, all vulnerable countries, should equip themselves with drought plans, in effect a practical strategy to protect people, lands and resources against future drought. Even if, the present harse conditions in so many countries are relieved by abundant rain, drought will undoubtedly return. The technical and institutional means of combating such drought are also fundamental to the struggle against desertification.

(c) The central part of any plan is to ensure that natural meteorological and hydrological services and international ventures such as AGRHYMET in Africa, have the needed resources to do their work.

It is a unfortunate fact that meteorological and hydrological services in many parts of the arid zone have seriously deteriorated. In part this arises from adverse climate, which has so weakened the economies of many countries that the latter feel unable to provide the money required. But one does not combat disease by firing the doctors. Neither can one expect to overcome drought by closing weather stations or by abandoning hydrological measurement of stream flow. The Expert Group on the Climatic Situation and Drought in Africa considered the latter problem (WMO 1983) and made following recommendations that are equally applicable to other stricken areas:

(a) Take prompt action to provide necessary resources to activate the national meteorological services.

(b) Necessary meteorological data are observed regularly and disseminated promptly to the operational and research institutions. Special emphasis should be placed on necessary observations of rainfall, the upper atmosphere and hydrological values.

(c) The necessary data are available to detect impeding drought conditions and to warn governments of the situation.

2. Interpreting the Information for Forecast

The problems of meteorological and hydrological testing agencies related to analysis of the atmospheric change leading to rainfall or its failure is never true. The atmospheric variations need to be carefully assessed along with its movement directions and possible results in terms of coverage of area. Collection of information from the space satellites is possible at a given time in a day but interpretation of the information keeping in view the wind direction, humidity, air depressions created and possible movement are not easy factors to be assessed. The interpretation of such information require detailed experience on the basis of past data trends and possible impact.

The information gathered through satellite imagery need to be analyzed with the present situation of atmospheric conditions gathered from various meteorological and hydrological agencies and the forecasts made on each day about the rainfall need to be matched with the forecasts and actual occurrence. The variations so observed need to be analyzed in terms of mistakes in prediction and reason for failures in the proper assessment help in arriving the correct system of forecasting to the maximum possible extent.

3. Importance of Weather Forecasts in Dryland Regions

Whole economy of dryland regions is dependent on agriculture and weather forecast play in important role at every stage of. monsoon occurrence. The arrival of monsoon helps cultivators to seed in their cultivated fields and general monsoon conditions and day-to-day situation helps in sowing the fields. In case of any deviation from the forecast keep the cultivation practices difficult and most crucial in rain-fed cultivation practices. The accuracy in forecast helps he cultivators to grow the selective crops as per monsoon conditions but changes in between are not possible once he seeds are sown. The loss to the cultivators in case of deviation put severe loss as most of the cultivators of the areas are utmost poor and possess no capacity to replicate the sowing process in case of sudden variations.

The accuracy in forecast system helps the cultivators to choose the variety from among the food grain; grass or tree saples as per monsoon conditions and the expectations are made on the basis of such forecasts. The economy of the area is mostly dependent on the monsoon conditions and their future course of action including migration in search of employment is based on such conditions. Therefore, for survival of people and ecology of the dryland areas, every developing country need to strengthen the system of monsoon forecasts to the most possible accuracy levels.

4. Strategy for Future

The critical moment of mankinds' relationship to climate on the threshold of human induced climatic change is the key factor. The human economy is releasing into the atmosphere substances that change its optical properties and hence the manner, in which it receives and uses solar energy. It is even possible that the most catastrophic consequences of all out nuclear war' might be a drastic cooling of climate, with consequent destruction of much of the world's biota (Crutzen and Birks 1982). Yet there is necessity to maintain a balanced view. The large-scale climate of the earth seems not have been greatly altered by mankind's century long removal of forest, wastage of soil carbon and release of carbon dioxide. Even though post glacial times, the past 10,000 years, have seen drastic reductions of rainfall in some sub-tropical and tropical areas, temperatures appear not to have fluctuated greatly. But the effect of a doubling of CO2 may well be greater than the post glacial fluctuations. Nature suppresses the expected effect by some heat storage process yet to be identified. Most authorities believe that the CO_2 warming will happen, perhaps delayed for a few decades by the heat capacity of the ocean.

Seen by this perspective of climatic stresses in the drylands also appear to need a balanced appraisal. One school of thought, certainly dominant among professionals, says that the high incidence and prolonged duration of recent droughts are simply aspects of natural fluctuations, due to some deep seated oscillation of the general circulation of the atmosphere and may be the ocean. Proponents of this view point out that prolonged desiccations have affected Africa before. On this view, the situation will right itself, the long term stability of climate will override the protracted drought and correction in Australia has already been witnessed. In many parts of the interior, notably around Alice Springs, the intense drought of 1972 was the culmination of a 20 year downward trend of rainfall. Local residents were gloomily predicting the abandonment of much rangeland. Yet the rains of 1973-74 reduced these same regions to a quagmire. Tropical drought is often ended in this exaggerated fashion.

Nevertheless, there is apprehension that human misuse of the desert fringe is indeed producing a lasting decay of the arid zone climates. There are plausible hypotheses, supported by modeling experiments, which blame the albedo and soil moisture feedbacks for the intensification of drought in Africa. There is yet no way in which climatologists can decide whether it will eventually be overridden by the general circulation. There are now many more climatologists, who are prepared to say that desiccation will continue than there were at UNCOD in 1977. This issue, though crucially important, does not really alter the need for intensified action to combat the effects of drought and to try to foresee the latter. Even if, the African droughts are indeed only an unusually prolonged fluctuation, they have still wrought untold damage and it is certain that they will recur time after time in future generations. The same is true of Brazil, Australia and inner Asia.

Hence, the drought-prone countries of the tropical world, especially those dependent on rain-fed agriculture and pastoralism, must at all costs preserve the scientific capability of monitoring and if possible predicting the needed rain. It is tragic that they have allowed their scientific services to deteriorate. They must continue their efforts to bring destructive forms of land use under control. They must contrive, somehow, to develop effective strategic plans to combat droughts, as they will occur again and again. It is vital that the return of adequate rains in some areas not weaken this effort. To combat against desertification will take generations to complete and even then the problem may need vigilance.

10.6 CONCLUSIONS

Carbon sequestration, the process of carbon stock protection and aggradations, is being looked upon as a viable option by a world increasingly worried about the potential impact of global warming. The world's drylands have the potential to reach an annual carbon sequestration rate of over 1.0 Gt (UNEP 1995). This is the figure cited as the minimum to be considered as relevant to the efforts to mitigate the build-up of atmospheric carbon.

There are, however, significant risks with carbon sequesterisation strategies. These arise because of the small differentials between the degraded and rehabilitated land and because of the vagaries of the climate. Carbon is easily gained but just as easily lost. The risks associated with carbon sequestration projects inevitably increase along the aridity gradient, due to increasing climatic variability and because of the complex land tenure and community structures. The only ways to remove carbon are the more efficient production and use of energy and in better land management.

The links between drylands and climate change mitigation are established. Stopping land degradation will slow the rate of release of CO_2 to the atmosphere; rehabilitating degraded lands can help sequester carbon in plants and soils (Figure 10.3). Carbon sequestration is compatible with techniques adopted to mitigate the effects of land degradation through increasing land cover and improving soil organic matter content to restore productivity to benefit the local inhabitants and the ecosystem (Squires *et. al.* 1997).

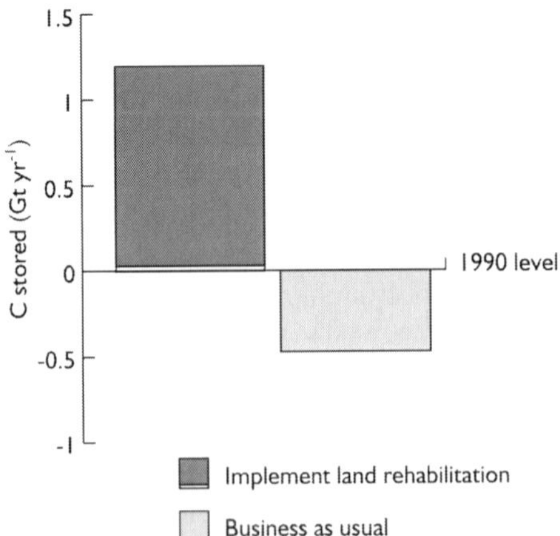

Figure 10.3 : *Projected carbon storage in the world's drylands to the year 2020 with and without active rehabilitation measures (Ojima et. al. 1993)*

Achieving carbon storage requires careful planning, however, as not all rehabilitation options increase soil carbon storage. Some dryland soils have a finite storage capacity for carbon, which cannot be increased by boosting biomass production alone. Improved carbon storage also requires careful consideration of local economic conditions and must be compatible with the needs and aspirations of the local peoples. Sequestration opportunities in the world's drylands may be categorized as follows:

- Increasing the standing biomass of dryland plant communities, either through more conservative grazing and browsing of rangelands, rehabilitation and revegetation of degradation lands, or afforestation of areas devoid of trees.
- Increasing the storage of carbon in long-lived woody plants, trees and shrubs.
- Substituting non-biomass fuels, solar stoves, biogas etc., for fuel woods.
- Utilizing biomass energy as replacement for fossil fuels.

Revegetation projects appear to be one of the most cost-effective ways to increase terrestrial carbon storage and many countries are considering or have implemented such programmes as part of their strategy to combat desertification. In addition, soil carbon can be increased too (Trexler and Meganck 1993).

Sequestration can occur in natural woodlands, grasslands and shrublands. Natural vegetation presents a wide variety of opportunities, both in terms of physical situation and the social change that may be needed in order to sequester and store additional carbon. Here, the challenge may be less with the physical context of the dryland region and more with the social and cultural changes needed to affect the carbon dynamics in the desired manner. These lands, while they are widely diverse, are usually home to people, who have cultural roots deep in the drylands and their method of use. Where those uses have grown unsustainable, or where carbon stocks are being needlessly depleted, it may be difficult to cause real change to occur.

Chapter 11

Saline Soils in the Drylands: Extent of the Problem and Prospects of Utilization

Natural, or primary, salt-affected soils are widespread in dryland, because the potential evaporation rate of water from the' soil exceeds the amount of water arriving as rainfall, allowing salts to accumulate near the surface as the soil dries (Seabolcs 1979, 1989). Vast tracts of saline soils limit the amount and quality of pasturage available in the drylands. Secondary, or human-induced salt-affected soils cover a smaller area than primary salt-affected soils (Oldeman *et. al.* 1991, Ghassemi *et. al.* 1995), but the secondary Salinization represents a more serious problem for the human use of drylands because it mainly affects cropland.

Arable land is a scarce and valuable resource in dryland regions, yet it is frequently abandoned, when it becomes salinized, due to the very high cost of repair (Dregne 1995). The problem is aggravated by the low salt tolerance of major agricultural crops (Abrol *et. al.* 1988) compared to wild salt-tolerant plants- halophytes, which are adapted to saline soils (Glenn 1995), as given in Figure 11.1. Salt buildup is the biggest enemy of irrigated agriculture in the drylands (Postel 1990, Van Schilfgaarde 1993, Ghassemi *et. al.* 1995, Gardner 1997). In the past several decades, progress has been made in understanding and avoiding the practices that lead to soil Salinization and in developing new crops and agronomic techniques for salt affected soils so they can remain in production without requiring complete restoration (Lieth and Masoom 1993). In recent years, even sea water has been used to produce halophyte crops on an experimental basis (Glenn *et. al.* 1997).

11.1 GLOBAL EXTENT OF SALT-AFFECTED SOILS

There are two main types of salt affected soils (Abrol *et. al.* 1988). The first are saline soils, which contain sufficiently high levels of NaCl, NaSO4 or more rarely, other neutral salts, to inhibit the growth of crop plants. All types of soils can become saline. The second types are sodic, or alkaline, clay soils. They generally have low levels of total salts, but contain enough of the alkali salt, Na_2CO_3, to damage heavy soils, Na_2CO_3, cause the clay particles in soil to disperse, or defloculate, by ion exchange processes, resulting in a deterioration of soil structure. A sodic soil has a low permeability to water and air and a pH above 8.2; plant growth is inhibited by negative effects of sodium on the soil and by the high pH, which reduces nutrient availability to the plant. Soils can become saline or sodic through natural or human-induced processes.

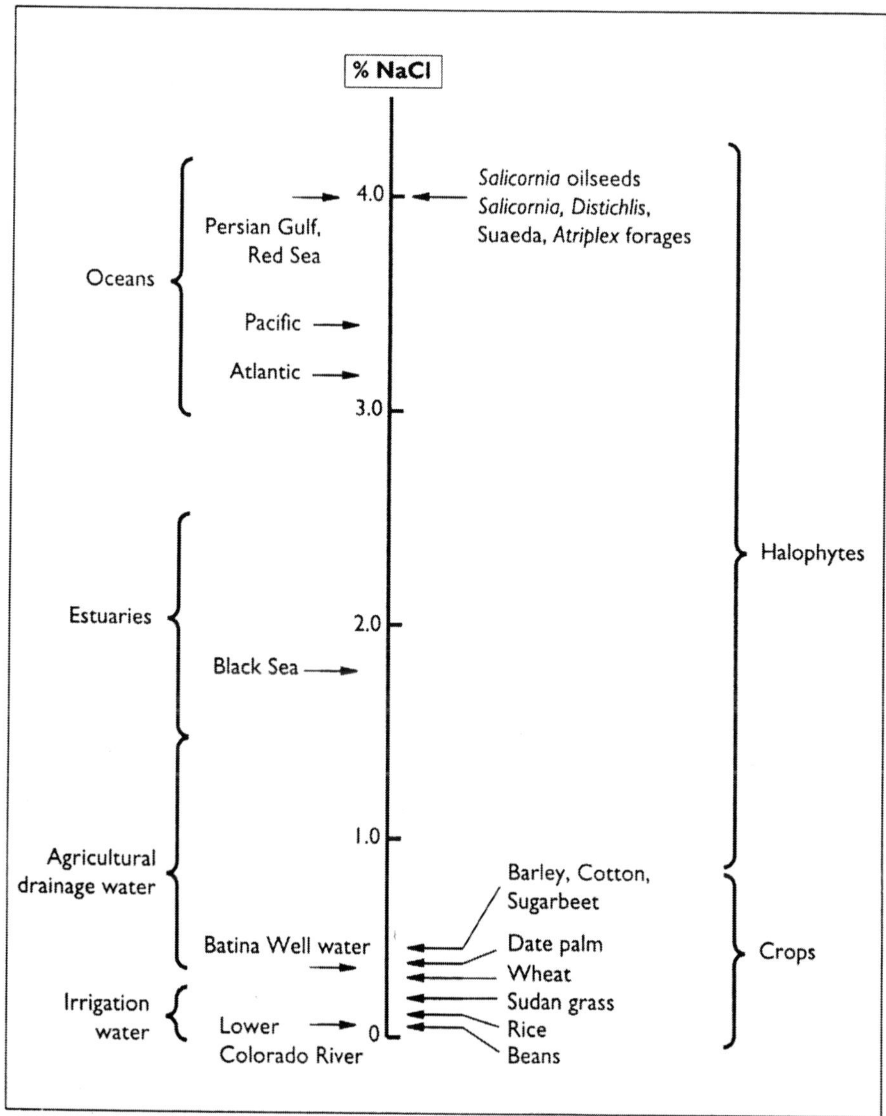

Figure 11.1 : *The salt tolerance limits for acceptable yields of crops and halophytes.*
Source: *Glenn, E.P., Environmental Research Laboratory, University of Arizona*

Accurate statistics on the extent of salt-affected soils are not available for all of the drylands, despite the importance of the problem, but some estimates of the extent of damage have been made. Figure 11.2 shows the broad distribution of saline and sodic soils around the world, both natural and human-induced (Ezabolcs 1979), while Figure 11.3 shows the distribution of human-induced or secondary saline soils as identified in the GLASOD survey (Oldeman *et. al.* 1991). Although salt-affected soils often have a patchy distribution, with salt problems affecting from 10% to 50% or more of the soils in a region, Table 11.1 indicates the best current estimates of all saline and sodic land areas, while Table No. 11.2 shows data taken from the GLASOD database on human-induced salt affected soils in the susceptible drylands.

Figure 11.2 : *Salt-affected soils.* Source: *Szabolcs (1979, 1989)*

Table 11.1 : World distribution of salt affected areas

(Million Hectares)

Continent /Sub-continent	Saline	Sodic	Total
1. Africa	122.9	86.7	209.6
2. South Asia	82.2	1.8	84.0
3. North and Central Asia	91.4	120.0	211.4
4. South-East Asia	20.0	—	20.0
5. South America	69.4	59.8	129.2
6. North America	6.2	9.6	15.8
7. Mexico/ Central America	2.0	—	2.0
8. Australasia	17.6	340.0	357.6
Global Total	**411.7**	**617.9**	**1029.6**

Source: Abrol *et. al.* 1988

Table 11.2 : Global extent of human induced salinization in the usceptible drylands (Million Hectares)

Continent	Light	Moderate	Strong	Extreme	Total
1. Africa	3.3	1.9	0.6	—	5.8
2. Asia	10.7	8.1	16.2	0.4	35.4
3. South America	0.9	0.1	—	—	1.0
4. North America	0.3	1.2	0.3	—	1.8
5. Europe	0.8	1.7	0.5	—	3.0
6. Australasia	—	0.5	—	0.4	0.9
Global Total	**16.0**	**13.5**	**17.6**	**0.8**	**47.9**

Source: GLASOD

The maps and tables illustrate four main points about the global salt problem:

1. Salt-affected soils cover very large area of land.
2. The dryland regions are disproportionately affected.
3. Saline and sodic soils pose approximately equal problems, and
4. Human-induced salinity problems occur throughout the dryland countries, both in irrigated and dryland farming districts.

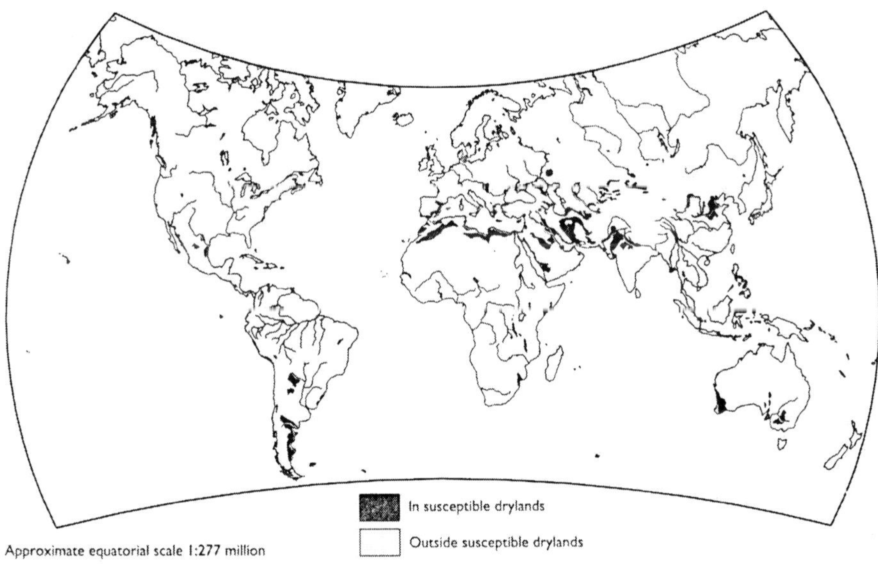

***Figure 11.3 :** Human-induced saline soils.* Source: *GLASOD*

Altogether, about 1 billion hectares of land, or 20% of the dryland area, have saline or sodic soils as mentioned in Table 11.2 (Thomas and Middleton 1993). Human-induced Salinization affects a much smaller area than natural salinity, but still affects approximately 77 million hectares (Oldeman *et. al.* 1991), 48 million hectares of which are in the susceptible drylands as given in Table 11.3. Salt problems affect rich and poor countries alike, but Africa and Asia are disproportionately affected. Figure 11.4 details out the process of land becoming salt affected.

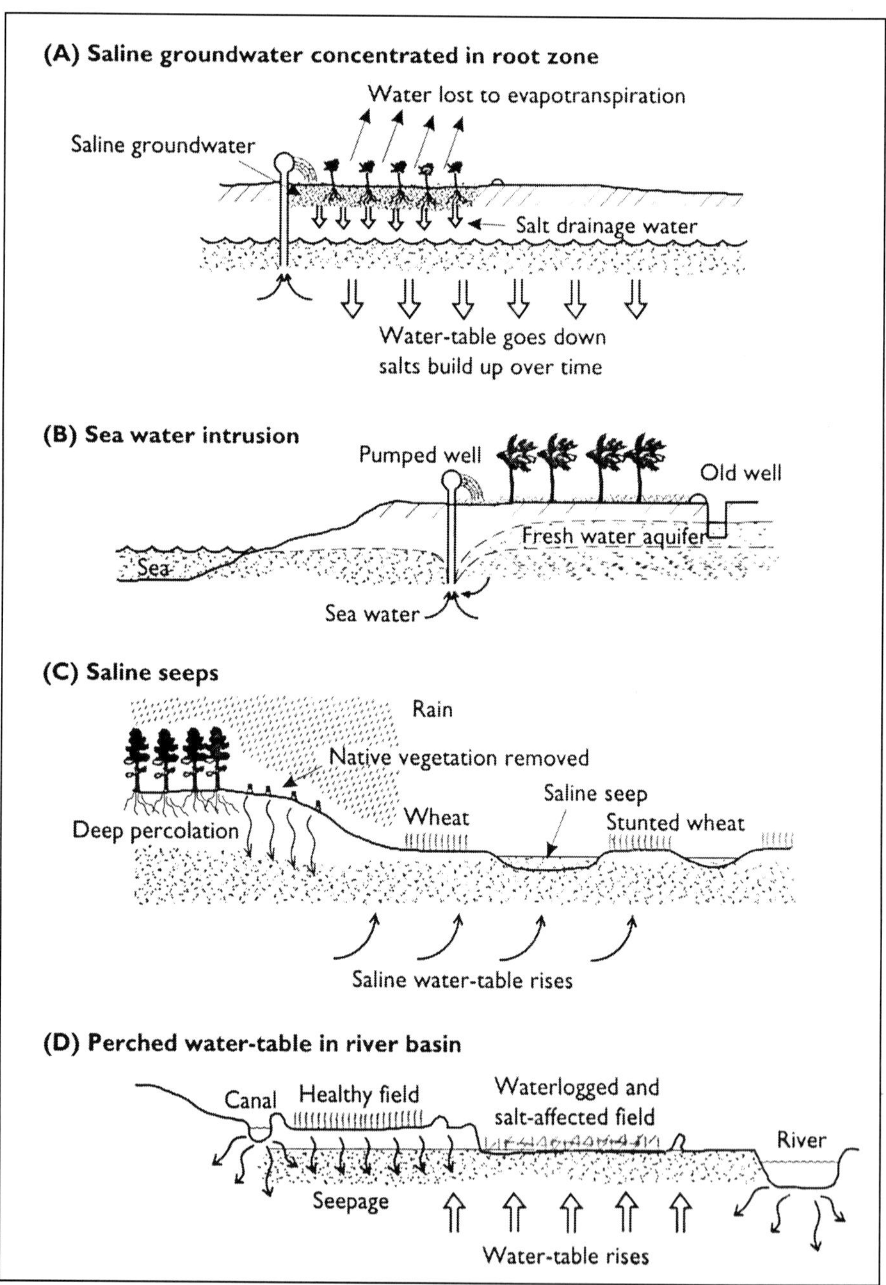

Figure 11.4 : *How soils become salt-affected. E.P., Environmental Research Laboratory, University of Arizona*

1. Salt Affected Soils: Their Development and Grouping

Although the properties and attributes of salt affected soils have been well-known for some time, it is useful to give a brief definition of this group of soils since the salinity, alkalinity and acidity of soils are substantial stress factors on land productivity. Salt affected soils can be characterized as formations that are predominantly influenced by different salts in their solid or liquid states, which eventually have a decisive influence on the development characteristics, physical, chemical and biological properties and eventually the fertility of the soil. Whenever and wherever this phenomena occurs, it produces specific formations of soils where the high electrolyte concentration and its consequences overshadow the former soil-forming processes or former soil properties and environmental conditions, often radically changing them. High electrolyte concentration is the only common feature of all the salt affected soils. Their chemistry, morphology, pH and many other properties may be different, depending on the character of Salinization and / or alkalinization. In the broader sense, salt affected soils can be divided into following groups:

(a) Saline soils that develop under the influence of electrolytes of sodium salts with nearly neutral reaction, dominantly Na_2SO_4, NaCl, seldom NaN03. These soils occur mainly in arid and semi-arid conditions and form a major part of all the salt affected soils of the world.

(b) The alkali soils that develop under the influence of electrolytes capable of alkaline hydrolysis, mainly Na_2SiO_3 and $NaHCO_3$ and seldom Na_3SiO_3 and $NaHSiO_3$. This group is extensive in practically all climatic regions, from the humid tropics to beyond the polar circles. Their total salt content is commonly lower than that of saline soils, sometimes even strongly alkaline.

(c) The salt affected soils that mostly develop owing to the presence of $CaSO_2$ gypsiferous soils or rarely in the presence of $CaCl_2$. Gypsiferous soils can mainly be found in the arid and semi-arid regions of North America, North Africa, the Near, Middle and Far East and also in Australia.

(d) Salt affected soils, which develop under the influence of magnesium salts. This group occurs and is particularly significant in arid, semi-arid and even semi- humid conditions, especially in soils with a heavy texture.

(e) Acid sulphate soils, whose salt content is composed mainly of $Al_2/SO_4/SO_3$ and $Fe_2/SO_4/SO_3$. These types of salt-affected soils are generally extensive in tidal marsh areas.

The halophytes and xerophytes growing on alkali soils differ in accordance with the different properties of these two subtypes. In solonetz soils particular, it is not the high salt concentration but the poor water physical properties, which are responsible for hindering plant life. Consequently, the vegetation on these soils only partly consists of halophytes with halo-frequent xerophytes accounting for a significant proportion. In sub-group (a) Salsola soda, Festuca ovina and Comphorosma ovata often occur together with Static gmelini and Puccinellia limosa. The latter also can be found in sub-group (b) in case of high alkalinity.

Together with some halophytic species, in solonetz soils xerophytes or halo frequent xerophytes prevail. First of all, different kinds of saga- Artemisia maritime, A.monogyma, A. alba, etc. Statice gmelini Fextuca ovina and F. pseudovina. In cases of low salt concentration, Achillea millefolium, Hardeum gussonearum, Astertripolium and Psa bulbosa, even Proa pretensis are found. Parallel with the lowering of salt concentration, the leguminae - Melilotus indica, Medicago hispida, Trifolium naritimum, etc., and herbs typical o the no-salt affected surrounding areas, also appear.

Among the halophytes, Puccinellia limosa, in particular, has a high value as fodder, despite its low yield, due to poor ecological conditions. Nevertheless, with irrigation and the application of nitrogen fertilizers, in several countries, such as Hungary, it helps considerably in solving fodder problems in dry areas. There have been similar experiences with a number of local xerophytes and halophytes in different countries, where efforts have been made to relieve the fodder problem that always arises in salt-affected areas. Evidently, in cases of low alkalinity and salinity, there are more possibilities for producing fodder crops.

(i) ***Magnesium Soils:*** In Magnesium soils, representatives of the two former groups can often be found, depending on the salt concentration and on the alkalinity of the soil. Apart from this, some specific plants associated are characteristic of magnesium soils, e.g., Spergularia salina and Suaeda maritime in soils containing magnesium carbonate or Ormenis praecox and Beta macrocarpa in saline magnesium soils.

(ii) ***Gypsiferous Soils:*** Gypsiferous soils are extensive in semi-arid and arid areas and in most cases, the gypsophilic xerophytes prevail among their natural plant associations. Lygneum spartum and Limomium delicatulum are specific of the calcic gypseous salt-affected soils in the Middle-East. Arthrocnemum glaucum cushions occur on micro relief. Inula crithmoides also occurs frequently on gypsiferous soils.

(iii) ***Acid Sulphate soils:*** On acid sulphate soils a diverse vegetation can be found, depending on local geographical conditions. The development of the various herbaceous species is also subject to the degree of salinity and acidity of the soil and they can be used as fodder. It is well-known that acid sulphate soils are well-developed in mangrove eco-systems, whose biota is determined by such circumstances. Reed vegetation in brackish swamps is also associated with acid sulphate soils in West Africa, where Racemose is common or in south America, where Rhizophora mangle has been studied (Brinkmann and Pons 1973).

In freshwater areas, imperate brasiliensis (white cedar) is characteristic of South America. In Eastern Asia, Melaleuca can be observed. The later has a wide ecological amplitude. The halophytic and xerophytic vegetation, typical of the different groups of salt affected soils has considerable significance in national parks and other environmentally-protected areas. Together with native ecological factors, the preservation of native xerophytic and halophytic vegetation on salt affected soils is a substantial part of keeping protected land undisturbed.

The close inter-relations between eco-systems and soils make it possible to differentiate between the soil types of vegetation. In this respect, salt affected soils are closely related to halophytes. The different groups of salt-affected soils can be associated with different groups of halophytes and xerophytes.. Several soils attributes. are pre-conditions to the development of certain plant associations.

There is a high diversity of halophytes, xerophytes and other plant associations in the different areas of the world covered by different types of salt affected soils. In order to promote the better knowledge and better utilization of the halophytic vegetation in salt affected areas, further studies are necessary so that the complex inter-relations between the coil conditions of salt affected areas and the vegetation growing on them can be revealed

11.2 LAND-USE PRACTICES THAT LEAD TO THE SALINIZATION OF CROPLAND

The major factors responsible for the development of human-induced saline and sodic soils are illustrated in Figure 11.2. Four main forms of poor management can be recognized (Abrol *et. al.*1988, Ghassemi *et. al.* 1995):

1. *Use of Saline Groundwater for Irrigation:* Dryland groundwater is frequently saline and prolonged irrigation with such water leads to the build-up of salts in the root zone, which can eventually render the land too saline for crop growth. If the groundwater is high in Na_2CO_3 rather than NaCl and the soil has a high clay content, sodic rather than saline soil conditions result. In dryland regions, aquifers are rarely recharged as fast as they are pumped, hence the water-table tends to be lowered and becomes more saline over time, because water is evaporated during crop production, but the salts in the water are returned to the soil and aquifer. This is a common problem in inland irrigation districts throughout the drylands.
2. *Ingress of Sea Water:* Coastal arid zones often have sandy soils overlying shallow, freshwater aquifers, which have accumulated over geological time spans. When these aquifers are pumped for irrigation or their uses, the water-table can be quickly lowered to below sea level, allowing sea water to seep into the aquifer through the sand. Coastal irrigation districts often develop unusually high salt levels in their well fields within 10-20 years of installation. On the Batina coast of Oman, hand-dug wells, which sustained date gardens for centuries, become

salinized within a few years after the introduction of motorized pumps, which extracted water from deeper levels, thereby lowering the water-table below sea level and permitting sea water to enter the wells (Speece and Wilkinson 1982).

3. *Saline Seeps in Non-irrigated Dryland Farming Regions:* When native shrubs, trees and grasses are replaced by annual grain crops, the evapo-transpiration rate of the landscape is lowered. Rainfall that have been used up by the native vegetation instead percolates through saline sub-surface sediments to impermeable horizontal layers and is conducted laterally to low spots in the landscape, producing extensive areas of salty, water-logged soil. Large areas of dryland wheat farms in Australia, the USA and Canada have been salinized by this process of saline seepage (Ghassemi *et. al.* 1995).
4. *Perched, Saline Water-tables Under Irrigated Fields:* The more intractable salinity problems occur in the large river basins that support vast dryland irrigation districts around the world. In its natural state, a river basin for irrigation, the natural balance is disrupted. Excess water seeps into soil from unlined canals, drainage ditches and from over-irrigation of the farm fields themselves. The excess water creates a high water-table throughout the basis, which rises into the root zone of the crops, water-logging the soil and inhibiting plant growth. Salts build up in the root zone over time because there is no drainage to carry them away. Glistening white salts patches in irrigated fields are, unfortunately, a common sight in many irrigation districts, This problem affects irrigation districts all over the world, from the lower Colorado River of the USA to the Nile and Indus Valley (Ghassemi *et. al.* 1995).

11.3 A CRISIS FOR DRYLAND IRRIGATED AGRICULTURE

Irrigated agriculture has become more and more important in feeding the world's growing population. The amount of irrigated land in the world doubled every 15-20 years from 1800 to 1980 and two-thirds of the world's irrigated area has been developed since 1950. (Postel 1990, Brown 1997). Figure 11.5 indicates the details in respective durations in terms of coverage. By 1990, one-third of the global food harvest came from irrigated fields, which represented only 17% of the total cropland. Many of the new irrigation districts have been developed in dryland countries through the madding of rivers to provide irrigation water and electricity.

Since 1980, however, the trend towards irrigated agriculture has been slowing. Per capita irrigated area actually peaked in 1978 at 53 hectares per 1000 people and has decreased since then, by 2000 it is estimated to decrease to 43 hectares. Lending by major international donors for new irrigation projects has fallen sharply.

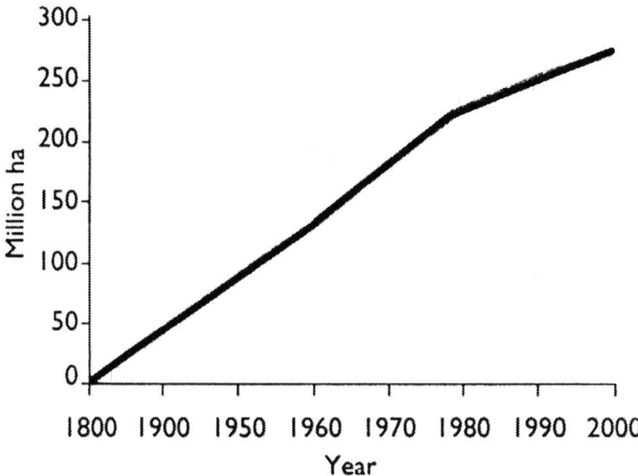

Figure 11.5 : *World gross irrigated area.*

Source: Glenn, E.P., Environmental Research Laboratory, University of Arizona

There are several explanations for the recent trend (Postel 1990, Brown 1997); the best sites for irrigation have already been developed; the costs of adding irrigation capacity have risen; and irrigated drylands have been found to be extremely vulnerable to degradation from water-logging and salt build-up, so further expansion of irrigation into new dryland districts is seen as a risky investment. The amount of salinized land in the irrigation districts can only be estimated since exact statistics are not kept for all countries. By one estimate, 38 million hectares or 24% of the irrigated land of the top five irrigating countries has already been damaged by salt and extrapolating to the world at large, 60 million hectares are thought to have been damaged, as mentioned in Table 11.6. Another estimate puts the amount at 45.4 million hectares, representing 20% of the total irrigated area (Ghassemi *et. al.* 1995). Other countries with large amounts of salt-damaged land in irrigation districts, not shown in Table 11.3, include Afghanistan, Egypt, Iraq, Mexico, Syria and Turkey.

Table 11.3 : Irrigated land damaged by salinization in top five irrigators and world estimates, Mid 1980s

Country	**Area damaged (Million Ha.)**	**Share of irrigated land damaged (%)**
1. India	20.0	36
2. Pakistan	3.2	20
3. China	7.0	15
4. USA	5.2	27
5. Soviet Union	2.5	12
Total	**37.9**	**24**
World	**60.9**	**24**

Source: Postel 1990

Salt damage is progressive, with more land lost to production each year in most irrigation districts. One estimate of the amount of new land that become salinized each year is 1-1.5 million hectares (Kovda 1983). The statistics for total irrigated acreage in Figure 11.5 above, do not deduct for the amount of land lost to salt damage. Hence, a point is reached, if it has not already, when the total effective area under irrigation may actually decrease. The lost income from salinized land in the irrigated drylands is estimated to be US$ 11.4 billion per year (Ghassemi *et. al.* 1995).

1. Areas of Concerns

The origin of all living things on earth can be linked, directly or indirectly, to soil resources of the world. Only 11% of the earth's total land area of 13.40 million hectare is currently being cultivated. The remaining potentially cultivable land is marginal for agricultural use because most of it is either inaccessible or is severely constrained by unfavourably steep terrain, shallow rooting depth, extremes of moisture and / or temperature regimes, shortages of essential inputs or is located in ecological sensitive regions.

The per capita arable land area of 0.3 hectare in 1990 has been decreased to 0.25 hectare in 2000 and is expected to decrease to 0.15 in the year 2050 and about 0.10 hectare in the year 2150. Human needs, with per capita arable land area of 0.10 hectare only can only be met with the assistance of science based and innovative technology. Consequently, there is an overwhelming interest in instigating sustainable land use systems that prevent or minimize soil degradation and restore productive capacity and life support processes of degraded lands.

Agricultural sustainability and soil degradation are complex concepts with subjective values. What these concepts are understood to mean varies according to the user's immediate and long-term concerns and needs. The scientific community needs to develop objective, consistent and operational definitions of both agricultural sustainability and soil degradation, which can then by used to create a rational and science based strategy for resources management. The objective is to have practical and operational ways to achieve agricultural sustainability and to prevent soil degradation and thereby to alleviate the pressing problems of modern agriculture. In other words, to increase per capita production and decrease the risks of soil and environmental degradation suitable measures are required to be undertaken. In this context, agricultural sustainability means " an increasing trend in per capita productivity per unit consumption of the non-renewable or the limiting resource or per unit degradation of soil and environmental characteristics or per unit reduction in soil's life support process." The aim is not only to increase per capita productivity but also to maximize production per unit of soil loss, per unit of energy input, per unit of reduction in soil organic carbon, per unit of afflux or radiatively active gases, per unit of consumption of ground water and per unit of increase in concentration of NO_3, PO_4 or other pollutants in natural water.

2. Soil Resilience

Soil is a dynamic entity and dynamism is life. Early civilizations originated and flourished on fertile soils, many ancient cultures developed reverence for the soil on which they depended. They believed that human originated from the land and land is a living entity worthy of worship. The Maori of New Zealand believe that "the land is the mother never dies." These beliefs are similar to the modern "Gaia hypothesis" that proposes that "earth is a living organism" and "is a complex entity involving the biosphere, atmosphere, oceans and soil; the totality constituting a feedback or cybernetic system, which seeks an optimal physical and chemical environment for life" (Lovelock 1979).

Soil dynamism is evidenced by continuous formative and evolutionary changes until a dynamic equilibrium is achieved. Anthropogenic perturbations alter this state of equilibrium and may change the rate and direction of principal processes (Rozanov 1990). With the advent of modern technology, human actions are becoming progressively drastic due to the development of bigger and faster machines and concentrated and highly active chemicals designed o enhance soil fertility and accentuate plant growth. The greater the human demands, the more potent are the technologies likely to have drastic effects on soil processes.

The "Magic of Mother Earth" is her ability to heal herself and the term soil resilience refers to the soil's ability to bounce or spring back into shape or position after being stressed. As a dynamic and an organic entity, soil has an in-built ability to restore its life-support processes, provided that the disturbance created by human intervention is not too drastic and sufficient time is allowed for the recovery processes to take effect. However, soil can undergo irreversible degradation, if it is drastically disturbed and its life support processes are severely jeopardized. Through constant misuse and mismanagement "man may sap the viability of Gaia by reducing productivity and by deleting key species in her life-support systems" (Lovelock 1979).

Soil resilience can, therefore, also be defined as the ability of soil to restore its life-support processes, after it has been degraded to natural or human-induced perturbations. Soil resilience depends on balance between soil restorative and soil degradative processes. It is a function of the harmonious and symbolic action of all living organisms in that it creates favourable environments for life to thrive. Although humans value only crops and livestock, other unwanted species are not necessarily pests, weeds or vermin- they too play an important role in the self-regulatory mechanisms of Mother Earth (Lovelock 1979).

Soil resilience depends on close inter-relationships between soil degradative and soil restorative processes and the factors ad processes that influence their magnitude and direction. Soil degradative processes are set in motion, when the soil structure starts to deteriorate, the carbon cycles are disrupted, nutrient reserves within the soil are depleted and nutrient recycling mechanisms are weakened. Soil restorative or self-regulatory processes are strengthened by science- based agriculture; discriminate and judicious land use, reduction of pressure on marginal. lands and fragile ecosystems and adoption of improved technology.

3. Rate of New Soil Formation

Soil is a renewable resource as long as the balance between the rate of soil formation and soil degradation is positive, otherwise it is a non-renewable entity, at least within a human lifetime. The timeframe is an important consideration in this analysis. Most available data are based on informed opinion and indirect evaluation and vary enormously. Some estimates put the rate of soil formation at about 2.5 cm in a thousand years. However Friend (1992) estimated that worldwide the rage of soil formation is about 2.5 cm per 150 years. In general, soils of volcanic origin develop faster than those developed or gneiss or basement complex rocks. The rate of new soil formation for Andisoils in the humid tropics is < 1 mm per year. In contrast, the rate of new soil formation for Alfisoils and Utisoils is < 0:01 mm per year. Under normal conditions, however, new soil is formed at the rate of about 2.5 cm in 300 to 1000 years, it would seem, therefore, that most soils are non-renewable within the human life span. The major exception to this rule is the formation of alluvial flood plains by rivers carrying heavy sediment loads, such as the Yellow River, the Ganges, etc.

4. Soil Degradation Rate

Soil degradation or the decline in the soil's capacity to produce goods of value to humans, has plagued the earth ever since human exploitation of land began. Many ancient civilizations thrived on 'good soil' and declined as soils became degraded through misuse. Typical examples are the Riparian and Harappan Kalibangan culture in the Indus Valley. Mesopotamian and Lydian kingdoms in the Mediterranean region and the Mayan civilization in Central America. These great civilizations declined along with depletion of the original mantle of fertile topsoil. The problem of soil degradation has been drastically accentuated by change in land use since the 18th century. It is estimated that out of the world's total land area of 1340 million hectare, about 200 million hectare is degraded land. There are several processes of soil degradation. Most prominent among these are wind and water erosion, chemical degradation and fertility depletion and physical degradation resulting from decline in soil structure.

Estimates of soil degradation in dry climate are equally alarming. Such statistics play an important role in creating awareness about the magnitude in severity of the problem and in formulating a global strategy in addressing it. Soil degradation presents on of the greatest challenges to the human race because soil resources are finite and to all intent, may be considered non-renewable. As explained earlier, the term 'soil degradation' is vague and highly subjective; to avoid ambiguity, it is important that soil degradation by different processes be defined quantitatively. To do so, is to delineate threshold values or critical limits of soil properties beyond which, soil's life support processes are severely jeopardized.

The important soil properties for which critical limits need to be defined are rooting depth, plant available water capacity, soil organic matter content, soil structural attributes, capacity and intensity factors and limiting levels of principal nutrients. These limits vary among soils, climatic conditions, farming systems, land uses and managerial inputs. There is a problem even with the data since ignorance about the critical limits and threshold values of soil properties often makes global data vague and unreliable. Soil degradation and productivity must be considered in relation to land use, inputs and management using improved technology.

11.4 METHODS OF PREVENTING AND REPAIRING SALT DAMAGED SOILS

Preventing salt damage is less expensive than repairing it. Most salt problems are caused by attempting to irrigate marginal land or by irrigating more land than the natural or manmade drainage system can handle (Abrol *et. al.* 1988). Low-lying heavy soil that lacks adequate drainage is usually the first to become damaged and should probably not is irrigated from the start. However, social and economic pressures often result in the creation of irrigation districts that initially contain more land than can be sustainably framed. The switch from seasonal to perennial irrigation after construction of Aswan High Dam has raised soil salinity levels throughout the Nile Valley (White 1988). In the Nile Delta, an ambitious irrigation scheme was developed for the reclaimed Mansour and Zawia polder area (Boumans and Mashali 1983). About half of the land that was intended for irrigation has either abandoned or never put into production due to salt problems.

Technically, both sodic and saline soils can be repaired (Abrol *et. al.* 1988), but this may not always be cost effective. Sodic soils can be reclaimed by applying gypsum ($CaSO_4$), calcium chloride ($LaC1_2$) or other calcium salts to the soil. Calcium replaces sodium on the clay particles and partially restores the soil structure. Sulphuric acid, elemental sulphur, iron sulphate and aluminum sulphate are also effective amendments, which react to produce calcium sulphate from calcium carbonate already in the soil. Adding the organic mulch such as straw or manure can also aid reclamation. To be effective, several tonnes per hectare of such inputs is required; it must be ploughed into the soil and irrigated prior to planting a crop. The restoration of soil structure is usually temporary and gypsum or other amendments may be needed every few years in some sodic soils. A local, low cost source of soil amendment is required for this method to be economical. An alternative method is to flood the land with high salinity water after gypsum application then gradually dilute it with fresh water; this keeps the soil structure open during reclamation (Oster 1993). The land can be used for salt-tolerant crops during reclamation.

Saline soils can present an even more serious challenge. The usual method used to reduce soil salinity is leaching, in which large quantities of water, in excess of the amount is needed by the crop, is added to the irrigation schedule or pounded onto the field between crops (Abrol *et. al.* 1988). In theory, the excess water dissolves salts in the root zone and moves them down the soil profile where they will not damage the crop. In practice, however, leaching requires good soil drainage and poor drainage is often on of the very reason for Salinization in the first place. Before leaching can be effective, therefore, a surface or sub-surface drainage system must be installed and a mechanism for the collection and disposal of the saline drainage water introduced. Irrigation canals must be lined to prevent seepage and control measures implemented to prevent over-irrigation of crops. This often requires reworking the infrastructure throughout an entire irrigation district, because the water table is likely to be high throughout the district.

Installing a sub-surface drain system costs from US$ 1000 to 2000 per hectare of farmland in USA and the disposal system for the drain water, requiring collection sumps, pumps and lined canals to carry away the brine, involves additional costs. Finding a place to deposit the brine has become a problem in its own right. Selenium and other toxic elements accumulate in the water (Presser 1994). If the water is discharged into surface ponds it can poison wildlife and even people and basin-wide drainage projects often succeed only in displacing salt damage to adjacent ecosystems. The Aral Sea, was once the world's fourth-largest freshwater lake, but is now saline and toxic from diversion of river water for agriculture and the deposit of irrigation drainage in the sea (Micklin 1988). In the USA. California's Salton Sea is also becoming hyper-saline and toxic from the deposit of agricultural waste water (Boyle 1996). Elsewhere in California, the San Joaquin Valley lacks drainage canal to the sea because of concerns that the drainage water might damage the estuarine ecosystems. Instead, saline drain water is being temporarily stored in evaporation ponds in the valley with no clear permanent solution in sight (Presser 1994).

Potential solutions to the toxic drainage dilemma are all expensive; chemical treatment, filtration or bioremediation would cost US$ 30-100 million per year in California alone and would be required for an indefinite period (Postel 1990). Because of expense and difficulty of installing drain systems and disposing of the output, only the most valuable land is likely to be rehabilitated this way. Much more frequently around the world, fields and sometimes whole irrigation districts are abandoned, when they become too saline to grow crops.

11.5 NEW SOLUTIONS TO SALINITY PROBLEMS

Increasing irrigation efficiency can prolong the life of an irrigation district, make water available for other uses and extend the salinity limit at which crops can be grown (Lerey 1993, Miyamoto 1993, Rhoades 1993, Van Schilfgaarde 1993). Farmers using traditional flood basin or furrow methods of irrigation often apply 50% more water than is actually sued by the crop due to the inefficiency of these application methods. It is difficult to add a precise amount of water, when an entire field must be flooded at once. The excess water is not just wasted, it contributes to raising the water-table under the fields, bringing up salts and creating the need for a drainage system. More efficient irrigation methods are now in widespread use in dryland irrigation districts all over the world. These include drip irrigation, in which a precise amount of water is trickled into the soil at the base of each plant through a plastic tube and moving boom irrigation, in which a motorized, wheeled irrigation machine travels over the crop, applying a predetermined amount of water as a simulated gentle rain from an overhead boom. Even flood irrigation has been improved by the use of precision, laser guided land plane to level fields so that water spreads over them more evenly.

More efficient irrigation techniques have been accompanied by a better understanding of crop water requirements and better methods of detecting water stress, by the use of infrared imaging of the crop. When the soil water supply becomes low, the plants close their stomata to conserve water. This results in an increase of leaf temperature, because the leaves are no longer cooled by water evaporating through stomata. The temperature increase can be detected by infrared sensors in the field or on satellite. A fully equipped, modern farmer potentially knows just when and how much to irrigate a field and has the irrigation technology to accomplish the task.

These technologies are effective but expensive. Surface irrigation methods such as furrows or basin cost US$ 100- 400 per hectare to install, compared to US$ 1000-2500 per hectare for self-moving systems or drip irrigation (Ghassemi *et. al.* 1995). Although, the economic costs of such technologies can be offset against their water savings and the benefits of continued crop production, they cannot be expected to solve all the salinity problems affecting the world's croplands. An alternate of complimentary approach is to shift towards more salt-tolerant crops. Crops that can tolerate high salinities can be grown on saline soils and water supplies, thereby extending the life of an irrigation district. They can also be used as secondary crops to be irrigated with drainage water from conventional crop fields. This second use of the irrigation water reduces the amount of drainage water that must be disposed of. an environmental benefit.

Intensive breeding and selection programmes have been carried out over the last two decades to improve the salt tolerance of rice, wheat, barley, alfalfa and other major crops (Shannon and Nobel 1990). A few improved cultivars have been released, but in general the results have been disappointing. Few conventional crops can be irrigated with water even 10% of salty as sea water and drainage water often exceeds this salinity. More rapid progress has been made in domesticating wild halophytes as domestic crops (National Reaserch Council 1990, Glenn *et. al.* 1997). Halophytes such as Atriplex and Destrichlis have been established on thousands of hectares of salinized land to provide forage for sheep and cattle. Animals often avoid halophytes in natural rangelands because they tend to be less palatable than grasses, but when halophyte biomass is incorporated into mixed ration, it can completely replace conventional grass hays in fattening diets. The animals need no special inducement to eat halophyte material if it is mixed with the other diet ingredients. They do, however, consume more feed and water per unit weight gain due to the high salt contents of the halophytes (Swingle *et. al.* 1996).

One of the most promising halophytes is an annual, succulent salt-marsh plant, Salicornia Bigelovii (Glenn et. al 1991). It produces an oilseed that can be pressed to provide human-edible oil and protein meal for animal diets, while the straw can be used for forage (Swingle *et. al.* 1996). It can be grown using undiluted seawater in coastal desert soils. Using sea water for irrigation requires special irrigation techniques, since the soil must be kept continuously moist and an excess of sea water, above consumption use, must be added to keep salts from accumulating in the root zone. Altogether, halophytes grown with sea water require about 35% more irrigation per unit of biomass production as conventional crops (Glenn *et. al.* 1997) (Figure 11.6).

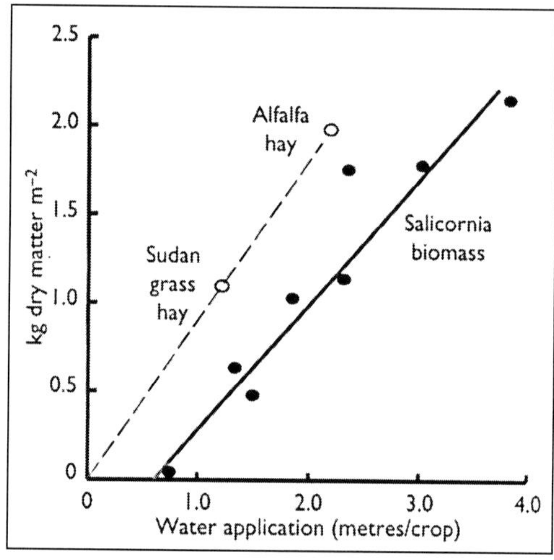

***Figure 11.6 :** Centre-pivot boom irrigating the oilseed halophyte Salicornia with seawater (Jim Riley, Environmental Research Laboratory, University of Arizona)*

In coastal regions, this disadvantage can be outweighed by the lower lift required to irrigate with sea water compared to deep wells. Although, sea water irrigation produces a large amount of saline drainage water, in coastal regions it can be drained back to sea by gravity, whereas inland irrigation basins often have no outlet. Seawater-irrigated halophytes crops might become cost-effective in coastal communities, which lack fresh water for crop production or wish to conserve their fresh water resources for human use. Halophytes may also become cost-effective as crops to reuse saline water and soils within existing irrigation districts.

1. Inputs and Improved Technology

It is true that the majority of soils in ecologically sensitive regions have undergone drastic changes in their properties due to intensive cultivation, land misuse and other anthropogenic perturbations. However, the impact of these changes on productivity potential has not been determined. The productivity of degraded soils can be greatly enhanced by improved management and proper land use. Soil degradation must, therefore, be determined scientifically according to its degree and the processes affecting it so that improved management and judicious land use can be implemented to set in motion the restorative processes, even on drastically disturbed lands.

Wolman (1967) developed schematics of sequence of land use changes in USA with sediment yield over a period of two centuries. As the forests were cleared and cultivated, grasslands ploughed and construction activity increased so die the sediment yield. Adoption of conservation practices reduced sediments yield for the same land use system. The pattern of sediment yield in relation to land use and conservation technology was similar for all soils, except that the amplitude of change deferred depending on soil properties, terrain characteristics and the climatic environment. In addition the benefits in crop yield, scientific management can also set in motion soil restorative processes. Vast areas of land that were once the 'dust bowl' of the south-central USA have been transformed into highly productive land by science-based agriculture. Similarly, vast areas of salt crusted barren land in Haryana and Punjab provinces in India have been transformed into highly productive lands by science based improved land use system.

Agriculture in North America, Western Europe and other developing economies has undergone several phases of development,. Per capita productivity has significantly increased with each phase, beginning with human power, moving on to horse power, improvements in farm implements, use of motorized equipment and chemical fertilizers and pesticides and use of improved and input-responsive varieties of crops. These technological innovations have been used most successfully on undegraded naturally resilient, prime agricultural land. Similar innovations need to be developed for soil degraded by different processes since improved technology and appropriate land use vary according to the processes of soil degradation and the extent to which they have progressed.

2. Time Scale

The issue of time scale is highly pertinent. Soil may be renewable resource over the geological time scale but not. over the human life span. However, with increasing demographic pressure, issue of productivity and sustainability must be addressed over the human life span, if not over the even shorter, common economic planning time span. In fact, there is tremendous pressure to produce needed goods and services over the economic planning time span of five or ten years or less. Therefore, it is appropriate to test the impact of innovative technologies on soil restoration or resilience over the human or scientific time span of 20 to 25 years.

3. Preventive Measures

The soil salinity is to be checked in the new areas where there is every possibility for turning an area into salinity affected. Since most crucial factor of this situation is the excessive use of water, cultivators are required to use water saving devices and selection of crops requiring lesser water. The cultivated area to be brought in to irrigation facilities through any source of watering, would change the soil structure as being a natural phenomena. The soil being cultivated for long with rain-fed system for years together, develop changes due to irrigation facilities. Every dose of water is inserted into the upper and lower soil structure and the depth it reaches start responding to the change.

In flow irrigation system through canals from a reservoir, silting start developing in the source and water transportation to fields creates salinity problems. In such a situation, removal of silting from the reservoir is necessary during the summer to keep the source clean. In addition, water-logging is the ultimate danger as cultivators start growing more crops and use excessive water to get maximum return from the available water. In case the water is drawn from underground water source, the indicator of legitimate use level is depletion level. In dryland areas, recharge is generally stopped in view of erratic monsoon and successive use creates such problems.

In all above situations the ultimate solution is the restrictive use of water to maintain the fertility of soil. The impact of water-logging can be stopped with sufficient drainage systems and plantation using more water. Eucalyptus plantation is suggested on the boundaries of the fields for diminishing the water logging problems to some extent. In case of underground irrigation, salinity occurs due to excessive water resulting in constant water depletion and the total area using such water is gripped with the problem. There is competition among the cultivators to get their source further deepened to get more water and every cultivator adopt similar practice.

In case of excessive withdrawal of water from underground sources, the area is categorized as 'dark zone' and further efforts for withdrawal of water for irrigating the crops further aggravate problem. This is possible to alarm the cultivators to warn further use of ground water by change in cropping pattern to save the fertility of the soil. Cultivated areas severely affected with salinity problems, crop practices adaptable to grow crops resistant in such system is also the remedy. Turning to biotechnology is also a sustainable remedy for such areas as excessive use of chemical fertilizers and pesticides further aggravate salinity. The fertilizers prepared through biotechnology consume lesser water and able to maintain the productivity level of similar extent on sustainable basis.

In areas severely affected with salinity problem, the ultimate solution is the treat areas with biological process with selection of crops possible to be grown and maintain the productivity levels. Sprinkler and drip irrigation system are the better solution as the transportation of water through field channels helps in water-logging and salinity. Use of water saving devices helps to check the impact to significant level.

4. Curative Measures

Treatment of soil and change in cropping pattern are the solution for sustainable use of land. Ayers (1977) proposed guidelines for the quality of water for irrigation to be used for growing various crops. Crop tolerance to soil salinity values for various crops for assessing the quality of water suitable for irrigation assuming a leaching fraction of 15-20% Maas and Hoffman (1977) suggested piecewise linear model to threshold electrical conductivity of irrigation water, slope which represents yield reduction per unit increase in salinity of the irrigation water and the salinity of the irrigation water at which yield would be 50% of the optimum value. It is apparent that both the threshold and slope would vary with different crops and performance of crops would differ significantly as a consequence of application of saline water. Thus selection of crops could be a primary tool to plan conjunctive use of water within an area.

To maintain the yield level under saline water irrigation, a number of varieties of different crops have been screened. Jain and Sas (1988) concluded that tolerance of salinity level in irrigation water was more in Sev variety of Jujube than in Gola and Mundia. Similarly, Khog variety of pomegranate was more tolerant. Irrigation with saline sodic water on medium textured soils under semi-arid climate revealed that soils become hard due to receding soil moisture at critical seedling stage. Although gypsum application yielded favourable results, for improvement of water infiltration (Goyal and Jain 1982).

The study revealed that soluble silica of ground waters resulted in increased crust force at low moisture regimes. (Goyal 1984). Integrated soil and irrigation management treatments which comprised soil treatment with gypsum, straw mulch and farm yard measures with advanced second irrigatipn after 7 days helped in maintaining higher seedling emergence which resulted in increased yield.

11.6 CONCLUSIONS

Saline soils represent a lost opportunity for crop production and pasturage throughout the world's dryland regions. Secondary Salinization of cropland is the most expensive form of land degradation to repair, costing thousands of US Dollars per hectare and frequently resulting in off-farm environmental damage through the disposal of saline drainage water. The best available estimates suggest that 45-60 million hectares of irrigated land have already been damaged by salt build-up (Postel 1990, Ghassemi *et. al.* 1995) and the GLASOD data indicate that the global extent of human induced Salinization on all farmland is 77 million hectares, including 48 million hectares in the susceptible drylands. The problem is getting attacked by developing better irrigation and drainage technologies, taking more care in the initial design of irrigation districts and by developing new crops that can be grown using highly saline water. Nevertheless, salinity will continue to be a serious problem throughout the arid regions for the foreseeable future. Population growth and the need for more food will put more, pressure on dryland cultivated areas in future. Finding ways to bring salinized soils into agricultural production must continue to be an important part of the international agricultural research agenda.

Chapter **12**

Population and Desertification

The growth of desertification is the consequence of four main direct causes-over cultivation, overgrazing, deforestation and mismanagement of irrigated cropland. These causes are greatly influenced by a variety of underlying social, economic and political factors and are not just a consequence of faults- in the particular land use, although such faults undoubtedly exist. Population growth and economic development are the two main driving forces, which lead to an expansion of agriculture and to changes in the types of agricultural practices. Economic development is generally accompanied by growth of urban population and serious degradation of natural resources is common around towns and cities.

The benefits of economic development are not shared equally among the habitants of a country and it is often the poorer people are forced to live on the worst lands, who are the most directly involved in causing desertification, the most serious affected by it and the least able to prevent it from happening. They are also often affected by famine, although that is no an inevitable consequence of drought or desertification and can occur when government policies constrain food production in particular areas and fail to alleviate poverty. In order to assess the causes of desertification correctly and devise programmes to bring it under control, it is necessary to take account the pattern of land use not only in the desertified areas themselves, but in the country as a whole and social and economic factors and government policies determine the pattern. Although, desertification is more regrettable symptom of underdevelopment, at the same time, it is also shown amenable to control by better land use planning and improved government policies.

Desertification is a consequence of peoples' efforts to use natural resources in environments that are highly susceptible to natural variability (Thomas and Middleton 1994). A full understanding of desertification, therefore, must incorporate an examination of issues that relate to the human populations of susceptible drylands. If solutions are to be found for desertification problems, attention must focus on human actions and human-environment relationships. To this end, the Convention to Combat Desertification commits its parties to the development and implementation of programmes for dryland management by focusing on the needs of affected populations. At the global level, an important first step towards achieving this goal is to gain a reliable picture of the distribution of populations in areas experiencing desertification. When combined with detained local studies of social issues that contribute to desertification and the positive actions taken by Non-Governmental Organizations, local communities and other parties, this will provide a valuable hierarchical input to addressing the desertification problem at scales ranging from global to individual communities.

To this end, UNDP / UNSO has undertaken a detailed global assessment of the distribution and density of human populations in susceptible drylands. The assessment has been undertaken in conjunction with the World Resources Institute, Washington DC and has included both a global and a detailed African Assessment.

12.1 EXPRESSING POPULATION CHARACTERISTICS

Populations are growing rapidly in most developing countries and the drylands are no exception. Since the mid-seventies, population growth rates have exceeded 2% per annum in four-fifth of the sixty-six countries containing appreciable areas of arid or semi-arid lands and exceeded 3% pet annum in a third of Those countries. They have also been greater than 3% per annum in over half of the thirty four predominantly dry countries, which possess more than three quarters of the total arid areas. In sudano-Sahelian region, the population increased by almost a quarter in seven years from 1977 to 1984 (Berry 1984-a)

Population growth is a major cause of land use change, as farmers try to grow more food to feed rising number of people by increasing either the yield per hectare or the area under cultivation. Food production has more than kept pace with population growth in Latin America and Asia but not in Africa. Within Asia, there is a great difference between the performance of agricultural sectors in the countries Far East, i. e, South and south-east Asia, which have adopted the intensive green revolution approach so enthusiastically and predominantly dry countries in the West Asia, which have lagged behind. Between 1961-65 and 1985-86, food production per capita fell by 13% in Africa, but rose by 17% Far East, by 8% in the Middle East and by 5% in Latin America (FAO 1975, 1987). Such comparisons are inevitably biased by the effects of drought on food production in Africa especially, but do illustrate general trends.

Population growth does not necessarily result in desertification (Blaikie 1985), but can help to induce it, if the only way to increase food production is to increase the area being cropped. Yield per hectare can be increased by intensive cropping, but these can only be sustained through investment in fertilizers and more productive cropping systems. Many dryland countries have insufficient capital to invest in agriculture and without fertilizers to replenish nutrients taken from the soil, more intensive cropping leads to declining fertility and falling yields. To maintain production, a larger area of land has to be cropped and since this is often marginal land, unsuitable for cropping, the result is further soil erosion. Agriculture production in the largely dry areas of Africa and Middle East (West Asia), has grown much more slowly in recent years, than that of Latin America and the Far East (South and South-East Asia). Between 1961-65 and 1985-86, cereal yield per hectare rose by 39% in Africa and 35% in the Middle East, but by 68% in the Far East and 58% in Latin America. In the Sahel region, the rise was only 8% (FAO 1975, 1987).

Often, the reason for over cropping and over use of natural resources generally, is not so much that of the population in the country has risen, but that there has been a growth in population density in certain areas. Many farmers are unable to move elsewhere, when their areas become over-crowded and are too poor to invest in more productive types of farming (Blaikie 1985). So, over-crowding causes farmers to crop their lands more intensively by reducing fallow periods or cultivating a particular patch of land for a longer time. Traditional cropping systems, such as the Acacia bush fallow system in Sudan, which have proved sustainable for hundreds of years because of their long fallow periods, break down, when the fallow periods are reduced. Yields decline and the land becomes degraded.

Population density can also rise because of land tenure constraints and political reasons. Land degradation in Lesotho and Zimbabwe, where large number of peasants are confined by long standing land tenure rules to relatively small areas of unproductive land. As their population rises, so does population density and degradation occurs. The Government of Tanzania has for some time settled people into new villages and the resulted concentration of population led to desertification because of over-grazing and over-cultivation in the vicinity. Desertification due to over-grazing also occurs around villages in which, nomadic herdsmen have been settled or sedentarized. Rings of rangeland degradation and deforestation are found around settlements, whether small towns or large cities, as their inhabitants graze livestock or gather fuel wood.

Migration is one way to alleviate over-crowding, although in Kenya the continuing movement of people from the highlands to arid and semi-arid lands caused great concern as to whether those lands were able to support an ever-increasing population. Despite government efforts to mitigate the situation, life in the Kenyan drylands has become ever more hazardous (Bernard 1985). Many migrants in the African drylands were refugees fled from drought, famine and war and were anxious to re-establish their livelihood elsewhere. But in the process, they created tremendous social and environmental problems. When the circumstances are very dire, migration even transcends national boundaries; the Ivory Coast had received a large number of migrants from Burkina Faso and Mali.

At its simplest level, human population can be expressed in terms of absolute numbers. From environmental point of view, however, it is also valuable to consider population pressure, a complex concept that is related directly or indirectly to the environment through the impact of land-use practices. A simple measure of population pressure is its density. This is by no means a full expression of population pressure, since it takes no account of land use, socio-economic factors or the carrying capacity of the land, which in itself varies through time in response to changes in background environmental conditions. Furthermore, pressure can be exerted on the land of an area by people residing elsewhere, by the need to grow food to support urban populations. Nonetheless, population density is a measurable spatial demographic factor, being the ratio of number of people per unit of area and can, therefore, be evaluated against environmental factors in order to assess human pressures thereon. If the population density is greater than the density supportable by the area's resources, degradation is likely to take place, the UNDP/UNSO (1997) population assessment is explained and presented for global drylands, along with a consideration of factors that cause dryland peoples to exert pressure on the environment.

1. Economic Development

One of the most worrying kinds of migration for government planners is the constant stream of people traveling from rural to urban areas in search of jobs in commerce and industry as the market economy grows. Urban population in dryland areas has been growing at 5-6% per annum, much faster than rural population (MacGregro and Valverde 1975, Berry 1984-a), and significant proportion of this growth is accounted for by migration from rural areas, in Sahel the growth was recorded as 50%. Rural urban migration is typical of countries undergoing economic development, but places a tremendous burden on governments, which have to provide costly urban services. Agricultural development also suffers, since the cities consume a growing proportion of the national budget, the rural labour force declines and farm prices have to be kept down by government to satisfy urban population and maintain political stability. Rise in food process can easily lead to riots, which may provide the excuse for a coup d'etat by ambitious army personnel.

The growth of urban centers represents a serious environmental problem for many developing countries because they lack sophisticated production technology and supply network found in western developed cities. Therefore, local forest and pasture resources usually have to be exploited for fuel-wood and fodder (Kates *et. al.* 1977). Other key resources, such as groundwater, also become depleted. The Sahel is one of the least urbanized regions in the developing world, with less than 20% of its total population living in towns and cities compared with an average of 29% of Africa as a whole and 31% for all developing countries. There are so few large towns and cities, urban growth in the region has particularly harmful environmental effects, because it is concentrated in the capital cities. The capitals of Chad, Mali, Niger and Burkina Faso contain more than half of all four nations' urban population. The population of Ouagadougou, the capital of Burkina Faso, nearly tripled between 1961 and 1979, while that of Khartoum (Sudan increased even faster).

Another consequence of economic development is that general pattern of land use undergoes changes. Some of these changes may be beneficial, for there is an increased ability to invest in more productive cropping techniques, improved seeds, fertilizers, pesticides, better crop storage etc. Other changes lead to mixed benefits. A developing country usually seeks to improve its international trading position by growing more cash crops for export and while this improves its foreign currency earnings, the expansion of cash crop cultivation displaces traditional land uses. Farmers practicing subsistence rain-fed cropping, then have to move to land previously considered too marginal for this use, which in turn shifts pastoralism to even more marginal areas.

The growth of the market economy also leads to a breakdown in traditional forms of social control over land use, the way in which nomadic herdsmen used to control the grazing of rangelands. The importance of such traditional mechanism may not be realized until they have largely been dismantled, by which time it is usually too late to restore them. Numerous attempts have been made to introduce mew forms of rangelands management in Africa, but all have failed.

The adoption of technological innovations, characteristic of economic development in the industrialized countries, can often take long time to occur in developing countries. In the case of energy, many urban dwellers remain committed to the use of fuel wood or charcoal for cooking and heating and this leads to tremendous pressure on local wood resources and therefore, leads to deforestation. On the other hand, technological change over-emphasized at the expense of traditional methods can also have detrimental effects. In the process of development, huge funds have been poured into expanding irrigated cropping, while subsistence rain-fed cropping has been neglected. Much of the money has in affect been wasted, because the poor design and management of irrigated cultivation .schemes have resulted in Salinization and water-logging.

The introduction of mechanized farming techniques from developed nations also caused problems. In Tunisia, where the disc ploughing of fragile soils has led to widespread soil erosion. The immunization of animals against debilitation diseases should, by reducing animal mortality, lead to an increase in the productivity of livestock raising, lessening pressure on rangelands. But. pastoralists continue to keep their herds as large as possible, to ensure that some survive the frequent droughts. This results into overgrazing and thereby moving towards desertification. Well meaning attempts to improve livestock health and production by sinking more boreholes also leads to desertification, because they tend to channel nomadic herds along small number of routes, which link the boreholes. Pastures along these routes become overgrazed and the soil becomes compacted after being trampled on by thousands of hooves.

2. Desertification and underdevelopment

Although population growth and economic development are the two basic driving forces, which lead to changes in land use, the nature of these changes depend very much on a particular country's economic status. Poorly developed countries usually lack the capital to invest in making agriculture more productive and sustainable, many people live in conditions of abject poverty and the skilled personnel needed to plan the wise uses of land and of natural resources, generally are in short supply. Most of the countries, which are experiencing serious desertification are economically underdeveloped and, therefore, lack the resource to respond to change in sustainable way (Caldwell 1984). Piers Blaikie (1985) of the University of East Angila described the dilemma: "environmental degradation is seen as a result of underdevelopment of poverty, inequality and exploitation, a symptom of underdevelopment and a cause of underdevelopment contributing to a failure to produce, invest and improve productivity."

Some evidence for a link between desertification and underdevelopment is apparent in the results of the UNEP survey of desertification, completed in 1984. It showed that desertification affected 70% of drylands of developing nations, but only 40% or less of drylands in developed nations (Mabbutt 1984). The wider Sudano-Sahelian region embracing the dry zones of both West and East Africa, contains some of the poorest countries in the world. This explains the desertification and related problems have affected its land and people so strongly in comparison with other regions. Mali, Burkina Faso, Niger and Chad, together with Sudan, Ethiopia and Somalia, each had a GNP per capita of less than $ 400 in 1984; for Mauritania and Senegal it was under $ 500. In contrast, the USA had a GNP per capital of $ 14080.

One of the most serious human consequences of economic under development is large number of people living in conditions of extreme poverty and deprivation. When population rise, many small farmers simply lack in means to increase food production without degrading land, for the only land to which they have access is of the poorest quality and therefore, vulnerable to. desertification. Governments are forced to make a choice between food security for the poor and the expansion of cash crop cultivation to improve foreign exchange earnings, often choose the latter. Desertification is not just affecting the least developed countries more than the industrialized; it selectively hurts the poorest sections of the population, the poorest of the world's poor (Kates *et. al.* 1977).

The social and economic dimensions of desertification indicate the need for utmost care in formulating strategies to bring it under control. Piers Blaikie (1985) has put it, "conservation is as much about social processes as physical one and the major constraints are not technical in the agricultural engineering sense, but social." The reality of such constraints becomes apparent while leading to pessimism about the feasibility of bringing desertification under control, if this appears also to require fundamental changes in the structure of society.

The need for care is perhaps most evident in the very delicate subject of family planning. Promoting this more vigorously and therefore reducing the rate of population growth have some impact on desertification. The likely future trends in desertification can be gauged, to some extent, from the predicted balance between population and resources. According to John Caldwell (1984), the Australian National University in Canberra, the problem may not be too serious in parts of Asian drylands, since the availability of water and generally good grassland soils allow them to absorb a greater fraction of their future population growth than is likely for the regions of non-arid lands. In Latin America, the proportion of population located in dryland areas, could fall as young people migrate to cities. However, he warns that in Africa, most rural population does not have access to irrigated land or to rich urban areas, so the majority of population growth occurs in rural areas. "Africa is the one continent, where the early stage of demographic transition, the relationship between falling mortality and rising or falling birth rates and low level of urbanization mean that a huge growth in rural population is still take place." Bearing in mind the economic dimensions of the problem of desertification, it is very doubtful whether controlling population growth can check desertification. Caldwell that family planning campaigns have so far not proved very successful in the African drylands and he argues that the only way to reduce the birth and death rates and likelihood of death during drought and famine is by continued economic development.

3. Role of Government

If desertification is partly a consequence of underdevelopment, then the role of the government is questioned, which is responsible for guiding the course of development. Discussions of land degradation often assume that the government's role is essentially neutral, but this is mistaken (Blaikie 1985). Government policies may actually promote desertification, albeit indirectly and unwittingly. One major problem with the governments of developing countries is the lack of resources, to invest in a phased manner for land upgradation. The poor people of drylands have neither money nor skill to treat their lands, though they are the most sufferers of the problem. The reality of the desertification and land degradation is the slow but recurrent process, which does not divert the attention of the government as done in case of drought or other natural calamities.

To promote national development, governments in most developing countries encourage the growth of industry and urban centers and the building of roads, electricity generation, transmission and distribution as well as other infrastructural facilities characteristic of a developed society. However, urban growth reduces the agricultural labour force and causes environmental degradation in the hinterlands. Fulfilling the aspirations of urban dwellers means keeping food prices low, but this makes agriculture unprofitable for many farmers, who therefore, have little incentive to improve productivity.

Foreign exchange earning is an important prerequisite for development because it enables a country to import the vast range of goods needed to built and maintain the fabric of a modern society. Governments, therefore, promote the cultivation of cash crops and the exploitation of minerals and other natural resources, which may be exported to earn the foreign currency (Schmidt-Wulffen 1985). Ironically, in doing so, they continue to export practices of their former colonial rulers (Bothomani 1984). Expanding cash crop cultivation displaces subsistence cropping and in turn livestock grazing, so that land uses become inappropriate to the land on which they are practiced and insults into degradation. Too great an emphasis on cash crops can also result in plummeting food crop production and even more cash crop export are then needed to pay for imports of food. When wood deficits first developed in Ghana and Nigeria, the governments increased imports to resolve the situation, but local production of staple crops fell, so even more cash crops had to be exported to earn sufficient foreign currency to pay for imported food (Anson-Meyer 1983). The import of wheat flour and rice increased the local demand for these grains and governments of developing countries find it difficult to maintain to increased quantum of imports thereby increasing the pressure of foreign exchange deficits. It can bias a country's whole agricultural development efforts towards establishing expenses irrigation cropping patterns. The governments of the developing countries have to select any of the options, whether they are in a position to import essential commodities on increasing pattern or they intend to invest more funds for creating more irrigation facilities.

The overall food policy need to be assessed in national perspective as the primary responsibility of the government is to provide nutritious food to its growing population either through imports or increases, production and productivity. Import and export items related to agricultural produce is the essential component of country's food policy. Growing cash crops for export to import cereals in not a wise step and unless a country is self-dependent in food production for its population, it cannot face the challenges of other developmental issues like infrastructure creation for betterment of its population. This calls for a fresh look over the country's population policy. The alarming growth of population disrupts total developmental efforts of the government of developing countries. In case the growth rate of population and the domestic product or national income. The development becomes stagnant and sometimes negative growth starts, which is not a healthy situation for the country's economy and development.

The fact is that food production for national requirement and for exports could not be increased while-meeting the challenge of growing population at the existing rate. Therefore, it is utmost essential to regulate the growing population rate to the level of its agricultural growth. The growth of agriculture production has various constraints in dryland regions. This problem is significantly serious countries, where dryland areas are large. The increase of irrigated area cannot increase immediately. The experience of performance of irrigation projects is quite unsatisfactory in terms of huge gap between project potential creation and achieved utilization. In addition, various problems of land degradation like salinity and water logging start occurring. In canal systems, cultivators of upper reach, middle reach and lower reach of water have different level of water availability. Resultant impact of land degradation is visible in upper reaches of the canal at the cost of the lower reach cultivators and generally the cultivators of upper reach suffer most from land degradation problems due to mismanagement of water use.

Growth of population causes desertification in view of catering needs of increasing population. The role of the government in the developing countries is significant in curbing the alarming growth of population. The population policy need to framed by the developing country to reach the growth level at par with the developing countries as it is most essential for the well-being of the people of the country as well as for the development process of the government. The national population policy must be framed keeping in view the following factors:

(1) For curbing the alarming growth of population the population policy must set a target to bring down the population growth at the rationale level within a specified period and must be adhered with legal enactments, awareness generation, motivation of the people to adopt small family norm and possible fixing the upper limit of children.

(2) Provide nutrients in sufficient quantity to prevent mother and child from diseases. This helps in removing fear of child deaths due to malnutrition and lack of proper medical facilities. Casualties from natural disasters are one of the significant components.

(3) In view of sufficient nutritional and medical facilities, death rate also declines significantly. The population proportion in the society misbalances due to decreasing of the working force. This is a temporal phenomenon but has significant impact on the economic conditions of the society and the country.

(4) Population policy must consider employment opportunities. Additional job creation in rural sectors help in maintaining the balance in rural urban culture and employment opportunity for each eligible person is helpful in expanding the process of development much faster.

(5) Literacy has a positive role in reducing the rate of population growth. Family planning measures must be essential aspect in education. The significant reduction in rate of population growth in developed countries is the impact of literacy. The education should be formal and informal both and according to the social requirement of the community. Creation of education facilities in rural areas is not sufficient but to bring all the children to schools is the ultimate goal of education.

(6) Capacity building of the people in village employment generation activities is necessary to produce marketable commodities and get remunerative return. Areas of skill development must be decided keeping in view the job opportunities. This will help in curbing rural migration for seeking jobs in cities or towns.

(7) Small family norms help the developing countries in settling lesser problems of food, employment, education, health, housing, nutrition and pressure on land.

Another development goal for many governments is social change. Due to deforestation, nomadic pastoralists were compelled to settle in villages and they are more easily regulated, taxed, educated and given proper health care. But such policies have led to widespread overgrazing around the new settlements. Many aspects of traditional society, such as subsistence cropping and tribal practices are regarded by governments as old fashioned and while they may not be deliberately dismantled or discriminated against, their further development is certainly not encouraged. Traditional customs and practices are in any case highly vulnerable to the encroachment of the market economy and may quickly become redundant, even without government intervention. In some countries, like Zimbabwe and Lesotho, people are constrained by land tenure regulations imposed during colonial times, but governments find it difficult to replace though they have resulted in high population densities and land degradation.

Governments of a Marxist persuasion, such as those of Ethiopia and Tanzania, often attempted to introduce central planning and community forms of social organization. Tanzania's villagization programme aimed to resettle scattered farm populations in villages but, as with the sedentarization of nomads, this had a detrimental effect on the local environment. In promoting cash crop cultivation, socialist countries are not very different from countries adhering capitalist philosophy. By favouring state-controlled cash crop cultivation at the expense of domestic food production and concentrating agricultural production in a small number of regions, the Ethiopian government increased the likelihood that the drought would lead to a major famine (Kelemen 1985; Griffin and Hay 1985).

There are a number of possible reasons why desertification is ignored in government policies. First, despite UNCOD and the activities, which flowed from it, there is still much confusion, misunderstanding and ignorance about the meaning and causes of desertification. Second, desertification and environmental conservation in general, do not fit easily within the "conventional development model", which only considers the environment and natural resources as source of income to be exploited, rather than elements vital to ensure the sustainability of the whole economic development process. The great achievement of the World Commission on Environment and Development, the "Brundtland Commission" was established a new frame of reference for development models and showed the development and environmental conservation are self enforcing, not contradictory (World Commission on Environment and Development 1987). It may, however, take time for the concept to permeate through the planners and policy makers. The third and least charitable reason is that governments do not care much about desertification unless; if directly affects the interests or the power of the ruling political groups and their friends (Blaikie 1985).

12.2 POPULATION IN GLOBAL DRYLANDS: DATA SOUCRES AND METHODS

To determine population densities in susceptible drylands, it is necessary to combine both data that determine the extent and distribution of dryland climatic zones and data on population distributions. In order to do this at the global level, the climate surface data used to delimit aridity zones are used together with the population surface produced by the US National Centre for Geographic Information and analysis (NCGIA). The latter uses census data collected at first and second sub-national administrative levels, for approximately 15,000 administrative units worldwide (Tobler *et. al.* 1995). Population counts were extrapolated to 1994 in order to standardize the year of assessment. Aridity surface, population distribution data and country boundary information were handled and analyzed in raster form using an ARC/INFO GIS system.

12.3 POPULATION IN GLOBAL SUSCEPTIBLE DRYLANDS

The data given in Table No. 12.1 show the total human population in each aridity zone for each continent. Data for hyper-arid areas are included for although they fall outside the susceptible drylands, they do support significant human populations in some parts of the world, albeit that they are usually heavily reliant upon resources imported from outside the hyper-arid zone. This is especially the case of Africa, where almost 10% of the population lives in the driest areas, in South America, where several urban centers are located in the Atacama Desert and in parts of the Arabian Peninsula and central Asia. African data are further broken down which reveals that 38% of the population of North Africa lives in hyper-arid areas. This figure includes the Nile Valley, where a special situation is created whereby large numbers of people are supported in this otherwise hyper-arid area by the perennial River Nile.

Table 12.1 : Population of the world's susceptible drylands

Continent	Hyper-arid	%	Arid	%	Semi Arid	%	Dry Subhumid	%
1. Africa	58175624	9	41366493	6	117573208	18	109038742	16
2. Asia	29506286	1	161556598	5	500695158	15	657899360	19
3. Australasia	0	0	275039	1	1352905	5	5318077	19
4. Europe	0	0	628630	<1	28811499	5	115146764	21
5. S. America	3877706	1	6330581	2	46851867	16	33777496	12
6. N. America	508971	<1	12750631	3	53900339	13	24342220	6

Source: UNDP/UNSO 1997

The importance of the susceptible drylands in supporting a significant component of the world's population is clear from table no. 13.1. Indeed, in Africa, 40% of the total population or nearly 268 million people live in these areas prone to drought and susceptible to desertification. Equivalent figures for Asia are 1320 million people i.e., 39% of the total continental population and

South America 30%. These figures clearly indicate that desertification is regarded as a problem of great significance.

Table 12.2 shows the population data expressed as densities, population per square kilometer. The range of average population densities is considerable both between different aridity zones and within the same zone in different continents. Asia has the highest population densities in all three susceptible dryland zones, while Australia has relatively low overall population densities in all the susceptible dryland zones, when compared to the other continents. Despite the limited number of people living in Australia's arid areas, the GLASOD survey demonstrated that significant soil degradation problem exists. This point shows that population numbers alone cannot always be used to explain the occurrence of land degradation or to predict future degradation problems to occur; in the interior of Australia, it is relatively extensive pastoral activities that are often associated with soil erosion problems (Pickup and Stafford Smith 1993).

Table 12.2 : Population density in drylands (Per sq. km.)

Continent	Hyper-arid	Arid	Semiarid	Dry pubhumid
1. Africa	8.6	8.1	22.9	40.6
2. Asia	10.6	25.9	72.2	186.5
3. Australasia	0	0.1	0.4	2.4
4. Europe	0	5.7	27.4	62.8
5. North America	16.4	15.6	12.9	2.9
6. South America	15.1	14.2	17.7	16.3

Source: UNDP/UNSO 1997

The greatest population densities occur in semiarid and dry sub-humid areas. This is expected in terms of the. susceptible drylands as a whole, they offer the greatest potential for human activities, particularly agriculture. As population and land use pressures grow in these areas, both people and the environment become more prone to the impacts of climatic variability (Glantz 1994). The high population densities in dry sub-humid and semi-arid zones of many continents mean that environmental pressures increase, during times of drought, there exists less opportunity for flexibility in human actions. Care, however, needs to be taken, while making generalization for population data as population totals and densities may serve as gross indicators of desertification risk, misleading pictures may emerge if taken alone (UNDP/UNSO 1997).

There are no clear relationships between population density and erosion (Blaikie 1985, Warren *et. al.* 1996). On the one hand, more people consume more resources and, therefore, may create more degrading processes, on the other, more people can make greater contribution to conservative measures. There are growing number of examples from around the world that detail situations, where rising population numbers and increasing population densities can contribute to declining or low rates of degradation (Mortimore 1989, Mainguet 1990, Tiffen *et. al.* 1994) and rural population decline can result in enhanced soil erosion (Millington 1989, Thomas and Middleton 1994). In such situations, it is necessary to exercise care in seeking the reasons behind them, on order in order that the full range of socio-economic factors such as change in land tenure and local governance, as well as the demographic issues, which can be incorporated in analyses that seek explanations (Tiffin *et. al.* 1994).

A number of reasons explain the. relationships between population and degradation which are so complex. One factor, that is certainly important is the range if socio-economic triggers or 'root cause' that can lead to land degradation. As UNDP/UNSO note, political and socio-economic factors are often the main drivers of land degradation. One issue noted as particularly important in recent decades is increased rates of migration caused by war, drought and famine (Westing 1994). Migrations can lead to rapid changes in population densities at local and regional scales and this occurs significant environmental pressures can ensue.

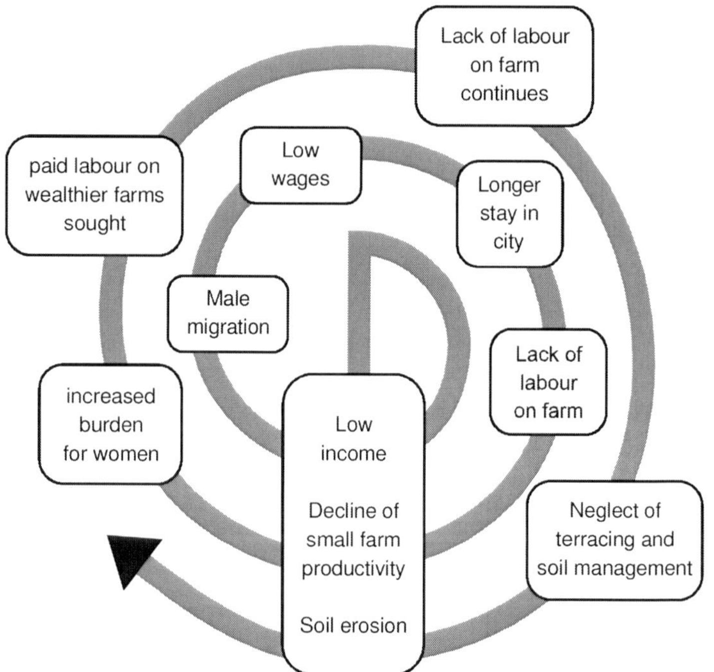

Figure 12.1 : *The implications of rural-urban migration for subsistence farming (after Millington et. al. 1989)*

Another issue is population growth rates that are important in determining the rate at which pressures are exerted on the land. leading to rapidly growing demands for food that result in agricultural expansion and intensification. In semiarid areas of Inner Mongolia, rapid population increases since the 1950s are believed to have made a marked contribution to enhanced land degradation (Takeuchi *et. al.* 1995). An important consequence of increased pressures to produce more food is that a number of studies show how traditional systems of agriculture that incorporate fallow periods are declining as more land comes under permanent cultivation to feed rising population (Bremen *et. al.* 1990).

Drylands in both the developed and developing countries now support many large urban centers (Cooke *et. al.* 1982, Beaumont 1989). Urban centers not only have high population densities, but exert a range of direct and indirect impacts on the environment that contribute to desertification. Direct impacts include enhanced demands on water resources leading to ground water depletion, demands for fuel that, in the developing countries contribute significantly to fuel-wood exploitation, the creation of waste products and demands for food. The most significant indirect factor is the attraction urban centers offer to rural populations in terms of employment opportunities, both real and imagined. Mortimore (1993) outlines how the displacement of poor rural people from the land for a variety of reasons makes them more dependent on off-farm income sources and these are frequently sought in urban areas. In some countries, this can contribute to a marked decline in agricultural labour force (Millington *et. al.* 1989), which in turn can have marked and complex impacts on degradation and rural population (Figure 12.1). One study notes how in parts of the Sudan, many families already have more land than labour available to cultivate it (Myers *et. al.* 1995).

Chapter 13

Poverty and Environmental Degradation

Poverty has been described as both the cause and the effect of environmental degradation. In some circumstances, poor people over-exploit the environment in their struggle to survive, while in other situations, environmental degradation leads to falling crop productivity or increasing poor health that in turn, create impoverishment. However, the extent of poverty and the state of the environment may be closely linked, this does not imply simple, clear lines of causation. Separating the environmental effects of poverty, population growth, inappropriate property rights and other factors is difficult. Some authors think it more appropriate to view poverty as the mechanism through other factors, lead to degradation (Pearce and Warford 1993).

Moreover, it is clear that poor people often rely directly on natural resources, or 'natural capital', for their living and very survival. Soil can provide a flow of food or income in the same way that a machine can produce a flow of manufacturing goods. Both are forms of capital, one provided by the earth, the other man-made. Poor people usually have very small amounts of man-made capital and this means, they have virtually no opportunities to generate income by producing manufactured goods. Poor people also have their own labour as a resource but employment possibilities are usually limited.

Poor people generally rely upon a very limited stock of natural capital, such as a small plot of land or access to a small communal common. Even those people neither own nor rent land and rely on employment for income, often supplement this income in an important way by access to some natural capital. Natural capital that can be accessed without a rental or ownership contract is called a "Common Property Resource" (CPR). Examples of CPRs are woodlands, grazing areas and rivers that can be used by members of the community. A study of rural villages in India found CPRs to account for 14-23 % of average household income (Jodha 1992). It is this characteristic of heavy reliance upon natural, capital-based sources of income that typifies many poverty-stricken people., therefore, that food imports may continue to rise and there is danger that increased emphasis may be placed by growing cash crops to pay for these imports. Furthermore, in times of drought, migrants are drawn to the cities in search of food and employment, depriving rural areas of the farm workers needed to grow crops and further reducing agricultural production (Glantz 1987-b).

Poverty is the result of various adversities in dryland areas. The most serious cause of the poverty is famine and drought. Adversities of monsoon is a recurring phenomena in dryland regions and impact of famine and drought further adds the problems of desertification and land degradation. The people residing in dryland areas have little livelihood resources and main resource is the land they cultivate is generally marginal land or marginalized in view of adversities caused by recurrent droughts. The worst affected are the people specially children and women in view of malnutrition. The livestock population decreases due to droughts, which is the supplemental source of their livelihood. The incidence and impact of poverty is very significant in dryland areas, which need immediate attention of the country government for initiating suitable measures to combat the impact of droughts.

Underdevelopment and misguided government policies can combine with drought and desertification to cause hunger or even famine. The United Nations Conference on Desertification owed its genesis more to worldwide horror at the terrible effects of famine during the first spell of drought in the Sahel in the early 1970s, than to concern about dryland environments alone. Famine generally occurs, when the poor crop yields lead to severe shortages of food in a particular area and market mechanisms or government interventions are insufficient to balance these shortages with supplies from elsewhere in the country or from overseas, but it does not have to be a consequence of desertification and drought.

The causes of famine have been debated at length. A period of prolonged, extended drought is obviously a contributing factor: Michel Glantz (1987-a). has referred to it as "setting the stage" for famine. Amartya Sen (1981), on the other hand has asked, "Why is it that producers of food are the first and most serious affected by drought or famine and why do so few town dwellers die from hunger, while rural areas are decimated by starvation and death." The implication is that famine is selective and causes of famine are heavily influenced by social and economic factors. Famine can arise, from misguided agricultural policies or the inadequacy of transport or ship for food from one part of a country to another. Thus a country may have an overall food surplus, yet some of its people can still starve. According to a report by Independent Commission on International Humanitarian Issues, famine is really a "man-made factor" and over-intensive agriculture, loss of topsoil and desertification help to establish the preconditions for it long before delivers the "coup de grace" (Brown 1985).

Numerous studies have linked famine with government policies that encourage the cultivation of cash crops at the expense of subsistence food crops, not necessarily with the laudable aim of preventing hunger. Government price policies, which forced small farmers to produce export crops at expense of food crops, were blamed for the famine in Mali from 1969 to 1973 (Schmidt-Wulffen

1985). These exports were needed to earn foreign currency to finance imports, create jobs in industry and increase the size of the government bureaucracy, but only farmers in favoured areas could buy enough food for their subsistence from the profits of cash crop production. In northern Mali, neither cash crop profits nor food supplies were adequate. No exchange of food was possible between different regions because the south of the country did not have a food surplus to meet the needs of north. A similar criticism has been expressed about policies in Kenya (Hegmar *et. al.* 1982).

Ghana and Nigeria had experienced great emphasis on cash crops, resulting in a vicious circle in which, domestic food crop production fell down and an ever-greater volume of cash crop had to be exported to pay for imports of food, which could easily grown in the country itself (Anson- Meyer 1983). Rising food imports, whether through normal trade or as "food aid" from developed nations concerned about famine, had changed consumers' taste away from locally grown staples and so depressed local food production. Per capita consumption of wheat and rice in Sahel grew by 4.6 % and 2.0% respectively between 1966-70 and 1976-80, while that of maize, millets and sorghum fell (Delgado and Miller 1985). Continuing demand for such "exotic" imported grains could bias a country's whole agricultural development efforts towards producing them, often by expensive irrigated cropping and this seriously hampers the rehabilitation of the subsistence farming sector. Rice production in the Sahel and other parts of West Africa has been expanding but not as rapidly as consumption. Only limited areas are suitable for wheat growing in the region.

Poverty may be severe as a consequence of serious political conflict inside the country. Michael Glantz had pointed out that although, there were major droughts and climate induced food shortages around the globe in 1982-84 in various parts of Africa, India, China, Indonesia, Brazil, famines occurred in Africa. Even a cursory review of the historical record of any drought plagued region shows that famine does not necessarily follow drought. Thirty one countries in sub-Saharan Africa were declared to have drought related food shortages, only five of them, only five of them, viz. Mozambique, Angola, Chad, Ethiopia and Sudan, were plagued by famine and each of them were subjected to internal wars (Glantz 1987-a).

The misery of the people of Ethiopia and Sudan provides a good illustration of how, even in countries affected by drought, civil war can make the difference between food shortages and famine. In Ethiopia, one million people died from starvation in 1984-85, because the government did not alert international agencies early enough to the food shortfalls likely to result from drought (MacKenzie 1987).

13.1 POVERTY, NATURAL CAPITAL AND INVESTMENT

Natural capital can be degraded by the actions of its users. Growing inappropriate crops can reduce soil fertility and erosion can result from excessive grazing pressure. The maintenance of the stock of natural capital requires sustainable management. In particular, current resources must be invested to ensure that the capital is productive in the future. The fertility of soil to ensure that it retains its productivity involves not using resources for immediate benefits e.g., the animal dung used as fertilizer could have been used as fuel and investing them for return to be received in future. Irrigation schemes, tree plantation and decisions to keep number of livestock are main issues of investment decisions; planning today to ensure future production levels.

In making any investment decision, people need some way of comparing future benefits e.g., improved soil fertility and crop yield to current costs i.e., the cost of fertilizer. Each person has a time preference that determines the worth of such investment undertaking. Economists use discount rates to indicate time preferences. As the discount rate increases, the level of total investment falls as few investment projects could provide a significant return. Higher discount rates also mean that short- term investments are favoured over long-term investments.

Poor people often have relatively high discount rates. Studies in India have shown that the rate of return required by poor farmers is 30 - 40% per annum (World Bank 1992). High discount rates arise from the uncertainty that poor people face about the future. Poor health, diet and living conditions mean that people are uncertain about their future survival and work capacity. In addition, the uncertainty caused by possible drought or flooding may mean investments, especially long-term investments, are not undertaken. Efforts to introduce soil conservation techniques in Burkina Faso, showed that adoption of new techniques was highest when returns occurred within two or three years (World Bank 1992). Another key factor in the investment decision is that poor people may view their property entitlement as uncertain and increase their discount rate accordingly. Future tenancy agreement changes and policy changes may move poor farmers from their land so that they will not be there to reap the benefits of previous investments.

Poverty means, therefore, that discount rates of individuals are high and the planning horizon is relatively short. But many investments in natural capital have relatively low rates of return that continue far into the future (Newcombe 1989). Planting trees, yield benefits in terms of fuel wood and prevention of erosion but these only occur after 5 and more years. A poor community may, therefore, invest only a small amount in its natural capital. This lack of investment may lead to degradation, which in turn implies that future productivity will fall (Vebator *et. al.* 1992).

It is important to note that poverty does not always lead to degradation. There may be strong social and cultural rules for controlling the use of natural capital. Such rules, which can over-ride individual preferences, may have been built up over long period of time to ensure survival of the community. Long-standing rules break down by migration to a new area or by changes in tenurial rights and degradation becomes a problem. This is especially the case with the management of common property resources. Traditionally CPRs are governed by social rules established over time but the joint effects of population growth, poverty and policy changes may break down these rules (Jodha 1992). The CPRs then become 'Open Access Regimes' (OARs), areas where people can extract as many resources as they are able. These high extraction rates occur because individuals have insufficient incentive to preserve a resource that is open to all. In essence, the distinction between a CPR and an OAR is those individuals producing under OAR act as though the discount rates were indefinitely high, due to the low probability of appropriating the benefits of foregone harvests. Under OAR, the incentive to invest in the natural resources is effectively eliminated.

1. Poverty and Population Growth

Poverty and population growth are closely related to each other. In dryland areas, the main cause of the desertification and land degradation is the population pressure, which compels people, generally poor, to remove the vegetation for cultivation, cutting of trees and bushes for fuel and grazing the livestock on the grassland. The population of livestock is almost equal or more than human population, as the people keep those for their survival base for using the animal products in difficult times like drought or failure of monsoon.

Growth of the population in dryland is significantly high as the people of reproductive age groups are tough to bear all the natural adversities and possess no means of entertainment to pass their leisure time. Furthermore, the lack of awareness for adopting the population control and keeping gap in child births. The social aspect of larger family to provide support in domestic activities like watch of the herds and support in cultivation. The children of early age are engaged in fuel wood collection or keeping vigil on the herds. Lack of medical facilities to prevent from diseases and malnutrition is another aspect of large family concept to keep living child even after natural calamities.

Population growth in poor social groups is a common phenomena and no efforts of the government are possible in far flung areas. Herdsmen of migratory nature in reproductive age groups witness addition of one more child after one year return from their long tours of thousand kilometers. After span of seven to ten years, the male child is also accompanied with the herds as a support person and it is treated as a necessity, which further gets one independent herd on maturity. Therefore, increase of population and livestock is inter-related issue and poverty is regular phenomena as the croplands are generally deserted due to failure of monsoon.

In the Thar Desert of Rajasthan, western state of India, the herdsmen move with their herds after July while sowing their crops and return after one year for sowing next crop. A large number of herdsmen move from these areas on fixed migratory routes and in the way they sell cattle product like milk and wool at the selected spots. Cultivators keep the herds in their fields during the night by providing the cost of their night halt as the herds provide them sufficient bio-fertilizer for their ensuing crop. The herdsmen generally attain their livelihood through these herds and their family members get some crop in case of favourable conditions. The herds increase generally on each trip as the reproductive capacity of goat and sheep is quite significant.

Herdsmen of Thar Desert are well aware of the circumstances of their native areas, where possibility of crop yield is quite insignificant and they are compelled to adopt the livestock rearing as the source of their livelihood. The poor people surviving in dryland areas have very little sources of livelihood as they have marginal cultivated lands dependent on rainfall. No other means of employment is possible in these areas as government assistance is only possible during droughts, where the people are engaged in wage employment activities.

2. Poverty and Desertification

Poverty and desertification in dryland areas are also related factors. Droughts are recurrent in these areas and each drought brings misery for them. In such a situation, the total vegetative cover is removed for fuel wood and cattle grazing. The land loses its fertility by human and livestock efforts and before next monsoon the complete areas is barren and deserted. The sand dune stabilization efforts are mitigated by cutting of trees by government machinery and the dunes are free to move with wind pressure.

Deficiency of water is also a common phenomenon in these areas. Actually, desertification occurs due to lack of water resources in these dryland areas. The situation is still worse in rain-fed areas and even the drinking water problems become severe for their survival. The people of these areas stay in their native areas as they have no access to cities in view of their inherent problems like cattle or hope of good rainfall in ensuing monsoon.

The situation of dryland areas is comparative better as the Thar Desert is the most populated place in the deserted areas of the world. The impact of green revolution has succeeded in procurement of food grains in the country, which is made available for schemes like food for work to create employment opportunities during droughts. In other drylands of the world, specially the Sahel are not so better as during droughts they still need international assistance of food for their survival. Prevalence of poverty in such areas is quite severe.

3. Poverty and Famine

Famine conditions in any underdeveloped country enhances the impact of poverty. The civil war in southern Sudan had led to a famine of massive proportions as millions of people were driven from their homes and lands by rebel troops or by inter' tribal conflict encouraged by both the government and rebels. It was estimated that during the summer of 1988, hundreds of people in the region died each day from hunger and disease. Large areas of crops were burnt so that people may unable to feed themselves. In October 1988, the Sudanese government ordered the UN Children's Fund (UNICEF) to stop distributing food in rebel-controlled areas (McElvoy 1988). The revels laid siege to towns like Torit that were held by government troops and prevented food supplies from reaching them (Coughlin 1988).

Famine is therefore, as much a result of inadequate government policies and underdevelopment as it is of drought. According to Bradford Morse (1987), the former administrator of United Nations Development Programme (UNDP), "drought itself is not the fundamental problem in sub-Saharan Africa. The present drought has, however, intensified the interaction of the factors impeding development in Africa; it has laid bare the African development crisis." The danger is that if famine is blamed on drought, then the deeper socio-economic and agricultural changes needed to ensure food security for all will not be made, since these require long term action and this is not attractive to politicians. Famine certainly exacerbates the land-use pressures, which cause desertification, for when food is in short supply, people become desperate and over-cultivate the land to get to possible extent. They know that their efforts would further deteriorate the fertility of the land to significant extent, but meeting the present crisis before them is the ultimate necessity for their survival.

Famine most seriously affects the poorer people and they are often forced on to the marginal lands and such lands are most vulnerable to being desertified . If desertification does not occur, then it would reduce the area of farmland available for growing food, total food production would decline and the people would be further compelled to use more marginalized land than before. The poverty is the resultant impact as the country governments have no plan to attend their conditions. The fact is that the expectations of the people of dryland areas from the government are very meagre and the solution of their problems even in phased manner could change their conditions, but the political system never thought it as their responsibility to do something for such people.

4. Linkage Between Desertification and Poverty in Drylands of the World

Drylands are inhabitated by approximately 2.56 billion people globally, accounting for more than 47 % of the world's population. In terms of the dryland areas consist of 53.5 million sq. km or 40% of the land area. According to the 1997 UNDP Human Development Report, poverty is worse in drier zones than it is in wetter zones. For 10 countries of the Sahel Region in Africa, the Human Poverty Index was 25% of the humid areas and 61% for arid zones. To gain better understanding in the link between drylands and the UNDP Human Development Index, which provides a measure of the socio-economic development. UNSO undertook an analysis of some 90 countries. On that basis, the countries were classified into three clusters:

Cluster 1 : Low HDI, PLVD > 35%.

Cluster 2 : Medium HDI, PLVD > 35%

Cluster 3 : Others

Where HDI = Human Development Index and PLVD= Productive Land Vulnerable to Desertification.

The classification of the countries based on their vulnerability to degradation and human development is given in Table 13.1, where cluster demarcation is given with respective countries.

Table 13.1 : Link between desertification and poverty in countries from various regions

Latin America and Caribbean	Europe and CIS	Africa	Arab States	Asia and Pacific
Argentina -3	Armenia-3	Angola-1	Algeria-2	Afganistan-1
Barbados-3	Bulgaria-3	Benin-1	Bahrain-2	China-2
Bolivia-1	Georgia-3	Botswana-2	Djibouti-1	Ondia-1
Brazil-1	Kazhaksthan-2	Burkina Faso-1	Egypt-1	Iran-2
Chile-3	Kyrguzstan-3	Cameroon-3	Iraq-2	Mongolia-2
Colombia-3	Turkey-2	Cape Verde-1	Jordan-2	Pakistan-1
Costa Rica-3	Tajikistan-3	Central	Kuwait-3	Sri lanka-3
Cuba-3	Turkmenistan-2	American	Lebanon-2	
Dominican	Uzbekistan-2	Republic -3	Libya-2	
Republic-3		Chad-1	Morocco-2	
Eucador-3		Coted' Ivoire-3	Oman-2	
El Salvador-3		Eritrea-1	Qatar-3	
Guatemala-3		Ethiopia-1	Saudi Arabia-2	
Haiti-3		Gambia-1	Somalia-1	
Honduras-3		Ghana-3	Sudan-1	
Jamaica-3		Guinea-3	Syria-2	
Mexico-3		Guinea Bissau-3	Tunisia-2	
Nicaragua-3		Kenya-1	UA Emirates-2	
Panama-3		Lesotho-1	Yemen-1	
Paraguay-2		Medagaskar-3		
Peru-3		Malawi-1		
Venezuela-3		Mali-1		
		Mauritania-1		
		Mozambique-1		
		Namibia-1		
		Niger-1		
		Nigeria-1		
		Senegal-1		
		South Africa-2		
		Switzerland-1		
		Tanzania-1		
		Togo-3		
		Uganda-3		
		Zambia-1		
		Zimbabwe-1		

Source: UNDP: Programme Activity Report 1995-1998

So far as dryland area coverage is concerned, in various continents, the arid, semi-arid and dry sub-humid regions in terms of area of respective continent and its percentage from the continental cultivated area of 134.79 million sq. kms. is given in Table 13.2 to specify regional variations.

Table 13.2 : Extent of drylands in different continents

(Area number in million sq. km.)

Continent	Arid	%	Semi-Arid	%	Dry Sub-Humid	%
Africa	5.052	3.75	5.073	3.76	2.808	2.08
Americas and Caribbean	1.201	0.89	7.113	5.28	4.556	3.38
Asia	6.164	4.57	7.649	5.67	4.558	3.38
Australia and Oceania	3.488	2.59	3.532	2.62	0.996	0.74
Europe	0.005	Neg.	0.373	0.28	0.961	0.72
World Total	**15.910**	**11.80**	**23.740**	**17.61**	**13.879**	**10.30**

Source: UNDP/UNSO- An Assessment of Population Levels of World's Drylands.

Concentration of population in dryland continental areas in comparison to total global population of 5395.26 million is given in Table 13.3. The percentage given in respective dryland areas is in relation to total population to reveal the situation in relationship with the dryland area.

Table 13.3 : Population of drylands in different continents of the world

(Population in Millions)

Continent	Arid	%	Semi-Arid	%	Dry Sub-Humid	%
Africa	40.50	0.75	117.65	2.18	109.37	2.03
Americas and Caribbean	19.08	0.36	100.76	1.87	581.20	10.77
Asia	161.55	2.99	625.41	11.59	657.90	12.19
Australia and Oceania	0.28	Neg.	1.34	0.02	5.32	0.10
Europe	0.63	0.01	28.72	0.54	115.22	2.14
World Total	**222.04**	**4.11**	**873.87**	**16.20**	**1469.01**	**27.23**

Source: UNDP/UNSO- An Assessment of Population Levels of World's Drylands.

A comprehensive conducted by UNEP to assess the status of desertification and implementation of United Nations Plan of Action to Combat Desertification in 1992 indicated that the dominant forms of land use and land cover in drylands are irrigated and rain-fed agriculture and rangelands, with forest and woodland occupying up to a quarter of drylands. The survey concluded that 73% of the rangelands, 47% of the rain-fed croplands and 30% of the irrigated areas showed some degree of desertification as given in Table 13.4. It is however mentioned that global assessments of this nature can vary from survey to survey based on criteria used.

Table 13.4 : Dominant types of land use in susceptible drylands

(In Million hectares)

Region	**Irrigated Croplands**	**Rain-fed Croplands**	**Range lands**	**Total**
Africa	10424	79822	1342345	1432591
Asia	92-21	218174	1571240	1881435
Australia	1870	42120	657223	701213
Europe	11898	22106	111570	145574
N. America	20867	74199	483141	587207
S. America	8415	21346	380901	410662
World Total	**145495**	**457767**	**4556420**	**5159682**

Source: UNEP, 1992-a.

According to Human Development Report of UNDP 2001, evaluation of 162 countries was conducted United States, Japan and Argentina are grouped under Category of countries showing high human development Brazil, China and India are grouped with countries showing medium Human Development and Pakistan and Bangladesh are grouped with countries showing low human development. This information has been depicted in Table 13.5 to show the comparative position of these countries along with the world average, high, medium and low human development index of the world countries.

Table 13.5 : Human development index of selected countries from different regions of the world

Human Development Parameters	United States	Japan	Arge-ntina	Brazil	China	India	Pakistan	Bangladesh	World Ave.	High HDI	Med. HDI	Low HDI
1	2		4	5	6	7	8	9	10	11	12	13
Life Expectancy at birth (years) 1999	76.8	80.8	73.2	67.5	70.2	62.9	64.4	59.6	66.7	77.3	66.8	52.6
Adult literacy rate (% age 15 and above) 1999	99.0	99.0	96.7	84.9	83.5	56.5	45.0	40.8	78.8	98.5	76.9	48.8
Population with access to improved water sources 1999			79.0	83.0	75.0	88.0	88	97	27			
Population with access to essential drugs(5) 1999	100	100	70	40	85	35	65	65				
Population with access to (%) adequate sanitation 1999			85	72	38	31	61	53				
Total population (Million) 1999	280	127	37	168	1265	993	138	135				
Expected population (2015) (Millions)	322	128	44	201	1410	1231	204	183				
Annual population growth rate (%)												
1975-1999	1.0	0.5	1.4	1.8	3.8	2.0	2.8	2.4	1.6	0.7	1.7	2.7
1998-2015	0.8	0.0	1.1	1.1	2.5	1.3	2.5	1.9	1.2	0.4	1.0	2.5
Female adult literacy (% age 15+) 1999	—	—	96.7	84.9	75.5	44.5	30.0	29.3	—	—	69.7	38.9
Total fertility rate (No/ Female)												
1995-2000	2.0	1.5	2.6	2.3	1.8	3.3	5.5	3.8	2.8	1.7	2.6	5.6

Human Development Parameters	United States	Japan	Arge-ntina	Brazil	China	India	Pakistan	Bangladesh	World Ave.	High HDI	Med. HDI	Low HDI
1	2		4	5	6	7	8	9	10	11	12	13
Infant mortality rate (Per 1000 live births) 1998	7	4	19	34	33	70	84	58	56	7	46	99
Public Education Exp. As % to Total Government Exp. 1995-97	14.4	9.9	12.6	—	12.2	11.6	7.1	13.8	—	—	—	—
GDP per capita (US$) 1999	31872	24898	12277	7037	3617	2248	1834	14830	6980	23410	3850	1200
Overall ranking in Human Development Index	6	9	34	69	87	115	127	132	—	—	—	—
Gender related Development Index	4	9	33	64	76	105	117	121				
Human Poverty Index				18	24	55	65	73				

Source: Human Development Report, UNDP, 2001

13.2 MINING THE FUTURE

Poor people may be able to live in harmony with the environment using traditional methods under normal conditions. However, subject to 'shocks' due to natural disasters like floods or droughts or human induced issues like land tenure changes, poor people either migrate from the affected area or remain in the degraded area while struggling for their survival.

Those who stay in the affected area need to extract sufficient resources to survive. Their poverty means they have a very limited range of resources, only their natural capital stock and their labour. The usefulness of 'human capital' under these circumstances are very limited. A shock could reduce employment opportunities, which means the poor will only have their natural resources for survival. People may be forced to over-exploit these resources; land may be grazed too intensively, crops may be planted on marginal soil or additional fuel wood may be collected from woodlands for sale. This type of resource exploitation is known as 'mining', which means that the stock of natural capital is being reduced for current consumption at the expense of future production. Mining ensures short-term survival, but at the expense of environmental degradation and future adverse impacts.

Occurrence of degradation depends on local circumstances. Environments can accommodate a certain level of exploitation and still regenerate themselves to their original state. However, when environments are pushed beyond certain threshold levels, a future decline in productivity may result. Land may be able to sustain some periods of heavy grazing pressure, but if the grazing is too intensive, the mix of species may be altered, after which water and wind erosion may remove soil and prevent plant regeneration (Perrings and Walker 1992).

It may be stressed that all shocks affecting a community need not necessarily lead to degradation. Whether or not degradation occurs, depends on the resources available and the social rules governing the community. One study in Nigeria found that the pressure of famine did not affect small-holders' tree conservation practices (Mortimer 1989). Further, Jagannathan (1989) stressed that if additional employment can be obtained in the event of a shock, and then poor people may restore their income level through work rather than mining the environment. Mining is one possible response to an environment shock, but it is by no means a necessary consequence.

1. Efforts by Government to treat degraded lands

Desertification is strongly linked with underdevelopment. This might give an impression that when dryland countries would become developed economically, desertification would go away. However, desertification is not confined to developing nations and poor land management can occur desertification everywhere, even in USA. More than 13 million hectare of US farmland is so highly vulnerable to erosion that the annual soil loss can only be kept to tolerate levels of the most restrictive land use practices. Half of all cropland requires conservation measures to keep soil loss within tolerable limits, 8% is so vulnerable that intensive cultivation leads to unacceptable rates of soil loss and only one-third is not inherently susceptible to erosion under any type of management (Bills and Heimlich 1984).

The consequences of poor management of drylands in the Great Plains became apparent in the 1930s, when huge areas were turned into the infamous Dust Bowl. The situation was saved only by a massive mobilization of manpower and funds by the federal government, which established the Soil Conservation Services and the Civilian Conservation Corps to advise farmers on conservative regimes, plant shelterbelts and do other conservation work. The fact that wheat yields in the Great Plains did eventually recover after the Dust Bowl and even exceeded former levels, given rise to optimism that desertification can be brought under control.

Soil conservation, however, demands continued vigilance and the USA dropped its guard. In the early 1970s, the Department of Agriculture told farmers to bring back in to production 24 million hectare of idled cropland to increase American exports. According to James Risser (1985) of Stanford university: " Soil conservation structures were destroyed so that all land could be put into row crops. Terraces and rows of trees were removed because they kept valuable land idle and because they got in the way of big new farm machinery that was being developed and hilly, erosive marginally usable land was ploughed and planted, some of it for the first time. Crop rotations were abandoned and farmers adopted monoculture cropping of corn, wheat or soyabeans."

The effects of this measure on soil erosion is not clear, although Riser claims that total US soil erosion is now greater than in the worst years of the Dust Bowl. Fortunately, at the same time, as some of the windbreaks were removed between 1970 and 1975, new windbreaks were established elsewhere, giving an apparent net gain of 1098 km of windbreaks and in 1985, the Department of Agriculture decided to introduce a "set aside" scheme, which aimed to take more than 18 million hectare of cropland out of production in order to reduce food surpluses. By 1988, about 11 million hectare had already set aside, although this was less than half of the area brought back into production a decade earlier.

The USA was not alone in slackening its grip on soil conservation. The expansion of croplands in to marginal areas had also increased the problem of wind erosion in Australia and the Soviet Union. In addition, Australia, USA and Canada had also experienced worsening trends in the desertification of rain-fed croplands in areas, where saline groundwater was seeping the surface (Dragne 1985). In 1988, a massive Salinization problem of different kind in Soviet Union was published for the first time. Since 1956, water has been taken by canal from the Amu-Dar'ya River in Soviet Central Asia to irrigate millions of hectares of land under cotton cultivation in the Karakum Desert several hundred miles away. The Amu-Dar'ya is one of the two rivers that are the main source of water for the Aral Sea, previously the fourth largest lake in the world. The diversion of water had reduced the area of the sea by two-thirds since 1960 and the exposed lake bed had been turned into salt desert said to be thousands of square kilometers in extent. Surface salt was swept up by winds and transported for hundreds of miles, causing considerable damage to farmland at its destination (Miller 1988; Lindley 1988). The fact that such desertification problems could occur in developed countries offers a warning that the future of the drylands may not be secure even if developing nations do succeed in dramatically improving their economic status.

2. Micro Level Efforts

Desertification and land degradation are the results of faulty approach of people residing in those areas and wrong policies of the country governments, but the fact finding or cursing the wrong doings is not the solution of the problem. Past efforts for combating desertification and land degradation through governmental or outside efforts could not succeed in view of various reasons comprising of lack of peoples support and cooperation and technical support. Planning for combating desertification were framed without considering local situations. However, the past events analysis is relevant to the extent to initiate efforts without repeating similar mistakes of the past. The experiences gained by respective government and the affected people reveal that measures for combating desertification and land degradation should be formulated at micro level keeping in view the level of degradation and expectations of the people.

Micro level schemes are cost effective and shorter gestation period. People of the dryland areas understand their problems better than anyone else including the country government. Similarly the people are well aware about the measure best suited to the problem. Therefore guidelines framed by country government should be flexible to accommodate local situations and benefiting the community. The ultimate objective of such measures should be land upgradation and increasing the productivity to the maximum possible extent, but the measures to be adopted at micro level must incorporate the local requirements. This process can yield success in short span and the country domestic product can be increased significantly while the aspirations of the community are met.

Technology support from any successful measure can be replicated in other countries, but with cautious efforts keeping in view the local conditions to be acceptable to ecology and environment. Technology import must be welcome from any part of the world irrespective of the developing or developed countries but its application in toto cannot succeed in achieving similar results and the past mistakes should not be repeated. The poverty issues are quite delicate and need patient approach to provide early results as this is linked with the starvation and misery. The strategy for future must be framed keeping in view of the geographical and ecological conditions on the one side and natural resources availability on the other hand.

13.3 POVERTY MITIGATION AND DEGRADATION

Another possible response open to poor people faced with increasing pressure on resources is to migrate to another region. The pressure on resources could arise from a 'shock' or from steadily worsening conditions. Historically, the process of development has led to movements of people from rural to urban areas, fuelling the rapid growth of cities in many developing countries. Table 13.6 shows details of migration trend in selected countries, illustrating that there has also been substantial rural to rural are migration in developing countries over the recent past. Such migration has environmental effects in both the regions of origin and destination. Lower population in the area being vacated can reduce pressure on natural resources and reduce degradation.

Table 13.6 : Movements of migrants by origin and destination in selected dryland countries

Country (Year)	Proportion of Total Annual Migrants (%)			
	Rural-rural	**Urban-urban**	**Rural-urban**	**Urban-rural**
1. Botswana (1985)*	60.0	8.0	29.0	3.0
2. Ivory Coast (1986)	14.8	44.2	20.3	20.7
3. Ecuador(1982)	16.0	46.0	18.0	21.0
4. Egypt (1976)	26.0	55.2	12.0	6.8
5. Ghana (1988)	4.6	25.6	9.5	37.3
6. India (1981)*	16.6	11.9	65.4	6.1
7. Pakistan (1973)	17.2	38.7	33.0	11.1
8. Peru (1986)	11.6	51.6	13.6	23.2

Note : Data refer to previous place of residence except countries with (*) where data refer to place of birth

Source: Bilsborrow (1992)

Migration is not simply a response of 'push factors' in the region of origin e.g. poverty, natural disasters, changes in land tenure, but is also dependent upon 'pull' factors from the destination region. Such factors may include the offer of free land and hence the land's resources of subsidies paid by governments. The movement of people to forest areas and the subsequent deforestation caused by land clearing for agriculture has been a major factor in the loss of forests. Leonard (1989) states that about two-thirds of tropical deforestation in developing countries occurs for agricultural reasons, with about 80% of this attributable to slash-and-burn agriculture being practiced by poor migrants.

New arrivals often clear land rapidly in order to make a legal claim. Indeed, in some countries, including Ecuador and the Sudan, formal property rights are only given when the land has been cleared (Southgate 1990, 1991). Deforested land is usually subject to soil erosion, by runoff or deflation and accelerated leaching. Some estimates suggest that 50% of a soil's fertility can be lost within 3 years by these processes (Leonard 1989). Such losses of fertility may cause migrants to clear more land locally or to move on to other areas. Although, natural environmental factors are often involved in the movements of people to new areas, much of the impact of migration is actually the conswquence of a wide range of governmental policies on such issues as land tenure, subsidies and infrastructure.

1. Poverty Eradication

Studies carried out for human resource development are evident that the people of the country need to be taken as asset and not a iiability. The government of dryland areas must understand the reality that every workless person is the liability on the state and providing them food from external aid is not the sustainable solution. In the developing world, the proportion of persons dependent on agriculture need to be reduced by creating employment opportunities in other sectors. The growing population and its pressure on land is unhealthy for the well being of a country and the shifting of people from cultivation to other areas within the same locality is the only solution of the problem. To begin with the land upgradation alone is capable to provide gainful and whole year employment to a significant number of the working force.

Other areas of shifting the people to alternative occupation is the animal husbandry sector, which is the ancillary but dependable source of livelihood and can be done as a supplementary activity of cultivation as input requirement are met from agricultural produce. But this is possible for good breed livestock. Dairy activities are successfully running in various parts of India providing remunerative return to the beneficiaries. This sector can be developed in any country specially in dryland areas for benefit of the larger area of the community, but with breed improvement programme of local breed. Experiences of imported breed would not succeed in any country and efforts for local breed improvement is the only solution though it may take reasonable time.

Another area of deployment of population is the biotechnology in agriculture production starting with bio fertilizers to save significant foreign exchange on import of chemical fertilizers and pesticides. Biogas is also an effective tool of fuel as a replacement of cutting of trees. Significant employment can be achieved in this sector. Other activities of the rural areas are the collection of medicinal products from unproductive land and its application through indigenous system of medicines. In India and China the export of medicinal products in raw shape and in he form of medicines is increasing for use within he country and abroad. This sector is a fast growing activity and significant employment can be given to the people through a well developed system through concerted efforts by the country government.

Use of technology can be done in low cost housing and other construction activities by replacement of local material rather than imports. These activities will yield significant employment in preparation and application of the activities on large-scale basis. Crafts and agro based processing activities are other areas of shifting of people from cultivation. In India, stone carving industry in village sector is developing fast and fetching good return in employment and activity execution. Similarly cutting of precious and semi-precious stones has developed fast providing employment to people with little skill upgradation. This activity is running through import of raw material and export of finished goods.

Information technology sector is fast developing by fulfilling country requirement as well technology assistance to other countries. Governments developing countries need to adopt similar activities in their countries for meeting the national requirement and information dissemination to requisite sectors including cultivation. The scope of fast expansion can economize the processing cost as well processing of the information very fast. There are innumerous areas of operation in this sector and its application.

Peoples' acceptability for the programmes is the prerequisite of success and skill upgradation of the people is the answer to this problem. The developed countries of the world have engaged their population in other sectors and are getting significant results. Shifting of the population to other areas is essential for fast developing the agriculture sector in terms of intensive and extensive approach. The intensive approach includes the upgradation of facilities in the areas of existing cultivation with focus on increasing the productivity. Extensive efforts include the utilization of unutilized land for productive purposes through suitable treatment. The demonstration impact of replication of activities is the process of entry of other persons in the areas of success.

The role of the country government is most essential in mitigating the poverty. The government is required to select few innovative schemes for combating desertification and land degradation through one single aspect and involve the people in total process. The empowerment in decision making process would help in attracting their active participation and the success would be replicated by others through demonstration impact. Dry farming practices in water deficient areas of Rajasthan State of India have succeeded through replication of successful results and presently, the production and productivity is not far behind the national average.

Role of Central Arid Zone Research Institute Jodhpur in India has helped in developing dry farm practices where the Institute provided various suitable varieties of seeds able to grow in adverse conditions. The success of the products multiplied the application and every important crop of coarse grain, pulses and oilseeds including cash crops are grown in Rajasthan in every possible area. The ultimate rate of growth of population in dryland areas of Rajasthan and other states of India is attraction of the people from other areas due to usefulness of the degraded lands for cultivation purposes. Therefore, the high rate of population growth in desert area of western Rajasthan is not only more births but in-migration from other parts of state and country.

An innovative approach for wasteland cultivation introduced by Government of India is Jetropha, used as bio diesel by mixing 20% bio-diesel in diesel for Railway diesel engines. Jetropha cultivation is promoted by the government to be grown in all the wastelands including railway side, road side and canal side. Jetropha can be cultivated in dryland areas and animal grazing is not feared. The tree can provide the bio-diesel up to fifty years and possesses medicinal value. This will help in providing employment to the poor on large scale and help in curbing the foreign exchange to significant extent. Similar activities can be introduced in other countries having large dryland areas.

2. Introduction of Self-employment Activities for Poverty Mitigation

There are numerous self-employment activities which can be introduced for the benefit of the poor for their social and economic upliftment. In this regard group based for activities suitable for women and youth are necessary to be introduced on cluster approach. The activity suitable for rural poor are local resource based activities where no significant amount is required as an incentive. Some machinery and equipment and skill upgradation can help in initiating activities as supplemental income to women and youth in spare time." Some activities are being carried out in each developing county, but there is necessity to develop a systematic approach for developing these activities on cluster-based approach. Activities like leather product, woodcarving, honeybee collection, preparation of medicines for acute diseases through local medicinal products, ready-made garments, agro-processing activities, wool based activities and micro-enterprises activities. For developing the activities, access to market is the necessity for sustainability. None of the activities would deteriorate the environment and ecology of the areas.

13.4 POVERTY AS A DISABLING FACTOR

The conditions of poverty also act to disable people from taking actions that benefit their environment. Efficient and sustainable management of natural resources often requires a range of other resources; human, natural, legal and informational. Poverty implies a lack of breadth as well as depth in such resource assets. The absence of any one of these assets may in itself lead to environmental degradation.

One disabling factor is the lack of credit availability to poor people. Even if such people want to invest in their natural capital, by irrigating or fertilizing land, borrowing facilities may be limited and interest rates high. Limited credit and high interest rates reflect several factors; the isolation of poor areas, high levels of uncertainty and poor people's lack of collateral. Poor people may also be dissuaded from investing in the environment because they lack any legal entitlement to it. Women in Sub-Saharan Africa often have no right of tenure to land despite the fact that they provide 50-80% of the labour (World Bank 1992). Consequently, the women have limited access to credit and may also have incentives to invest due to uncertainty over their rights in the future.

Illiteracy and low levels of education among the poor means that it may be difficult to obtain information about the environmental implications of certain actions. A new pesticide may cause degradation, if used improperly, but literacy may prevent this knowledge being known. Insufficient knowledge of application method leads to soil and water pollution and to the destruction of wildlife.

Environmental problems of this nature are most likely to occur, when traditional methods have been replaced by new methods or when people have migrated to new agricultural areas, where soils are different and hence there is no historical experience to rely upon. When this is the case and information in costly, environmental degradation may be the result.

1. Accelerating the Economy

The most crucial factor of the governments of developing countries specially having huge dryland areas is the idle economy. Large crowd of unemployed persons out of total working force is the ultimate cause of poverty who consume the produce without any work. The necessity of creation of employment opportunities is the ultimate solution for accelerating the economy as this could help in creation of demand for more goods and activization of existing industries and opening of new industries with added purchasing power of more employment avenues. The process activates the economy within the country and in short time the economy is activated in full swing. Poverty eradication is the only mode and with handful of money the avenues can be created with a systematic approach.

Social forestry for combating deforestation is another area, where the people of rural areas especially the poor can be put to job with some incentive support. This adds to environmental upgradation as well as production of fruits, timber, fuel-wood and fodder from the existing deforested lands. Social forestry is the solution of various problems of desertification and land degradation, but require careful planning by using the areas for selective trees to be grown in particular environment. The indigenous species with high yielding varieties can be successfully grown and short gestation trees is the ultimate requirement of the people of the area.

2. Idle Working Force

Generally poor people intend to get some work for livelihood. This is the minimum necessity of the poor to get atleast the earnings to subsistence level. The people without job are the curse to social system as various unlawful activities like theft, decoity, looting, etc. generate including illicit trading and immoral trafficking. The country government of developing countries need more force to keep safety and security of the people. Therefore, in the best interest of the peace and harmony, job creation atleast to minimum subsistence level is essential and responsibility of the government. Unless the government think in this direction, there is no end to the problem.

In the situation of mass unemployment and poverty, various gangs are organized creating various problems to government, society and people. The ultimate solution is the curbing such activities at its root level, otherwise it becomes difficult to handle the situation. There are many international linkages of gangs creating various problem for the country government. It is advisable to curb such acts for the benefit of the country and the society. In general practice, vacant lands of common property resources, wasteland or government lands are encroached by the people through unlawful measures and the country government find helpless for curbing such activities and ultimate solution in such cases is the regularization of such lands in possession of person or group of persons. In case such lands are allotted to needy persons as country policy, the government and people get benefit from such lands and its productive use is ensured.

13.5 CONCLUSION

Poverty and degradation have particularly close links, most obviously because the poor people of the world tend to live in ecologically marginal areas. Viewing the environment as a form of natural capital used by poor people highlights a number of reasons why degradation might occur.

1. The discount rate used by poor people may be high, with the result that investment in natural capital is often low and biased towards short-term projects.
2. Poverty means people have no surplus resources and this implies that any shocks may result in the mining of the environment.

Poverty also acts as a disabling factor: poor health, poor education, illiteracy and lack of legal rights may all act to reduce disability and incentive to prevent degradation. The fact that degradation may lead to more poverty, falling crop yields as soils are eroded, means that a vicious circle can develop with disastrous consequences.

Despite the close links between poverty and degradation the two are not always connected. Population pressure, land tenure arrangements and government policies act in conjunction with poverty to cause degradation. Even in conditions of poverty, the existence of traditional methods and forms of co-operation may prevent degradation if there is an absence of other malign factors.

The poor people know the productive use of degraded lands provided they are given opportunities to use that. The poverty is a curse to the incumbent, society and government. The ultimate necessity to create awareness in people to adopt best possible measures for treatment and use for maximum production and productivity. The ultimate objective of the national government and the people is best use of the land but require support and incentive from the government which poor lacks.

Chapter 14

Programmes for Combating Desertification and Land Degradation

Desertification is the degradation of lands in dry areas. There is conceptual difference between desertification and natural deserts and differentiated in view of physical characteristics of degradation of soil and vegetation. Desertification is not the desert expansion of popular imagination, instead, it is essentially a subtle, dispersed and continuous process, which mainly occurs far away from desert fringes, with the outright conversion of fertile land into desert only taking place in extreme cases.

The direct cause of desertification and land degradation is poor land use. A long-term change in climate could make an area more desert-like without human intervention, but so far, there is no firm evidence that desertification and land degradation are taking place. Poor land use is not unique to dry areas and consequently soil erosion and the removal of vegetative cover are major problems throughout tropics. However, it is severe environmental and human effects and the influence of low and erratic rainfall, which makes desertification such a distinctive and worrying problem.

There are four main types of poor land use in the drylands: over-cultivation, overgrazing, mismanagement of irrigated cropland and deforestation. All these issues need to be taken up to reverse the process of desertification and land degradation. Since desertification and land degradation are the universal problems, it is not possible to provide a general yardstick for combating it the accepted factors for land degradation and desertification are varying in view of geographical, biological, physical, social and economic conditions and accordingly the package of programme for combating it is a general conception, which can be adopted in any specific country after ascertaining its conditions as well as extent and magnitude of the problem. The programmes for combating desertification and land degradation can be broadly classified under the following areas:

(a) Efforts of UNCCD for combating desertification and land degradation.

(b) Activities of UNCCD at Global and inter-regional levels.
(c) New initiatives for combating desertification and land degradation.
(d) Information dissemination and awareness raising.
(e) Social Sector and community development programmes.
(f) Programmes specifically for poverty eradication.
(g) Technology for combating desertification.
(h) Role of Non-governmental organizations in combating desertification and land degradation.

14.1 EFFORTS OF UNCCD SO FAR FOR COMBATING DESERTIFICATION AND LAND DEGRADATION

The United Nations Convention for Combating Desertification was established with the following mandate:

(i) Regularly review the implementation of the convention and functioning of its institutional arrangements.
(ii) Promote and facilitate the exchange of information on measures adopted.
(iii) Establish such subsidiary bodies as are deemed necessary.
(iv) Review reports submitted by its subsidiary bodies.
(v) Rules of procedure for itself and for the subsidiary bodies.
(vi) Adopt amendments wherever necessary.
(vii) Approve programme of budget.
(viii) Utilize services and information by competent bodies or agencies, whether national, international, inter-governmental or non-governmental.
(ix) Promote and strengthen the relationship with other relevant conventions.
(x) Exercise such other functions as may be necessary for the achievement of the objective of the convention.

All countries, which are signatories to the Convention, are party to the Convention and participate in the Conference of Parties (COP) for regular planning and review of activities under the Convention. In addition, a number of international organizations, international donors. Research and Development Institutions and NGOs also participate in the Conference of Parties, depending upon the bloc they belong to. As per definition in the UN, they are "developed" or 'developing' country parties. Until 2001, the COP held every year but later it was decided to convene meeting every subsequent year. Therefore, next Cop was convened in 2003.The summary of the activities of COP held so far, is given in Table 14.1.

Table 14.1 : Summary of activities of the cop to the UNCCD

COP	Year	Date	Venue	Main Issues Taken Up
COP-I	1997	September 29 Oct., 2011	Rome, Italy	Work Programme of CST- Roster of independent Experts. Programme and budget for the convention and for the Sect.
COP-II	1998	November, 30 Dec., 2010	Dakar, Senegal	Establishment of a Permanent Secretariat, Regional Action Programme. Preliminary discussion on NAP preparation. Review report of the Global Mechanism. CST related issues. Ad-hoc panel of Traditional Knowledge and its linking with modern technology.
COP-III	1999	November, 15-26	Recife, Brazil	Presentation of national Reports of African Country Parties. Review work of the Global mechanism. Establishment of an Ad-hoc Working Group under the COP. Establish synergies amongst Convention. CSP related issues. Recife initiative.
COP-IV	2000	December, 11-22	Bonn, Germany	Discussion on National Reports of Asia, Africa, Latin America and northern Mediterranean in the AHWG. CST related issues. Discussion on constitution of CRIC
COP-V	2001	October, 1-12	Geneva, Switzerland	National Reports of all regions. More detailed discussion on Africa. Proposals on procedures and modalities for CRIS. CST discussion on EWS. Constitution of Ad-hoc panels, if required.

Source: UNEP-2001.

UNCCD was initially serviced by an Interim Secretariat located in Geneva. Switzerland. Since 1998, the UNCCD is supported by a permanent secretariat located in Bonn, Germany. The functions of the permanent secretariat established by Article 23 of the Convention includes:

(i) To make arrangement for sessions of the COP and its subsidiary bodies.

(ii) To facilitate assistance to affected developing countries for compilation and communication of information.

(iii) To co-ordinate it's activities with the secretariats of other relevant conventions.

(iv) To prepare reports on the execution of its functions and to present to COP.

14.2 ACTIVITIES OF UNCCD AT GLOBAL AND INTER-REGIONAL LEVEL

The UNCCD after its constitution has done significant work for combating desertification and land degradation at global and inter-regional levels. The development of general framework for all the drylands of the world and various measures specific to continents of the world are the key factors of addressal. Some important issues are narrated hereunder:

1. Early Warning Systems (EWS)

Early Warning Systems is one of the main issues that has been reviewed at the inter-regional activities under the UNCCD and has resulted in the organization of two Afro-Asia Forums of Early Warning systems and a Technical Workshop following it. The first Afro-Asia Forum of EWS was held in Beijing during August 1996 to exchange experiences and explore areas of collaboration in combating desertification and mitigating the effects of the two continents. The Second Forum was held in Niamey, Niger during September 1997 with the objective of providing the Committee on Science and Technology an evaluation of the prospects of integrating EWS with environmental information on desertification. The workshop also recommended the development and testing of practical examples of such integration.

The Asia-Africa Technical Workshop on Early Warning systems was held in July 1999 at Beijing, China. These were aimed towards exchange of practical experience, training, research cooperation, collection, analysis and exchange of data in some specific field of selected areas. A great amount of potential was identified in preventing desertification and drought mitigation. The major recommendation was that affected countries of the region, may strengthen and / or establish information, evaluation and follow-up EWS taking into account the climatic, meteorological, hydrological, biological and other relevant factors.

2. Assessment of Status of land Degradation in Dryland (LADA) and Millennium eco-system Assessment (MA)

The necessity for having basic standardized information and methodological tools for land degradation assessment at various levels has been echoed by parties and other relevant actors. These include an assessment of the impact of land degradation on sensitive eco-systems and international waters, such as shared river, lake basins, watershed and coastal eco-systems. The UNCCD is collaborating with the UNEP and FAO amongst other key players with support of GEF to facilitate a project on dryland land degradation assessment, with a view to assessing both the level and trends of degradation of natural resources and environmental driving factors that cause this degradation. This is a global initiative that has to target areas affected by drought and desertification.

The Millennium Eco-system Assessment is a four year process designed to improve the management of world's natural and managed economic by helping to meet the needs of decision makers with scientific information on the conditions of eco-systems, consequences of eco-system change and options for response. Southern Africa and South-East Asia have been selected as two of the local regions for the sub-global component of the MA process. The MA has a core budget of US $ 21.1 million contributed by GEF, UN Foundation. Packard Foundation, Government of Norway, Rockefeller foundation, world Bank, UNEP, FAO, UNESCO, UNDP and NASA. Amongst the important activities proposed is Remote Sensing and the Millennium Assessment. Summary of activities proposed at national, regional and global levels are given in Table 14.2.

Table 14.2 : Activities of UNCCD at various levels in the context of combating desertification in dryland regions

National	Sub-regional	Regional	Inter-regional	Global
1. Ongoing schemes, programmes and projects - Desertification, Monitoring and assessment. - Soil and water conservation - Natural resource management - Local area development programmes - Awareness raising - Peoples participation - RandD activities - Institutional strengthening and capacity building 2. Programmes of State Governments 3. Externally aided projects-ongoing and new 4. Projects arising from NAP 5. Developmental activities and measures formulated at the local level 6. Various measures identified and decided from time to time	1. Concerns sub-regions such as South Asia, West Asia, Central Asia, South-east Asia etc. 2. Areas not formulate programmes on CCD related issues. 3. Areas that could be considered include, establishment of sub-regional programmes, monitoring and assessment on desertification and drought phenomena. 4. Sub-regional co-operation and eco-restoration 5. Exchange of RandD for drought prone and desert areas 6. Traditional technologies used in the region 7. Local capacity building	1. Concerns to continents as a whole 2. Establishment of six thematic programme network - Desertification monitoring and assessment - Agro-forestry and soil conservation - Sand dune stabilization in rangeland and pasturelands - Water conservation in agriculture - Drought preparedness - Local area development programmes	1. Concerns two or more regions. 2. Co-operation on studies on early warning systems, through inter-regional forums on EWS etc.	1. Concerns the whole world and global effects/ phenomena such as El-Nino, ENSO, La-nina etc. 2. Global monitoring and assessments on land degradation 3. CST related activities 4. Synergies amongst conventions 5. Establish-ment of EWS on drought

Source: UNCCD, 2001

3. CST: Sue of Benchmarks and Indicators

The Committee on Science and Technology has general observations that most countries do not use a scientifically based and verifiable set of benchmarks and indicators for monitoring and assessing desertification and land degradation. It is for this reason that the CCD and annexes had made provisions for countries to report their progress in solving the problem. Monitoring of the CCD demands two major types of indicators-implementation and impact indicators. Implementation indicators are those that can be used to track the progress of programmes and activities put in place by the countries that have ratified the Convention. Impact indicators follow the effects of programmes and activities on desertification in general. They measure the effects on natural resources, the economy, society and institutions. They may cover bio-physical, socio-economic and institutional aspects of the problem. Use of implementation and impact indicators is given in Table 14.3

Table 14.3 : Use of implementation and impact indicators

Type of indicator	Characteristics	Users
Implementation	- Reflect progress and steps in implementing the CCD - Measure the development and implementation of NAP. - Show progress in partnership arrangements, the involvement of local people and NGOs. - Indicate appropriate regulations or legislation. - Can be applied at level of Convention, NAP or project.	National and Sub-regional Coordination Committees, political decision makers, Conference of Parties, Donors, NGOs etc.
Impact	- Measure the impact of CCD. - The state of natural resources, reflecting the containment of desertification. - The state of economy. - The strength of society. - The presence and capacity of institutions. - Can be applied at global, regional, sub-regional, national and local level.	Political decision makers, national Ministries and Departments, Technical Experts, economists, Donors, Local Communities and other partners.

Source: UNEP-2001

Selection of indicators for assessing the impact and severity of desertification, following indicators were suggested by UNCCD to member countries:

(a) Reliable and scientifically valid.

(b) Independent of sample size.

(c) Measurable - standardized, accurate method and analysis with low measurement error.

(d) Biologically and socially relevant.

(e) Sensitive to stress factors without high natural variability and have an early warning function.

(f) Easy and cost effective.

(g) Able to assess trends over time-benchmarks.

(h) Based on readily available data of known quality.

Keeping in view of above criteria, a suggested list of impact indicators is given in Table 14.4.

Table 14.4 : List of impact indicators of desertification

Physical	**Biological**
Climate: Indices of temperature and rainfall levels Evapo-transpiration Precipitation tendencies Albedo Soil: Surface run-off Visual erosion of soil loss. Soil compaction Soil fertility Organic matter Water: Water table potential and use. Ground water decline Salinization Volume of water bodies Land Use: Crop and livestock productivity	Flora: Changes in vertical structure- percent cover Changes in horizontal structure- Strata Changes in dominant species - species dominance Changes in richness and species diversity- index of diversity Changes in phyto-volume Fauna: Density, abundance and rarity of species Livestock pressure Socio-Economic and Demographic Indices: Income percentage Infant mortality Agriculture income/ total income Collection time for fuel- fodder-water needs of the day Nutrient status by age and sex Age specific literacy rate by gender Migration from degraded lands Percentage of population aware of desertification phenomenon

Source: UNEP, 2001

4. Desertification Monitoring and Assessment and Use of Early Warning Systems:

For participation in UNCCD activities at the regional, inter-regional and global levels, it is important that the key institutions, organizations, experts, professionals and others are involved in various disciplines and aspects of the desertification governing the physical, biological, socio-economic and climatic aspects of desertification aspects are identified. A roster of experts requires to be prepared for the active participation of these experts in the various activities of the UNCCD both at the national as well as at the international level. A logical framework matrix for addressal all the related aspects are given in Table 14.5.

Table 14.5 : Logical framework matrix for institution building of a project on combating desertification with peoples participation

Intervention Component	Indicator Group (OVI)	Means of Verification	Risk / Assumption
Strategic Level: Long Term Development Objectives: Sustainable arid equitable management of rehabilitated common lands with the active participation of village-level institutions and local communities	Establishment of Village Level Institutions: Awareness level of the institution and villagers. Functioning of village level institutions	6 months progress reports	Target population willingness to participate. Stable project environment with special respect to staffing.
Tactical level: immediate Objectives Hectare of common lands rehabilitated and managed by number of village committees with extension provided by the project authority	Establishment of village committees Awareness generation level of the village committees and villagers Functioning of village committees	6 months progress reports Forester/ SDO/ DFO records Village committee interviews	Target population to participate Stable project environment with special respect to staffing
Tactical Operational Level: Output Village committees established which manage common lands on the basis of approved village management plans	No. of village committees established	6 months progress reports	Stable project environment with special respect to staffing
Tactical Operational Level: Information and mobilization activities Community organizing activities Community extension and counseling activities	Sample indicators: No. of campaigns Publicity material distributed No. of village management Plans Frequency of community/ project staff meetings at different levels etc.	6 months progress reports	Stable project environment with special respect to staffing

Source: R.N. Kaul, 2001

5. Synergies of UNCCD With Other Conventions

There are number of financial institutions, RandD institutions , organizations and agencies of international, regional, global, which are involved in the formulation and implementation of the Convention. These include UNDP, UNSO, WMO.

FAO. IF AD, World Bank and regional banks such as the Asian Development Bank. Article 8 of the UNCCD refers to the linkages with other conventions. In addition, in every COP, efforts to strengthen the collaboration with the Rio convention and with other conventions related to sustainable development have been emphasized.

(a) Rationale for Synergies Amongst RIO Conventions

- They share a common sustainable approach and the same basic principles enshrined in the RIO Declaration.
- The conventions have a ramification on not only ecological but also the socio- economic and institutional aspects.
- Several of the policies and measures suggested for achieving the objectives of one convention may also contribute to objectives of others.
- Often, decision and policy makers of one convention belong to the same governmental organization dealing with closely relationed portfolios such as the other conventions or directly play a role in its implementation.
- Joint efforts that could be mutually beneficial in achieving sustainable development.

(b) Linkage Between Desertification and Bio-diversity

Topsoil erosion leading to loss of nutrients, organic matter. Vegetative cover protects the topsoil from both water and wind erosion. The critical living organisms in a pastureland are the pasture plant species. The variability amongst these is essential for the sustainability of livestock farming. Different livestock have different forage preferences and therefore, greater diversity of forage plant species results in greater stability and profitability of their livestock economy. The within-species variability is also critical for breeding and improved vigour. Loss of dryland bio-diversity can impact on loss of livestock, especially during drought and famine.

(c) Linkage Between Desertification and Climate Change

Desertification is closely inter-dependent with reductions in global carbon sinks and reservoirs. Desertification and the loss of vegetative cover increase the albedo-reflectance properties of the soil surface. Under specific circumstances, increased albedo reduces local precipitation, exacerbating the process of desertification.

(d) Linkage Between Desertification and Ramsar Convention

Protection of wetlands is not only" important from the point of conservation of natural resources i.e., flora and fauna, but also is an important indicator of stability of the eco-system.

(e) Linkage Between Desertification and Forest Principles

Forests being an important natural resource and have helped in regulating climate. They are also important for meeting the needs of fuel and fodder.

Conservation and protection of forests is an important component towards arresting desertification.

6. Funding Mechanism Under UNCCD

Prevention and control of land degradation, primarily desertification and deforestation, are critical to achieving sustainable development. The consequences in terms of loss of bio-diversity, reduced atmospheric and sub-terranean carbon sequestration and pollution of international waters can be significant and have global implications. For addressing the problem of desertification occurring across the globe, the magnitude of extent of funding required is enormous. The estimated average annual global cost of implementing the activities for combating desertification are given in Table 14.6.

Table 14.6 : Cost of implementing activities for combating desertification in dryland regions

Activity	Major Programme Areas	Major Global Activities	Annual Global Cost (in US $)
Combating Desertification and Drought	1. Strengthening know-ledge base and develop information and monitoring systems for regions prone to desert and drought including socio-economic aspects	Management related activities Data and Information International and regional cooperation	300 million.
	2. Combating land degradation through intensified soil conservation, Afforestation activities	Direct preventive measures. Accelerated Afforestation programmes. Direct corrective measures. Promote and introduce improved land/water/ crop management systems, combat Salinization in irrigated crop lands, promote incentive measures, use of energy, which lessens pressure including alternate sources and improved stoves.	6 billion.
	3. Strengthening integrated development programmes for eradication of poverty and promotion of alternate livelihood systems in areas prone to desertification	Adopt policies at national level for decentralized approach to land resource management. Create/strengthen rural organizations in charge of village pastoral land management. Establish intersectoral mechanism	Included below
	4. Develop comprehensive antidesertification programme and integrating them into national development plans and national environmental planning	To handle environmental and developmental consequences of land use and land ownership, particularly to property rights of women Promote rural credit	Included below
	5. Develop comprehensive drought preparedness and relief Schemes including self help arrangements for drought prone areas and designing programmes to cope with environment refugees	Develop infrastructure by involving local people to alternate livelihood systems and poverty alleviation programmes. Data and information. International and regional co-operation and coordination	Included below

Activity	Major Programme Areas	Major Global Activities	Annual Global Cost (in US $)
Combating Poverty Enabling the poor to achieve sustainable livelihood	1. Empowerment of local communities 2. Alleviate poverty and to attain sustainability 3. Long term strategy aimed at elimination of poverty know-how Assist the most disadvantaged groups-women, children, youth and refugees	Promote sustainable livelihoods. Participation of local communities in sustainable management Establish grass-roots mechanisms for sharing/ exchange of and enquiries	30 billion
Promoting sustainable development through agriculture and rural development	1. Review of agriculture policy planning and integrating programming in the light of multi-functional aspects of agriculture, particularly water resources technology, food security and sustainable development	National policy reviews on food security. Review national and regional agricultural policies, water resource technology, foreign trade, price, exchange rate, agricultural subsidies and taxes as well as organizational and economic integration	3 billion
	2. Ensuring people's participation and promoting human resource development for sustainable agriculture	Implement policies to influence land tenure and property rights. Consider demographic trends and identify critical areas for agricultural production	4.4 billion
	3. Improving farm production and farming systems through diversification of farm and non-farm employment and infrastructure development	Low-input sustainable agricultural systems.	10 billion
	4. Land resource planning, information and education for agriculture	Review policies, water resource technology, improving harvesting, processing, distribution and marketing of products. 1.7 billion	1.7 billion
	5. Land conservation and rehabilitation	Formulate and implement integrated agricultural projects that include other natural resource activities such as management of rangelands, forests, wild life	5 billion

Activity	Major Programme Areas	Major Global Activities	Annual Global Cost (in US $)
	6. Water for sustainable food production and sustainable rural development	Promote social and economic research and policies that encourage sustainable agriculture. Identify storage and distribution affecting food availability. Data and information systems. International, regional cooperation and co-ordination	600 million
	Total Cost		61 billion annual

Source: United Nations Conference on Environment and Development 1992.

7. Global mechanism

The Convention has prescribed for a Global Mechanism, a multi-source, multi-channel financing mechanism for facilitating and mobilizing financial resources. The Convention under Article 21 states, "In order to increase the effectiveness and efficiency of existing financial mechanisms, a Global Mechanism to promote actions leading to the mobilization and channeling of substantial financial resources, including transfer of technology, on a grant basis, and/ or concessional terms, to affected developing country parties, is hereby established." The Convention also provides for partnerships with international organizations and funding agencies as a source for generating funding. At COP-1 held in Rome in September 1997, it was agreed that the Global mechanism would be housed in International Fund for Agriculture Development (IFAD). Subsequently, the Managing Director for the GM was identified and a secretariat for the functioning of the GM was established. A Facilitation Committee of the GM consisting of representatives of various international organizations and agencies such as the UNDP, IFAD, World Bank was constituted.

The GM's main task so far has been to collect and disseminate information on the financial needs of affected developing country parties for carrying out action programmes and other relevant activities related to the implementation of UNCCD. The GM is also identifying and drawing up an inventory of relevant bilateral and multi-lateral co-operation programmes, as well as, other sources and channels of funding patterns and eligibility criteria. The GM has with these guiding principles established a demand driven information system called "Financial Information Engine on Land Degradation" in May 2000. This system is intended to be instrumental to the financial functioning of the GM and to its efforts to foster partnerships, improve communications and co-ordination between sources and recipients of funding and technical assistance related to the implementation of the UNCCD as well as with other interested institutions, organizations, bilateral and multi-lateral agencies, foundations, academic institutions, networks, bodies dealing with land degradation issues.

8. Global Environment Facility (GEF)

Unlike the other environmental conventions that emerged from Earth Summit (UNFCC, CBD), the UNCCD did not establish any particular fund for financing desertification programmes such as the GEF. Funding from GEF are available, in so far as land degradation relates to the GEF Focal areas, namely Climatic Change, Bio-diversity and International Waters and have agreed to fund the incremental costs of activities concerning land degradation within the GEF outline.

The strategy approach to support activities, which keep focus on global benefits of the GEF focal areas arise from the undertaking of the fund-assisted projects. The activities geared by achieving global environmental objectives through merging of projects into the mainstream of ongoing and planned developmental efforts, activities and programmes such as proper land use planning and implementation, which often requires national, sub-regional and regional efforts. Activities relating to land degradation and GEF focal areas are given in Box No. 1.

BOX 1 : GEF Related activities of land degradation

GEF PROJECTS ON LAND DEGRADATION AND CLIMATE CHANGE

GEF funded activities in the interface between carbon emission and reduction and prevention and control of land degradation could include:

- Sustainable land use planning and management.
- Regional cooperation to achieve them, to enable countries to secure better correspondence.
- Between land capability and use.
- Furthering the use of solar, wind and biogas energy for. lighting, water heating, cooking, water pumping and other needs and at increasing the efficiency of wood burning stoves and charcoal producing kilns to reduce net carbon dioxide emissions.
- Strengthening national and regional capabilities of information networking, technical assistance and extension services to introduce viable land management practices.

GEF PROJECTS ON LAND DEGRADATION AND BIO-DIVERSITY

Developmental assistance activities could include:

- Land use planning and management, on a catchment and regional basis to facilitate the integration of conservation and production-oriented management of biodiversity, in areas susceptible to land degradation.
- Development and application of integrated land management, resource monitoring and ecosystem management, in relevant institutions and local communities, improvement of management practices, institutional arrangements, policies and incentives, in sectors of agriculture, forestry, urban development and water use, all of which impact land degradation and through it the global environment.
- In-situ conservation of significant components of bio-diversity, particularly native vegetation and forests.
- Rehabilitation of degraded land and water resources with a view to conserving ecosystems, species and genetic diversity and with appropriate participation of the local communities.
- Measures relating to sustainable use of biological resources in degraded lands, which avoid or minimize adverse impacts on biodiversity.

GEF PROJECTS ON LAND DEGRADATION AND INTERNATIONAL WATERS:

- Management of watersheds, which.come under the jurisdiction of more than one country, with a view to promote afforestation, reforestation, sustainable management of forests, soil and water conservation.
- Strengthening regional co-operation to control sediment pollution and salt intrusion in international waters, including aquifers and of co-ordinated implementation of action programmes for drainage basins and regional seas, to control pollution and to protect aquatic ecosystems.

Source: UNDP, New Delhi.

9. Establishment of National Desertification Fund (NDF)

An important element of the Convention is the means of enhancing the participatory process in the decentralized governance, to enable communities to decide, where resources should be spent in recognition of this, Article 21 of the Convention calls for "the establishment, as appropriate, of mechanisms such as the national desertification funds, including those involving the participation of non-governmental organizations, to channel financial resources rapidly to the main source of funding for actions at the national and local levels. The funds are envisaged as national mechanisms, which make small grants and loans accessible to communities. They are a potentially powerful instrument that could provide a major boost to translating the participatory approach into practice and could empower local communities to advance their own initiatives through a participatory mechanism in decision making. The NDF channels funds to a local Area Development Fund (LADF) is the most direct, simple and effective way, as given in Box 2.

BOX 2 : National desertification fund- key elements suggested by UNDP/UNSO

Governance / Institutional Structure:

- Ensure transparency and efficiency
- Internal checks and balances
- Embody/ access to special expertise.

Operational Modalities:

- Activities should be community driven
- Simple operational procedures
- Access should be limited to community driven projects
- Low ceilings should be adopted on grants
- Community input for resource mobilization urged (part of the funding is borne by the communities themselves)

NDFs could support:

- Alternate livelihood systems in drought-prone areas
- Participatory technology development
- Management of common property resources
- Awareness raising
- Strengthening local coping strategies for drought
- Enhancing food security at the household level

Fund Capitalization:

- Domestic Resources: Government budget, taxes, stamps, levies, lotteries, reallocation of savings, private donations etc.
- External sources: Bi/multi-lateral grants, loans, debt conversions and swaps, endowments, foundations, etc.

Source: UNDP/UNSO 2000

14.3 NEW INITIATIVES FOR COMBATING DESERTIFICATION

Certain new initiatives have been evolved during the past decade for the sustainable development and addressal of the problems of desertification, specifically conservation and eco-restoration, community based programmes, integrated watershed approach, decentralization of governance, participatory approach towards implementation of programmes and preparation of long-term integrated programmes for meeting the food, fuel, fodder and water requirement of dryland areas.

1. Integrated Watershed Programmes

The sector-based programmes formulated during seventies and eighties, although benefited the affected areas, it was felt deficient in achieving objectives. Also, a sharp and clear focus on achieving drought-proofing and combating desertification was not properly enunciated. As a result, the directions of the programme got-diluted depending upon various factors such as, the perception of the governments, capabilities of sectoral departments and pressures exerted by different interest groups. In the process, each activity under the programme was implemented in an isolated and segmented manner and watershed unit of area development was not duly respected. It is therefore, needed to integrate and co-ordinate all programmes related with land, water conservation and development of degraded lands based on a watershed basis. Since watershed in a geo-hydrologically delineated natural unit that is drained by a water system, its adoption as the basis for soil and water conservation includes bio-physical, socio-economic and sometimes political intervention for its planning, management and execution of schemes for conservation of natural resources at the micro-level. Essentials of watershed management are as under:

(i) Integration of land, water and natural resources and development of degraded lands.

(ii) Community driven projects and schemes.

(iii) Agencies for implementation can be local communities, non-governmental organizations, village associations and development line departments of the government or agencies also involved in planning, development, execution of programmes including awareness raising.

The watershed approach is an appropriate strategy for grass-root level development, where every element can be thoroughly kept in view and programmes can be developed with specific modifications. This is the right approach for development of degraded lands. There is necessity to develop elements of common guidelines for watershed development comprising following issues:

(i) Focus on village common land

(ii) Institutionalization of community participation

(iii) An emphasis on enabling the creation of sustainable rural livelihood support systems

(iv) Recognition of capacity building measures as a vital component of the watershed project design

(v) Institution of committee systems of the at various levels starting from the village up to top government decision making level

(vi) Decentralization of the processes of planning and bringing user-communities to centre stage

(vii) Securing the participation of women

(viii) Putting in place arrangements that facilitate equitable sharing of the usufruct

(ix) Pilot projects for reclamation of saline and waterlogged areas

(x) Creation of watershed development fund

(xi) Active participation of beneficiary group including cost sharing by the members to develop ownership in decision making and created assets

(xii) Much greater degree of flexibility in choice of technology

(xiii) Greater thrust on research support, innovative activities specific to local conditions

(xiv) Transfer of technology to beneficiary groups and its application with local circumstances

(xv) Better linkages with institutions including credit linkages

(xvi) Efficient management of common property resources with people's participation.

Watershed is a slow process where the process need to be started from awareness generation about such practices. People coming forward for becoming members should be taught about the benefits of collective working and their strength. Once the process is developed, its benefits would continue to remain them associated. However, the conflict resolution should be a regular process while monitoring on timely basis. For agriculture research in soil and water management and conservation programmes, it is essential to derive Natural Resource Management, which include addressal of resource inventory, soil management, nutrient management, water management and cropping system management.

In India, watershed programme was launched on a large scale by unifying and simplifying the implementation as a single national initiative. An Inter-Ministerial Committee was constituted in 1999 for formulation of common approach for watershed development. The mandate of the Inter-Ministerial Committee was to examine the existing guidelines of watershed development projects and programmes of the Ministry of Agriculture and the Ministry of Rural Development with a view to identify the areas of convergence and commonalities in approach in respect of specific criteria for selection of rain-fed areas for treatment, programme components and interventions, institutional framework and modalities of implementation.

In 2000, a "Common Approach for Watershed Development" was jointly formulated and adopted by Ministry of Agriculture and Ministry of Rural Development, Government of India incorporating the strength of the earlier first generation based watershed programmes. The Common Approach for Watershed Development Projects is now accepted by the Planning Commission. Based on the common approach, the Ministry of Agriculture has restructured National Watershed Development Programme in Rain-fed Areas providing for decentralized procedure in planning, implementation and evaluation of the programmes to make it sustainable (Ministry of Agriculture 2000).

Realizing the effectiveness of watershed approach adopted for controlling land degradation and increasing productivity. Watershed Development Fund was established at the National Agriculture Bank for Rural Development with the objective of spreading the message of participatory watershed development. The fund is utilized to create the necessary framework conditions and consolidate the isolated successful initiatives under different programmes in the government, semi-government and non-governmental sectors. The total corpus of the fund is Rs. 2 billion envisages coverage of 100 districts in three years. The choice of the district is to be determined in consultation with the concerned state government. The districts and blocks selected must be predominantly backward, with significant proportion of Scheduled Castes and Scheduled Tribe population, having irrigation percentage preferably less than 30% i.e., national average and where the potential of watershed development is large.

The watershed development is a unique micro approach for water conservation, which helps in retaining the fertile soil in fields, increases water level in surrounding areas, development of arable and non arable areas, livestock development, pasture development and tree plantation. It is accordingly significantly helpful in combating desertification and land degradation in rain-fed areas.

2. People's Participation in Conservation and Eco-restoration of Degraded Lands

For combating the desertification and land degradation in drylands, people's participation is an essential tool for the success in conservation and eco-restoration. People's participation is essential in designing the project, planning and implementation of the programme, use of indigenous technology, monitoring, demonstration and dissemination, sharing the benefits and replication of success experiences for sustenance of the programme.

Joint forestry management is to be implemented with involvement of village communities in regeneration of degraded forest lands. This is a process of bringing together the forest department and local people in the management of forest and the degraded land resources. For sustenance of the programme and keeping transparency in the process following issues need to be adhered:

(a) Register all the forest protection committees under prevalent country act like Societies Registration Act of 1860 in India. This is necessary to make the committee into a legal framework for functional requirements where relationships with the forest department are clearly defined.

(b) Give uniform nomenclature to such committees viz. Joint Forest Management Committees.

(c) Ensure that at least, 50% of the members in the JFM general body and at least 30% in the executive committee are women.

(d) Bring good forests with crown density above 40% except the protected areas, under the scope of JFM. However, in the first instance, implementation of JFM in good forest areas should be done in a phased manner on a pilot basis. The activities should concentrate on non-timber forest products, which can be given free or on concessional rates as per prevalent system or agreed procedure in degraded forest areas under JFM.

(e) Felling of trees and harvesting of timber is to be done under the provisions agreed between forest department and JFM.

(f) Provide overlapping working circle in new areas for incorporating broad provisions of micro plans.

(g) Emphasize a well-established marketing linkage for better return to the members.

(h) Ensure that no less than 25% of the share of village community be contributed to the village development fund for long-term sustainability of resources and ensure that the accumulated funds are used for village development activities with no diversion to other villages or areas.

Community involvement in Social Forestry Programmes for development of Common Property Resources like village woodlots, grazing lands, tank beds, forest shores and tree plantation for multiple use play an important role in supporting not only land holding families, but also the landless in meeting their basic needs such as food, fodder, fuel-wood and small timber Various Afforestation programmes like social forestry and farm forestry are under implementation to deal with degraded common property resources. In spite of some failure, the social forestry programme has witnessed success. However, some problems do persist in ensuring people's participation in the management of CPRs, which are due to lack of awareness of people about the social forestry programme. There are evidence of some operational problems e.g., local people were not consulted in choosing in choosing plantation sites and seedlings to be raised on the proposed plantations which resulted in conflict between officials and local people at a later stage. Also, the available grazing lands in most sample villages are not sufficient to feed the existing livestock and this compelled the villagers to graze their livestock in plantation areas in the initial years of plantation resulting destruction of saplings and frequent quarrels between the rural community and forest authorities. Above all, the wages to villagers are less than prevailing wage rates.

The above examples are the procedural laps and lack of coordination of forest department people with the village community. The social forestry projects, where peoples' participation is taken in transparent manner, the projects have succeeded and the village people have been benefited through meeting their deficiencies. The poor people having no land resources or very little land are the main beneficiaries of these programmes.

3. Eco-Development Schemes

The communities residing in fringes of National Parks and Sanctuaries including Tiger Reserves, schemes to provide alternate sources of sustenance can be initiated, which help to improve the ecological productivity of buffer zones of protected areas through the involvement of these communities in keeping security of the protected areas. The projects can be designed through a package of activities aimed at providing sustenance to the forest side and ameliorating their hardships to minimize conflicts between them and protection staff. The various activities proposed are (i) Habitat improvement, (ii) Alternate sources of energy, (iii) Infrastructure building and roads, (iv) Small welfare measures, and (v) Collection of minor forest produce.

4. Involvement of NGOs

Non-Governmental Organizations form a very important interface between the government and the local community. While, the government machinery is not able to handle all sort of activities in view of their administrative and coordination problems, NGOs are better option to implement the activities with peoples participation. Activities to be assigned to NGOs can essentially cover the following major programmes and areas:

(a) Common Property Resource Management.

(b) Creating environmental awareness.

(c) Family planning and welfare.

(d) Sanitation.

(e) Gender equality development.

(f) Child welfare.

(g) Self-sufficiency programmes.

(h) Alternative livelihood programmes.

(i) Evaluation and impact of agricultural and rural development projects.

A number of successful examples are available from different parts of the world on the involvement of NGOs in eco-reforestation and land degraded watershed and wastelands. These success stories indicate that problems of desertification and land degradation are continuing in such areas, where the micro-watershed programmes have not been implemented seriously. They were also weak, where community participation has not been vigorous. Following the recent droughts, NGOs and local communities in many drought-affected areas have come together for building and reviving traditional water harvesting systems. This indicates that the local communities turn to the traditional systems in serious natural disasters. The ultimate success in the continuation of a measure or scheme lies in the acceptance and ownership of those activities there under and it is vital that the communities are involved from the beginning of the programmes.

5. Management Through Co-operative of the Members

The National Tree Growers' Cooperative of the Members in India is an example of management of degraded land through a cooperative system. Its objective is to restore and protect the biological productivity of marginally productive and unproductive degraded land or wasteland for sustaining fuel wood and fodder plantation by organizing Village Level Tree Grower' Cooperative Society in the states of Andhra Pradesh, Gujarat, Karnataka, Orissa, Rajasthan and Uttar Pradesh. The NTGCF has assisted the VLTGCD in making available plantation materials, financial assistance, carrying out technical and socio-economic research and development activities through its spearhead team located in the above states.

The VLTGCS plantation site at Mallenahally in the Kolar district of Karnataka, is no more a wasteland as the federation has been very successful in achieving the objective of regenerating the wastelands in the country (Reddy *et. al.* 1997). This site covers an area of 20 hectares of land surrounded predominantly by agricultural land and few barren or being used for grazing purpose. The land was fully barren with about 10% of the total root stock underneath the soil cover. With the inception of plantation since 1989, there has been a lot of developments in and around this small watershed. The Millennahally VLTGCS land, which was once an under utilized degraded wasteland, is now filled with different grass and plant species. The main forest plants include babul, acacia and eucalyptus. Besides the society has undertaken soil conservation measures such as making contour buildings, mulching, and plantation of grasses along the draining streams. This has considerably reduced the soil loss and considerably improved the quality of the soil in-situ. Therefore, this micro watershed has been developed and protected from being degraded further.

The Samal Tree Growers Cooperative Society, which was organized by the people of village Samal located at the left bank of river Mahi in district Kheda in Gujarat is another clear example of efforts in regeneration of degraded wasteland (Saxena and Agarwal 1997). The society had acquired about 35 hectare of revenue wasteland on lease from the Government of Gujarat in 1987 for the purpose of tree plantation. The soil of this plot was sandy loam but highly eroded. There were many gullies in the plot, some of these as deep as 30 metres. Various soil conservation measures were taken up on the land immediately after acquisition and the land was put under tree plantation. There were 60,000 trees of nineteen species on the cooperative plot in 1994-95.

NTGCF initiated Energy Conservation programmes in its project areas and in other villages through 60 milk unions across the country, in order to reduce the pressure on the common lands and thereby consolidating the afforestation efforts and help to improve the quality of life of the rural people. These programmes include installation of biogas plants, improved chullhas and the distribution of solar and pressure cookers. The fuel efficiency studies conducted in Kolar district of Karnataka and Gujarat corroborated the fact that the improved chullhas save 30 to 40% of the fuel wood as compared to the conventional types of chullhas. In a classic case of energy consumption 90 out of 97 family biogas plants installed in the Namanar VLTGCS of Gujarat are functioning well for the past four years resulting in an estimated fuel wood saving of 90 kg per day (G.K.Kadekodi 2000).

6. RIOD – An NGO Network Created Under the UNCCD for Combating Desertification and Land Degradation

The UN Convention to Combat Desertification is one of the International conventions, which lays emphasis on people's participation, involvement of local communities and partnership building in the implementation of the Convention. NGOs and CBOs have an important role for improving the living standards of the local populations by working as an interface between the State and local population in areas such as raising awareness, assisting in local development programmes and schemes, capacity building and directly empowering their employment opportunities and livelihood.

RIOD is an international network of NGOs formed as a part of the UNCCD process and their international headquarters is in Nairobi. RIOD network has many NGOs from different countries as Members. RIOD India was formed in 1995 and its Secretariat is located in Youth for Action, Hyderabad and is a vast network of NGOs from all States divided into 5 regions- Northern, Eastern, Western, North-eastern and Southern and has a large number of NGOs from these regions as its members. It is actively involved in empowerment of women, assisting in providing micro-credits for alternate employment, generating and assisting with seed money for taking up eco-regeneration programmes, promoting micro-watershed schemes, generating awareness on issues concerning desertification. Success story of activities of RIO India is given below:

Food Security: Example of Becoming Self-reliant

Chennamma, 48 of Ullemkonda village in Mahaboobhagar district in Andhra Pradesh has experienced recurrent drought in her area. With 5 acres of dryland and no other avenues of income except to earn a living as labourer, the option available to her was to migrate elsewhere from the hostile living conditions. But with the initiative of Youth for Action, RIOD India, Chennamma and other members of the village have been involved in regeneration of the soil fertility and productivity of their land through soil and water conservation measures such as gully checking, bunding, construction of percolation tank and deepening of wells, which has helped her rebuild her life in her village.

Chennamma was provided loans by Sangham and with her own contribution, bought fertilizer and high yielding varieties of seeds of jowar, castor, roster and ground nut. Her husband who was working in another place began to work in their own field. Through their efforts, the yield per acre from various crops increased by more than 50%. The improvement in the yield and the income level increased in 3 years and even during lean periods. This helped her to provide better nutritious food to her children, the rice eaten earlier was supplemented with Jowar and ground nut, fulfilling the basic nutritional requirement. Chennamma extended her skills to raising of species such as Tamarind, Cheema. Chinta, Subabul, Glaricidia and Acacia in nurseries. These are especially grown in semi-arid tropics and drought resistant varieties. Through appropriate utilization of her land, she was able to obtain fodder, fuel, biomass, and timber. She started vegetation cultivation programme in her garden such as tomato, brinjal, ladies finger etc. Her physical health improved due to better nutritional diet intake. She presently helps other women in the neighbouring villages about the skills, she has acquired through her efforts of the past 3 years. The case of Chennamma is an example of how one can attain food security and self-sustenance, through simple practices of land conservation measures and provision of micro-credit.

7. Participation of NGOs and Civil Society in Decision Making Process

Involvement of NGOs and members of the civil society in policy making, planning and programme implementation and review is through their membership in various committees, Task Forces set up by the Government under various programmes. Distinguished members of the civil society are also nominated to the Parliament to represent the people in come countries. CCD definition of Civil Society is "Individuals or organizations from a wide range-non-governmental, private sectors, professional associations, local organization of economic, social and cultural nature."

8. New Initiatives of Science and Technology for Women

Over the last decade, special efforts have been made to demonstrate the application of science and technology focusing on women as a specific target group. The objectives of such scheme are: (a) to promote research, development and adaptation of technology with a view to improve the living and working conditions and opportunities for gainful employment of women especially in rural areas and (b) to increase the contribution of women to science and technology and development. Priority areas can be identified under the scheme could include following:

(a) Specific science and technology application programmes to solve the problems of women in different regions such as hill areas, coastal and arid zones.

(b) Research and development on post-harvest technology and agricultural implements used by women.

(c) Involvement of women in low-cost, production and preservation of horticulture produce.

(d) Utilization of agricultural and animal residues through such technologies, which can provide opportunities for income generation.

(e) Preparation and association of skill manuals, training aids in areas of agriculture and animal husbandry.

(f) Involvement of women in indigenous health practices and cultivation of medicinal plants.

(g) Research, development and demonstration programmes to minimize occupational hazards of women in agriculture and industries.

(h) Design, fabrication and improvement of equipment, furniture and instruments used by women in different occupations. (i) Identification and popularization of technologies relating to water management, health and sanitation.

(j) Adaptation of innovative technologies developed in centers abroad.

14.4 INFORMATION DISSEMINATION AND AWARENESS RAISING

Information dissemination is necessary for communication of technology to cultivators and others engaged in regeneration of arid, semi-arid and sub-humid areas in the process of combating desertification and land degradation. The technology developed by agronomists and other experts for treatment of degraded lands is essential to replicate in the field for removal of the problems faced by rural people. This area is multiple disciplinary and the expertise need to be disseminated keeping in view the environment, ecology, natural conditions and soil and water situations.

The information dissemination is two way approach for application in dryland areas. First thing of the system ins information which heed to be disseminated. In this connection the local situations need to be analyzed before building up the information system for specific area. The information technology need to be prepared keeping in view the soil and water conditions, level of deterioration of land, rainfall conditions, geo¬physical situations and inclination of the people to address those issues.

It is essential that information dissemination centers need to be established keeping in view the different climatic zones and soil strata. Level of desertification and land degradation need to be ascertained for each area in moderate, significant and severe category classification. One way need to be developed is the information communication to local levels, where the technology to be adopted keeping in local situations. This is the principle aspect to communicate the people about the technology to be adopted in any specific area as a measure to combat desertification and land degradation. The activities to be taken up at the local level need to be suitable designed to make it practicable and acceptable to local cultivators, cattle herders, government departments and specialists individual persons and institutions.

Another aspect of the information dissemination is the response to local enquiries. Such enquiries are area specific and problem is representative faced by a single person or community as a whole. The response of such enquiries need to be focused and very specific keeping in view the environment, ecology, geo-physical conditions and other factors of local nature. The person responsible to information dissemination must be quite capable and skilled with complete knowledge of representative area.

The awareness generation is the key factor of information dissemination, where people living in drylands and facing various problems need to kept aware about the land degradation and desertification situation in terms of severity of the problem. Through awareness generation, the people living in dryland areas need to be made aware about the actions they are doing are harmful for environment degradation and need to be suitably amended to retain the vegetable cover intact. At the local level, there are conflicting interests of various sections of societies carrying out their activities without thinking their impact on the total community. Awareness generation brings the total community to think over collectively in the interest in totality and take remedial measures at their own levels.

INFOTERA Network is a global information network of the United Nations Environmental Programme to exchange the world wide information in the field of science and technology developed in any one of the area in the world and successfully implemented to combat desertification and land degradation. Such exchanges or information sharing is quite significant and can be replicated at local level in any country after ascertaining its viability and adaptability.

14.5 SOCIAL SECTOR AND COMMUNITY DEVELOPMENT PROGRAMMES

Impact of desertification and land degradation lies most on the society of any country and the most deprived class in particular. Since, it is not possible to eliminate the difference of "haves" and "have- nots" in the society, the most affected people of the society from the problems of desertification and land degradation are the most poor, resource-less people, children and women facing the problems of mal-nutrition, old and the victims of natural calamities particularly in drylands in view of adversities.

Land based programmes for their upgradation are essential but the social sector programmes are the priority issues to be addressed by any government for mitigating the impact of the people of the society. Therefore, it is the foremost responsibility of the governments of the countries of drylands in particular and world-wide in general. Various programmes under implementation in India through the Department of Family Welfare, Department of Education, Department of Women and Child Development, Ministry of Social Justice and Empowerment, Ministry of Rural Development etc. are model to be adopted in other countries facing the problems of desertification and land degradation as a measure to combat it. The programmes are broadly divided in four main areas:

(a) Programmes for social awareness, Security and empowerment of women and weaker sections of the society.

(b) Programmes for infrastructure development and employment generation.

(c) Programmes on health, nutrition, sanitation and safe drinking water, and

(d) Programmes of literacy and pre-school.

1. Programmes for Social Awareness, Security and Empowerment of Women, Weaker Sections of Society and Poverty Alleviation

In January 1992, the Government of India set up National Commission for Women as a statutory body with specific mandate to the constitutional and legal safeguards. A landmark initiative to promote gender equality even in political sphere was the reservation of one third elected seats for women in all local self governments i.e., urban bodies as municipalities and rural bodies as Panchayati Raj Institutions. Through this measure, about one million women emerged as leader at grass-root level with 75,000 of them being chairpersons. Similarly National Plan of Action for the Girl Child 1991-2000 was enacted to ensure survival, protection and development of the girl child with the ultimate objective of building a better future for girl child and maintaining rationale sex ratio in the country.

With a view to raise the status of girl child for changing social attitude, towards the family of girl child, to improve enrolment and retention of girl child in school. For empowerment of women in the society, their rights in family, community and government with increased access to control over material, social and political resources. The overall objective must be to strengthen the processes and create environment for empowerment of women at all levels.

For helping the people of weaker sections of the society, there is need to implement schemes like availability of essential commodities at reasonable rates and if possible, at subsidized rates to provide sustainable livelihood. Awareness generation about the schemes being implemented for the benefit of weaker sections for their upliftment, social harmony and prevention of atrocities.

For poverty eradication various self-employment and wage employment schemes should be initiated to help the agricultural labour and non-agricultural labour, women etc. for attaining their livelihood. In dryland areas, cultivation is possible for few months and the people of such areas feel problems of employment opportunities during most part of the year and the situation is further aggravated in case of failure of monsoon, which is a common phenomena in such areas.

2. Programmes for Infrastructure Development and Employment Generation

Creation of infrastructure facilities is utmost essential for developing forward and backward linkages for marketing the surplus of agriculture sector and input availability to poor through credit support. Infrastructure creation is helpful for generation of wage employment opportunities for local population for some time. In addition, it helps in mobility of the poor in developing access to employment in distant areas. This is also helpful in accelerating rural economy, which is generally idle for want of infrastructure facilities.

For community development activities in rural areas, it is necessary to allow the people of rural areas to create demand based activities rather target approached activities. Empowerment of the people to select the activities as per their community requirement with their involvement in preparation and implementation of schemes. Generally, government departments and agencies have little access to work in remote areas in view of their skeleton staff and other problems, implement of community activities through village association is better option for creation of such facilities.

This is an accepted fact that there is vast variation in the community facilities in rural and urban areas even in the most developed countries. There must be a phased time bound programme by the government to provide at least minimum facilities like village roads, electricity, education and health institutions, safe drinking water facilities and dryland areas should be taken on priority basis. Government of India has already fixed norms different from other areas to provide slight relief to the people living in hard-hit areas.

Employment generation is the basic requirement, which helps in increasing the purchasing power of the poor for meeting essential requirement of livelihood. The governmental efforts required for this purpose are the identification of areas for rural small sector. Such schemes need to emphasize on local resources and capacity building for skill development, soft credit linkages at reasonable terms and access to market are the conditions for creating employment opportunities in the villages. This helps in removing unnecessary burden on urban centers through mobility of rural poor in search of employment.

3. Health, Nutrition, Sanitation and Drinking Water Supplies

Health facilities are the general requirement and rural remote areas are generally deprived by such facilities. Access to such facilities at a reasonable distance helps people to avail these facilities and this helps in releasing the burden over the city facilities. In addition, reproductive and child health care is necessary for improving health status of young women and children. This is essential for child survival, safe motherhood, reproductive health both for men and women.

Integrated Child Development Services (ICDS) is the world's largest and a unique community based outreach integrated programme for holistic early child care and development. The programme provides an integrated approach for converging basic services for improved child care, early stimulation and learning, health and nutrition, water and environmental sanitation , targeting young children, expectant and nursing mothers and women's/ adolescent girls' groups, supportive community structures, women's groups, through anganwadi centers. They are reached through nearly 0.5 million-trained community workers. The necessity in implementation is constant supervision and reach to needy people of the society for which the scheme is designed.

Rural drinking water and sanitation programmes should be designed to provide facilities to all the areas with special emphasis to dryland areas. Safe drinking water facilities helps in health care in areas affected with salinity, chemical infested, fluoride etc. affected areas. The strategy should be to cover all habitations, especially non-covered and partially covered areas with sale drinking water, ensure sustainability of the system and sources, to tackle water quality problems in affected habitations and preserve good quality by institutionalizing water quality. The Government of India has regulated facilities in dryland areas by relaxing the general norms by providing :

(a) 40 litres per capita per day drinking water for human beings.

(b) 30 lpcd additional water for cattle in the Desert Development Programme areas

(c) One hand-pump or stand-post for every 250 souls.

(d) Water sources within the habitation within 1.6 km. in plain areas and 100 m in hilly areas.

(e) Rural sanitation programme need to be implemented in all rural areas for demand based activities in the villages.

4. Literacy Programme

For literacy at least for elementary education, formal and non-formal education system is essential. The children of elementary level are unable to move to far distant locations and therefore, educational institutions need to be kept at minimum possible distance. In India present norm of availability of such facilities is 1.6 km. from the habitation and this is a reasonable distance to be covered by a child. Another problem in rural areas is the availability of child for learning. This is in view of involvement of children for meeting domestic requirements like collection of fuel-wood, carrying small livestock for grazing or similar activities. In such cases, an equilibrium is necessary to operate non-formal literacy centers to run as per local requirement. Such schools are already running in India.

Centres for adult literacy are also must for keeping aware with the programmes of community and individual benefits. With inception of formal and non-formal literacy centers, literacy level has been increased during 1991-2001 from 52.11 % to 65.38%, with male and female literacy rates of 75.85 and 54.16 percent respectively.

14.6 PROGRAMMES SPECIFICALLY FOR POVERTY ERADICATION

People living in degraded regions, drought prone and desert areas have very limited sources of income. The income from agriculture is not assured in view of uncertainty of monsoon. Scarcity of water is acute to maintain cattle for dairy activities. Power supply is also not certain. There are problems about transportation, availability of raw materials and reliable market for sale of finished goods. Despite efforts made for the past few decades, rural poverty continues to be significant. For raising the economic status of the people living in poverty, in the drought prone and desert areas and eradication of poverty, there is need to implement skill building and income generation schemes in an integrated manner.

1. Credit Assistance for Rural and Agro-based Activities and Schemes

There is significant problem for credit availability for rural sector, particularly for the input requirement of agricultural activities. Mostly rural people have to depend on moneylenders. For micro credit requirements role of the banks need to suitably amended as "Change Agents" for rural development with specific target to achieve, assigning about 40% total credit to priority sectors comprising of agriculture and allied activities, small scale industries, self employment retail trade and transport business etc. Banks provide differential interest rate ranging from 4 to 19% for rural credit, particularly agriculture and animal husbandry, aimed at the economically and socially weaker sections. The commercial banks are often rigid, unwilling and often take away the advantages of the government's socially oriented schemes and of low and differential interest rates. Co-operative banks are more sympathetic to rural credits and are more flexible and accessible to credit seekers. The traditional money lenders are also well entrenched form of credit institution and their influence among the rural population is still formidable. It is estimated that they service the credit needs of almost 30-50% of the rural producers and often with interest rates, which could be as high as 16 to 30%. (Ministry of Agriculture, 1996).

Poverty estimates for India and its various States are made by using a poverty line cut-off after fixing a price for the minimum required consumption levels of food, clothing, shelter and fuel. Allocation of health and education are included since it is presumed that the government provides these facilities free of cost. A norm of 2400 calories per day for rural and 2100 calories per day for urban areas is used as minimum. The methodology focuses on the purchasing power needed to meet the specific calorie intake standard with some margin for non-food consumption (Planning Commission, Government of India).

Sustained and broad-based growth of agriculture is essential for alleviation of poverty, generating income and employment, assuring food security and sustaining a buoyant domestic market for industry and services. In order to strengthen the rural economy, credit flow to agriculture through institutional channels of commercial banks, co-operative banks and regional rural development banks must be in easy access to cultivators with little barriers. The problem with the credit system linkages to agriculture sector is the crucial factor of uncertainty of monsoon, where the crop failure is recurrent phenomena, especially in dryland areas. Banking sector provides credit on assured return basis and there are varied problems for credit support to agriculture sector. The requirements of credit for agriculture sector are for short-term requirements related to inputs, medium requirement for land up-gradation measures, water saving devices, electric and diesel pump-sets and long term requirement for lift irrigation, water harvesting structures, wells etc. Since agriculture activities are quite uncertain specially in drylands, problems faced by cultivators are as under:

1. In dryland areas, costs of cultivation practices are quite high in view of adverse conditions. Small and marginal farmers require credit support for their short, medium and long term requirement. Though every country government is quite vigilant about the interest of small cultivators as they are quite large in number but possess very small area out of the total cultivated land of the country.
2. Credit requirement can be met from banking institutions and every government has managed easy credit flow to cultivators, but due to erratic monsoon conditions, repayment of credit is very serious problem of the cultivators and generally, banks avoid to provide credit to cultivators as realization is the responsibility of respective bank.
3. In adverse monsoon conditions, cultivators, face severe problem of their own survival and repayment of loans becomes impossible, which put them in defaulters list, debarring further credit facilities. In such a situation, continuations of cultivation practices become difficult. The cultivators are compelled to divert at private moneylenders and once they are gripped in the clutches of moneylenders, they loose their land and ultimately become agricultural labour.
4. For the sake of benefiting small cultivators, Government of India has introduced a system of conversion of short term loans to medium term loans or defer the payment till the cultivators are able to get a good crop in the monsoon cycle of five years, where at least two years are average or good from rainfall point of view. This system helps cultivators to continue credit support for their agricultural requirements.
5. Deferment of loans is possible in case of medium and long term loans and National Bank for Agriculture and Rural Development (NABARD) provide credit support to banks lending to cultivators. Similar facilities are required to be managed in all the developing countries for assured and timely availability of credit to small cultivators.
6. Another hindrance in cultivation is farm harvest price of agricultural produce. Generally, the prices are Shoot up in case of monsoon failure and vice versa, decrease significantly when the crop yield is quite high. Cultivators suffer in both the situations. For benefiting the total cultivator community, minimum support price of agriculture produce need to be announced by country government before sowing the crop. This helps cultivators to grow requisite crop beneficial in every monsoon conditions. This helps cultivators to get remunerative price of their produce as country government purchases all the surplus produce from the market at the given support price.

7. In India, agriculture marketing centres network has been spread in the country, where cultivators can sell their agricultural produce at the market price. In case, cultivator feel after reaching to market that the price is not favourable and decides to wait till enhancement of prices, they are facilitated to keep their crop in agricultural marketing centre and able to get credit of 75% of the market price. The amount is met when the cultivator sells its crop in favourable conditions and the balance amount is given.
8. In addition to cultivation of food crops, credit requirement is necessary for forestry sector activities, where small cultivators need to adopt the activity. In addition to credit issues, long gestation period deprives them to adopt such practices. In such a situation, credit requirement is necessary for input requirement as well as livelihood support for the gestation period. Such provisions can help the cultivators to divert to forestry sector.
9. Collection and processing of minor forest produce and medicinal contents from mountains, hills and wasteland credit support becomes necessary but poor people are unable to get credit as this component is not included in credit list of developing countries. This is a source of livelihood to many poor and need to be encouraged.
10. Credit support to small cultivators is required for non-farm activities like illness and social customary occasions. For such credit, cultivator has to reach private moneylender and the repayment of loans become impossible in many situations. For ascertaining the safety and security of the cultivated land of small farmers such credit requirements need to be managed by the government of developing countries.

2. Grameen Bank, Bangladesh - A Success Story

The emergence of "Grameen (Rural) Bank" as a key factor has been successful in developing the rural economy and bringing out the positive social changes. Prof. Mohammad Yunus launched an action-research- programme in 1976 at Jobra village to explore possibilities of credit access by rural poor-landless and women, which finally got the shape of GRAMEEN BANK. As per this, loan recipients and landless, who get loans, organizing themselves into groups of like-minded persons with almost persons with almost the same socio-economic status. A person is eligible to get loan, if his / her family owns less than half an acre of cultivable land and the value of whose family assets does not exceed the market value of one acre of land. The cumulative repayment rate in 1995 was about 99%. The success of Grameen Bank is due to following factors:

(A) *An Appropriate Banking System*

(i) Intensive supervision and monitoring of repayment rather than enforcing tough rules and regulations.

(ii) Continual efforts to become self-sustainable and autonomous.

(iii) Hard work to manage the sanction to small loans and weekly repayment nstalments.

(B) *Engagement in Social Development*

(i) Women empowerment, where 93% of borrowers are women, motivating as well as monitoring people to get married without dowry, no loan to a man with two wives, intensive and extensive workshops organized for women awareness.

(ii) Promoting children education.

(C) *Successful Organizational Pattern*

(i) Decentralized flow of information.

(ii) Staffing network.

(iii) Ensuring the feedback on innovative ideas enforced by Grameen Bank.

The concept of micro-financing, started by Prof. Mohammad Yunus is presently replicated in 52 countries. Many international NGOs such as the Foundation for International Community Assistance (FINCS), SCCION, Women World Banking, Freedom From Hunger etc. are executing projects through their partners in developing countries. Prof. Yunus has estimated that of the 1.2 billion poor in the world. Grameen has reached 2 million, other services on similar lines covering another 1.5 million in Bangla Desh. International replicates have about 2.5 million borrowers. It is estimated that this has helped overall 36.million people worldwide. (Kurukshetra 2000).

3. Water Management

In dryland areas, scarcity of water is the severe constraint in development. Rural sector is generally neglected for basic amenities and the people in dryland face serious problem in irrigation, drinking water for human beings and livestock. Availability of water is dependent of the geo-physical conditions of the area and such efforts can be taken up efficiently at the micro level. For management of water for all-out requirement of the people in dryland areas, dependent source need to decided for long term availability and similar arrangement can be made after taking appropriate decisions.

Safe drinking water facilities for human beings and livestock is the primary requirement for survival of the people of the drylands. In this regard growing population trends of human and livestock demographic trends help in taking appropriate decision regarding sustainable source for availability of safe and potable drinking water. Suitable norms need to be decided for provision of drinking water at least at the minimum level. In India, criteria for human and livestock in rural and urban areas has been decided for all type of geographical areas and decided to provide drinking water at 100 litre per capita per day in urban areas and 40 litre per capita per day for rural areas. This is minimum level to be managed through out the year. In some problematic areas, water is provided through rail or truck born tankers in summer. The general matter of principle is availability of safe drinking water source within same habitation to enable people to save time in carrying water.

Making available irrigation facilities is not an easy task as almost all the drylands face problems of water. There is positive correlation between non-availability of sufficient water and desertification / land degradation conditions. Similarly, it is also true that provision of water for irrigation create many problems of land degradation due to lack of proper management. Retaining the fertility of soil and irrigation facilities on sustainable basis is an ideal situation, which generally lacks in all the irrigated lands. This is resultant impact of source of irrigation and its management for providing irrigation facilities.

Awareness generation can be helpful in proper water management for irrigation. The cultivators require specialists' support for each type of irrigation facilities to guide the cultivators for use the quantity of water, time and its adverse impacts due to mismanagement. Generally preventive measures are more reliable and cost effective than curative measures. Water-logging, salinity, alkalinity etc. are the problems of land degradation cause due to water mis-management and ultimate sufferers are the individual cultivators, village community, society and the country. Awareness generation can be helpful in tackling these problems in the beginning, when their remedies are possible.

Another area of enhancement of crop yield is water saving devices to maximization of irrigated area and increase in productivity. Water is scare commodity especially in dryland areas and awareness for rationale use helps in land up-gradation, increase in soil fertility and increase in irrigated area without much efforts from the same quantum of water.

4. Micro Assistance to Women

In dryland areas, there is very severe problem of employment opportunities as rain-fed cultivation is generally possible. In such situations rural people are unemployed for most part of the year. The women unemployment is quite high. The solution of this problem for poverty alleviation is the micro credit for self-employment activities possible within the village itself. Sustainability of these activities depend on selection of activities based on local resources but possible through credit support. Norwegian Agency for Development Cooperation (NORAD) is an example for external assistance to women for self-employment activities. Under this programme, financial assistance is given to Women's Development Corporations, Public Sector Corporations, autonomous bodies and voluntary organizations to train poor women, mostly in non-traditional trades and to ensure their employment in these areas. Some of the trades are computer programming, electronic watch assembling, radio and television repairs, garments making, secretarial practice, community health work, embroidery, weaving etc.

In such micro enterprises activities, there is necessity of capacity development through skill up-gradation, marketing linkages, marketable produce and remunerative return for sustainable functioning of the activities. The agency responsible needs to possess capabilities to train the beneficiary group in selection of activity, input availability at cheaper rates and processing capabilities of the product in terms of cost and time are the areas of concerted efforts. In rural areas, women can handle the task in spare time and get some income for their family support.

14.7 TECHNOLOGIES FOR COMBATING DESERTIFICATION

Technology packages for dryland areas need to be developed by associating different research agencies functioning in the field of agriculture, forestry, and community awareness for combating desertification and for mitigating the effects of drought, which are common phenomena in dryland areas. In addition, traditional practices, which are relevant, need to be revived, improved and incorporated in the programme. A number of technologies to combat the process of land degradation have been evolved and there are various traditional technologies under application in various developing countries. Technology packages can be broadly categorized as follows:

(a) Technology for conservation of soils, water and vegetation.
(b) Technologies specifically to control, land degradation in different bio-climatic regions: Arid, Semi-arid and Dry-sub-humid.
(c) Technologies for management and reclamation of degraded lands: Mine Spoils, Alkaline/ saline Soils, Water logging, Industrial effluents.
(d) Traditional technology for combating desertification.
(e) Technologies for mitigating the effects of drought.

1. Technologies for Conservation of Soil, Water and Vegetation

Degradation of soils, water and vegetation effects severely in dryland areas. Technology has been developed for addressing all the degradation problems.

(a) *Integrated Soil Fertility Management*

Soils in the semi arid tropics are poor in fertility. Though the present materials largely influence the inherent fertility of the soils, climate and crop management have significant role in building up and maintenance of soil fertility. Soil organic matter status regulates the available nutrient status of the soil not only at source, but also in buffering the soil reaction (Katyal and Reddy 1997). Nearly 95% of nitrogen and sulpher are contained in soil organic matter. Bulk of the zinc and copper are also in organic farm.

Therefore, decline in soil organic matter levels due to soil erosion by run-off or poor vegetal cover also multiplies the nutrient deficiencies. Management of soil organic matter at its equilibrium level is one of the foremost frontal challenges of resource management in arid and semi-arid tropics.

(b) *Integrated Nutrient Management*

Since no immediate improvement in the use of chemical fertilizers is likely to take place in rain-fed areas, integrated nutrient management is the key to maintain the productivity of soils on a sustainable basis. Globalization may eventually result in steep rise in fertilizer prices in the near future, making them beyond the reach of small farm-holders. Large number of experiments at farm stations and on farmers' fields need to be demonstrated to emphasize the importance of organics. Farm Yard manure, compost and bio-fertilizers in supplementing the nutrient requirements of crops and providing stability to yields in rain-fed areas (Subha Rao and Rao 1980, Singh *et. al.* 1999, Venkateswarlu and Wani 1999). Fifty percent of the fertilizer nutrient could be replaced with the use of farm yard manure / compost in a variety of soils. Use of organic manure not only reduces the use of chemical fertilizer nutrient requirement substantially in addition to supplementary important primary and secondary nutrients. The use of compost and farmyard manure also improves the soils' physical condition and crop yield on a long-term basis. In addition, it also improves the moisture holding capacity of spoils.

The other approach to enhance the use of organics could be the compost. This way animal manure, which is in short supply for use in rain-fed regions, can be converted into compost, increasing the bulk several fold, yet the benefits remain the same or a shade better than farm yard manure. Moreover, decomposed manure could make the nutrients more readily available to the existing crop. Considering the local market price of farmyard manure, the cost of composting organic wastes may be on par with that of pure farmyard manure as the quality gets reduced substantially. It has also been reported that termite infestation is much less, when compost is used, compared to farmyard manure.

(c) *Soil Quality Monitoring*

Maintenance of soil health is critical for sustained agricultural productivity more in rain-fed areas. In view of the problems on a continuous basis, rainfall variability among years further compounds this relationship. Nevertheless, soil quality monitoring on a regular basis under different cropping systems and land uses is necessary to capture significant enhancements or deterioration in physical, chemical and biological properties over time. Changes in physio-chemical properties affecting soil quality have been reported adequately through land use experiments. Recent studies by Venkateswarlu (1998) across several dryland locations, highlighted the significant efforts of cropping systems and manuring practices on reversible and irreversible changes in soil micro flora and its diversity.

(d) *Permanent Vegetative Cover through Alternate land-use Systems*

Several locations in semi-arid tropics eco-system indicate that annual crops cultivated on lower capability land are prone to lower yields and risks and lack of response to inputs. Soils in these capability categories can be best utilized for alternative land uses, where self generating grasses, legumes and perennial woody trees constitute the major components. Agro-forestry approach including systems of agri-silvi-culture, agri-horticulture, horti-pasture and silvi-pasture can be successful as an alternate land use. Management of lands of lower capability through such interventions is the best way of integrating livestock production in rainfed areas, thus contributing to sustainability of the production system. Alternate land uses not only provide fodder, fuel wood, fruits and timber but also enhance the quality of resource base through greater biomass production and providing a land cover for most part of the year, which constitutes the basic step for control of soil erosion by wind and run-off. Off-season rainfall, which otherwise goes unutilized in single rainfed cropping areas, can thus be best utilized with such production systems.

Silvi-pasture has proved as an ideal land use system for marginal degraded lands, where a perennial tree component. Ley farming is another approach, where legume or non-legume forage is rotated with cereals. This system improves soil quality besides providing fodder. Stylo or grasses can be grown in two to four years period followed by arable crops.

(e) *Alternate Land Use Systems for Different Agro-ecological Regions*

In order to provide stability to farm income and also utilize the marginal lands for production of fodder, fuel wood and fibre, a number of alternate land use systems have been evolved based on location specific experimentation and cafeteria studies (Singh 1988). The general recommendations for alternate land use systems based on annual rainfall and land capability are outlined in Table 14.7.

Table 14.7 : Recommended alternate land use system options for different agro-climatic conditions

Annual Rainfall (mm)	Soil Type	Land Use System	Suitable Tree/ Grass/ Legume Species
<500	Shallow (0-0.30m)	Tree farming	Prosopis cineraria P.juliflora, Acacia Aneura, A. Nilotica, A. fortilis, Pitheceliobium dulce.
	Medium (0-0.45m)	Pasture Management	Lasiurus sindicus (light textured soil), Cenchrus, Sehima nervosum, Stylosanthes scabra, Clitoria ternatea.
500-750	Shallow (0-0.30 m)	Silvi-pastoral system	Acacia nilotica, Colophosphermum mopane, Dalbergia sissoo, hardwicia binata, Cassia sturti, Albizia amara, Leucaena leucocephala, Censhrus ciliris, C. setigerus, Dicanthium ennutalum. Panicum antidotala, Stylosanthes hamala, macroptillum atropurpureum.
	Medium (0-0.45 m)	Horti-pastoral system	Annona squamosa, Ziziphus mauritiana, Syzigium cuminil, Emblica officialinis, Tamarindus indica, Feronia limonia, Aegle marmelos, Cenchrus ciliris, Panicum antidotale, Urchloa mosambicensis, Stylosanthes hamata, Macroptilum atropureum, Clitoria tenatea.

Annual Rainfall (mm)	Soil Type	Land Use System	Suitable Tree/ Grass/ Legume Species
>750	Shallow (0-0.39 m)	Ley farming or silvi-pastoral system	3 years stylosanthes hamata and 4th year arable crop (sorghum on heavier soils, pearl millet on lighter soils) Silvi-pastoral system as above.
	Medium (0.30-0.45 m)	Ley farming or horti-pastoral system	Ley farming as above. Mangifera indica, Achras zapota, Psidium guajava, Embica officinalis, Stylosanthes hamata/ Macroptilium atropurpureum.
	Deep (>0.45 m)	Agri-silvi or agri-horti system	Acacia ferruginea, Prosopis cineraria, Tectona grandis, Hardwickia binata, Dalbergia sissoo + arable crops. Mangifera indica, Achras zapota, Psidium guajava + arable crops.

Source: Singh 1988.

(f) *Technologies for Soil and Water conservation*

Soil and rainwater conservation and management research has received maximum emphasis during last three decades with the objectives to:

(a) integrate conservation and production technologies taking watershed as the unit of management,

(b) Ensure maximum in situ rain water conservation and

(c) harvest run-off water for recycling as come up supplemental irrigation and supporting establishment of tree crops.

High intensity rainstorms during monsoon cause significant loss of topsoil both in Alfisols and Vertisols. Gully and sheet erosion are the most common. A combination of mechanical, agronomic and vegetative practices can help in arresting soil loss and run-off in cropped lands. Usually, resource loss is more when lands are kept fallow. Even crops differ significantly on the canopy and the root system.

Mechanical measures like contour and graded bunding and bench terracing are designed for lands of different slopes as permanent structures, which can be termed as 'hardware' treatments. Contour bunding in cultivated lands intercepts the run-off, reduces soil loss and provides increased opportunity time for water intake. This practice is useful in low rainfall areas of less than 600 mm, having soils with high infiltration/ permeability rates. In Alfisols, contour bunding helps not only in controlling run-off but also increases crop yields, while it is not suitable for deep black soils due to prolonged water stagnation. Graded bunding is recommended for areas having rainfall higher than 700 mm for safe run-off disposal. Adoption of these practices by the farmers has been dismal due to inherent limitations of small holders and in fact that such bunds cut across their field boundaries/ holdings. As an indigenous system, the farmers install small holder bunds and mud-cum-pebble bunds across the slope to control moisture and to control erosion.

Inter-terrace land treatments are of semi-permanent nature, primarily useful to minimize the velocity of overland flow. These practices have a significant role in checking of soil loss and ensure better utilization of rainwater for crop growth. Land treatments like ridging, compartmental bunding", conservation, furrows, broad-bed and furrow widely tried across the country are some examples (Singh 1998). While the advantage of such practices in resource conservation are demonstrated unequivocally, the yield gains depend on the quantum and distribution of rainfall in a particular year. Compared to seed based technologies, the acceptance of these practices by farmers has been much less.

2. Technologies to Control land Degradation in Different Bio-climatic Regions

For combating desertification and land degradation region specific strategy need to be evolved for application of technology, as universal strategy cannot be effective for every area. The drylands have their own problems varying from area to area and region to region and there is vast difference in geographical, ecological and environmental conditions.

(a) *Technology for Management of land Degradation in Arid Region*

The problem of desertification is severe in the arid zone especially in places, which receive rainfall less than 250 mm annually. Vast stretches of land are already out of cultivation, however, tremendous scope to arrest the process of degradation by adopting appropriate technologies. Rainwater harvested by induced catchment can be judiciously used to raise nurseries, to manage livestock and for life saving irrigation to tree crops. Since the harvested rainwater in arid regions is very precious, it should be used with utmost care. Inter-dunal plains in some places are covered with local species of grass, which are either over-grazed or dominated by non-palatable grasses. Such areas can be rejuvenated by using improved grass species, which help serve as fodder and also arrest erosion.

Control of wind erosion through mechanical and chemical methods of sand stabilization or through vegetative measures. Sand dune stabilization through planting trees is common to protect from shifting and creating disruption. Management of pastures and rangelands is suggested in rainfed agro-ecosystem. The approach should be integrated with research and technology in complementary sectors. Dynamic methods need to be developed that help to identify the appropriate balance between the resilience of vegetation to short term overgrazing, the variability in the weather overtime, the indebtedness of herdsmen and their attitude to risk.

(b) *Semi-arid and Sub-humid Regions*

Conservation of soil, efficient management and utilization of rainwater and management of the non-arable lands for fodder, fruit and fuel wood production in the watershed perspective are the core strategies of natural resources management in the semi-arid tropics and in the dry sub-humid regions. Other aspects as suggested in arid lands are equally significant to these regions.

3. Technologies For management and Reclamation of Degraded lands

Areas covered with shallow ravines can be utilized for silvi-pasture, which also encourages) livestock enterprise. Aerial seeding of grasses of improved strains can also be tried to reclaim the shallow ravines. Erecting mechanical checks and stabilization of ravine slopes, wherever possible should be taken up to prevent further degradation (Singh and Singh 1999).

(a) *Technologies for Rehabilitation of Mine Spoils*

One of the major constraints in the rehabilitation of mine spoils particularly in arid regions is the very poor status of plant nutrients in the over-burden mine dump. Higher pH and exchangeable sodium, magnesium, sulpher and phosphorus, as well as salinity-alkalinity build up are the other constraints in many mine spoils (Saxena and Chatterjee 1995). Revegeta.tion of the mine spoils is, therefore, a challenging task. CAZRI has attempted to revegetate a limestone mine spoil area and a gypsum mine spoil area with some success.

In areas under mine spoils and quarry based, which are permanently out of cultivation, the erosion rate of the mine-spoils is known to be about 1000 times more than normal soils. Mine-fillings and quarry wastes are a hazard as the surface runoff brings the mine and quarry wastes into the territory of fertile cultivated soils, thereby rendering them unfit for cultivation.

(b) *Technologies for Management / Utilization of Industrial Effluents*

Management of salt affected areas needs to be addressed with proper technologies to regain their lost productivity. Salt affected land lying in low rainfall areas should be focused for silvi-pasture development, while those lying in high rainfall areas can be considered for rice cultivation with suitable amendments and varieties. Much of water logging and salinity-alkalinity hazard is associated with appropriate water management. This can be partly countered by vertical and horizontal sub-surface drainage. Lining of canals, judicious use of canal water using sprinkler and drip irrigation techniques, and efficient management of command areas are necessary to arrest further degradation of land resources.

The sodic soils can be reclaimed or moderated by the application of gypsum. The requirement of gypsum has been standardized. With proper choice of crops, only the upper layer of 15 cm of soil needs to be-amended by application of gypsum (Khosla 1973). Other amendments useful for reclaiming salt affected soils include pyrite and organic manures. Network research has further resulted in identification of crops and varieties that can tolerate salinity and alkalinity. Usually the crops withstand excess moisture conditions, can also tolerate sodicity. Highly degraded salt rich soils of 15 tonne/ hectare gypsum is required for rice-wheat rotation, in which the cultivation of rice by itself is a reclamation measure (Abrol and Gupta 1990). Number of Afforestation and agro-forestry techniques are developed for rehabilitating the salt affected soils (Gill *et. al.* 1990). Tree species of Prosopis Chilensis, P. juliflora, Tamarix troupii, Tamarix aphylia, A.

nilotica, Acacia auriculaefromis, Casuarina obesa, C. equisatifolia and Eucalyptus camaldulensis and highly tolerant of soil salinity thus suitable for plantation in such areas.

(c) *Technologies for Reclamation of Water-logged Saline Soils*

Saline soils are often associated with water logged areas due to rise in the water table. Surface stagnation of water in these areas is a serious problem during monsoon. Cost effective measures for managing surface drainage are the key to minimize the adverse effects of water logging / soil salinity. Following are some of the measures for reclaiming water logged saline soils:

(i) *Drainage Management:* Design criteria for' sub-surface drainage for water logged saline soil representing alluvial plains have been standardized. Desalinization of soils under sub-surface drainage can be achieved through the rainwater conserved in the field by providing strong bunds (Singh and Pandey 1991).

(ii) *Irrigation System Management:* Drip and sprinkler irrigation .can increase the water use efficiency by 70-80% in contrast to conventional system of surface irrigation with unlined field channels, which have an efficiency of 20-30% only. Improvement in the conventional methods by borders / furrow irrigation and brick lined water courses has an efficiency of 40-60%. These improvements can bring down the rate of secondary Salinization by irrigation water.

(iii) *Disposal of Drainage Affluent:* Sub-surface drainage systems produce poor quality drainage water, which needs to be disposed off carefully. This problem is imposing restrictions on the reclamation plans, using sub-surface drainage system on about 0.4 m ha, which is potentially productive but for water logging and salinity. Feasible alternatives for use of high salinity drainage water by blending it with canal water need to be worked out as a strategy for managing saline drainage effluents.

(iv) *Agronomic Practices for Managing Salinity/ Alkalinity:* About 25% higher seed rate over the normal is recommended to account for mortality of young seedlings and poor tillage of crops in salt affected soils. Heavy irrigation is recommended before sowing to each down the accumulated salts so as to improve germination and initial growth of the crops in such soil conditions.

(v) *Alternative Land Uses for Salt Affected Soils:* Suitable tree species such as Prosopis Juliflora, Acacia nilotica and Tamarix articulata are recommended for plantation in salt affected soils. Long-term field studies indicate that growing leguminous tree species such as Prosopis Acacia, Casuarina etc. can help ameliorate alkali soils at much faster rate than non-leguminous trees because of farmers inability to build up soil N/organic matter status. Growing legume trees is highly sodic soils can contribute in their amelioration for crop production. Grasses in general are more tolerant to alkali conditions than most field crops.

Promising grasses in alkali soils are Leptochloa fusca, Chloris gayana, Brachiaria mutica and cynodon dacylon (Kumar and Abrol 1986). Other promising grasses for utilization of saline soils are Aeluropus lagopoides, Chloris barbata, Echinocloa colonum, Decanthium annulatum, Phragites and Sida spp.

(vi) *Agro-forestry for Moderately alkaline Soils/Reclaimed Soils:* Promising trees for agro-forestry on moderately sodic soils or the reclaimed ones are Populus deltoids. Eucalyptus tereticornis, Acacia nilotica and Tectona grandis. Populus-based agro-forestry system has proved more remunerative due to faster growth and better market price. Intercrops such as rice-berseem and rice-wheat can also be taken during the initial period of tree growth.

(vii) *Horticulture:* Promising fruit tree species such as Ziziphus mauritiana (ber), Psidium guajava (Guava), Punica granatum (pomegranate), Emblica officinalis (amla), Carissa carandum (karaunds), Tamarindus indicus (imli) and Syzygium cumini (jamoon) can be grown in highly alkali soils by adopting proper site modification technology including the use of organic and inorganic amendments.

4. Traditional/ Indigenous Technologies for Combating Desertification

Traditional knowledge and practices have their own importance as they have stood the test of time and have proved to be efficacious to the local people. Some of these traditional practices in fields of agriculture are crop production, mixed farming, water harvesting, conservation of forage, combined production system, biodiversity conservation of forage combined production system, biodiversity conservation, forestry and domestic energy.

Technology is not neutral. Political arid economic forces both drive it at the macro and micro level of development planning. Most development projects have involved the injection of technologies and on traditional knowledge of the people. Technology choice has usually been influenced by external factors i.e., national policies vs. policies of external donor agencies. Village eco-system planning should try to integrate the best of traditional knowledge with the best of modern science and technology. Traditional technologies and practices are usually rooted in ecological wisdom and the social traditions of local communities. Modern technologies, however, aim to increase productivity, usually through an intensive use of external inputs. Accordingly, modern technologies are often are very capital intensive and demand new skills generally lacking in local area and thus tend more to mystify than enlighten.

The high productivity wrought by modern technology can be sustained only if it is based on production systems ecologically sound. To integrate the best element of modern and traditional technologies offer promising entry points for developing package on community based dryland resource management technologies. The participation of local communities can lead to greater use of traditional practices.

5. Technologies to Mitigate the Effects of Drought

Droughts driven land degradation is severe and efforts should be made to create the situations that impact of droughts is minimized through specific management strategies. Deficit of rainfall during the onset, progress and withdrawal of monsoons can cause early, mid and late seasonal droughts. Measures to tackle early season droughts include transplantation and changing the crops and varieties, whereas mid season droughts can be controlled through reducing the leaf area by regulating the plant population or management strategies like mulching and keeping the field weeds free. In situ water conservation, water harvesting and recycling as supplemental irrigation obviously can mitigate drought to a great extent, but need more efforts and inputs. There are very few options to manage terminal droughts due to early withdrawal of monsoon. Harvesting the fodder crops in the only way to slightly reduce the damage.

(a) *Early Warning Systems*

Long term whether data can be used for forecasting of the rainfall and behaviour of monsoons. While long term forecasts are still not possible in developing countries specially for dryland areas, short term range and medium term forecasts of the likely aberrations of the spatial and temporal distribution of rainfall are now possible. An effective and timely information delivery system needs to be developed and put in place for forecasts for region specific trends.

(b) *Crop-Weather Modelling*

With the availability of high speed computers, it has become possible to include large number of variables and data sets from many years and locations to predict the behaviour of the rainfall for particular locations. Cultivators need to select drought resistant crops capable to be grown with monsoon variations. The cultivators need to be motivated to manage the cropping pattern mix of fodder trees, fruit trees and food grain to minimize the loss in adverse conditions.

(c) *Mid Season Corrections*

Even in timely sown crop, long dry "spells during season can cause significant drop in yields. Since such drought spells occur during standing crop, the farmer has limited options to prevent the negative impact. However, certain specific corrective measures can be taken to minimize the effect of mid-season drought. Reduction of leaf area either by rationing or thinning can mitigate the effect to some extent. Weed control and mulching are other short term measures, which can mitigate the drought by conversing the scarce moisture. In case of long duration crops like castor, pigeon pea and sorghum. 2% urea spray is useful after a good rain wets the foliage and the plant starts recouping. Long term measures for improving the soil moisture storage through appropriate land configurations also contribute to minimize the mid-season drought effect. However, these measures are to be planned well in advance at least in areas vulnerable to recurring droughts. In short, the main strategy of mid-season correction involves around extending the moisture availability period in the soil.

14.8 ROLE OF NON-GOVERNMENTAL ORGANIZATION IN COMBATING DESERTIFICATION AND LAND DEGRADATION

In past decades, role of non-governmental organizations has been increased significantly. NGOs have various- success stories of working with the people for their social and economic upliftment. Combating desertification and land degradation are the areas, where NGOs are involved for betterment. Special features of NGO system is as under:

(a) Non-Governmental Organizations are voluntary organizations and are ready to accept the responsibilities in the desired fields.

(b) NGOs are non-profit and charitable organizations, where profit earning in not the motto in response to the services rendered for the benefit of the community.

(c) NGOs are dedicated organizations to wide range of community service activities at home and overseas. They have access to work in remotest areas and most deficient areas of the world, facing all the adversities.

(d) They function with the help of from voluntary organizations, donations from their supporters and general public for the assigned cause.

(e) Their services are cost effective and their ability to handle the tasks according to their organizational set up. Services for the assigned tasks are comparatively cheaper than the governmental organizations.

(f) While working with the community, they gain confidence of local community as they work for their betterment and people also have faith in them in view of their dedication and access to reach the vulnerable areas.

(g) NGOs have their in-built organizational set up, which can be suitably strengthened to handle the assigned tasks and possess functional liberty in comparison to governmental organizations, where procedural and administrative barriers are numerous.

(h) NGOs perform in significantly better with result oriented approach. While providing services to governmental schemes for implementation, their performance is suitably adjudged by governmental organizations in terms of quality and cost effectiveness.

(i) In view of their working with the people, various functions of the programme as awareness generation, participatory rural appraisal, community mobilization, group cohesiveness and peoples' participation are functional tools of their working system.

(j) Raising resources from the communities for their own benefit is much easier to NGOs and people happily contribute their share in the desired terms in wages, material and even in cash.

(k) After completion of programmes, they hand over created assets to the beneficiary groups and remain associated for sustenance of the projects while motivating them for maintenance of assets for long term benefits.

(1) NGOs associate the people in preparation and implementation of the projects and with flexibility in their approach, make suitable corrective measures as per local requirements and aspirations of the people of the area.

1. Global network of NGOs

These range in size from small church groups active only within a single city in a developing country to huge organizations like OXFAM that support development and humanitarian work all over the world. What NGOs lack in funds and technical knowledge, they make up for in enthusiasm and grassroots contracts that sometimes result quite extensive tree planting campaigns (Grainger 1984-b). A number of projects involving NGOs, such as CARE, have already been referred to. although some NGOs have been established specifically to promote or undertake tree planting, often the most prominent organizations in this field are church groups, women's organizations and youth clubs for which tree planting is just one of a number of activities. Gujarat state of India has cooperated with a variety of NGOs in its social forestry programme (Karamchandani 1982). According to Fred Weber (1982), the success of NGOs Afforestation schemes in the Sahel, compared with the poor performance of village woodlot schemes organized by government forest departments, suggests that foreign donors could make more effective use of their money by working through NGOs.

The National Council of Women of Kenya (NCWK) has thirty-five affiliated women's organizations and is a powerful example of NGO achievements. In 1977, it founded a national Green Belt Movement to help communities establish greenbelts of at least 1000 trees on open spaces, school grounds and roadsides. The first greenbelt was planted at the time of UNCOD, in September 1977, but the number of greenbelts had risen to more than 1000. This does not include many smaller greenbelts, which farmers had been inspired to plant on their own properties. The organization had associated communities to choose the sites and was made responsible for organize the planting and management. In early years, all the seedlings were supplied free of cost by the Ministry of Natural Resources, but demand outstripped supply and member groups had to establish sixty five own nurseries to increase seedling production. The practical job of tending the greenbelts and supervising the nurseries was undertaken by Green Belt Rangers. The NCWK raised money for its work from member groups, local companies and individuals all over the world (Harrison 1986; Mathai 1985; Anon 1982).

The actions of the NGOs in the field of combating desertification and land degradation, while not unnoticed, were mainly unsupported and undocumented. Information on the extent of Afforestation work by NGOs based in developing nations is, therefore, very sparse, even though in some countries, it is known to exceed that performed by government forest departments. In Senegal, 5100 hectare of trees were planted by community efforts compared with 4700 hectare planted by government forest department (Sene 1984). Involvement of NGOs in tree planting was given a fresh focus in 1985, as part of International Youth

Year, The Tree. Project was established at the United Nations under the auspices of the UN Liaison Service for Non- Governmental Organizations. The International Tree Project Clearinghouse was set up to collect and disseminate information on NGO tree-planting activities and later published the first detailed directory of African NGOs involved in the field (NGLS 1986).

Indigenous NGOs often receive funding for their work from NGOs based in developed nations. The contribution of these NGOs to development assistance as a whole is quite substantial. The number of NGOs based in developed nations, which have a significant forestry component in their programmes is still small, mainly due to lack of awareness of the need for action in this field. NGOs involvement can do much to help to motivate local people, but the success of the projects may still be affected by organizational, technical and financial constraints. There is often poor coordination between the NGOs on the one hand and governments and international organizations like FAO on the other. This can mean that NGOs are not consulted about or involved in the planning of national and international programmes and receive only limited access to necessary technical information.

2. Importance of NGOs in Combating Desertification and Land Degradation

Governments of the developing countries are facing serious constraints in functioning in view of increased responsibilities and aspirations of the people for their social and economic development. The awareness created in the people has significantly increased their aspirations and the gap between rich and poor has increased significantly. Situation of people in rural areas and particularly in dryland areas is quite miserable due to access to infrastructure facilities and social amenities. The financial burden of government setup is increasing significantly and developing countries are generally unable to shoulder all the responsibilities. Sharing the responsibilities by NGOs is the need of present situation and accordingly the importance of NGOs is increasing very fast.

It is also true that NGOs number in developing capable to undertake the responsibilities is quite meagre and therefore, it is necessary to make NGOs viable to bear the responsibilities. Combating desertification and land degradation is quite difficult task in view of its multi-facet problems and constraints. It is therefore, necessary to identify some areas, where the NGOs can function in view of their capability.

While selecting the areas of operations of NGOs, it is advisable to identify programmes which generate resources to meet the expenses of the services of the NGOs. Though the NGOs function on non-profit motive, but at least the actual expenses of their services need to be provided for their sustenance and perennial support in the programmes. Another aspect need to be considered here is the benefit from the activities assigned to NGOs. Invest made for development of degraded lands can be helpful in providing significant employment opportunities to the people of the drylands. Increase in production and productivity of degraded lands help in attaining the livelihood of the people and resultant benefit the governments by net addition to national economy. Therefore, this aspect is equally thoughtful in selection of the list of activities to be assigned to NGO sector for betterment of the area, people, environment, ecology. The suggested list of activities to be shouldered to NGO sector is as under:

(a) Social forestry programme on government, community and barren lands, keeping provision of plantation of trees able to provide fruits, fodder, fuel wood and timber on regular basis.

(b) Rail, road and canal side plantation of trees, bushes and grass with involvement of local community along with right to use the benefits.

(c) Development of waste land with proper or requisite treatment and sharing the benefits by the community.

(d) Water harvesting structures and watershed development for the long term benefit of the local community.

(e) Awareness generation for population control, water saving devices, application of biotechnology, farm forestry, development of environment and ecology.

(f) Community mobilization, participatory rural appraisal, motivation of community for sharing the cost and benefits of created assets.

(g) Collection of minor forest produce through community and ensuring for remunerative return.

(h) Cultivation of medicinal plants in farmer's fields, community land and wasteland and collection of medicinal products from mountains, hills, wastelands and uncultivated lands.

(i) Assistance in establishing micro enterprises units for self employment activities and establishing marketing linkages for remunerative return.

(j) Awareness generation for conflict resolution in common property resources and sharing the benefits from group based activities.

(k) Transfer of technology related to dry farm practices, improved seeds, selection of crops befitted to local conditions and drought resistant varieties of seeds, trees and grass.

(1) Livestock development and stall feeding practices to save land deterioration and loss of grown fodder and grass.

Other aspects of NGO functioning are credit linkages for rural activities relating to development of degraded lands, keeping vegetative cover, use of fallow lands etc.

3. Capacity Building of NGOs

For optimal utilization of the services of NGOs, capacity building is necessary. NGOs lack in specialized knowledge of each and every aspect of operational activities. Capacity building can be helpful in their result-oriented approach. It is advisable to keep them aware with the technical aspects of the projects and train them about the objectives and methodology of implementation of the projects. Capacity building also includes specialized support in case they lack in requisite spheres. In case the NGOs are working on the government projects, involvement in preparation of projects is beneficial and in case it is not possible, communication of all the related issues to the maximum possible extent can help in effective implementation of the projects. In case the NGOs are involved in activities directly funded by donor agencies, the concerned state department needs to coordinate for better results as they are working for the community benefit.

4. Cordial Relations Between Government and NGOs

While associating NGOs in government activities specially for combating desertification and land degradation, active and perennial support of the concerned government department is essential for getting good results. Generally, projects suffer in view of indifferent behaviour of the government machinery as they treat their responsibility over while assigning the task to NGOs. The project suffered mostly for the reason for lack of government support or lack of cordial relations with the government functionaries.

For this purpose, while deciding the task to be assigned to NGOs, government department should set a line of action where NGOs must be free to operate their activities and the functional responsibility for coordination need to be assigned to nominated officers for transfer of technology, technical supervision in terms of quality and time over runs, NGOs should be given functional liberty at. the field level. Their filed level operational problems need to be sorted out effectively so that NGOs function smoothly.

Innovative ideas of NGOs for field application need to be attended minutely with its consequences and in case the approach is applicable, need to be appreciated. The guidelines of project activities should be kept flexible as per local conditions and decisions over such issues need to be taken quickly. Appreciation for better implementation through cost and time saving devices is necessary for encouragement and those lacking in performance need to be assisted to subsidize the deficiencies within time span with suitable corrective measures.

Bibliography

Abel NOJ and Blaikie, PM (1989); Land Degradation rates and conservation policies in the communal rangelands of Botswana and Zimbabwe. Land degradation and Rehabilitation. 1. 101-23.

Abrahams, A.D. Parsons, A.J. and Wainwright, J. (1995). Effects of vegetation change on inter-rill run-off and erosion. Walnut Gulch, Southern Arizona, Geomorphology 13, 37-48.

Abrol, IP and Benkateswarlu, .1 (1995), Sustainable development of arid areas in India with particular reference to Rajasthan. In AK Sen and Amal Kar (eds). Land degradation and desertification in Asia and the Pacific region. Jodhpur, Scientific Publishers, 135-53.

Abrol, IP, Yadav, JPS and Massoud, FI, (1988); Salt affected soils and their management, FAO Soils Bulletin, 39.

Abu-Zaid, M (1989); Environmental Impacts of Aswan High Dam. Water Resources Development 5, 147-57.

ADB (1994): The environment programme; Past, present and future. Manila; ADB. Adams, JM, Faure, H, Faure- Dencard.n L, McGlade. JM and Woodward, FI 1990: Increases in terrestrial carbon storage from the Last Glacial maximum to the present, Nature 348, 711 -44.

Adams M. (1996); When is eco-system change land degradation? Pastoral Development Network Paper 39e, ODI. London, 1-9.

Agnew, CT 1990 Spatial aspects of drought in the Sahel, Journal of Arid Environments 18, 279-93.

Agnew CT and Anderson E (1992), Water Resources in the arid realm, London, Rourledge.

Agnew CT and Warren A (1996): A framework for tackling drought and desertification, Journal of Arid Environment 33, 309-20.

Abaldejo J (1990); Impact of the degradation processes on soil quality in arid Mediterranean environment. In Rubio JL and Rickson RL (eds). Strategies to combat desertification in Mediterranean Europe. Commission of the European Communities, Brussels, Report EUR 11175 EN/Es, 193-215.

Allen JA (1984): Oases. In Coudsley- Thompson JL (ed), Sahara Desert. Oxford: Pergamon Press 32-33.

Alison FE (1973). Soul organic matter and its role in crop production. New York: Elsevier.

Anton Imeson (2012) Desertification Land degradation and Sustainability Published by John Wilcy & Sons.

Arar A. (1993): Water issues in the Palestine Occupied Territories. Middle East Water Forum, February 7-9, 1993, Cairo, 1-29.

Archaeomedes ND, Archaeomedes: A DG XII research programme to understand the natural and anthropogenic causes of land degradation and desertification in the Mediterranean Basin. University de Paris 1, Paris.

Arshad M, Rao AR and Akbar G (1995): Cholistan Desert in the state of flux. Desertification Control Bulletin 26, 55-58

Ash JE and Wasson RJ (1983): Vegetation and sand mobility in the Australian desert dune field. Zeitschrift for Ceomorphologie 45, 7-25.

Aubreville A. (1949): Climate, forests et desertification de. Pafrique tropicale. Paris: Societe d'Editions Geographiques maritimes et Coloniales.

Aweto AO and Adejumobi. DO (1991): Impact of grazing on soil in the southern Guinea savanna zone of Nigeria. Environmentalist 11, 27-32.

Babaev AG, Kharin NG and Orlovsky NS (1993): Assessment and mapping of desertification processes, a methodological guide. Desert Research Institute, Academy of Sciences of Turkmenistan, Ashgabad.

Ballin-Cortes J (1990): Desertification Comentarios y Posibilidades de Control, Universidad autonoma de San Luis Potosi.

Balling RC (1994b): Analysis of historical temperature and precipitation trends in the drylands of Mexico, the United States and Canada. Report to UN IPED/INCD, Nairobi.

Bari MA and Schofield NJ (1992): Lowering of a shallow, saline water table by extensive eucalypt reforestation. Journal of Hydrology 133, 273-91.

Barraclough S (1995): Social dimensions of desertification; a review of key issues. In Stiles D. (ed). Social aspects of sustainable dryland management, Chichester: Wiley, 21-79.

Barrow CJ. (1987): Water resources and agriculture in the tropics. Harlow: Longman.

Barrow CJ. (1991): Land degradation. Cambridge University Press.

Basnet, K. (1992) Conservation practices in Nepal: Past and present. Ambio 21. 390-93.

Bathhurst, JC, Kilsby, C and White S, (1996): Modelling the impacts of climate and land-use change on basin hydrology and soil erosion in Mediterranean Europe. In Brandt, CJ and Thornes JB (eds), Mediterranean desertification and land use. Chichester: Wiley, 355-88.

Batjes NJ (1996): Total carbon and nitrogen in the soils of the world. European Journal of soil Science 47, 151-63.

Bharara LP : (1999): Man in the desert: Drought, desertification and indigenous knowledge for sustainable development. Scientific publishers Jodhpur, India.

Behnke RH and Scoones 1. (1993): Rethinking range ecology: Implications for range management in Africa. In Behnke RH, Scoones, I. and Kerven, C. (eds), Range ecology at disequilibrium. London: Overseas Development Institute, 1-30.

Behnke RH. Scpones I and Karveen C. (eds) (1993): Range ecology in disequilibrium:New models of variability and pastoral adaptation in African Savannas. London: Overseas Development Institute.

Belnap J. (1995): Surface disturbances: Their role in accelerating desertification. Environmental Monitoring and Assessment 37, 39-57.

Bennet HH. (1938): Soil Conservation. New York: McGraw Hill.

Berg WA., Naney JW and Smith SJ. (1991): Salinity, nitrate and water equality in rangeland and terraced Wheatland above saline sweeps. Journal of Environmental Quality 20, 8-11.

Beschorner N. (1992): Water and instability in the Middle East. Adelphi Paper, International Institute for Strategic Studies, London. 273, 8-26.

Bhalla AS. (1992): Environmental degradation in rural areas. In Bhalla AS (ed), Environment, employment and development. Geneva: International Labour Office.

Bilsborrow RE. (1992); Rural poverty, migration and the environment in developing countries. World Bank Policy Research working Paper 1017, Washington DC.

Biswas MR. (1994): Agriculture and environment: a review, 1972-92. Ambio 23, 192-97.

Biswas A. and Tewatia RK. (1991): Nutrient balance in agro-climatic regions of India-an overview. Fertilizer News 36(6), 13-18.

Blaikie,P. (1985): The political economy and soil erosion. London: Longman. Blaikie, PM. And Brookfield, HC. 1987: Land degradation and society. London: Routledge.

Boumans J. and Mashali, A. (1983): Seepage from Lake Burullus into the reclaimed monsour and Zawia polder area. Agricultural Water Management 7, 411-24.

Bowonder, B. Prasad, SSR and Unni, NVM (1988): Dynamics of fuel wood prices in India. World Development 16, 1213-229.

Boyle, R. (1996): Life or death for the Salton Sea. Smithsonian 27, 86-96.

Breemen, H. Ketellaars, SG and Traore, N. (1990): Une remede contre le manque de terre? Secheresse 1, 109-17.

Brown, L. (1997): Facing the prospect of food scarcity. In brown, L. Flevin, C. and French, H. (eds), State of the world 1997. New York: Norton and Co, 23-41.

Brown, LR. And Wolf, EC. (1984): soil erosion: Quiet crisis in the world economy. World watch, Washington DC.

Brown RB. and Wolf EC, (1986): Restoring African soils. World watch Paper 65, Worldwatch, Washington Dc.

Bryan, RB and Harvey LE. (1985): Observation on the geomorphic significance of tunnel erosion in a semi-arid ephemeral drainage system. Geografiska Annaler 67A, 257-72.

Buckley, R. Wasson, R. and Gubb, a. (1987): Phosphorous and potassium atatus of arid dunefield soils in central Australia and southern Africa. Journal of Arid Environments 13,211-16.

Bullard, JE, Thomas, DSG, Livingtone, I. And Wiggs. GFS. (1997): Dune field activity and interactions with climatic variability in the south-west Kalahari Desert. Earth Surface Processes and Land forms 22.

Cardy, F. (1993). Desertification- A fresh approach. Desertification Control Bulletin 22, 4-8.

Carreira, JA and Neill, FX. (1995: Mobilization of nutrients by fire in a semi-arid gorse scrub land ecosystem of southern Spain. Arid Soil Research and Rehabilitation 9, 73-89.

CAZRI. (1991): Indo Soviet integrated long-term programme of cooperation in science and technology Discussion Paper 4-8-3-1991 on Ecology of arid zones and control of desertification, Jodhpur, India.

Charney JG. (1975): Dynamics of deserts and drought in the Sahel. Quarterly Journal of the Royal Meteorological Society 101, 193-202.

Choudhary, SA (1995): NGOs participation in he development of rain-fed areas of Pakistan. In Sen, AK and Kar, A. (eds_ Land Degradation and desertification in Asia and the Pacific Region. Jodhpur: Scientific Publishers, 353-54.

Chepil, WS and Woodruff, NP. (1963): The physics of wind erosion and its control. Advances of Agronomy 15, 211-302.

Child, G. and Grainger, J. (1990): A system plan for protected areas for wild life conservation and sustainable rural development in Saudi Arabia. Riyadh: national Commission of Wildlife Conservation and Development.

Chinese Academy of Sciences (1993): Wondrous Taklimakan. Beijing and New York: Science Press.

Choksi, A. and Dyer, C. (1996): Pastoralism in a changing world: Patterns of adaption among the Rabaris of Kutch, Gujarat, IIED Drylands Programme. Issue Paper 69, London.

Chopra, H. (1989): Land degradation: dimensions and casualties. Indian Journal of Agricultural Economics 44, 45-54.

Chouhan T.S. (1992): Desertification in the world and its control. Published by Scientific Publishers (India), Jodhpur.

Chouhan T.S. (1998): Global Status of Land Degradation and Desertification: A study of causative natural and anthropogenic factors. Printwell Publishers Distributors, Jaipur, India.

Chouhan T.S. (2018): Geo-informatics for combating Land Degradation and Desertification. Published by Scientific Publishers (India), Jodhpur.

Chouhan T.S. (2018): Rajasthan ka Bhugol (Geography of Rajasthan) (Hindi) Published by Scientific Publishers (India), Jodhpur.

Chouhan T.S. (2018): Remote Sensing and GIS GPS Based Resource Management. Published by Scientific Publishers (India), Jodhpur.

Chouhan T.S. (2018): Space Technology and GIS for Disaster Monitoring and Mitigation. Published by Scientific Publishers (India), Jodhpur.

Chu, PS, (1983): Diagnostic studies of rainfall anomalies in northeast Brazil. Monthly Weather Review 111, 1655-664.

Ciofolo, I. (1995): West Africa's last giraffes: The conflict between development and conservation. Journal of Tropical Ecology 1 1, 577-88.

Clark, WC. And Munn, Re (eds) (1986): sustainable development of the biosphere. Cambridge: Cambridge University Press.

Cook FEM (1995): Economic botany data collection standard. Kew: Royal Botanic Gardens.

Cooke, RU, Brunsden, D. Doornkemp, JC and Jones DKC (1982): Urban geomorphology in drylands. Oxford: Oxford University Press , UN University. Coppen, JJW, 1995:Gums, resins and latexes of plant origin. Rome: Food and Agriculture Organization of United Nations.

Corbett JD. (1996): The changing face of agro-eco-system characteristization. Paper presented at the Third International Conference on Integrating GIS and Environmental Monitoring, Santa Fe, New Mexico, NCGIA.

Coronato FR and Bertiller, MB (1996), Precipitation and landscape related effects on soil moisture in semi-arid rangelands of Patagonia. Journal of Arid Environments 34, 1-9.

Cox, J. and Dougill, A. (1995). Land degradation and grazing in the Kalahari: New analysis and alternative perspectives. Pastoral Development Network Paper 38c, ODI, London.

Cox J. and Dougill, a. (1996): Ecosystem dynamics and degradation in the Kalahari-a reply. Pastoral Development Network Paper 39e ODI London, 10-15.

Crawford, RMM. (1989). Studies in plant survival, ecological case histories of plant adaptation to adversity. Oxford: Blackwell Scientific.

Critchley, W., Reij. C. and Wilcocks, T. (1994): Indigenous soil and water conservation: A review of the state knowledge and prospects for building on traditions. Land Degradation and Rehabilitation 5, 293-314.

Cross, H. and Sandos, J. (1996): Across the border: Rural development in Mexico and recent migration to the United States. Berkeley: University of California, Institute of Government Studies.

Cunninghum, AB. And Milton, SJ. (1987): Effects of basket weaving on Mokola Palm and dye plants in North-western Botswana, Economic Botany 41, 386-402.

Davidson, EA. (1994): Climate change and soil microbial processes: Secondary effects are hypothesised from better known interacting primary effects. In Rounsevell, MDA and Loveland, PJ (eds). Soil response in climate change. Amsterdam: Springer-Verlag, 156-68.

Davis SD. Droop SJM, Gregerson, P. *et. al.* (1986): Plants in danger. What to de know" Gland and Cambridge: IUCN.

Davis, SD, Ambrose, JR, Cook, FEM and Prendergast, HDV, (1997): The survey of Economic Plants for Arid and Semi-arid Lands. What's in it for you? In Etkin, N, Harris, DR, Hather, JG, Houghton, PJ and Prendergast HDV. Plants for food and medicine. Kew: Royal Botanic Gardens.

Dean, WRJ, Hoffman, MT, Meadows, Milton and SJ, ME. (1995): Desertification in the semi-arid Karoo, south Africa: Review and reassessment. Journal of Arid Environments 30, 247-64.

Dean, WRJ, Hoffman, MT and Willis, CK, (1996): The light and the way in south African Desertification Research. South African Journal of Science 92, 170-71.

Department of Science and Technology, Government of India, New Delhi (1998): Science and Technology for Women II- Some Experiences in Green Farming Practices and Cover Management.

Dhir, RP. (1995): Problems of desertification in the arid zone of Rajasthan: A view. Desertification Control Bulletin 27, 45-52.

Dhir, RP., D. C. Joshi and S. Kathja (2018) "Thar desert in reteospect and Prospect" Published by Scientific Publications, Jodhpur.

Dhruva Narayana VV (1993): Soil and Water Conservation Research in India. New Delhi, India. Indian Council of Agricultural Research 454.

Doornkamp, JC (1993): Clay shrinkage induced subsidence. Geographical Journal 159, 196-202.

Dougill, AJ. Heathwaite, AL and Thomas DSG (1997). Soil water movement and nutrient cycling in semi-arid rangeland: Vegetation change and system resilience. Hydrological Processes.

Douglas, I. (1967): Man, vegetation and sediment yield of rivers. Nature 215, 925- 28.

Dowdeswell, E. 1995): The way ahead for UNEP. Desertification Control Bulletin 27.3

Dregne HE, (1976): Soils of arid regions. Amsterdam: Elsewiev.

Dregne, HE (1992). Erosion and soil productivity in Asia. Journal of Soul and Water Conservation 47, 8-13.

Dregne HE (1995): Desertification control: a framework for action. Environmental Monitoring and Assessment 37, 111 .22.

Dregne, HE. And Tuckler, CJ. (1988): Desert encroachment. Desertification Control Bulletin 16, 16-19.

Dregne He. Zhixun Xiong and Siyu Xiong (1996): Soil salinity in China. Desertification Control Bulletin 28,28-33.

Dresch J. (1986). Degradation of natural ecosystems in the countries of the Meghreb as a result of human impact. In Glantz MH. (ed). Arid land development and combat against desertification. Moscow: UNICEF, 65-7.

Dunkerley, DL and Brown, KJ. (1997). Desert Soils. In Thomas DSG (ed), Arid zone geomorphology: process, form and change in drylands. 2nd edition. Chichester: Wiley, 55-68.

Dunne T., Deitrich, WE and Brunengo, MJ, (1978): Recent and past erosion rates in semi-arid Kenya. Zeitschrift fur Geomorphologie Supplement band 29, 130-40.

Easterling, W. Rosenberg, N. Mckenney, M. and Jones C. (1992). An introduction to the methodology, the region of study and a historical analog of climate change. Agricultural and Forestry Meteorology 59, 3-15.

El-Baz, F. (1994): Gulf war disruption of the desert surface of Kuwait. In El-Baz, F. and Makharita RM (eds). The gulf War and the environment. Lausanne: Gordon and Breach, 131-61.

Eldridge, DJ and Greene RSB. (1994): Assessment of sediment yield by splash erosion on a semi-arid soil with varying cryptogam cover. Journal of Arid Environments 26, 221-32.

Ellis JE and Swift DM. (1988): Stability of African pastoral ecosystem: Alternate paradigms and implications for development. Journal of Range Management 4`, 450-59.

Ellis JE, Coughenour, MB. And Swift, DM (1991): Climatic variability, ecosystem stability and the implications for range and livestock development. In Cincotta, RP, Gay, CW and Perrier, GK (eds). New concepts in international rangeland development: Theories and applications. Proceedings of the 1991 international Rangeland Development Symposium, Society of Range Management. Logan: Utah State University.

Ervin RT and Lee JA (1994): Impact of conservation practices on airborne dust in the southern High Plains of Texas. Journal of Soil and Water Conservation 49,430-38.

Evernari, M. (1985). The desert environment. In Evernari, M., Noy-Meir, I. And Goodall, DW (eds), Ecosystems of the World 12A. New York: Elsevier, 1 -22.

Eyre SR. (1968): Vegetation and soils: A world picture Chicago: Aldine.

FAO (1979): A provisional methodology for soil degradation assessment. Rome: FAO.

FAO (1993a). World soil resources. World Soil Resources Report 66 Rev. 1. Rome FAO.

FAO (1994): Land degradation in south Asia: Its severity, causes and effects upon he people. World Soul Report, 78, FAO, Rome.

FAO (1995): Forest resource assessment 1990: Global synthesis. FAO Forestry Paper No. 124 Rome, FAO.

FAO/UNEP (1984): Provisional methodology for the evaluation of soil degradation. Rome: FAO.

FAO/ UNEP (1983). Provisional methodology for assessment and mapping of desertification. Rome: FAO.

FAO (2000) Asia Regional Workshop on "Approaches to Resource Mobilization for UNCCD" by the Food and Agricultural Organization. Workshop supported by the global mechanism. International Fund for Agricultural Development (IFAD) and the UNCCD.

Farres, P. (1978). The role of time and aggregate size in the crusting process. Earth Surface Processes 3, 243-54,

Fearnside, PM. (1979): O processo de desertificcacao e or riscos de sua oceuriencia no Brasil (Nordeses). Acta Amazonica Brazil 9, 393-400.

Feng,L. Lu, J. and Di, Y. (1994): Review on the prevention of sand damage to railway lines in desert areas of China. Journal of Desert Research 14, 47-53.

Flour, D. (1991). Energy and economic evaluation of C02 removal from fossil-fuel power plants. Report IE 7365. Palo Alto, CA: Electric Power Research Institute.

Forest Survey of India, Dehra Dun, ministry of Environment and Forests, Government of India, New Delhi, 2000. The State of Forest Report -113.

Francis, CF and Thornes, JB. (1990): run off hydrographs from three Mediterranean vegetation cover types. In Thornes, JB. (ed), Vegetation and geomorphology. Chichester: Wiley, 363-85.

Franco, AC and Nobel PS. (1989): Effect of nurse plants on the micro-habit and growth of cacti. Journal of Ecology 77, 870-86.

Fryrear, DW, (1981): Long-term effect of erosion and cropping on soil productivity. In Pewe, TL. (ed). Desert dust, origins, characteristics and effects on man. Geological society of America Special Paper 186, 253-59.

Fryrear, DW, Stubbendieck, J. and McCully. WG, (1973): Grass seedling response to wind and wind blown sand. Crop Science 113, 622-25.

Fullen, MA. And Mitchell, DJ. (1994): Desertification and reclamation in north-central China. Ambio 23, 131-35.

Gairon, S. and Hadas, a. (1973). Measurement of water status of soil and sooil characteristics relevant to irrigation. In Yaron, B., Danfora, E. ad Vaadia, Y. (eds). Arid zone irrigation. London: Chapman and Hall, 215-40.

Gardner, G. (1997): Preserving global cropland. In Brown L., Flavin, C. and French, H. (eds), State of the world 1997. New York: Norton and Co., 42-59.

George, RJ. And Conacher, AJ. (1993): Mechanisms responsible for stream flow generation on a salt affected and deeply weathered hill slope. Earth Surface Processes and Landforms 18, 291-309.

Ghassemi, F. Jakeman, AJ. And Nix, HA. (1995): Salinization of land and water resources: human causes, extent, management and case studies. Sydney: University of New South Wales Press.

Ghazanfer, SA. (1992): An annotated catalogue of the vascular plants of Oman, Meise: National Botanic Garden of Belgium.

Gibbens, R.Tromble, J. Hennessy, J. and Cardenas, M. (1983): Soil movement in mesquite dune-lands and former grasslands of New Mexico from 1933-1980. Journal of Range Management 36, 145-48.

Gifford, RM, Cheney, NP., Noble, JC, Russell, RS, Wellington, AB and Zaammir, C. (1990): Australian land use, primary production of vegetation and carbon pools in relations to atmospheric carbon dioxide concentration. In Gifford RM. And Barson, M. (eds), Australlia's renewable resources: Sustainability and global change. IGBP symposium. Canberra: Bur. Of Rural Resources, 151-88.

Gillette, DA, Adams, J, Endo, A. Smith, D. and Kihl, R. (1980): Threshold velocities for input of soil particles into the air by desert soils. Journal of Geophysical Research, 85C. 5621-630.

Gisser, M. and Pohoryles, S. (1977: Water shortage in Israel: Long-run policy for the farm sector. Water Resources Research 13(6): 865-72.

Glantz, MH. (1977): Water and inappropriate technology: Deep wells in the Sahel. In VP Nanda (ed). Water needs for the future. Westview Press, Boulder: 305-18.

Glantz MH(ed) (1994): Drought follows the plow. Cambridge: Cambridge University Press.

Glantz MH, Katz, RW and Nicholls, N. (eds), (1991): Telecommunications linking worldwide climatic anomalies. Cambridge: Cambridge University Press.

Glantz and Orlovsky, N. (1983): Desertification : A review of the concept. Desertification Conrol bulletin 9, 15-22.

Glantz, MH, Rubinstein, AZ and Zonn, I. (1993): Tragedy in the Aral Sea basin: Looking back to plan ahead. Global Environmental Change 3; 174-98.

Glozovslcy, NF. (1995): The Aral Seas basin. In Kesperson, JX, Kasperson, Re and Turner II, BL (eds), Regions at risk: Comparisons of threatened environments. Tokyo: United Nations University Press, 92-139.

Glenn, E. (1995): Halophytes. In Encyclopedia of environmental biology, Volume 3, New York: Academic Press, 403-18.

Glenn, EP, Brown, JJ and O'Leary, J. (1997): Seawater agriculture. Scientific American.

Glenn, E. Miyamoto, S., Moors, d. Brown, JJ, Thompson, TL, and Brown, P. (1997): Water requirements for cultivating Salicornia Bigelovii Torr. With seawater on sand in a coastal desert environment. Journal of Arid Environments 36, 711-30.

Glenn, EP. O' Leary, JW, Watson, MC, Thompson, TL and Kuehl, RO. (1991). Salicornia bigelovii Toll: An oilseed halophyte for seawater irrigation. Science. 251, 1065-67.

Glenn EP, Squires, VR, Olsen, M. and Frye, R. (1993): Potential for carbon sequestration in the drylands. Water, Air and Soil Pollution 70, 341-55.

Goldstein, B. (1985). Spirulina- an new high-value, saline water crop for energy, food and feed production in arid lands. In Whitehead, EE, Hutchinson, EF, Timmermann, BN and Varady, RG (eds) Arid lands. Today and tomorrow. Boulder: Westview Press and London: Belhaven Press, 755-63.

Gonzalez Loyarte, MM. (1996): Climate fluctuations as a source of desertification in a semi-arid region of Argentina. Desertification Control Bulletin 28,8-14.

Gonzalez- Villareal and Garduno, H. (1994): Water resources planning and management in Mexico. International Journal of Water Resources 10, 239-55.

Govil DP and Daima ML (2000): National action plan for combating the process of desertification: Experience and perspective plan of action, Paper prepared for the Working Group No. 4 of NAP, New Delhi. Ministry of Environment and Forests Government of India, 12.

Graigner, A. (1990): The threatening desert. London: Earthscan Publishers London in association with UNEP, Nairobi..

Great Plains Committee (1937): The Future of the Great Plains. 75the Congress, 1st Session, HR doc. 144. Washington DC: US Senate.

Grieve, AM. (1987): Salinity and water-logging in the Murray-Darling basin. Search 18, 72-74.

Grossman, R. Ahrens, R, Gile, L., Montoya, C. and Chadwick, A. (1995): Arial evaluation of organic and carbonic and carbonate carbon in a desert area of southern New Mexico. In Lai, R., Kimble, JM, Levine, e. and Stewart, BA (eds), Soils and Global Change. Boca Raton: CRC Press, 81-91.

Grove, AT. (1977): The geography of semi-arid lands. Philosophical Transactions of the Royal Society of London, B. 278, 457-75.

Grove, AT, (1996): The historical context: Before 1850. In Brandt CJ. And Thornes, JB (eds), Mediterranean Desertification and land use. Chichester: Wiley, 13-28.

Grunblatt, J. (1990): Kenya Pilot Study for Desertification Assessment and Mapping systems. Analyst Report. UNEP DEDC-PAC.

Grunblatt, J., Ottichilo, WK, Sinange, RK, Mwendwa, HA and Kinuthia, JH. (1992): Kenya: Pilot study for desertification Assessment and Mapping. Baringo. In UNEP World Atlas of Desertification. London: Edward Arnold, 54- 58.

Gruchkke, A. (1991): The opening of new land in the dry areas of the People's Republic of China. Consolidation and new orientation since the beginning of the 1980. Geographische Rundschau 43, 672-80.

Gueye, MB. (1995): The active method of participatory research and planning (MARP) as a natural resource management tool. In Stiles, D. (ed). Social Aspects of dryland management. Chichester: Wiley, 83-92.

Gupta JP, Kar A and Faroda As (1997): Desertification : Problems and possible remedies Yojna August 1997 , 55-50.

Guo, H. Wu, D. and Zhu, H. (1989): Land restoration in China. Journal of applied Ecology 26, 787-92.

Gupta RK and Saxena SK. (1971): Ecological studies on the overgrazed rangelands in the arid zone of west Rajasthan. Journal of he Indian botanical society 50, 289-300.

Guyot, L. (1990). Development of a new method for assessment and monitoring of desertification in Sahelian and Sudanian regions (1957-87): Presentation of the results obtained on the first transect: Mauritania- Mali. In Desertification revisited: Proceedings of an ad-hoc meeting on the assessment of desertification. Nairobi: UNEP DC/ PAC, 295-97.

Hagmann, J. and Murwira, K. (1996): Indigenous soil and water conservation in southern Zimbabwe: A study on techniques, historical changes and recent development under participatory research and extension. IIED Dryland Programme Issue Paper 63.

Hakkeling RTA (1989): Global assessment of soil degradation- eastern and southern Africa. Main Report (2 volumes). Washington STIBOKA.

Hanan, NP and Prevost, Y. Diouf, A. and Diallo, O. (1991): Assessment of desertification around deep wells in the Sahel using satellite imagery: Journal of applied Ecology 28, 173-86

Hardin, G. (1968): The tragedy of the commons. Science 162, 1243- 248.

Hare, FK, (1987): Drought and desiccation: Twin hazards in a variable climate. In Wilhite, D. and Easterby, W. (eds), Planning for drought. London: Westview Press, 3-10.

HarteAJ, (1984): Effects of tillage on the stability of three red soils of the northern wheat belt. Journal of soil conservation New south Wales 40, 94-101.

Harzallah, A., Aragao, JOR and Sadourny, R. (1996). International rainfall variability in north=west Brazil. Observation and model simulation. International Journal of Climatology 16, 861-78.

Heathcote, RL. (1983). The arid lands; lands and their use and abuse. London and New York: Longman.

Hecht, A. (1983): Drought in the Great Plains: History of societal response. Journal of Climate and Applied Meteorology 22, 51-56.

Hellden, U. (1991): Desertification- time for an assessment? Ambio 20-372- 83.

Helms, D. and Woodman, H. (1990): Conserving the plains: The soil conservation service in the Great Plains. Agricultural History 64, 58-73.

Hidore, JJ and Oliver, JE. (1993): Climatology: an atmospheric science. New York: Macmillan.

Hills ES (ed). (1964: Arid lands: A geographical appraisal. Methuen and Co. London, Paris: UNESCO.

Hinchcliffe, F. (1995): The hidden harvest: Wild resources and agricultural systems, Summary of a review and planning workshop, June 7-8 1995.: International Institute for Environment and Development, London.

Holburt MB. (1982): international problems on the Colorado River. Water supply and management 6, 105-14.Hong TD, Linington, S. and Ellis, RH. 1996: Seed storage behaviour: A compendium handbook for gene banks. No. 4: International Plant Genetic Resources Institute, Rome.

Hooke, Janet, Sandercock and Peter (2017) "Combating Desertification and Land Degradation Strategies using vegeetation, Springer.

Houghton, RA (1995): Changes in the storage of terrestrial carbon since 1850. In Lal, R., Kimble, J. Levine, E. and Stewart, Ba (eds), soils and Global Change. Boca Raton: Lewis Publishers, 45-65.

Hseung, Y., Zhu. S. and Wang, Z. (1981): Ecological regionalization of the Huang-Huai-Hai Plain in china. Nanjing, Institute of soil Science, Academica Sinica, Soil research Report No. 4.

Hulme, M. (1992). Rainfall changes in Africa: 1931-60 to 1961-90. International Journal of Climatology 12, 685-99.

Hulme, M. (1996 a): Recent change in the world's drylands. Geophysical research Letters 23, 61-64.

Hulme, M. (1996 b): Global warming. Progress in Physical Geography 20, 216-23.

Hulme M and Kelly, M. (1993): Exploring the links between desertification and climate change. Environment 35, 4-45.

Hulme, M. and Marsh, R. (1990): Global mean monthly humidity surface for 1930-59, 1960-89 and projected for 2030. Report to UNEP / GEMS/ GRID. Climatic Research Unit, University of East Angila, Norwich.

Hulme, M. Marsh, R. and Jones, PD (1992): Global changes in the humidity index between 1931-60 and 1961-90. Climate Research 2, 1-22.

Humphrey, C. and Sneath, d. (1996): Pastoralism and institutional change in Inner Asia: comparative perspectives from the MECCIA research project. Pastoral Network Paper 39-b, London: ODI.

Hurt, R. (1981): The Dust Bowl and Agricultural and social History, Chicago: Nelson-Hall.

ICAR (Indian Council of Agricultural Research) (2000): International Conference on managing natural resources for sustainable agricultural production in the 21st century. Yadan JSP and Singh GB (eds), ICAT, New Delhi.

Idso, KE and Idso SB. (1994): Plant responses to atmospheric C02 enrichment in the face of environmental constraints: A review of the past 10 years' research. Geoderma 69, 159-203.

Iglesias, AN. (1992): Erosion eolica, desertification Y crisis de rentabilidad de la economica garadera en Patogonia. El caso de la provincia de Santa Croz. Estudios Geograficos 53, 447-79.

IGN (1992): Mali transet methodology to assess ecosystem change. In UNEP, World atlas of desertification. London. Edward Arnold, 62-65.

IIED (International Institute for Environment and Development) (1995): The hidden harvest : The value of wild resources in agricultural systems. Compiled by Gujit, I., Hinchcliffe, f. and Melnyk, M.: IIED, London.

Ilaiwi, M., Abdelgawad, G. and Jabour, E. (1992): Syria: Human induced soil degradation. In UNEP, World atlas of desertification. London: Edward Arnold, 42-45.

Imeson, AC. (1995): The physical, chemical and biological degradation of the soil. In Fantechi, R. Peter, D, Balabanis, P. and Rubio, JL (eds), socio- economic aspects. Proceedings of the European School of Climatology and Natural hazards course, Alicante, Spain, 6-13 October 1993. Brussels: European Commission, 153-68.

Imeson, AC, Perez- Trejo, F. and Commeraat, LH. (1996): The response of landscape units to desertification. In Brandt, CJ and Thornes JB (eds), Mediterranean desertification and land use. Chichester: Wiley, 447-69.

Inegi (Instituto Nacional Estadistica Geografia Y Informatica, Maxico) (1991): X Censo General de Poblacion Y Vivienda (1990). Mexico DF: INEGI.

Ingram, J. (1991): Soil and water processes. In Scoones, I. (ed), Wetlands in drylands: the agro-ecology of savanna systems in Africa. London: IIED.

Inter-governmental Authority on Drought and Development (IGADD) (1990): Forum on environmental protection and development of sub-regional strategy to combat desertification. IGADD/ NORAGRIC.

IPCC (International Panel on Climatic Change) (1990): Climate change: The IPCC scientific assessment. Cambridge: Cambridge University Press.

Iraq, Government of (1980). Desertification in the Greater Mussayeb Project, Iraq. In UNESCO/ UNEP/ UNDP, Case studies on desertification. Natural Resources Research Series XVIII, Paris: UNESCO, 176-213.

Jackson LL, McAuliffe, JR and Roundy, BA. (1991): Desert restoration, Restoration and Management Notes 9, 71-79.

Jacobson, T. and Adams, RM (1958): Salt and silt in ancient Mesopotamian agriculture. Science 128, 1251-258.

Jagannathan, NV. (1989): Poverty, public policies and the environment. World Bank Environment Department Working Paper 1990-8. World Bank: Washington DC.

Jain JK (ed) (1986): Combating Desertification in developing countries. Scientific Review on Arid Zone Research Vol 4. Scientific Publishers, Jodhpur, India.

Jessop, JP (ed-in-chief), (1985). Flora of Central Australia, French Forest. Reed Books.

Jodha, NS (1992): Common Property Resources. World Bank Discussion Paper, August 1992. Washington DC.

Johnson DL. (1993): Nomadism and desertification in Africa and the Middle East. Geo-Journal 31, 51-66.

Justice, CE, Townshend, JRG, Hollen, BN and Tucker CJ. (1985): analysis of the phenology of global vegetation using meteorological satellite data. International Journal of Remote Sensing 6, 1271-318.

Kadomura, H. (1994): Climatic changes, droughts, desertification and land degradation in the Sudano- Sahelian region. In Kadomura, H. (ed), Savannization processes in tropical Africa II. Toyko: Toyko Metropolitan University, 203-28.

Kar A. (1996): Land surface processes in the evaluation of potential land degradation. In Kolarkar, AS, Joshi, DC, and Kar, A. (eds), land resources and their management for sustainability in arid regions. Jodhpur: Scientific Publishers, 29-47.

Kar, A. and Joshi, DC. (1992): Prediction of land degradation hazards and risk assessment: the experience from Indian arid zone. Proceedings, World Congress on Natural Hazard Reduction 2, New Delhi: 11-23.

Kassas, M. (1987): Drought and desertification. Land use Policy 4, 389-400.

Kassas, M. (1995): A review of desertification. Journal of Arid Environments 30, 115-28.

Kassas, M., Ahmad, YJ. And Rozonov, B. (1991): Desertification and drought: An ecological and economic analysis. Desertification Control Bulletin 20, 19-29.

Katyal JC and Reddy KCK (1997). Plant nutrient supply needs: Rain fed food crops. In plant nutrient needs, supply, efficiency and policy issues 2000- 2025. In Kanwar JS and Katyal JC (eds), NAAS New Delhi 91-113.

Kaushalya, R. (1992): Monitoring the impact of desertification in western Rajasthan using remote sensing. Journal of Arid Environments 22, 293-304.

Kay BK, Graves, WL. And young, JA. (1988): Long-term storage of desert shrub seed. Mohave Revegetation Notes No. 23. Davis, University of California.

Kellogg, CE, (1956): The role of science in Man's struggle on arid lands. In White, GF. (ed), the future of arid lands. Washington DC: American Association for the Advancement of Science, 26-46.

Kemp, DD. (1994): global environmental issues: A climatological approach. 2nd edition. London: Routledge.

Kerley, GIH, Knight, MH. And De-Kock, M. (1995): Desertification of sub- tropical thicket in the eastern Cape. South Africa: Are there alternatives? Environmental Monitoring and Assessment 37, 211-30.

Khan, AU. (1994): History of decline and present status of natural tropical thorn forest in Punjab. Biological Conservation 67, 205-10.

Kharin, NG. and Tsolmon, P. (1994): Effect of social and economic factors on desertification in Mongolia. Problemy Osvoeniya Pustyn 6, 17-23 (In Russian).

Khagoli, MM. (1991); Desertification, famine and the 1988 rainfall- the case of Umm Ruwaba District in the northern Kordofan region. Geo- Journal 25, 81-89.

Kinsman, JD and Trexler MC. (1991): Terrestrial carbon management and electric utilities. Water, Air and Soil Pollution 70, 545-60.

Kirkby, MJ. (1980): the problem. In Kirkby, MJ and Morgan RPC (eds), soil erosion. Chichester: Wiley, 1-16.

Kirkby MJ, Arkinson, K. and Lockwood, J. (1990): The inter-action of erosional and vegetational dynamics in land degradation: Spatial outcomes. In Thrones, JB. (ed). Vegetation and geomorphology. Chichester: Wiley, 25-40.

Kishk, MA (1986): Land degradation in the Nile Valley. Ambio 15. 226-30.

Kiss A. (ed), (1990): Living with wildlife: Wildlife resource management with local participation in Africa. World Bank Technical Paper. 130 Africa Technical Department Series. World Bank. Washington DC.

Klein, JC and Lacoste, A. (1994): The subalpine meadows of central Elbruz (Iran): South-eastern most distribution limit of Festuco-Brometea Br.-Bl and Tx. 1943. Phytocoenologia 24, 401-21.

Konstant, TL, Sullivan, S. and Cunningham, AB. (1995). The effects of utilization by people and livestock on Hyphaene petersians (Arecaceae) basketry resources in the palm savanna of North-Central Namibia, Economic Botany 49, 345-56.

Kovda, V. (1983): Loss of productive land due to salinization, Ambio 12, 91-93.

KREMU, (1983): Land use map of Kenya. Kenya Rangeland Ecological Monitoring Unit, Ministry of Environment and Natural Resources, Nairobi.

Kumar, M. and Bhandari. MM. (1993 a): Human use of the sand dune eco-system in the semi-arid zone of the Rajasthan Desert, India. Land Degradation and Rehabilitation 4, 21-36.

Kumar, M and Bhandari. MM. (1993 b): Impact of human activities on the pattern and process of sand dune vegetation in the Rajasthan Desert. Desertification Control Bulletin 22, 45-54.

Kumar, M, Goossens, E. and Goossens, R. (1993): Assessment of sand dune change detection in Rajasthan Thar Desert, India. International Journal of Remote Sensing 14, 1689-703.

Lal, R. (1984): Soil erosion from tropical lands and its control. Advances in Agronomy 37, 183-248.

Lal RA, (1995): Global soil erosion by water and carbon dynamics. In Lal, R., Kimble, JM, Levine, E. and Stewart, BA (eds), soils and global change. Boca Raton: Lewis Publishers, 13 1-42.

Lal, R. (1990): Water erosion and conservation: An assessment of the water erosion problem and the techniques available for soil conservation. In Goudie, AS. (ed) Techniques for desert reclamation. Chichester Wiley, 161-98.

Lal R and Cummings, DJ. (1979): Clearing a tropical forest I. Effects on soil and micro-climate. Field Corps Research 2, 91-107.

Lal, R and Stewart, BA (1990). Need for action: Research and development priorities. Advances in Soil Science 11, 331-36.

Lamb PJ and Peppier RA. (1991): West Africa. In Glantz MH, Katz, RW and Nicholls, N (eds), Teleconnections linking worldwide climate anomalies. Cambridge: Cambridge University Press, 121-91.

Lambin, EF, (1993): Spatial scales and desertification. Desertification Control Bulletin 23, 20-23.

Lamprey, HF (1975): Report on the desert encroachment reconnaissance in northern Sudan, October 21- November 10, 1975. National Council for Research, Ministry of Agriculture. Food and Resources, Khartoum.

Land Management Society (1991): Tree and water combinations in five countries: A farmer's report. Perth: Land Management Society.

Langbein, WB and Schunnn, Sa. (1958). Yield of sediment in relation to mean annual precipitation. Transactions of the American Geophysical Union 39, 1076-84.

Lavee H, Poesen J and Uair, A. (1997): Evidence of high efficiency water harvesting by ancient farming in the Negev Desert, Israel. Journal of Arid Environments 35, 341-48.

Le Houerou, HN (1958): Ecologie, phyto-sociologie et productivitie de Poliver en Tunisie meriodionale, Bulletin Cartographie et Phytogeographe B CHRS, Paris IV, 7-22.

Le Houerou, HN (1979): La desertisation des regions arides. La Recherche 99, 336-44

Le Houerou, FIN, (1990): The grazing land ecosystems of the African Sahel. Ecological Studies 75, Berlin: Springer-Verlag.

Le Flouerou, FIN, (1993): Changements climatiques et desertisation. Impact of Science on Societies 116, 183-201.

Le Houerou, HN (1996): Climate change, drought and desertification. Journal of Arid Environments 34. 133-85.

Le Prune, JC (1981): A eroso a conservacao e o manejo do solo no nordeste Brasileiro: Balanca, diagnostico e novas linhas de pasquisas. SUDENE, Recife, 15.

Leonard, HJ. (1989): Environment and the poor: Development strategies for a common agenda.: Overseas Development Council, Washington DC.

Lewis, M. (1989): National grasslands in the Dust Bowl. Geographical Review 79, 161-71.

Lieth H and Masson A. (eds) (11993): Towards the rational use of high salinity tolerant plants, Volumes 1 and 2. Dorbrecht: Kluwer.

Lin Y, Chen CK and Chen CT (1983): A discussion of the direction of productive development and way of halting desertification in the Mu Us sany land, Ordos Plateay- based on its physical conditions. Acta Geographica sinica 38, 230-40.

Liniger HP and Thomas DB (1996): Grass- Ground cover for Restoration of Arid and Semi-arid Soils: Paper for the 9th International Soil Conservation Organization (ISCO) Conference in Bonn, Germany, 26-30.8.1996.

Liniger HP, Hurni H and Thomas DB (1996): WOCAT- World Overview of Conservation Approaches and Technologies- Results from Africa. Paper for the 9th International Soil Conservation Organization (ISCO) Conference in Bonn, Germany, 26-30.8.1996.

Lockertz W (1978): Lessons of the Dust Bowl. American Scientist 66, 560-69.

Lockwood JG (1984): The southern oscillation and El Nino. Progress in Physical Geography 8, 102-10.

Loneragan, OW (1990): Historical review of Sandalwood (Santalum spicatum) research in Western Australia. Research Bulletin No. 4.: Department of Conservation and Land Management, Western Australia.

Lopez Bermudez F and Albaladejo J (1990): Factores ambientales de la degradacion del suelo en al area Meditteranea. In Albaladejo, J. Stocking M and Diaz E (eds), soil Degradation and rehabilitation in Mediterranean environmental conditions: Murcia: CEBAS-CSIC, 15-42.

Maas EV (1986): Salt tolerance of plants. Applied Agricultural Research 1, 12-26.

Magalhaes A and Magee P (1994): The Brazilian Nordeste (Northeast). In Glantz MH (ed), Drought follows the plow. Cambridge: Cambridge University Press 59-76.

Mainguet M (1985): le Sahel, un laboratoire naturel pour letude du vent, mecanisme principal de la desertification. Proceedings of an International Workshop on the physics of wind blown sand, Arhuus: University of Arhuus, 545-61.

Mainguet M (1991): Desertification. Natural background and human management. Berlin. Springer-Verlag.

Maley J (1989): Chronologie des proncipales fluctuations du lac Tchad an cours du dernier millenaire. In Barretan D. and Von Graffenried C (eds), Dataion et chronologie dans le Bassin duLac Tehad. Compates Rendus du Seminaire Mega-Tchad, Paris: Ostron. 161-63.

Margaris NS, Koutsidou E and Giourga C (1996): Changes in traditional Mediterranean land-use systems. In Brandt, CJ and Thomas JB (eds) Mediterranean desertification and land use. Chichester, Wiley 29-42.

Margaris M and Loyarte G (1996): Climatic fluctuations as a source of desertification in a semi-arid region of Argentina. Desertification Control Bulletin 28, 8-14.

Mark S. Reed and Lindsay c. stringer (2016) "Land degradation, Desertification and climatic change: Anticipating, assessing and adapting to future change "published by Routledge.

Marshall JK (1977): The below-ground ecosystem: A synthesis of plant- associated process IBP. Fort Collins, CO: Colorado State University.

Martin PL (1990): The outlook for agricultural labor in he 1990's. UC Davis Law Review 23, 499.

Martin PL (1993): Trade and migration: NAFTA and agriculture.: Institute for International Economics, Washington DC.

Martin PL (1996: Emigration dynamics in Mexico and Central America: The case of agriculture. Paper prepared for the International Organization for Migration, April.

Mather JR and Ambroziak RA (1986): A search for understanding potential evapotranspiration. Geographical Review 76, 355-470.

Matheson W and Ringrose S (1994): Assessment of degradation features and their development into the post-drought period in the west-central Sahel using Landsat MSS. Journal of Arid Environments 26, 181-99.

McCabe JT (1990): Turkana pastoralism: A case against the tragedy of the commons. Human Ecology 18, 81-103.

McCauley JF, Breed CS, Grolier MJ and Machinnon DJ (1981): The US oust storm of February 1977. In Pewe TL (ed) Desert Dust: origins, characteristics and effects on man. Geological Society of America Special Paper, 186, 123-47.

Meigs P (1953): World distribution and arid and semi-arid homoclimates. Arid zone hydrology, UNESCO Arid Zone research Series 1, 203-09.

Meteorological Office (1980): Tables of temperature, relative humidity, precipitation and sunshine for the world, Part 1. North America and Greenland (including Hawaii and Bermuda). London: HMSO.

Millington AC, Critchley, RW, Douglas TD and Ryan P, (1994): Estimating woody biomass in Sub-Saharan Africa. World Bank, Washington DC.

Millington AC, Mutiso SK, Kirby J and O'Keefe P (1989). African soil erosion- nature undone and the limits of technology. Land degradation and rehabilitation 1, 279-90.

Ministry of Environment and Forests, Government of India, (2001): India: national Action Programme to combat desertification: in the context of UNCCD Volume I: Status of Desertification.

Ministry of Rural Development, Government of India, (2000). Annual Report for 1999-2000, Ministry of Rural Development, 211, New Delhi.

Mistry JF (1989): Sainity ingress in the coastal area of Saurashtra- Gujarat State (India). ICID Bulletin 32, 52-60.

Mitchell DJ and Fullen MA (1994): Soil-forming processes on reclaimed desertified land in north-central China. In Pye, K 9ed), Environmental change in drylands: Bio-geographical and geomorphological perspectives. Chichester: Wiley, 393-412.

Molina JS (1993): Soil erosion and conservation in Argentina. In Pimental D (ed), World soil erosion and conservation. Cambridge: Cambridge University Press, 171-92.

Morgan, RPC (1995): Soil erosion and its conservation, 2nd edition. Harlow: Longman.

Mortin J, Benyamini Y and Michaeli A. (1981): The effect of raindrop impact on the dynamics of soil surface crusting and water movement in profile. Journal on Hydrology 52, 321-26.

Mortimore M (1993): Population growth and land degradation. Geo-Journal 31.15-21.

Moss AJ, Green P and Hurka J (1982): Small channels: Their formation, nature and significance. Earth Surface Processes and landforms 7, 401-15.

National Research Council (1990): Saline agriculture. Washington Dc: national Academy Press.

Neilson RP and Marks D (1994): A global perspective of regional vegetation and hydrologic sensitivities from climatic change. Journal of Vegetation Science 5, 715-30.

Nepstad DC, Carvalho E, Jipp P *et. al.* (1994): The role of deep roots in hydrological and carbon cycles of Amazonian forests and pastures. Nature 372, 666-69

Newcombe, K (1989): An economic justification for rural Afforestation: The case of Ethiopia. In Schramm Gunter and Warford (9eds) Environmental management and economic development. Bltimore: Johns Hopkins University Press.

Nicholls N and Krtz RW (1991): Teleconnections and their impacts for long range forecasts. In Glantz MS, Katz RW and Nicholls N (eds) Teleconnections linking worldwide climatic anomalies. Cambridge: Cambridge University Press, 511-25.

Nicholson SE (1978): Climatic variations in the Sahel and other African regions during the past five centuries. Journal of Arid Environments 1, 3-24.

Nicholson SE (1989): Long-term changes in African rainfall. Weather 44, 47-56.

Nickling WG and Wolfe Sa (1994): The morphology and origin of nabkhas, region of Mopti, Mali, West Africa. Journal of Arid Environments 28, 13-30.

Norse, D. (1994): Multiple threats to regional food production: environment, economy, population? Food Policy 19, 113-48.

OIES (1991): Arid and semi-arid regions: Response to climate change, Boulder, Co: Office of Interdisciplinary Science.

Ojima DS, Dirke BOM, Glenn EP, Owensby CE and Scurlock, JMO (1993): Assessment of C budget for grasslands and drylands of the world. Water, Air and Soil Pollution 70, 643-57.

Oldeman LR (1988): Guidelines for general assessment of the status of human induced soil degradation. Wageningen: ISRIC.

Oldeman LR Hekkeling RTA and Sombroek, WG (1990): World map of the status of human-induced soil degradation. An explanatory note. Wageningen: ISRIC and Nairobi: UNEP.

Oldeman LR, Hakkeling RTA and Sombroek, WA (1991): World map of the status of human-induced land degradation: An explanatory note (end revised edition) Wageningen: ISRIC and Nairobi: UNEP.

Oldeman LR, Van Engelen VWP and Pulles JHM (1991): The extent of human induced soil degradation. In Lodeman LR, Hakkeling RTA and Sombroek WG (eds). World map of the status of human induced soil degradation: an explanatory note: Wageningen: ISRIC, 27-33.

Olsson L (1983): Desertification or Climate? Investigation regarding the relationship between land degradation and climate in he Central Sudan: Lund Series in Geography, series A, 60, Lund.

Olsson L, (1993): On the causes of famine, drought, desertification and market failure in the Sudan. Ambio 22, 395-403.

Oostwoud Wijdenes DJ and Bryan RB (1994): Gully head cuts as sediment sources on the Njemps Flats and initial low cost gully control measures. In Bryan RB (ed), Soil erosion, land degradation and social transition. Advances in Geo-ecology 27, Cremlingen Destedt: Catena Verlag, 205-29.

Otterman J (1974): Baring high-albedo soils by overgrazing: A hypothesized desertification mechanism. Science 186, 531-33.

Owensby CE (1993): Potential impacts of elevated CO_2 on above and below-ground litter quality of a tall grass. Water, Air and Soil Pollution 70, 413-24.

Palutikof, JP (1986): Drought strategies in east Africa: The climatologists role. Climatic change 9, 67-78

Palutikof, JP, Conte M, Casimiro Mendes, J, Goodess, CM and Espirito Santo, F (1996): Climate and climatic change. In Brandt CJ and Thornes JB (eds), Mediterranean desertification and land use. Chichester: Wiley, 43-86.

Palutikof, JP, Farmer G, Wigley TML (1982): Strategies for the amelioration of agricultural drought in Africa. Proceedings of the Technical Conference on Climate in Africa. WMO Geneva 596, 222-48.

Pandi zdruli marcello pagliai, Selim kapur and Angle Faz cano (Eds) (2010) "Land Degradation and Desertification: Assessment, mitigation and Remediation" published by Springer.

PartonWJ, Scurlock, JMO, Ojima DS *et. al.* (1993): Observations and modeling of biomass and soil organic matter dynamics for the grassland biome worldwide. Global Biogeochemical Cycles 7, 785-809.

Patel J, (1997): Story of a Rivulet Aravalli: From Death to Re-birth: Taarun Bharat Sangh, Bhimakpura, Thanagazi, Alwar, India -48.

Paye M (1990): Senegal : La desertification du basin arachidier. Reflets Sahelians 13, 35-39.

Penman HL (1948): Natural evaporation from open water, bare soil and grass: Proceedings of the Royal Society, Series A 193, 120-45.

Perkins JS and Thomas DSG (1993 a): Environmental responses and sensitivity to permanent cattle renching, semi-arid western central Botswana. In Thomas DSG and Allison RJ (eds), Landscape sensitivity. Chichester: Wiley, 273-86.

Perkins JS and Thomas DSG (1993 b): Spreading deserts or spatially confined environmental impacts? Land degradation and cattle renching in the Kalahari desert of Botswana. Journal of Land Degradation and Rehabilitation 4, 179-94.

Perrings C and Walker B (1992). Biodiversity loss and the economics of discontinuous change in semi arid rangelands. Discussion Paper at the second Conference on the Ecology and Economics of Biodiversity Loss, Beijer International Institute of Ecological Economics: Stockholm, Sweden.

Petrova AV (1982): Changes in humus and nitrogen contents in calcareous chernozems caused by wind erosion. Agrokhhimiya 1, 76-80.

Pickup G and Chewings VH (1988): Forecasting patterns of soil erosion in arid lands from Landsat MSS data. International Journal of Remote Sensing 9, 69-84.

Pickup G and Chewings VH (1988): Soil erosion in arid lands from Landsat MSS data. International Journal of Remote Sensing 9, 1469-490.

Pickup G , Bastin GN and Chewings VH (1994): Remote sensing based condition assessment for non-equilibrium rangelands under large scale commercial grazing: Ecological Applications 4, 497-517.

Piper S (1989): Measuring he particulate pollution damage from wind erosion in the western United States. Journal of Soil and Water Conservation 44, 70-75.

Poeson J and Bunte K (1993): Systematic investigation of soils containing rock fragments. In Mediterranean desertification and land use> MEDALUS 1 Final report: Brussels: Commission of the European Communities Directorate, 378-404.

Polis GA (ed) (1991): The ecology of desert communities. Tucson: University of Arizona Press.

Postel S (1990): Saving water for agriculture. In Brown L (ed) State of the World 1990. New York: Norton 39-58.

Proctor J (ed) (1989): Mineral nutrients in tropical forest and savanna ecosystems. British Ecological Society. Special Publication 9, Oxford: Blackwell Scientific.

Purvis JC (1977): Satellite photos help in dust episode in south Carolina. Information Note 77/8. Washington: US National Weather Service.

Raich JW and Schlesinger WH (1992): The global carbon dioxide flux in soil respiration and its relationship to vegetation and climate. Tellus 44B, 81-99.

Rao VB, Satyamurry P and De Brito JI (1986): On the 1983 drought in north-east Brazil. Journal of Climatology 6, 43-51.

Raul Daussa Fausto pedrazzini (2009) Desertification in the Mediterranean Region Water Scarcity, Land Degradation Published by Springer, The Netherlands.

Ray JG, Travleev A and Belova NA (1993): The extinction of some perennial grass vegetation and the degradation of chenozem due to athropo- zoogenial factors, in some steppes of Ukrainian SSR. In Wicherek S. (ed) Farm Erosion in temperate plains environment and hills. Symposium proceedings, Saint- Cloud, Paris: Elsevier, 241-51.

Reining P. (1978): handbook of desertification indicators: Based on the Science Associations Nairobi Seminar on Desertification. American Association for the Advancement of Science. Washington DC.

Rhoades JD (1990): Soil salinity- causes and controls. In Goudie AS (ed), Techniques for desert reclamation. Chichester: Wiley, 109-34.

Rhoades JD (1993): Practices to control salinity in irrigated soils. In Lieth, H and Masson A (eds) Towards the rational use of high salinity tolerant plants, Vol 2. Dorbrecht: Kluwer, 379-88.

Ringrose S and Matheson W (1992): The use of Landsat MSS imagery to determine the aerial extent of woody vegetation cover change in the west- central Sahel. Global Ecology and Biogeography Letters 2, 16-25.

Ringrose S, Matheson W, Tempest F and Boyle T (1990): The development and causes of range degradation features in southeast Botswana using multi- temporal Landsat MSS imagery. Photogrammetric Engineering and Remote Sensing 56, 1253-262.

Ropelewski CF and Halpert MS (1987): Global and regional scale precipitation patterns associated with the El Nino/ Southern Oscillation. Monthly Weather Review 115, 1606-626.

Rosewell CJ (1970): Investigations into the control of earth-work tunneling. Journal Soil Conservation New South Wales 26, 188-203.

San Jose JJ and Farinas M (1991): Temporal changes in the structure of a Trachypogen savannah protected for 25 years. Acta Oecologica 12, 237-47.

Saxena SK and Chatterji PC (1995): Mine spoils of arid and semi-arid areas and their rehabilitation 233-244. In Land Degradation and Desertification in Asia and the Pacific Region (Sen AK and Kar A, eds.), Scientific Publishers, Jodhpur, India.

Schillhorn van Veen TW (1995): The Kyrgyz sheep herders at a crossroads. Pastoral Development Network Paper 38d, ODI: London.

Schlesinger WH, Reynolds JF, Cunningham GL *et. al.* (1990): Biological feedbacks in global desertification. Science 247, 1043-48.

Scoging FI. (1989): Runoff generation and sediment mobilization by water. In Thomas, DGS (ed), Arid zone geomorphology (1st edition). London: Belhaven Press.

Scoones I (1992): Land degradation and livestock production in Zimbabwe's Communal Areas. Land Degradation and Rehabilitation 3, 99-114.

Scoones I and Graham O (1994): New directions for pastoral development in Africa. Development in Practice 4, 188-98.

Seely M (1978): The Namib Desert: An unusual ecosystem. Journal of Arid Environments 1, 117-28.

Seelt MK and Jacobson KM (1994): Desertification and Namibia: A perspective. Journal of African Zoology 108, 21-36.

Seitz F, Jastrow R and Nierenberg WA (1989): Scientific Perspectives on the greenhouse problem. New York: George C. Marshall Institute.

Sen AK and Kar A. (eds) (1995): Land degradation and desertification in Asia and the Pacific Region. Scientific Publishers , Jodhpur, India.

Sghaier M and Seiwert WD (1993): Winds of change and the threat of desertification: Case study from the Tunisian Sahara. Geo-Journal 31, 95-99.

Sharma BR and Paul DK (1999). Water Resources of India . In 50 years of natural resource management planning with special reference to Sevan grasslands. Monograph 21, Jodhpur, India, CAZRI-92.

Shantz HL (1956): History and problems of arid hands development. In GF White (ed) the future of arid lands. Washington DC: Americal Association for the Advancement of Science, 3-25.

Singh AN (1994): Monitoring change in the extent of salt-affected soils in northern India. International Journal of Remote Sensing 15, 3173-182.

Singh G (1995): Watershed management and concepts. In Sen AK and Kar A (eds) Land degradation and desertification in Asia and the Pacific region. Scientific Publishers, Jodhpur, India, 209-18.

Singh S, Kar A, Joshi Dc, Ram B, Kumar S, Vats PC, Singh N, Raina P, Kolarkar As and Dhir RP (1992): Desertification mapping in western Rajasthan. Annals of the Arid Zone 31, 237-46.

Singh S and kar A. (1997): Desertification control in the arid ecosystem of India for sustainable development. Agro Botanical Publishers New Delhi, India.

Skarpe C and Bergstrom, R (1986): Nutrient content and digestibility of forage plants in relation to plant phenology and rainfall in the Kalahari, Botswana. Journal of Arid Environments 11, 147-64.

Skarpe C (1991): Impact of grazing in savanna ecosystems. Ambio 20, 351-56.

Smith DR (1992): Salinization in Uzbekistan. Post-Soviet Geography 33, 21-33.

Smith EL (1988): Successional concepts in relation to range condition assessment. In Tueller PT (ed). Vegetation science applications for rangeland science and management, Dordrecht: Kluwer.

Smith R. (1951): Pedogenesis in the Frankland River Valley, Western Australia, CSIRO Bulletin 265, Melbourne Australia.

Soloman AM. Prentice IC, Leemans and Cramer WP (1993): The interaction of climate and land use in future terrestrial carbon storage and release. Air, Water and Soil Pollution 70, 595-614.

Sombroek WG, Nachtergael FO and Habel A (1993): Amounts, dynamics and sequestrations of carbon in tropical and sub-tropical soils. Ambio 22, 417-26.

Southgate D (1990): The causes of land degradation along 'spontaneously' expanding agricultural frontiers in the Third World. Land Economics 66(1), 93- 101.

Southgate D (1991): Tropical deforestation and agricultural development in Latin America. Discussion Paper. Department of Agricultural Economics, Ohio State University.

Spaeth HJ (1996): land degradation in West Niger. Desertification Control Bulletin 28, 34-41.

Squires VR and Glenn EP (1995): Creating an economic linkage between fossil fuel burning, climate change and rangeland restoration. Proceedings of the 5th International Rangelands Congr. Salt Lake City, 531-32.

Squires VR, Glenn EP and Ayoub AT. Land management, desertification and the global carbon cycle. Tueson, AZ: University of Arizona Press.

Stiles C (1995): Desertification is not a myth. Desertification Control Bulletin 26, 29-36.

Sud YC and Molod A (1988): A GCM simulation study of the influence of Saharan evapo-transpiration and surface-alvedo anomalies on July circulation and rainfall. Monthly Weather Review 116, 2388-400.

Suliman MM (1988): Dynamics of range plants and desertification monitoring in the Sudan. Desertification Control Bulletin 16, 27-31.

Suzuki FI (1981): The transcendent arid environments: A historico- geographical study of world religions. Addis Abeba Sha, Yokohama.

Swingle R, Glenn E and Squires V (1996): Growth performance of lamba fed mixed diets containing halophyte ingredients. Animal Feed Science and Technology 63, 137-48.

Szavolcs I (1976): Present and potential salt affected soils. FAO Soils Bulletin 31, 9-13.

Szavolcs I (1979): Review of Research on Salt affected soils. Natural Resources Research XV, Paris: UNESCO.

Szavolcs I (1987): The global problem of salt affected soils. Acta Agronomica Hungaria 36, 159-72.

Szavolcs I (1989): Salt affected soils. Boca Raton, FL: CRC Press.

Takeuchi R, Katoh K, Nan Y and Kou Z (1995): Vegetation cover change in desertified kerqin sandy lands. Inner Mongolia. Geographical Reports, Tokyo Metropolitan University 30. 1-24.

Taylor PJ, Walker GR, Hodgson G, Hatton TJ and Correll RL (1996): Testing of a GIS model of Euclyptus largiflorens health on a semi-arid, saline floodplain. Environmental Management 20, 553-64.

Thiemeyer H (1992): Desertification in the ancient erg of NE Nieria. Zeitschrift fur Geomorphologie Supplementhand 91, 197-208.

Thomas DSG (1997-a): Arid environments: Their nature and extent. In Thomas DSG(ed), Arid zone geomorphology: process, form and change in drylands, 2nd edition, Chichester: Wilet, 3-12.

Thomas DSG (1997 b): Reconstructing ancient arid environments. In Thomas DSG(ed). Arid zone geomorphology: Process, form and change in drylands. 2nd edition. Chichenster, Wiley, 577-606.

Thomas DSG and Middleton NJ (1993): Salinization: New perspectives on a major desertification issue. Journal of Arid Environments 24, 95-105.

Thomas DSG and Middleton NJ (1994): Desertification: Exploding the myth. Chichester: Wiley.

Thornes JB: Mediterranean desertification and vegetation cover. In Fantechi R, Peter D, Balabanis P and Rubio JL (eds) Desertification in a European context; Physical and socio-economic aspects. Proceedings of the European School of Climatology and Natural Hazards course, Alicante, Spain, 6- 13 October 1993, Brussels: European Commission, 169-96.

Thornes JB (1996): Summary prospects. In Brandt CJ and Thornes JB (eds), Mediterranean desertification and land use. Chichester: Wiley, 543-48.

Thornes JB, Fransis CF, Loez Bermudez F and Romaro Diaz A (1990): Particular overland flow with coarse particles and vegetation roughness under Mediterranean conditions. In Rubio JL and Rickson RL (eds), Strategies to combat desertification in Mediterranean Europe. Commission of the European Communities, Brussels. Report EUR 11175 EN/ES, 228.43.

Thornes JB, Shao JX, Diaz E, Roldan A, McMahon m and hawkers JC (1996): Testing the MEDALUS hill-slope model, Catena 26, 137-60.

Thornewaite CW (1948): An approach towards a rational classification of climate. Geographical Review 38, 55-94.

Tinker PB and Ineson P (1990): Soil organic matter and biology in relation to climate change. In Scharpenseel HW Schomaker M and Ayoub, A (eds), Soils on a Warmer Earth. Amsterdam: Elsevier, 71-78.

Tolsma DJ Ernst WHO and Verwey RA (1987): Nutrients in soil and vegetation around two artificial waterholes in eastern Botswana. Journal of Applied Ecology 24, 991-1000.

Trexler MC (1991): Minding the carbon store: Weighing US Forestry strategies to slow global warming. World Resources Institute, Washington Dc..

Trexler MC (1993): Manipulating the biotic carbon sources and sinks for climate change mitigation: Can science keep up with practice? Water, Air and Soil Pollution 70, 579-93.

Trexler MC and Meganch R (1993): Biotic carbon offset programs: Sponsors or impediments to economic development? Climate Research 3, 129-36.

Tucker CJ (1980): A critical review of remote sensing and other methods for non-destructive estimation of standing crop biomass. Gross and Forage Science 35, 115-82.

Tusker CJ, Dregne HE and Newcomb WW (1991): Expansion and contraction of the Sahara Desert from 1980 to 1990. Science 253, 299-301.

Tucker CJ, Hielkema JU and Roffrey J (1985 a). the potential of satellite remote sensing of ecological conditions for survey and forecasting desert locust activity. International Journal of Remote Sensing 6, 127-38.

Tucker CJ, Townshend JRG and Goff, TE (1985 b): African land cover classification using satellite data. Science 227, 369-75.

Tucker CJ, Vanprae CL, Sharman MJ and Van Ittersum G (1985 c). Satellite remote sensing of total herbaceous biomass production in the Senegalese Sahel: 1980-84. Remote Sensing of Environment 17, 233-49.

Tyson PD (1986): Climatic change and variability in southern Africa. Cape Town: Oxford University Press.

UN (1977 a): Status of desertification in the hot arid regions, climatic index map and experimental world scheme of aridity and drought probability, at a scale of 1: 25,000,000, Explanatory note. Conference on Desertification A/ CONF. 74/31. New York.

UN (1977 b): Desertification, its causes and consequences. Oxford. Pergamon Press.

Un (1993): United Nations Demographic Yearbook. New York.

UN (1994): Elaboration of an international convention to combat desertification in countries experiencing serious drought and / or desertification, particularly in Africa, New York: General Assembly A/ AC -241/15/ Rev 5.

UN (1997 b): Desertification, its causes and consequences, Oxford: Pergamon Press.

UN (1987): State of the Environment Report: The West Bank and Gaza. New York.

UN (1989): Case studies in population policy: Mexico. New York.

UN (1993): United Nations Demographic Yearbook. New York.

Un (1994): Elaboration of an international convention to combat desertification in countries experiencing serious drought and / or desertification, particularly in Africa. General Assembly A/AC. 241/15/ Rev.5, New York.

UNCED (1992): Earth Summit, Rio de Janeiro. London: Regency Press.

UNDP (1995): The Aral in crisis. UNDP: Tashkent.

UNDP / UNSO (1997): Aridity zones and dryland populations: An assessment of population levels in the world's drylands with particular reference to Africa. New York: UNDP Office to Combat Desertification and Drought (UNSO).

UNEP (1982): A world soils policy. UNEP, Nairobi.

UNEP (1984): General Assessment of progress in the implementation of the plan of action to combat desertification 1978-84: UNEP, Nairobi.

UNEP (1992): The Aral Sea: Diagnostic study for the development of an action plan for the conservation of the Aral Sea. UNEP, Nairobi.

UNEP (1994 a): Success story evaluation mission to China: Techniques of cultivating Tara mix shrubs in sandy and saline- alkaline soils by use of floods. UNEP, Nairobi.

UNEP (1994 b): Success story evaluation mission report. Integrated whole farm- whole landscaping planning- 'Pynehem', Frankland, Western Australia. UNEP: Nairobi.

UNEP (1994 c): Success story evaluation report on WVI restoration and protection of the environment in Lounga region, Senegal. UNEP, Nairobi.

UNEP (1995 a): combating global climate change by combating land degradation. Desertification Control Bulletin, No. 29.

UNEP (1995 b)" Global Biodiversity Assessment. Heywood, VF1 and Watson, RT (eds), Cambridge: Cambridge University Press.

UNEP (1996): Success stories: Saving the drylands. 1995 award winners. UNEP, Nairobi.

UNEPCOM (USSR Commission for UNEP) (1994): Desertification assessment in southern and eastern pre-Aral region: Final report on the UNEP/ USSR project. UNEPCOM, Moscow.

UNEP/ EAP (1995 a): Land cover assessment and monitoring, Nepal. Volume 6-A, UNEP Environment Assessment Programme for Asia and the Pacific, Bangkok.

UNESCO (1973): International classification and mapping of vegetation. UNESCO, Paris.

UNESCO (1977): World map of arid zones 1/ 25,000,000, Explanatory notes. MAB Technical notes No. 7, UNESCO, Paris.

UNESCO (1980): Case studies on desertification. Natural Resources Research XVIII.

UNOSTD (1987): Report of the panel of experts on 'The application of science and technology to the study, prevention, monitoring and combating of drought and desertification'. New York.

Van Schilfgaarde, J (1993): Water management strategies for salinity control. In Lieth H and Masson A (eds), towards the rational use of high salinity tolerant plants, Vol. 2. Dorbrecht: Kluwer, 371-78.

Venator C, Glaeser J and Soto R (1992): A silvi-pastoral strategy. In Downing, TE (ed), Development or destruction. Colorado: Westveiw.

Walker JM and Rowntree PR (1977): The effect of soil moisture on circulation and rainfall in a tropical model. Quarterly Journal of the Royal Meteorological society 103, 29-46.

Warren A (1995): Changing understandings of African pastoralism and environmental paradigms. Transactions of the Institute of British Geographers 20, 193-203.

Warren A and Agnew C (1988): An assessment of desertification and land degradation in arid and semi-arid areas. Paper 2. IIED, London.

Warren and Khagoli M (1992): Assessment of Desertification and Drought in the Sudano-Sahelian Region 1985-1991. UN Sudano-Sahelian Office, New York.

Warren A, Sud YC and Rozanov B (1996): The future of deserts; Journal of Arid Environments 32, 75-89.

Watson A (1990): The control of blowing sand and mobile dunes. In Goudie AS (ed), Techniques for desert reclamation. Chichester; Wiley, 35-85.

WCED World Commission on Environment and Development (1987): Our common future. Oxford University Press, Oxford.

West JG (1984): A revision of Dodonaea Miller (Sapindaceae) in Australia. Brunonia 7, 1-194.

West NE, Stark Jm, Johnson, DW *et. al.* (1994: Effects of climatic change on the edaphic features in arid and semi-arid lands of western North America. Arid soil Research and Rehabilitation 8, 307-51.

Westing AH (1994): Population, desertification and mitigation. Environmental Conservation 21, 110-14.

Whitford WG (1991): Subterranean termites and long term productivity of desert rangelands. Socio-biology 19, 235-43.

Whittaker R (1970): Communities and ecosystems. London: Macmillan.

Wilhite D and Glantz M (1985): Understanding the drought phenomenon: The role of definitions. Water International 10, 111-20.

Williams MAJ and Balling RC (1996): Interactions between desertification and climate. Edward Arnold, London.

Williams MAJ, Dunkerley DL, De Dekker P, Kershaw AP and Strokes T (1993): Quarternary environments. London: Edward Arnold.

Wulliams WD and Aladin NV (1991): The Aral Sea: recent limnological changes and their conservation significance. Aquatic Conservation 1, 3,24.

WMO (World Meteorological Organization) (1975): Drought and agriculture. WMO Technical Note 138. Geneva, Switzerland.

World Bank (1992): World Development report: Development and environment. Oxford University Press, New York.

Worster D (1979): Dust Bowl. Oxford University Press, New York.

WrightJL (1982): New evaporation crop coefficients. Proceedings of the American Society of Civil Engineers Irrigation and Drainage Division 108, 57-74.

Yair A and Rutin J (1981): Some aspects of the regional variation in the amount of available sediment produced by isopods and porcupines, northern Negev, Israel. Earth Surface Processes and Landforms 6, 221-34.

Yan ZW, Ye DZ and Wang C (1992): Climatic jumps in the flood/ drought historical chronology of central China. Climate Dynamics 6, 153-60.

Yariv S (1976): Comments on the mechanism of soil detachment by rainfall. Geoderma 15, 393-99.

Yassoglou NJ (1987): The production potential of soils: Part II- sensitivity of the soil systems in southern Europe to degrading influxes. In Barth H and L'Hermite P (eds), Scientific basis for soil protection in the European Community. Elsevier, 87-122, London.

Yassoglou NJ (1995): Land and desertification in Pantechi R, Peter D, Balabanis P and Rubio JL (eds). Desertification in a European Context; physical; and socio-economic aspects. Proceedings of the European school of Climatology and Natural Hazards course, Alicante, Spain, 6-13 October 1993. Brussels: European Commission, 169-94.Zhao Songqiao 1988: Human impact on northwest China arid area: Desertification or de-desertification? Chinese journal of Arid Land Research 1, 105-16.

Zhu Y and Zheng X (1983): Shallow groundwater resources of the Huang- Huai-hai Plain. In Biswas AK, Dakang Z, Nickum JE and Changming L (eds), Long-distance water transfer: A Chinese case study and international experience. Dublin Tycooly International, 257-70.

Zonn IS (1995): Desertification in Russia: problems and solutions (an example in the republic of Kalmykia-Khalmg Trangch), Environmental Monitoring and assessment 37, 347-63.

■